Anti-infectives

Recent Advances in Chemistry and Structure–Activity Relationships

Anti-infectives

Recent Advances in Chemistry and Structure–Activity Relationships

Edited by

P.H. Bentley
Smith & Nephew plc, Heslington, York, UK

P.J. O'Hanlon
SmithKline Beecham Pharmaceuticals, Betchworth, Surrey, UK

THE ROYAL
SOCIETY OF
CHEMISTRY
Information
Services

The Proceedings of the 2nd International Symposium on Recent Advances in the Chemistry of Anti-infective Agents held at Churchill College Cambridge on 7–10 July 1996.

The cover illustration shows: left, *Escherichia coli* showing disruption resulting from antibiotic treatment; top right, *Candida albicans* hyphae infecting epithelial tissue; bottom right, *Varicella zoster* virus releasing from an infected cell.

Special Publication No. 198

ISBN 0-85404-707-7

A catalogue record for this book is available from the British Library.

Published by The Royal Society of Chemistry,
Thomas Graham House, Science Park, Milton Road,
Cambridge CB4 4WF, UK

Printed in Great Britain by Hartnolls Ltd, Bodmin, UK

Preface

The Second International Symposium on Recent Advances in the Chemistry of Anti-Infective Agents followed a similar event held in 1992 and earlier symposia (from 1976) which were devoted solely to developments in β-lactams. As before, presentations were provided which covered antibacterials/antibiotics, antifungals and antivirals.

The antibacterial/antibiotic contributions to the Symposium began with a reminder of how quickly bacteria can become resistant to agents. Dr. Williams went on to describe his work on elucidating the molecular mechanism of action of the glycopeptide antibiotics and progress using this knowledge towards identifying compounds active against vancomycin-resistant bacteria. The need for new agents active against novel targets to overcome multi-drug resistant bacteria was a consistent theme. Presentations described the design, synthesis and antibacterial evaluation of novel oxazolidinone antibacterial agents (Dr. Barbachyn), and the synergistic properties of protein synthesis inhibitors streptogramins and development of quinupristin/dalfopristin for treating severe Gram positive infections (Dr. Paris). Professor Kocienski described his work on the total synthesis of the polyether antibiotic salinomycin and the pivotal role that long bonded interactions can play in controlling stereochemistry at acetal centres. Bacterial signal peptidases are a novel target and inhibition by penem derivatives was described (Dr. Burton). Professor Burke presented his elegant work on the total synthesis of antimitotic agent halichondrin B, demonstrating two-directional chain synthesis and termination strategy. Developments in β-lactams were represented by the trinems, a new class of carbapenems and recent developments in the synthesis and SAR of these β-lactams was presented (Dr. Biondi). The production of β-lactamases is the predominant mechanisms of resistance for β-lactam antibiotics and Dr. Pflieger described the discovery of bridged carbacephems which are potent β-lactamase inhibitors possessing antibacterial activity. Professor Miller's lecture linked all three anti-infective areas covered by the Symposium by describing his latest work on developing novel agents using siderophore-mediated drug delivery.

The range of antifungal topics covered at the meeting embraced targets in the cell wall as well as intracellular mechanisms. The therapeutic target which continues to receive most attention is that of cell wall synthesis and the metabolism of its sterol components. The most recent successes from the industrial pipeline were described in presentations by Drs Street and Saksena which focused on improvements to the existing therapies based on azole structures and showed that refinement of existing knowledge can still make a significant contribution to improving treatment of fungal infections. This work has resulted in two outstanding compounds which are undergoing clinical trials namely UK 109496 (voriconazole) and Sch 56592. Research on alternative biochemical targets is at a much earlier stage. The potential of chitin synthetase as a fungal target was reviewed by Dr Georgopapadakou, whilst the likelihood of evolving ideas currently of interest in the agrochemical field into human therapies was discussed with respect to inhibition of mitochondrial respiration and interference with intracellular organisation via tubulin inhibition, (Drs Clough and Hollomon respectively). Clearly these approaches still have some way to go but the drive to find new therapies continues to be provided by the increasing incidence of systemic fungal infections among the immunocompromised patient population.

The antiviral lectures covered a wide range of antiviral research, from biosynthesis studies (Dr Turner), through the use of crystallography, to the development of novel antiviral agents and clearly illustrated the considerable advances made since the last symposium in this series. A number of approaches to anti-HIV agents were presented including novel nucleoside analogues (Professors Walker and Herdewijn), prodrugs of anti-HIV nucleoside phosphates (Professor McGuigan), non nucleoside HIV-1 reverse transcriptase (RT) inhibitors (Dr Camarasa) and an overview of the development of the HIV protease inhibitor Crixivan™ (Dr Dorsey). The use of crystallography to investigate the mode of action of non-nucleoside HIV-1 RT inhibitors was also discussed (Professor Arnold). These studies were expanded upon to postulate the mechanism of resistance to these agents and how this information could be used to design second generation inhibitors.

Viral proteases as targets in other viruses were reviewed (Dr Mills) and the approach was illustrated by the investigation of inhibitors of viral cysteine proteases (Professor Vederas). The breadth of the potential targets for antiviral agents was further illustrated by the use of sialidase inhibitors as anti-influenza agents (Dr Smith).

Once again this Symposium has highlighted the seriousness of microbial resistance developing to our mainline antimicrobials. It is hoped that the work presented will contribute to combatting this resistance in the years to come.

We thank all the speakers for their excellent presentations, the participants in the poster programme, the session chairpersons and the staff of Churchill College for contributing to an effective and enjoyable symposium. Finally, we gratefully acknowledge the sustained help of our fellow organisers, Dick Challand, Paul Wyatt, Richard Taylor and Elaine Wellingham.

P H Bentley, Smith & Nephew Group Research Centre

P J O'Hanlon, SmithKline Beecham Pharmaceuticals

October 1996

Editor's Note

Dr N Turner's presentation is not reported in these proceedings.

Contents

Advances in Antibacterials / Antibiotics

Advances in Antifungals

Advances in Antivirals

Abbreviations

ACCA	amino-1-cyclopropyanelcarboxylic acid
AIBN	2,2'-azobisisobutyronitrile
AIDS	acquired immune deficiency syndrome
All	allyl
AUC	area under curve
Bn	benzyl
Boc	t-butyloxycarbonyl
BOM	benzyloxymethyl
BOP	benzotriazol-1-yloxytris(dimethylamino)phosphonium hexafluorophosphate
Bs	brosyl
Bz	benzyl or benzoyl
CAN	ceric ammonium nitrate
Cbz/Z	benzyloxycarbonyl
CC_{50}	concentration cytotoxic to 50% host cells
CD_{50}	curative dose 50% survival
cfu	colony forming units
CIC_{95}	concentration cytotoxic to 95% host cells
CRO	ceftriaxone
CSA	camphorsulphonic acid
DBU	1,8-diazabicyclo[5.4.0] undec-7-ene
DCC	1,3-dicyclohexylcarbodiimide
DDQ	2,3-dichloro-5,6-dicyano-1,4-benzoquinone
DEAD	diethyl azodicarboxylate
DET	diethyl tartrate
DHP	dihydropyran
DIAD	diisopropyl azodicarboxylate
DIBAL	di-isobutyl aluminium hydride
DMAP	4-dimethylaminopyridine
DMB	2,4- or 3,4-dimethoxybenzyl
DMP	Dess-Martin periodinane
ED_{50}	effective dose 50% survival
EDC	1-(3-dimethylaminopropyl)-3-ethylcarbodiimide
5-FC	5-fluorocytosine
HIV	human immunodeficiency virus
HOBt	N-hydroxybenztriazole
HPβCD	hydroxypropyl β-cyclodextrin (vehicle)
IC_{50}	concentration inhibiting 50% of microbial growth or 50% of enzyme activity
Im	imidazole
LAH	lithium aluminium hydride
LDA	lithium N,N-diisopropylamide
LG	leaving group
LHMDS	lithium hexamethyldisilazide
LP	leader peptidase

MC	methylcellulose (vehicle)
mCPBA	m-chloroperbenzoic acid
MG/KG	MPK = mpk = $mgKg^{-1}$
MIC	minimum inhibitory concentration (in mcg/ml)
MMTr	monomethoxyltrityl
MOM	methoxymethyl
MOPS	3-N-morpholinopropanesulphonic acid
MRI	magnetic resonance imaging
MRSA	methicillin-resistant *Staphylococcus aureus*
MRSE	methicillin-resistant *Staphylococcus epidermidis*
Ms	mesyl (= methanesulfonyl)
MS	molecular sieves
NBS	N-bromosuccinimide
NCS	N-chlorosuccinimide
NHS/HOSu	N-hydroxysuccinimide
NOE	nuclear Overhauser effect
PAE	post antibiotic effect
PBP	penicillin binding protein
PCC	pyridinium chlorochromate
PFP	pentafluorophenyl
Piv/Pv	pivaloyl
PMB	p-methoxybenzyl
pNBz/PNB	p-nitrobenzoyl
PPTS	pyridinium p-toluenesulphonate
PSE	penicillin sensitive enzyme
pTSA	p-toluenesulphonic acid
RT	reverse transcriptase
SATE	S-acetylthioethanol
SEM	trimethylsilylethoxymethyl
SI	selectivity index = CC_{50}/EC_{50}
TBAF	t-butylammonium fluoride
TBDPS	t-butyldiphenylsilyl
TBS/TBDMS	t-butyldimethylsilyl
TES	triethylsilyl
Tf/Tfl	triflyl/trifluoromethylsulfonyl
THP	tetrahydropyran
TMS	trimethylsilyl
Trityl	triphenylmethyl
troc	trichloroethoxycarbonyl
Ts	tosyl
VRE	vancomycin resistant enterococci

Advances in Antibacterials/Antibiotics

1
The Glycopeptide Story – How to Kill the Deadly "Superbugs"

Dudley H. Williams, Martin S. Westwell, Daniel A. Beauregard, Gary J. Sharman, Robert J. Dancer, Andrew C. Try and Ben Bardsley
CAMBRIDGE CENTRE FOR MOLECULAR RECOGNITION, DEPARTMENT OF CHEMISTRY, LENSFIELD ROAD, UNIVERSITY OF CAMBRIDGE, CAMBRIDGE CB2 1EW, UK

1 INTRODUCTION

In 1967, the Surgeon General of the United States told a meeting in the White House that "the time has come to close the book on infectious diseases." The Surgeon General could perhaps be excused for his optimism, as antibiotics and vaccines appeared to be sweeping all before them. In making this statement, the Surgeon General failed to appreciate the power of natural selection and, in particular, he (among most others) did not appreciate how such selectional processes can work relatively quickly in bacteria. Bacteria are capable of passing genetic information between species, and additionally have a relatively short generation time. Suppose two bacteria can come into close proximity. If the first bacterium has a feature which renders it resistant to a given type of antibiotic, then that feature can be transmitted to the second species. These two variables have been important in leading to the current situation in which we are losing the fight against pathogenic bacteria; alarming numbers of people are losing their lives.

It is against the above background that I wish to present the role that the glycopeptide antibiotics have played in therapy during the last 40 years and how, in my research group in Cambridge, we were able to determine the molecular basis for their mode of action. I will then discuss how, in recent times, resistance of bacterial pathogens against the most important member of glycopeptide group, vancomycin, has emerged. Finally, I will illustrate how the principles of action of a new semi-synthetic glycopeptide - which is showing remarkable activity against bacteria which currently cause lethal infections - have emerged from our recent work.

2 VANCOMYCIN AND GLYCOPEPTIDE STRUCTURE DETERMINATION

Vancomycin was discovered in a soil sample from the jungles of Borneo during a research programme carried out by the American pharmaceutical giant Eli Lilly in the mid-1950s. It is produced by a microorganism as a secondary metabolite. That is, a compound which does not appear to be essential in the internal economy of the producing organism. It was first used clinically in 1959, although its early use was somewhat limited by side effects. For example, when vancomycin was administered intravenously, phlebitis would on occasions occur near the point of injection, and also its use was not recommended in those with any history of hearing difficulties. These side-effects were much reduced when purer vancomycin became available in later years.

The first structural work on vancomycin was carried out by Marshall and reported in 1965.[1] The work indicated the presence of *N*-methylleucine, glucose, and chlorophenols. However, progress was slow at this time because the methods needed to solve such a complex structure were not yet available. An important finding, even before the structure of vancomycin was known, was that of Perkins who showed that the antibiotic binds to

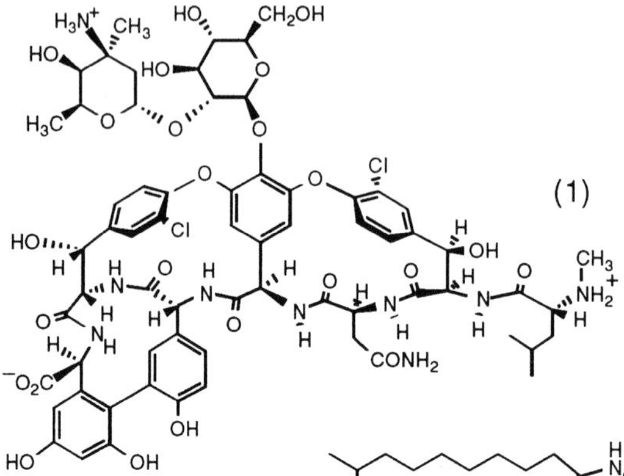

(1)

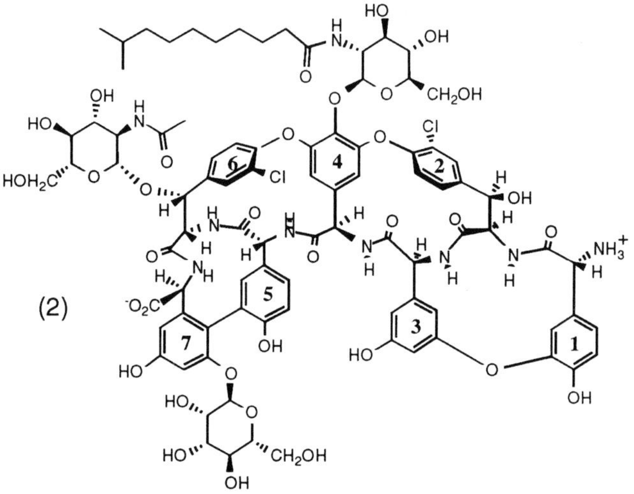

(2)

bacterial cell-wall mucopeptide precursors terminating in the sequence –L-lysyl-D-alanyl-D-alanine.[2] At the time of this finding (1969), little about the molecular basis of action of vancomycin could be inferred because the structure of the antibiotic was as yet unknown.

My own group carried out intensive researches on the vancomycin structure in the period 1972-81. This may have been the first work to use the negative nuclear Overhauser effect (NOE) in proton NMR to derive extensive structural information on a relatively large secondary metabolite of truly unknown structure. The use of the NOE represented a major advance at the time, because this method, though not so precise as X-ray crystallography, could be applied in solution and give approximate distances between hydrogen atoms which were close together in the vancomycin structure. We were able to obtain a partial structure by 1977.[3] Crucial information on the stereochemistry of the molecule was then derived from an X-ray study of a degradation product of vancomycin (CDP-1).[4] Early structures were incorrect insofar as they incorporated an iso-aspartic acid (rather than an aspartic acid) residue, and in misrepresenting the stereochemistry of a chlorinated aromatic ring. The former error was corrected by Harris & Harris,[5] and the latter by ourselves,[6] to give the now accepted structure (1). In a shorter, but overlapping period, we were also able to elucidate the structure of a second glycopeptide, ristocetin A.[7] Although much less important than vancomycin, this material is used in the diagnosis of von Willebrandt's disease (characterised by the absence of a blood clotting factor). By the early 1980s, proton NMR spectroscopy had become so powerful that our third structure elucidation - that of teicoplanin - could be carried out relatively quickly.[8] Although teicoplanin is not quite as important clinically as vancomycin, it is extensively used in a similar manner and has the advantage of being cleared more slowly from the body. Its structure (2), incorporates a

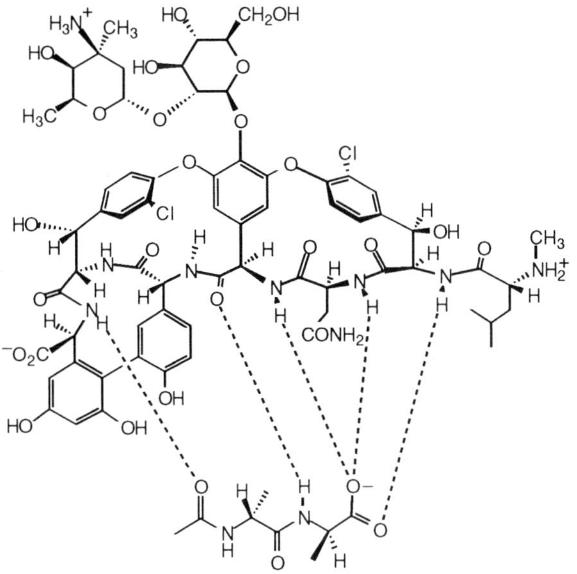

Figure 1 *The binding interaction between antibiotic and bacterial cell-wall analogue. Hydrogen bonds between the two are represented by dashed lines.*

hydrocarbon sidechain attached to a glucosamine residue on the fourth amino acid of teicoplanin. At the time of writing, the structures of about 100 glycopeptides are established.

3 DETERMINATION OF THE BINDING SITE OF GLYCOPEPTIDES FOR BACTERIAL CELL-WALL PRECURSORS

With the structure of vancomycin in hand, it was then possible to elucidate the molecular basis for binding of the mucopeptide precursors terminating in –L-Lys-D-Ala-D-Ala to the antibiotic. We did this for ristocetin A in 1980,[9] for vancomycin in 1983,[10] and for teicoplanin in 1985.[11] In all cases, the method used was proton NMR spectroscopy, carried out on the complex formed between the antibiotics and *N*-acetyl-D-Ala-D-Ala (or di-*N*-acetyl-L-Lys-D-Ala-D-Ala). Four key methods were used. First, we examined the chemical shifts of the alanine methyl groups of *N*-acetyl-D-Ala-D-Ala when this bacterial cell-wall analogue was both bound to the antibiotic and isolated from it. The changes in these chemical shifts indicated the location of these methyl groups over substituted benzene rings of the antibiotics. Second, we used the changes in chemical shifts of amide NH resonances of the antibiotic to indicate which of these were involved in binding the carboxylate group of the cell-wall analogue. Third, we determined which amide NH groups of the antibiotic were protected from exposure to water in the antibiotic/cell-wall analogue complex. This experiment allowed us to determine all of those NHs which were involved in binding the antibiotic to the cell-wall analogue. Fourth, we were able to determine which protons of the antibiotic were close in space to identified protons of the cell-wall analogue. This involved using the NOE as described above, although this time to estimate distances between protons of the two different molecules of the complex (rather than inter-proton distances within one molecule). These methods led to the model of the binding interaction which is reproduced in Figure 1.

4 NEW TWISTS IN THE MECHANISM OF ACTION OF THE GLYCOPEPTIDES

The work described at the end of the preceding section seemed to solve the problem of the molecular basis of the mode of action of the glycopeptides. However, in 1989, we discovered that the antibiotic ristocetin A forms a dimer, and we were able to elucidate the structure of the peptide portion of the dimer (although not at this stage the orientation of the saccharide portions of the molecule with respect to peptide within this dimer). It is interesting to look back and note that even in 1989, we suggested that the dimer may be of physiological significance.[12] We suggested this since delivery of a dimer to the site of bacterial cell-wall biosynthesis, with which the glycopeptides interfere, would simultaneously place two molecules of the antibiotic at the crucial site for action. By 1993/94 we had in fact showed that all the glycopeptides which we were able to examine, with the exception of teicoplanin, formed dimers and measurement of their dimerisation constants showed that they were very large for some of the antibiotics (see Figure 4).[13,14] These experiments also allowed us to detail many of the structural features which were important for strong dimerisation. The generality of the phenomenon of dimerisation led us to propose that if a dimer became bound to one –L-Lys-D-Ala-D-Ala fragment of growing cell wall, then the binding event in which a second fragment of growing cell wall could bind to the same dimer would effectively be intramolecular (Figure 2B, to be contrasted with Figure 2A).[15] Additionally, we noted that since teicoplanin has a hydrocarbon sidechain then it should, in principle, be able to target itself to the site of biosynthesis with which it interferes by using this hydrocarbon sidechain as a membrane anchor (Figure 2C).[15] Thus, in this latter case, dimerisation is unnecessary because both the growing cell-wall fragment and the antibiotic are attached to the membrane making binding an intramolecular event. Hence, in the mode of action of the vancomycin group of antibiotics, either dimerisation (Figure 2B) or membrane anchoring (Figure 2C) preferentially locate antibiotic at the crucial site for action.

It was not until early 1995, that we were able to demonstrate experimentally that the hypotheses with regard to the above locating devices seemed to be correct. Our experiments involved trying to antagonise the action of the antibiotics in killing bacteria on agar plates by the addition of external di-*N*-acetyl-L-Lys-D-Ala-D-Ala. Figure 2 also shows the basis for these experiments. Where binding is simply bimolecular, the probability of locating the antibiotic at the appropriate site is relatively low (Figure 2A). In contrast, whilst the first antibiotic molecule of a dimer has been located to bind to cell-wall precursor, the second antibiotic molecule in this dimer is already in an appropriate place to bind another growing piece of cell wall. It will therefore be more difficult to antagonise this kind of binding event by externally added tripeptide (Figure 2B). Finally, where teicoplanin is located at the site

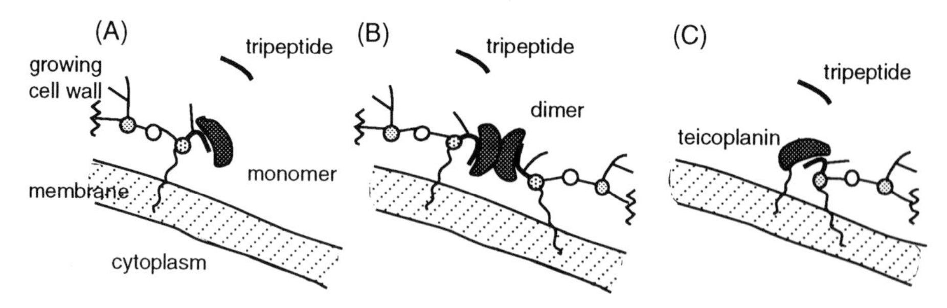

Figure 2 *Antibiotics binding as (A) monomer, (B) dimer & (C) with a membrane anchor. (A) The binding of single antibiotic molecule to the growing cell wall is a simple biomolecular association [such that externally added ligand can replace the cell-wall peptide]. (B) The antibiotic dimer can benefit from an essentially intramolecular association at the surface of a bacterium [and it is more difficult to disrupt by the externally added ligand]. (C) The benefit from intramolecularity can also be exploited if the antibiotic has a membrane anchor [again, the complex is more difficult to antagonise].*

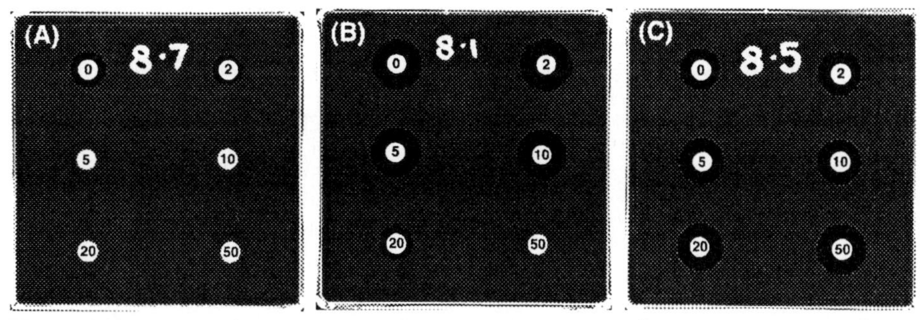

Figure 3 *Three agar diffusion plates illustrating the effect of di-*N*-acetyl-L-Lys-D-Ala-D-Ala on the potencies of three representative antibiotics in inhibiting the growth of* Bacillus subtilis. *Each dark circle represents an area where the bacteria are killed by the antibiotic placed on the paper disk (white circles). The size of the dark circle is a measure of the potency of the antibiotic. Each paper disk contained 1 μg of antibiotic. (A) teicoplanin A_3-1 (does not dimerise or possess a membrane anchor), (B) vancomycin (dimerises weakly), & (C) eremomycin (dimerises strongly). The number superimposed on each disk is the amount of di-*N*-acetyl-L-Lys-D-Ala-D-Ala (μg) added to the antibiotics.*

of biosynthesis through use of a membrane anchor, the binding of membrane-bound cell-wall precursors to antibiotic should be favoured because of proximity effects, and again subject to more difficult antagonism by externally added tripeptide (Figure 2C). In fact, approximately five times as much externally added tripeptide is required to antagonise the action of teicoplanin relative to teicoplanin lacking a membrane anchor (TA_3-1). More strikingly, the strongly dimerising antibiotic eremomycin requires roughly a thousand times as much externally added tripeptide to antagonise its action to a similar level as TA_3-1.[16] Representations of the actual inhibitions zones demonstrated by the various antibiotics strikingly confirm these remarkable effects (Figure 3). Indeed, there is a striking correlation between the amount of externally added tripeptide which is needed to antagonise an antibiotic and its dimerisation constant (Figure 4).

5 SIMULATING THE COOPERATIVE BINDING OF ANTIBIOTIC TO SIMPLE MODELS OF BACTERIAL CELLS

The experiments described in the previous section give strong support to the promotion of glycopeptide antibiotic action by means of devices which preferentially locate the antibiotic at the site where it can interfere with cell-wall biosynthesis. However, it is desirable to demonstrate directly that when cell-wall precursors bind to antibiotic in the previously described manner (Figure 2B), then the binding constant for the association is greater than in a simple bimolecular association (Figure 2A). This was shown by using a simple model system which can act as an analogue of a cell membrane.

Molecules of the surfactant sodium dodecyl sulphate (SDS) aggregate in aqueous solution to give a structure known as a micelle. In this micelle, the hydrocarbon tails of the molecule have a tendency to aggregate towards the centre of a roughly spherical structure, in which the sulphate groups reside near the surface - where they can be solvated by water molecules. The structure thus generated (Figure 5A) is a crude model of a bacterial

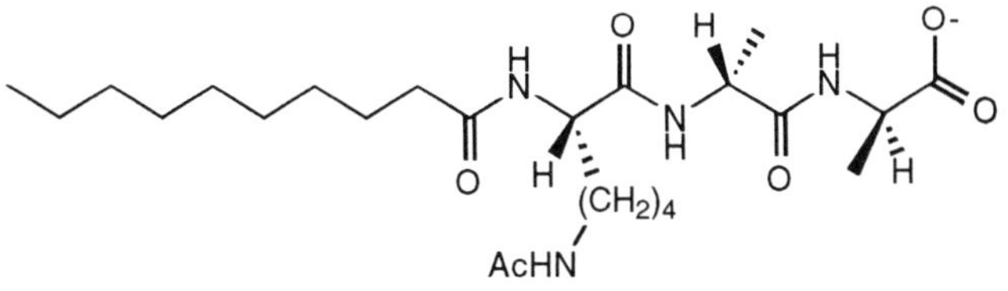

(3)

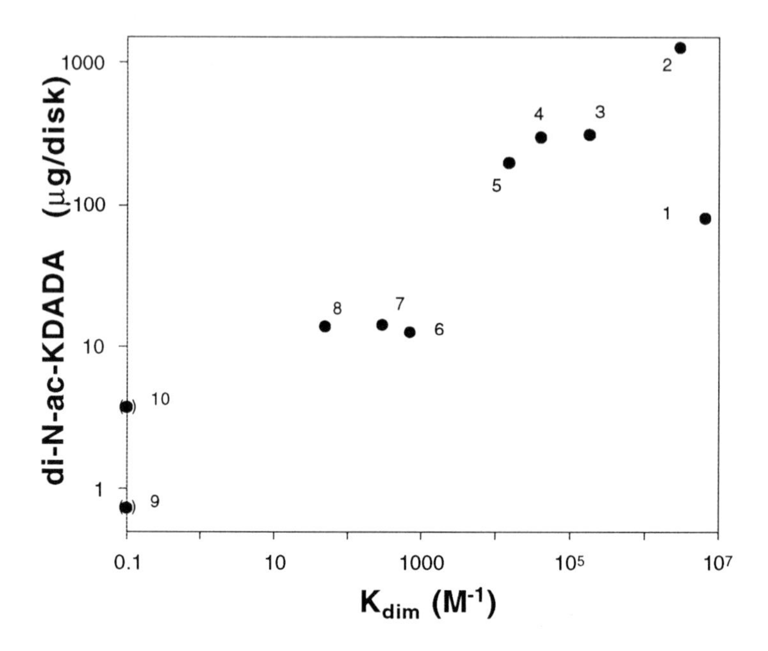

Figure 4 *Plot of amount of di-*N*-acetyl-*L*-Lys-*D*-Ala-*D*-Ala required to give a 50% reduction in the potency of glycopeptide antibiotic versus dimerisation constant, K_{dim}, of the antibiotic. Teicoplanin, which possesses a membrane anchor, (10) and teicoplanin A_3-1, without a membrane anchor, (9) show no evidence for dimerisation (i.e., $K_{dim} < 1\ M^{-1}$).*

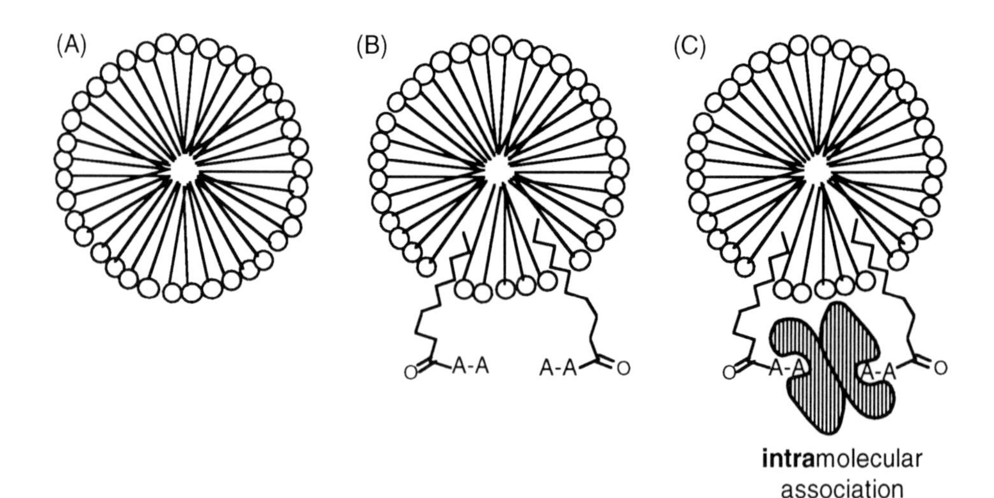

Figure 5 *Schematic representation of (A) a micelle, (B)* N-α-*decanoyl-*N-ε-*acetyl-*L*-Lys-*D*-Ala-*D*-Ala using its membrane anchor, and (C) an antibiotic dimer bound to the ligand at the surface of a micelle.*

membrane. Next, a fatty acid chain was put onto a cell-wall mucopeptide precursor, to give *N*-α-decanoyl-*N*-ε-acetyl-L-Lys-D-Ala-D-Ala (3). This compound, when in the presence of a micelle in aqueous solution, can lower its free energy by inserting its hydrocarbon tail in the hydrocarbon portion of the micelle, to give the crude mimic of a growing bacterium (Figure 5B). Several peptide structures can be incorporated into one pseudo-membrane in this way, and since the hydrocarbon portion of the micelle has the fluidity characteristic of a liquid crystal, two of these peptide structures can take up a geometry which might be appropriate to bind to a dimer without too much strain (Figure 5C). If our hypothesis is correct, then the binding affinity of a dimer glycopeptide to such an organised assembly should be greater than that of a simple bimolecular association in solution (Figure 2A). We were excited to find that when the binding of di-*N*-acetyl-D-Ala-D-Ala to ristocetin A in the presence of micelles is compared to the binding of *N*-decanoyl-D-Ala-D-Ala to ristocetin A also in the presence of micelles, the latter binding is greater by a factor of 100.[17] Thus, the stronger binding of an antibiotic dimer to a simple analogue of the bacterial cell system has been directly demonstrated.

6 THE OPERATION OF TARGETING DEVICES IN THE ACTION OF A NEW ANTIBIOTIC

Illnesses due to methicillin resistant *Staphylococcus aureus* (MRSA) and vancomycin resistant enterococci (VRE) are on the increase.[18-20] These illnesses are often lethal. The VRE problem arises from bacteria which would not ordinarily be pathogenic to a healthy individual. However, it is commonplace these days for hospitalised individuals to be immuno-deficient. Immuno-deficiency may arise due to cancer chemotherapy, AIDS, or simply the impaired function that often goes hand-in-hand with weakness following operations or old age.

The vancomycin-resistant enterococci were first reported in 1989. It appears that these enterococci have been able to obtain genes from other bacteria such that the precursor from which their cell wall is built no longer terminates in –D-alanyl-D-alanine, but rather terminates in –D-alanyl-D-lactate (–D-Ala-D-Lac).[21,22] As a consequence, the hydrogen bond which is normally made between the NH of the terminal D-alanine group and a carbonyl group of the antibiotic can no longer be made. Instead, it is replaced by a repulsive interaction between the oxygen of the *C*-terminal D-lactate group and the carbonyl group of the antibiotic (Figure 6). The affinity of the glycopeptide antibiotics for the vancomycin resistant enterococci precursors which terminate in –D-Ala-D-Lac is hence decreased by a factor of the order of 1000 (relative to the binding to precursors terminating in –D-Ala-D-Ala). Due to this markedly reduced binding constant, vancomycin has little activity against bacteria using such precursors. Significant inhibition of VRE requires 100-1000 times as much vancomycin as is required in the treatment of sensitive organisms. Thus, vancomycin

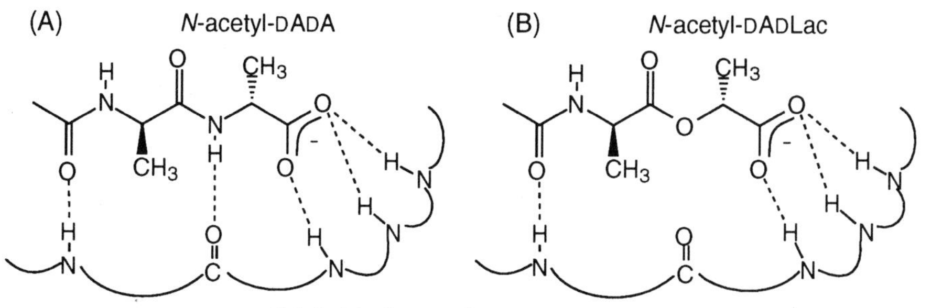

(A) *N*-acetyl-DADA (B) *N*-acetyl-DADLac

antibiotic binding pocket

Figure 6 *In vancomycin resistant enterococci, the terminal D-alanine of the immature cell wall has been replaced by D-lactate. This change drastically reduces the binding constant to vancomycin, as an NH which can form a hydrogen bond is replaced by an O which cannot.*

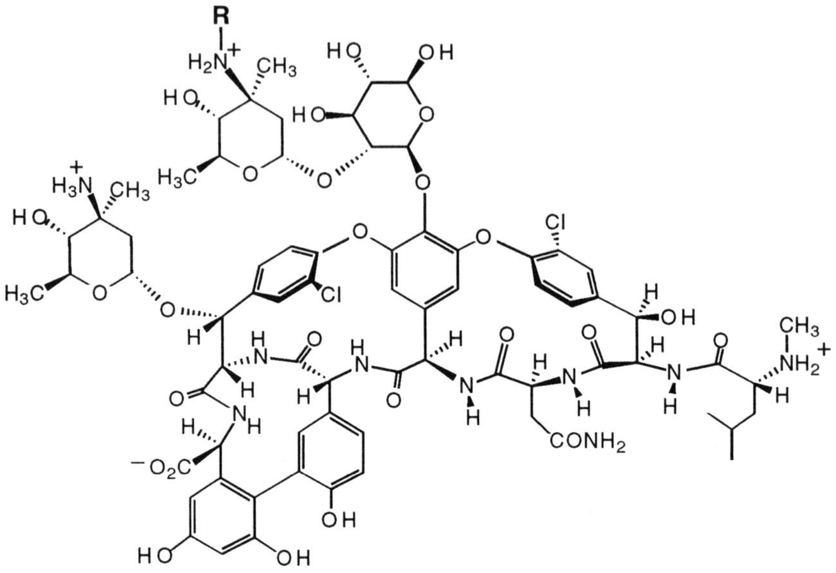

(4) R= Cl-biphenyl

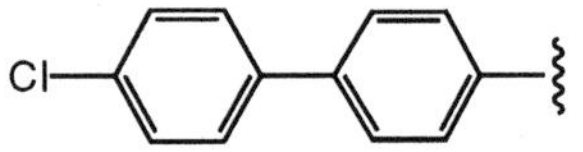

(5) R= H

is not a viable antibiotic in these cases, and an initial conclusion might be that there is no hope for useful activity of glycopeptides against these serious pathogens.

The above conclusion, however reasonable it might seem at the outset, is incorrect. Scientists at Eli Lilly in Indianapolis have recently taken the antibiotic chloroeremomycin,[23] (5), and added to this the hydrophobic tail which is shown in the resulting structure (4).[24] This molecule has remarkable activity against vancomycin resistant enterococci.[25] For example, in a mouse model, the effective dose in reducing colonisation and clearing infection is typically 50 times lower than for vancomycin.[26] The new compound also has good *in vitro* activity against MRSA, being generally about 8 times more active than vancomycin.[24] It also has exceptional *in vitro* activity against penicillin-resistant *Streptococcus pneumoniae*, and in general its *in vivo* activity is commensurate with its *in vitro* potency.[24]

The above facts regarding the activity of (4) raise a crucial question. How can it be that a compound that is not expected to bind strongly to –L-Lys-D-Ala-D-Lac precursors has remarkable activity? The apparent dilemma is made more clear by the observation that the new glycopeptide (4) and vancomycin both have low affinity for di-*N*-acetyl-L-Lys-D-Ala-D-Lac (both bimolecular binding constants lie in the range 500-1800 M^{-1}, in contrast to vancomycin affinity of ca. 10^6 M^{-1} for the cell-wall analogue di-*N*-acetyl-L-Lys-D-Ala-D-Ala). One possible approach to try to understand this apparent anomaly would be to seek a completely new mechanism of action for the binding of glycopeptides which are active against bacteria utilising D-Ala-D-Lac precursors. However, it seemed to us that we should first consider that the binding of glycopeptides to –D-Ala-D-Lac precursors might be strengthened by subtle cooperative phenomena. This viewpoint was reinforced by our earlier work on the origins of binding affinities of cell-wall precursors to glycopeptides.

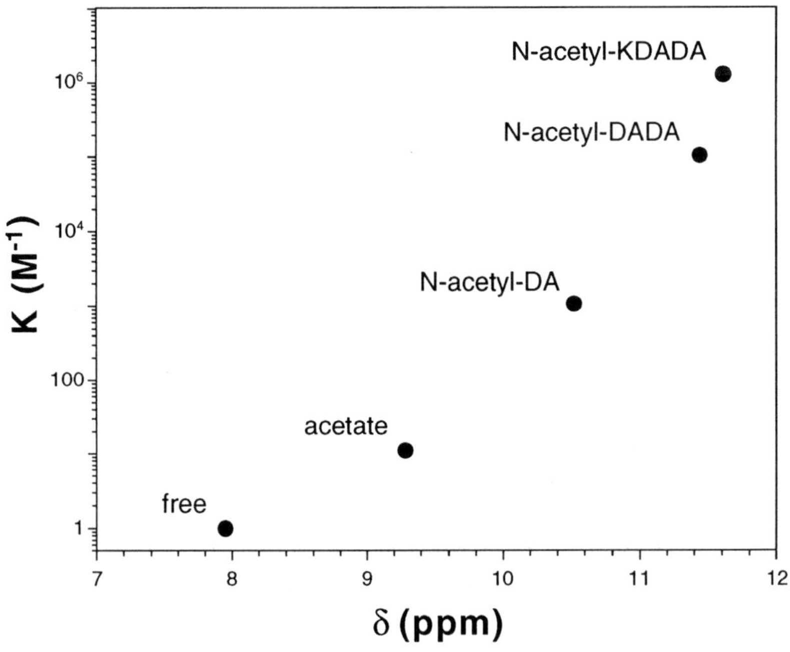

Figure 7 *Plot of the binding constant of peptide ligands to a vancomycin group antibiotic vs. the chemical shift of the NH proton w_2. As the ligand becomes progressively longer the binding constant increases and the chemical shift of w_2 appears further downfield; an indication that the hydrogen bond between w_2 and the carboxylate is becoming stronger.*

This work indicates that the main source of binding affinity lies in the binding of the carboxyl group into a pocket containing 3 NHs (Figure 1). This conclusion is by no means self-evident. Acetate ion binds into the aforementioned pocket of 3 NHs with a binding constant of only 10 M^{-1}, and it is the further interactions with other portions of di-*N*-acetyl-L-Lys-D-Ala-D-Ala that increase the binding constant to ca. 10^6 M^{-1}. Despite the fact that the latter number is dramatically larger than the former, it is in the interaction of the carboxylate anion with the NH binding pocket (Figure 1) that much of the adverse entropy of binding is removed, and it is therefore the intrinsic strength of this interaction that is greatest. Once this basic affinity is present, then it may be increased by neighbouring interactions in two ways. First, the neighbouring interactions add what is formally their own affinity to the binding. Second, and crucially, the neighbouring interactions strengthen the binding of the carboxylate into its binding pocket. This effect is shown in Figure 7: each point on the graph is for a cell-wall analogue, starting with the simplest (CH_3COO^-) to have a terminal carboxylate (analogous to that of the terminal -D-Ala of a piece of bacterial cell wall), and proceeding along the series to *N*-acetyl-D-Ala, thence to *N*-acetyl-D-Ala-D-Ala, and finally to di-*N*-acetyl-L-Lys-D-Ala-D-Ala. In Figure 7, along the horizontal axis is plotted the chemical shift of one of the NH protons of the antibiotic (w_2, that of the second amino acid from its *N*-terminus) which binds the carboxylate anion of the bacterial cell-wall peptide (Figure 1). It is important to realise that the plotted chemical shift corresponds to fully bound antibiotic in each case, such that in every antibiotic molecule the NH w_2 forms a hydrogen bond to a carboxylate anion. Along the vertical axis is plotted the equilibrium constant for the binding of the cell-wall analogues to the antibiotic as measured by UV spectrophotometry. It can be seen (Figure 7) that as the binding constants between the antibiotic and the cell-wall analogues increase (as the cell-wall analogues increase in length), the carboxylates of the various cell-wall analogues hydrogen bond more and more

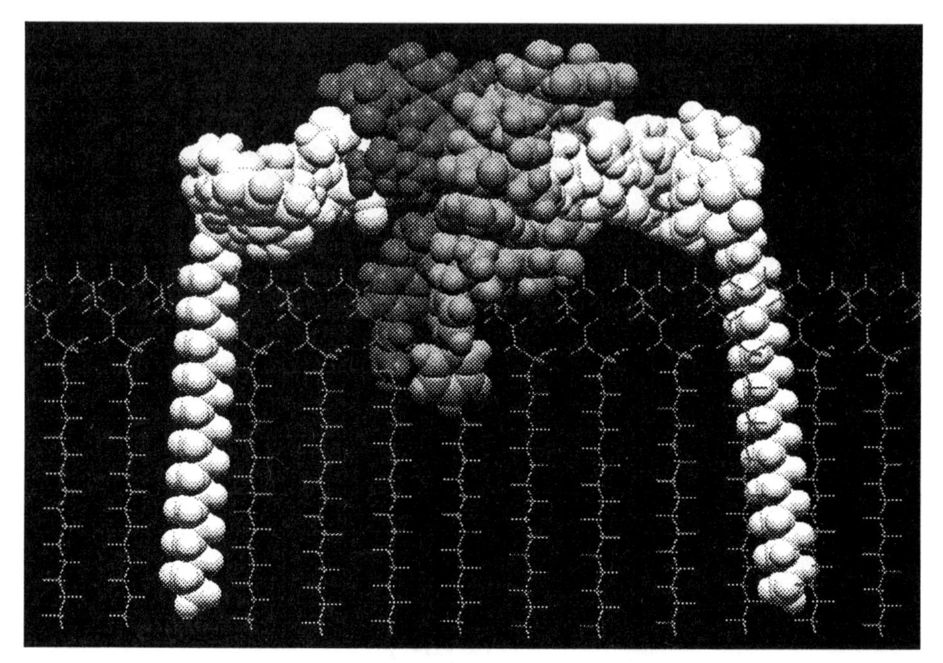

Figure 8 *Schematic representation of compound (4) utilising both a membrane anchor and dimerisation to bind to cell wall terminating in* D*-lactate at the surface of a bacterium. The molecules constituting the bacterial cell membrane are shown as stick structures. The immature bacterial cell wall is shown in white, and the two halves of the dimer in light and dark grey. Note the potential of the biphenyl groups (lower part of the antibiotic molecules) to act as a membrane anchor.*

strongly to the antibiotic [this conclusion is based simply on the assumption that a greater downfield shift of the NH reflects stronger hydrogen bond formation]. Thus, the experiments show how the tightness of binding of the carboxylate group, into the antibiotic pocket which accepts it, is increased by the cooperative binding provided by adjacent groups.[27]

The above demonstration that the binding of a functional group can be increased by other neighbouring groups which help to reduce the motion of the first in its binding site led us to the hypothesis that the affinity for –L-Lys-D-Ala-D-Lac precursors might be increased by alternative sources of cooperative interactions. It was felt that it should be possible to build on the strength of the weak interaction between the NH binding pocket and the carboxylate anion by dimerisation of the antibiotic, and by membrane attachment of the antibiotic, as outlined earlier in this article. In particular, it was felt that if these two features could operate simultaneously, then the largest cooperative increase in binding would be observed.

We hypothesise that (4) can bind to bacterial cell wall using the locating, and partially immobilising, devices **both** of membrane anchoring and dimerisation (Figure 8).

7 COMMENTS ON THE COOPERATIVITY AND CONCLUSIONS: THE GULLIVER EFFECT

We have recently commented upon the fact that the decreased motion of a ligand in a binding site works to improve the electrostatic interactions that are formed in that binding site.[28] In a reciprocal manner, features which improve the electrostatics of binding in a ligand/receptor interaction likewise reduce the degree of motion in the binding site. The two

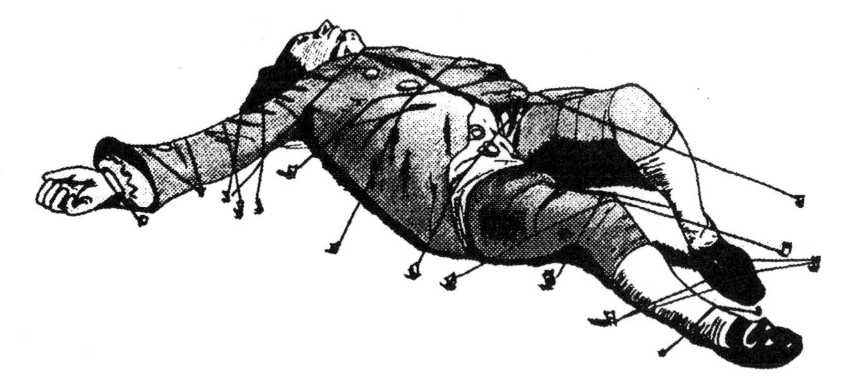

Figure 9 *Gulliver was tied down by many weak bindings. As with the association of molecules, an array of weak interactions is an effective way to restrict motion and ensure strong net binding.*

effects can be regarded as working iteratively on each other. For example, if **A** binds in isolation to part of a binding site with an exothermicity **X**, and **B** binds in isolation to another part of the binding site with an exothermicity **Y**, then attachment of **A** to **B** so that they can simultaneously bind into the binding site in a cooperative manner is not only advantageous through the operation of the classical chelate effect, but leads to an exothermicity of binding greater than **X** + **Y**. A useful way of looking at this phenomenon is in terms of what one might call the "Gulliver effect". If you wish to keep a close interaction between Gulliver's shoulders and the ground, then it is clear that an agitated Gulliver would be better constrained if he were also held at the waist and feet (Figure 9). The figure gives a crude, but perhaps useful, insight as to how the carboxylate group of a –D-Ala-D-Lac precursor can make a stronger interaction when other features that may reduce local motion (dimerisation and membrane anchor) act at the same time. The work indicates how subtle locating devices can strengthen adjacent interactions and how important antibacterial activity can be recouped from a situation that may initially appear hopeless.

REFERENCES AND FOOTNOTES

1. F. J. Marshall *J. Med. Chem.*, 1965, **8**, 18.
2. H. R. Perkins *Biochem. J.*, 1969, **111**, 195.
3. D. H. Williams and J. R. Kalman *J. Am. Chem. Soc.*, 1977, **99**, 2768.
4. G. M. Sheldrick, P. G. Jones, O. Kennard, D. H. Williams and G. A. Smith *Nature*, 1978, **271**, 223.
5. C. M. Harris and T. M. Harris *J. Am. Chem. Soc.*, 1982, **104**, 4293.
6. M. P. Williamson and D. H. Williams *J. Am. Chem. Soc.*, 1981, **103**, 6580.
7. J. R. Kalman and D. H. Williams *J. Am. Chem. Soc.*, 1980, **102**, 897.
8. J. C. J. Barna, D. H. Williams, D. J. M. Stone, T. C. Leung and D. M. Doddrell *J. Am. Chem. Soc.*, 1984, **106**, 4895.
9. J. R. Kalman and D. H. Williams *J. Am. Chem. Soc.*, 1980, **102**, 906.
10. D. H. Williams, M. P. Williamson, D. W. Butcher and S. J. Hammond *J. Am. Chem. Soc.*, 1983, **105**, 1332.
11. J. C. J. Barna, D. H. Williams and M. L. Williamson *J. Chem. Soc., Chem. Commun.*, 1985, 254.
12. J. P. Waltho and D. H. Williams *J. Am. Chem. Soc.*, 1989, **111**, 2475.
13. J. P. Mackay, U. Gerhard, D. A. Beauregard, R. A. Maplestone and D. H. Williams *J. Am. Chem. Soc.*, 1994, **116**, 4573.
14. U. Gerhard, J. P. Mackay, R. A. Maplestone and D. H. Williams *J. Am. Chem.*

Soc., 1993, **115**, 232.
15. J. P. Mackay, U. Gerhard, D. A. Beauregard, M. S. Westwell, M. S. Searle and D. H. Williams *J. Am. Chem. Soc.*, 1994, **116**, 4581.
16. D. A. Beauregard, D. H. Williams, M. N. Gwynn and D. J. Knowles *Antimicrob. Agents Chemother.*, 1995, **39**, 781.
17. M. S. Westwell, B. Bardsley, A. C. Try, R. J. Dancer and D. H. Williams *J. Chem. Soc., Chem. Commun.*, 1996, in press.
18. P. Courvalin *Antimicrob. Agents Chemother.*, 1990, **34**, 2291.
19. G. M. Eliopoulos *Eur. J. Clin. Microbiol. Infect. Dis.*, 1993, **12**, 409.
20. A. P. Johnson, A. H. C. Uttley, N. Woodford and R. C. George *Clin. Microbiol. Rev.*, 1990, **3**, 280.
21. M. Arthur, C. Molinas, T. D. H. Bugg, G. D. Wright, C. T. Walsh and P. Courvalin *Antimicrob. Agents Chemother.*, 1992, **36**, 867.
22. C. T. Walsh *Science*, 1994, **261**, 308.
23. Our nomenclature; also known as chloroorienticin and LY264826B.
24. T. I. Nicas, J. E. Flokowitsch, D. A. Preston, D. L. Mullen, J. Grissom-Arnold, N. J. Snyder, M. J. Zweifel, S. C. Wilkie, M. J. Rodriguez, R. C. Thompson and R. D. G. Cooper In *ICAAC, Session 152, New glycopeptides*; San Fransisco, 1995; F248.
25. T. I. Nicas, D. L. Mullen, J. Grissom-Arnold, N. J. Snyder, M. J. Zweifel, S. C. Wilkie, M. J. Rodriguez, R. C. Thompson and R. D. G. Cooper In *ICAAC, Session 152, New glycopeptides*; San Fransisco, 1995; F249.
26. C. J. Boylan, T. I. Nicas, D. A. Preston, D. L. Zeckner, B. J. Boyll, R. A. Raab, D. L. Mullen, N. J. Snyder, L. L. Zornes, R. E. Stratford, M. J. Zweifel, S. C. Wilkie, M. J. Rodriguez, R. C. Thompson and R. D. G. Cooper In *ICAAC, Session 152, New glycopeptides*; San Fransisco, 1995; F255.
27. P. Groves, M. S. Searle, M. S. Westwell and D. H. Williams *J. Chem. Soc., Chem. Commun.*, 1994, 1519.
28. M. S. Westwell, M. S. Searle and D. H. Williams *J. Mol. Recog.*, 1996, in press.

2

Design and Synthesis of Novel Oxazolidinones Active against Multidrug-resistant Bacteria

M. R. Barbachyn, S. J. Brickner, G. J. Cleek, R. C. Gadwood, K. C. Grega, S. K. Hendges, D. K. Hutchinson, P. R. Manninen, K. Munesada, R. C. Thomas, L. M. Thomasco, D. S. Toops and D. A. Ulanowicz

MEDICINAL CHEMISTRY RESEARCH, PHARMACIA & UPJOHN, INC., KALAMAZOO, MI 49001, USA

1 INTRODUCTION

The development of bacterial resistance to currently available therapeutic agents has been viewed with alarm by the medical community. Some particularly problematic organisms include multidrug-resistant strains of *Mycobacterium tuberculosis*, penicillin- and cephalosporin-resistant *Streptococcus pneumoniae*, methicillin-resistant *Staphylococcus aureus* (MRSA) and *Staphylococcus epidermidis* (MRSE), and vancomycin-resistant *Enterococcus faecalis* and *Enterococcus faecium*. Of even greater concern is the expectation of the vancomycin-resistant enterococci transferring their resistance determinants to the more virulent MRSA. When this occurs there will be no effective treatment for infections caused by such a bacterium.

A number of solutions have been suggested for overcoming problems associated with bacterial resistance.[1] Some examples include structural modification of known drugs, combination drug therapy, new therapeutic approaches such as the use of narrow spectrum drugs with improved diagnostics, and identification of novel agents with a unique mechanism of action.

The antibacterial oxazolidinones, exemplified by DuP 721, were discovered by workers at DuPont.[2] DuP 721 is a totally synthetic substance, possesses a unique mechanism of action involving the inhibition of protein synthesis at the initiation phase,[3] has activity against multiply resistant Gram-positive bacteria,[4] and exhibits good oral activity and satisfactory pharmacokinetic parameters. Some less attractive attributes of DuP-721 include the bacteriostatic nature of its action and, more importantly, the demonstration of significant toxicity for (±)-DuP 721 in a chronic rat model.[5]

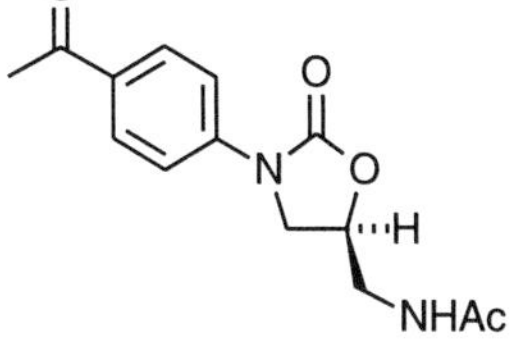

DuP 721

Subsequent synthetic efforts at DuPont led to more potent oxazolidinone analogs such as (1) and (2).[6]

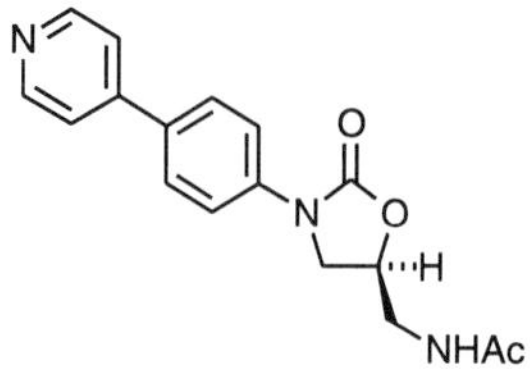

(1)

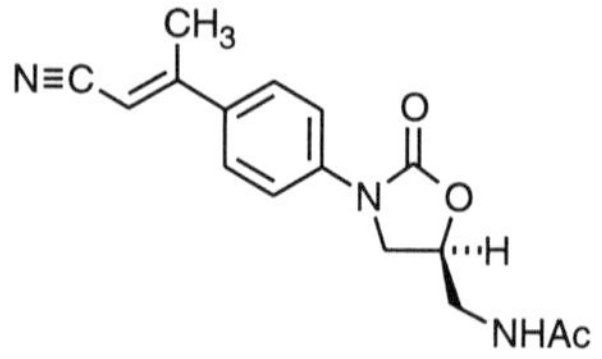

(2)

We speculated that the pyridyl ring system of oxazolidinones (1) could be replaced with a piperazine bioisostere. This strategy had been employed in a reverse sense in the quinolone antibacterial agent area.[7] Reduction of this strategy to practice eventually led to a new class of oxazolidinone antibacterial agents characterized by a substituted piperazine moiety, as shown in the generic structure (3). Ultimately, optimization of this subclass led to the selection of eperezolid (U-100592) as a clinical candidate.[8]

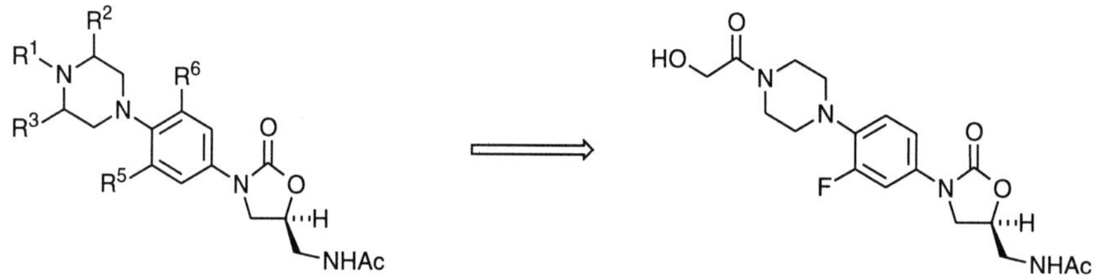

(3)

eperezolid (U-100592)

R^1 = alkyl, acyl, carboalkoxy, sulfonyl
R^2, R^3 = H, CH_3
R^5, R^6 = H, F

In a further extension of this bioisosteric strategy, we envisioned the incorporation of cyclic oxazine and thiazine appendages onto the phenyloxazolidinone template (4).

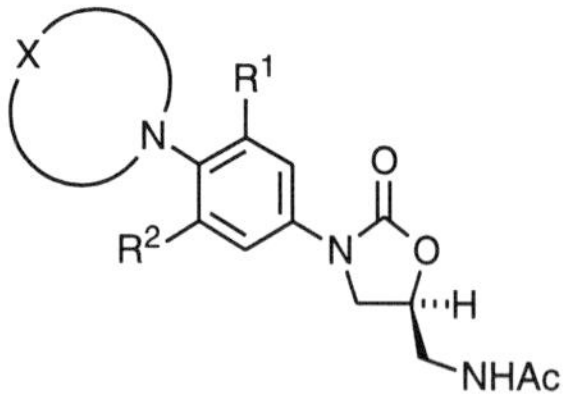

(4)

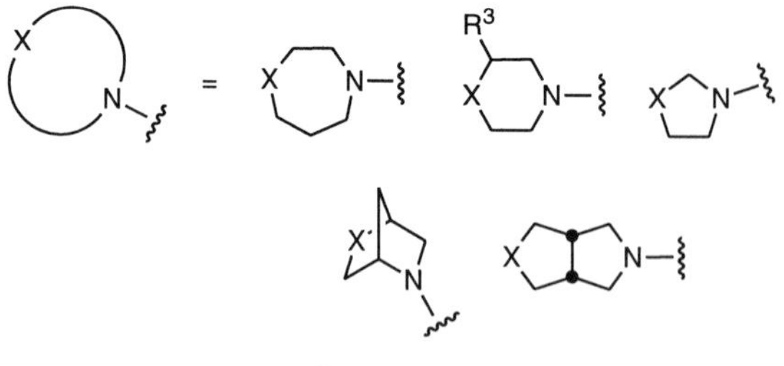

X = O, S, SO, SO_2
R^1, R^2 = H, F

R^3 = CH_3OCH_2-, $CH_3OCH_2CH_2$-

A retrosynthetic analysis for the targeted compounds (4) is shown in Scheme 1. The The C-5 acetamidomethyl sidechain of (4) should be readily available from the corresponding 5-(hydroxymethyl)oxazolidinone (5) through established methodology.[2] In what can be regarded as the key step of the planned synthetic protocol, an aryl carbamate (6) would be deprotonated with *n*-butyllithium and reacted with (*R*)-glycidyl butyrate, a strategy employed previously to prepare enantiomerically enriched 5-(hydroxymethyl)oxazolidinones,[9] to directly generate (5). Carbamate (6), in turn, would be readily available from starting materials (7) and (8), via a three step reaction sequence involving nucleophilic substitution (LG = halogen, triflate, or alternative leaving group), reduction, and Schotten-Baumann functionalization.

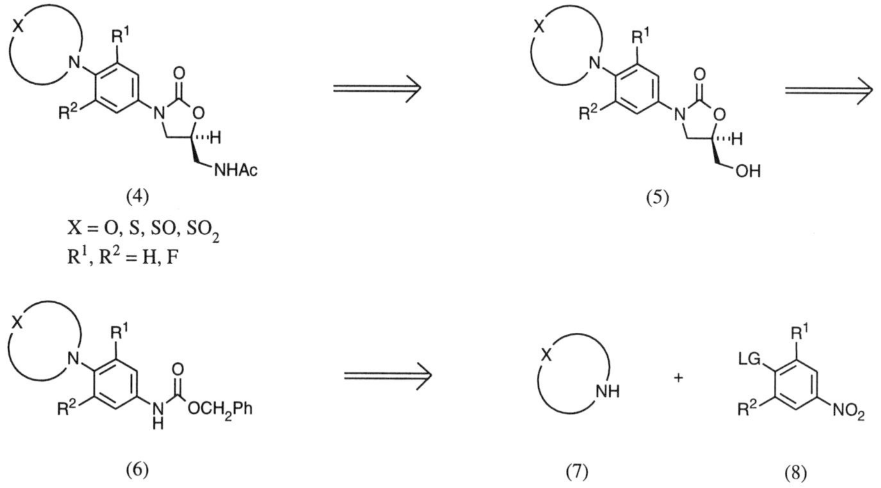

(4)

X = O, S, SO, SO_2
R^1, R^2 = H, F

(5)

(6)

(7)

(8)

Scheme 1

2 SYNTHETIC CHEMISTRY

A representative procedure for the assembly of oxazinyl- and thiazinyl-phenyloxazolidinones is shown in Scheme 2. Starting compound 3,4-difluoronitrobenzene was reacted with morpholine to give high yields of the adduct (9). Reduction of the nitro group and Schotten-Baumann reaction of the resultant aniline then afforded the Cbz derivative (10). The stage was then set for the pivotal oxazolidinone ring-forming reaction. Carbamate (10) was deprotonated at low temperature and treated with (*R*)-glycidyl butyrate. Warming to ambient temperature then generated the important 5-(hydroxymethyl)oxazolidinone intermediate (11). The hydroxyl group of (11) was then activated by conversion to the corresponding mesylate. Introduction of the acetamidomethyl sidechain was then achieved by treating the mesylate with sodium azide, followed by catalytic hydrogenation and acetylation to generate the morpholinylphenyloxazolidinone linezolid (U-100766).[8]

In the difluoro series we utilized 2,6-difluoro-4-nitrophenol (12)[10] as the starting material. Phenol (12) was first converted to the triflate (13) and then reacted with morpholine to generate a good yield of the addition product (14). Triflate transfer to morpholine was a competing side reaction. Elaboration of intermediate (14) to the

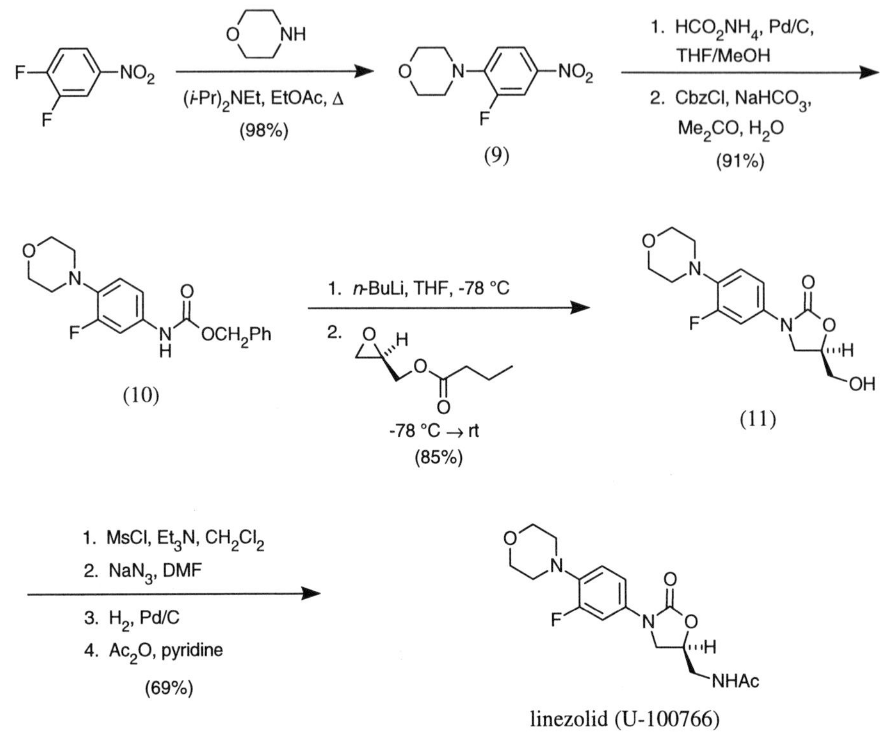

F F NO_2 O NH

(i-Pr)$_2$NEt, EtOAc, Δ
(98%)

O N F NO_2

(9)

1. HCO_2NH_4, Pd/C, THF/MeOH
2. CbzCl, $NaHCO_3$, Me_2CO, H_2O
(91%)

O N F N H O OCH_2Ph

(10)

1. *n*-BuLi, THF, -78 °C
2. O H O O
-78 °C → rt
(85%)

O N F N O O H OH

(11)

1. MsCl, Et_3N, CH_2Cl_2
2. NaN_3, DMF
3. H_2, Pd/C
4. Ac_2O, pyridine
(69%)

O N F N O O H NHAc

linezolid (U-100766)

Scheme 2

difluorophenyloxazolidinone U-99512 was accomplished via a synthetic procedure essentially identical to that employed for the monofluoro congener (*vide supra*).

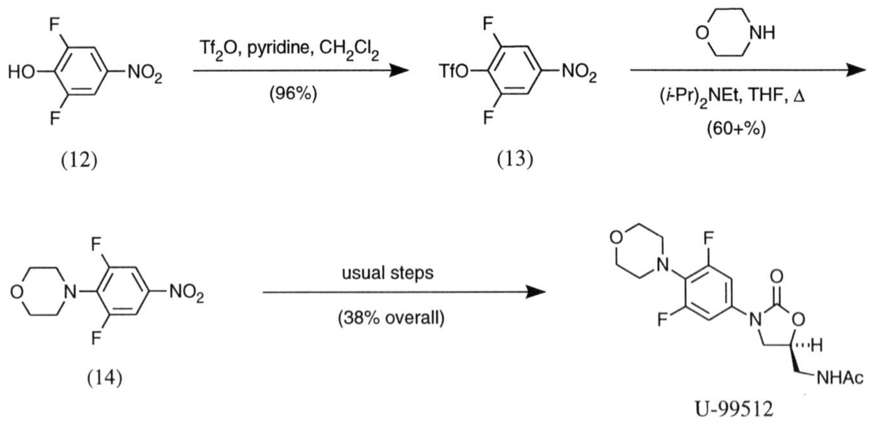

F HO F NO_2

(12)

Tf_2O, pyridine, CH_2Cl_2
(96%)

F TfO F NO_2

(13)

O NH
(i-Pr)$_2$NEt, THF, Δ
(60+%)

O N F F NO_2

(14)

usual steps
(38% overall)

O N F F N O O H NHAc

U-99512

Scheme 3

Another focus was on examples of (4) wherein X is sulfur. As shown in Scheme 4 for the thiomorpholine congener U-100480, the methodology used to prepare these compounds closely parallels that described previously. However, due to the presence of the sulfur moiety some changes to the usual reaction conditions were necessary. For example, reduction of the nitro group of intermediate (15) to give the aniline (16) required the development of alternative procedures. Interestingly, the reduction/Schotten-Baumann reaction sequence could be conducted in one pot to furnish the carbamate (17). Elaboration of (17) to the 5-(azidomethyl)oxazolidinone (18) proceeded uneventfully. Reduction of the azide to the corresponding amine was accomplished by application of the procedure of Vaultier and coworkers.[11] Acetylation then afforded the targeted analog U-100480.[12]

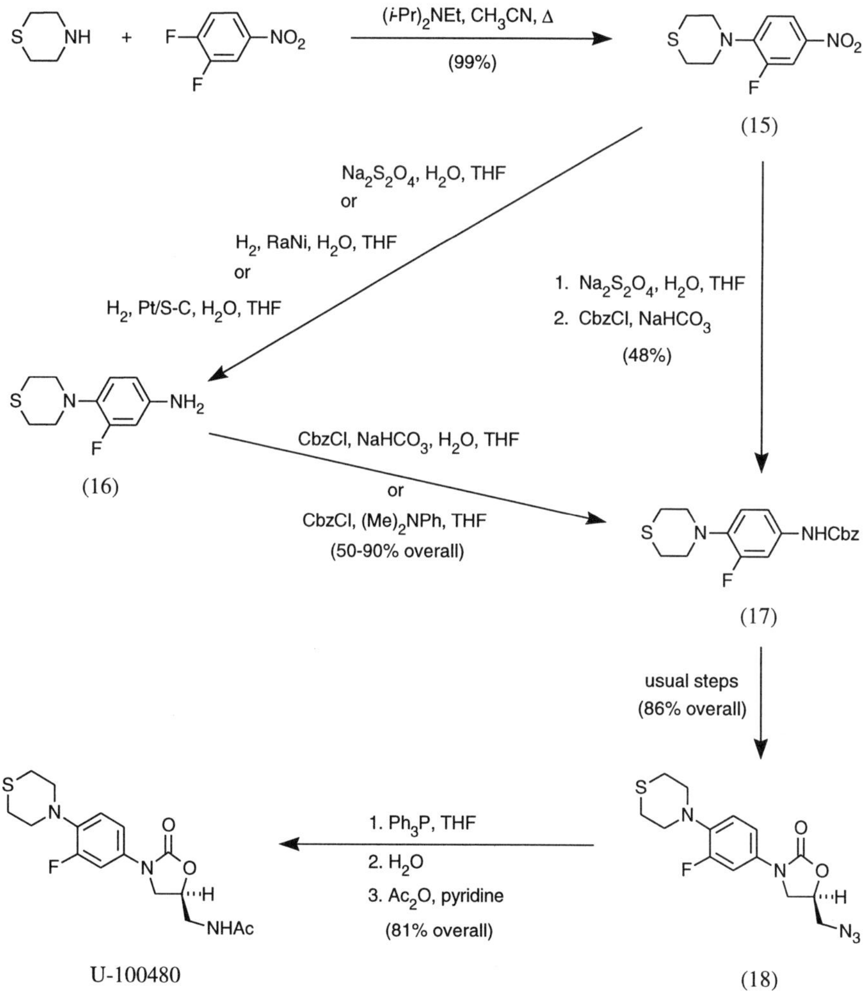

Scheme 4

An alternative and preferable method for introducing the requisite 5-aminomethyl sidechain is illustrated in Scheme 5. The oxazolidinone mesylate (19), bearing an interesting 2-thia-5-azabicyclo[2.2.1]heptan-5-yl appendage, was subjected to direct ammonolysis conditions to generate the amine (20). Acetylation under standard conditions provided the desired 5-(acetamidomethyl)oxazolidinone U-107120 in excellent overall yield.

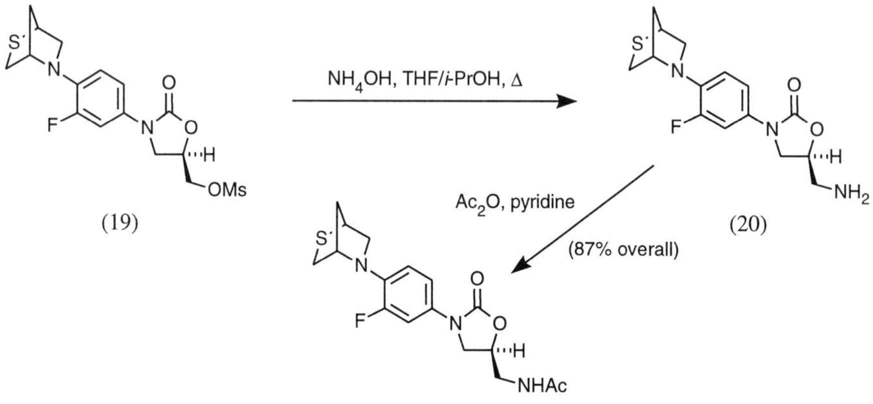

NH$_4$OH, THF/*i*-PrOH, Δ

(19) (20)

Ac$_2$O, pyridine

(87% overall)

U-107120

Scheme 5

Thioether linkages are amenable to further elaboration. We were especially interested in oxidizing selected sulfide-containing oxazolidinone analogs to the corresponding sulfoxide or sulfone to probe the effect of such a modification on antibacterial activity. As shown in Scheme 6, we were pleased to find that sulfide U-141235 could be readily oxidized to the sulfoxide U-141541, employing classical sodium metaperiodate conditions.[13] The corresponding sulfone analog U-141604 was prepared by treating U-141235 with catalytic osmium tetroxide in the presence of *N*-methylmorpholine *N*-oxide.[14]

3 BIOLOGICAL RESULTS AND DISCUSSION

The effects of ring substitution on antibacterial activity are highlighted in Table 1, using the clinical agent vancomycin as the control antibiotic. With regard to phenyl ring substituents, the primary emphasis was on fluorinated derivatives, as earlier work at Pharmacia & Upjohn had identified the pronounced potentiating effect of such appendages.[15] One trend that is clearly evident from the data is that increasingly large R^3 substituents on the morpholine ring have a detrimental effect on both the *in vitro* activity (MIC) and *in vivo* efficacy (ED_{50}) of the analogs against *S. aureus*.

A comparison of oxazinyl and thiazinyl ring systems is shown in Table 2. In general, the *in vitro* activity of the oxazines is very similar to that of the corresponding thiazine derivatives. Interestingly, with the exception of the six-membered ring series (U-100766

Table 1 *Effect of Ring Substitution on Antibacterial Activity*

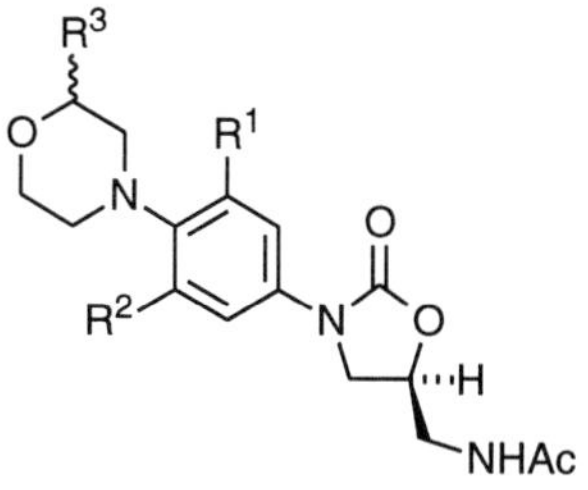

U-Number	R^1	R^2	R^3	*Staphylococcus aureus* UC 9213	
				MIC[a] (μg/mL)	ED_{50}[b] (mg/kg)
141960	H	H	H	4	—
100766	H	F	H	2	5.6 (3.9)
99512	F	F	H	2	5.5 (2.0)
104070	H	F	CH_3OCH_2	16	17.6 (2.0)
104159	H	F	$CH_3OCH_2CH_2$	32	15.0 (3.1)

[a]Minimum inhibitory concentration: the lowest concentration of drug that inhibits visible growth of the organism; vancomycin MIC 1 μg/mL. [b]Effective dose $_{50}$: the amount of drug required to cure 50% of mice subjected to a lethal systemic infection; oxazolidinones administered PO; SC administered vancomycin ED_{50} in parentheses.

Table 2 *Oxazine to Thiazine Comparison*

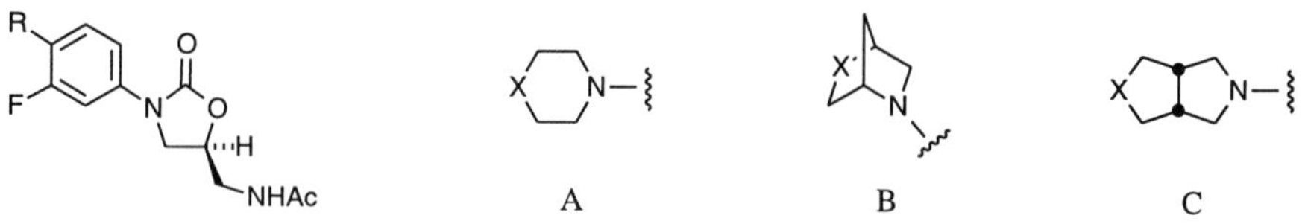

A B C

U-Number	R	X	*Staphylococcus aureus* UC 9213	
			MIC (μg/mL)[a]	ED_{50}, PO (mg/kg)
100766	A	O	2	5.6 (3.9)[b]
100480	A	S	4	1.5 (1.7)[b]
106647	B	O	4	13.8 (2.9)[b]
107120	B	S	2	4.2 (4.0)[b]
141258	C	O	4	>20 (3.8)[c]
141235	C	S	2	4.3 (2.1)[c]

[a]Minimum inhibitory concentration; vancomycin MIC 1 μg/mL. [b]Vancomycin ED_{50} (SC) in parentheses. [c]Eperezolid (U-100592) ED_{50} (PO) in parentheses.

and U-100480), the oxygen-substituted compounds are generally less active *in vivo* than their sulfur congeners.

The effect of ring size was explored. As shown in Table 3, the seven- and especially six-membered ring analogs are more active *in vivo* than the corresponding five-membered ring analog.

Table 3 *Effect of Ring Size on Antibacterial Activity*

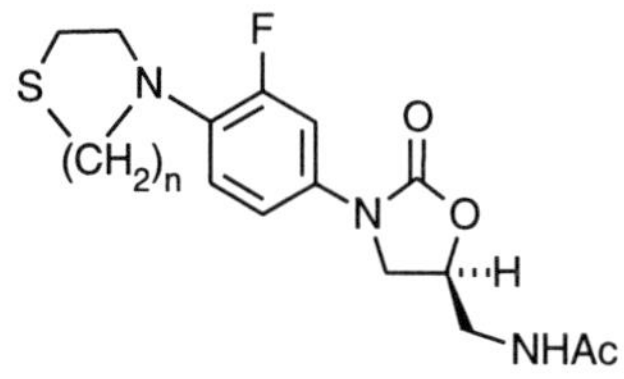

U-Number	n	*Staphylococcus aureus* UC 9213	
		MIC (μg/mL)[a]	ED_{50}, PO (mg/kg)
107296	1	2	11.8 (3.0)[b]
100480	2	4	1.5 (1.7)[b]
141259	3	4	8.8 (6.3)[c]

[a]Minimum inhibitory concentration; vancomycin MIC 1 μg/mL. [b]Vancomycin ED_{50} (SC) in parentheses. [c]Eperezolid (U-100592) ED_{50} (PO) in parentheses.

An examination of the effect of sulfur oxidation state on antibacterial activity proved interesting (see Table 4). In general, with the exception of the 3-thia-7-aza-bicyclo[3.3.0]octan-7-yl sulfoxide derivative U-141541, the sulfides, sulfoxides, and sulfones displayed comparable levels of *in vitro* and *in vivo* activity against *S. aureus*.

The thiomorpholine-substituted oxazolidinone U-100480 was submitted for a preliminary evaluation of its *in vitro* antimycobacterial activity (see Table 5).[12] The observed activity against a screening strain of *Mycobacterium tuberculosis* (H37Rv) was comparable to that of the clinical comparator isoniazid. The antimycobacterial activity of U-100480 was also found to extend to other species, including *Mycobacterium avium* complex, an opportunistic pathogen associated with the acquired immune deficiency syndrome (AIDS). The activity levels compared favorably with those of the comparator azithromycin. While the data is not shown, U-100480 has proven equally active against a panel of ten drug-sensitive and drug-resistant strains of *M. tuberculosis* and also multiple clinical isolates of *M. avium* complex.[12]

Linezolid (U-100766), with an appended morpholine moiety as its salient structural feature, underwent an *in vitro* activity evaluation against bacteria of interest (see Table 6 for selected examples).[16] Linezolid demonstrated good activity against methicillin-susceptible and methicillin-resistant strains of *S. aureus*. This activity extended to vancomycin-resistant *Enterococcus faecalis* and penicillin-resistant *Streptococcus pneumoniae.* Interestingly, linezolid also exhibited respectable activity levels against the important fastidious Gram-negative organisms *Haemophilus influenzae* and *Moraxella catarrhalis.* In addition, the anaerobe *Bacteroides fragilis* was susceptible to linezolid.

The *in vitro* activity of linezolid correlated quite well with the *in vivo* efficacy seen in lethal systemic mouse models (see Table 7).[17] Linezolid dosed orally or subcutaneously was equipotent with subcutaneously administered vancomycin against MRSA. Good efficacy against vancomycin-resistant *Enterococcus faecium* and penicillin- and cefaclor-resistant *S. pneumoniae* was also observed.

Table 4 *Effect of Sulfur Oxidation State on Antibacterial Activity*

R F N O O H NHAc

$(O)_n$=S N— A

$(O)_n$=S N— B

$(O)_n$=S N— C

$(O)_n$=S N D

U-Number	R	n	*Staphylococcus aureus* UC 9213	
			MIC (μg/mL)[a]	ED_{50}, PO (mg/kg)
100480	A	0	4	1.5 (1.7)[b]
101603	A	1	8	3.4 (2.5)[b]
101244	A	2	8	2.0 (2.3)[b]
141259	B	0	4	8.8 (6.3)[c]
141303	B	1	4	7.0 (5.2)[c]
141453	B	2	4	—
141235	C	0	2	4.3 (2.1)[c]
141541	C	1	8	10.0 (1.4)[c]
141604	C	2	2	3.5 (2.9)[c]
107120	D	0	2	4.2 (4.0)[b]
143610	D	1	4	—
107723	D	2	4	4.5 (1.5)[b]

[a]Minimum inhibitory concentration; vancomycin MIC 1 μg/mL. [b]Vancomycin ED_{50} (SC) in parentheses. [c]Eperezolid (U-100592) ED_{50} (PO) in parentheses.

Table 5 *In Vitro Activity of U-100480 Against Mycobacteria*

Organism	MIC (μg/mL)[a]		
	U-100480	isoniazid	azithromycin
M. tuberculosis H37Rv	≤0.125	0.2	—
M. avium ATCC 49601	4	—	4
M. avium 101	0.5	—	4
M. simiae	2	—	4
M. xenopi	2	—	1
M. malmoense	2	—	16
M. fortuitum	8	—	>64

[a]Minimum inhibitory concentration.

Table 6 *Selected In Vitro Activities of Linezolid (U-100766)*

Organism	No. Strains	MIC_{90} (μg/mL)[a]	
		U-100766	vancomycin
S. aureus, methicillin-susceptible	12	4	1
S. aureus, methicillin-resistant	41	4	2
E. faecalis, vancomycin-susceptible	14	4	2
E. faecalis, Van B	10	4	>16
S. pneumoniae, penicillin-resistant	10	1	≤0.25
H. influenzae	9	11.8[b]	>16
M. catarrhalis	10	4	>16
B. fragilis	10	4	>16

[a]MIC_{90}: concentration at which 90% of the isolates are inhibited. [b]Weighted MIC_{90}

4 CONCLUSIONS

The design, synthesis, and antibacterial evaluation of a series of oxazine- and thiazine-substituted phenyloxazolidinone antibacterial agents was described. Many of the analogs displayed potent *in vitro* and *in vivo* activity against aerobic Gram-positive bacteria. The thiomorpholine analog U-100480 also demonstrated excellent *in vitro*

Table 7 *Selected In Vivo Activities of Linezolid (U-100766)*

Organism	ED_{50} (mg/kg)	
	U-100766[a]	vancomycin[b]
S. aureus UC 9271[c]	2.9	13.2
S. aureus UC 9213[d]	5.6	3.9
	2.0[e]	3.9
S. aureus UC 6685[f]	3.8	2.6
E. faecium UC 15090[g]	24.0	>100
S. pneumoniae UC 15087[h]	3.8	—[i]

[a]PO administration unless noted otherwise. [b]SC administration. [c]Methicillin-susceptible. [d]MRSA. [e]SC administration. [f]Multidrug-resistant MRSA. [g]Vancomycin-resistant. [h]Penicillin- and cephalosporin-resistant. [i]Penicillin G and cefaclor ED_{50}s >20 mg/kg.

activity against acid-fast organisms such as *M. tuberculosis* and *M. avium*, including multidrug-resistant strains. Linezolid (U-100766), characterized by an appended morpholine ring system, exhibited *in vitro* and *in vivo* activity against drug-sensitive and drug-resistant Gram-positive bacteria, as well as some of the fastidious Gram-negative organisms. Good activity was also observed against anaerobic bacteria, exemplified by *B. fragilis*. On the basis of these and other considerations beyond the scope of this discussion, linezolid was selected for further evaluation and is currently in Phase II clinical trials.

Acknowledgements

The authors are indebted to J. W. Allison, C. W. Ford, J. C. Hamel, J. K. Moerman, R. D. Schaadt, D. Stapert, D. M. Wilson, B. H. Yagi, R. J. Yancey, Jr., and G. E Zurenko, all of Pharmacia & Upjohn, Inc., for the *in vitro* and *in vivo* test results. We also thank M. H. Cynamon and S. P. Klemens of the Veteran Affairs Medical Center and SUNY Health Science Center, and S. E Glickman and J. O. Kilburn of the Centers for Disease Control and Prevention, for the *in vitro* mycobacteria test results.

References

1. L. L. Silver and K. A. Bostian, *Antimicrob. Agents Chemother.*, 1993, **37**, 377.
2. W. A. Gregory, D. R. Brittelli, C.-L. J. Wang, M. A. Wuonola, R. J. McRipley, D. C. Eustice, V. S. Eberly, P. T. Bartholomew, A. M. Slee and M. Forbes, *J. Med. Chem.*, 1989, **32**, 1673, and references cited therein.
3. D. C. Eustice, P. A. Feldman, I. Zajac and A. M. Slee, *Antimicrob. Agents Chemother.*, 1988, **32**, 1218.
4. A. M. Slee, M. A. Wuonola, R. J. McRipley, I. Zajac, M. J. Zawada, P. T. Bartholomew, W. A. Gregory and M. Forbes, *Antimicrob. Agents Chemother.*,

1987, **31**, 1791.
5. R. C. Piper, T. F. Platte and J. R. Palmer, unpublished results, The Upjohn Company.
6. R. K. Carlson, C.-H. Park and W. A. Gregory, U.S. Patent 4,948,801, 1990. D. R. Brittelli, W. A. Gregory, P. F. Corless, C.-H. Park, European Patent Application 0,316,594, 1989.
7. For a general discussion of quinolone SAR see L. A. Mitscher and P. Devasthale, 'Quinolone Antimicrobial Agents', Second Edition, D. C. Hooper and J. S. Wolfson, Editors, American Society for Microbiology, Washington, D.C., 1993, Chapter 2.
8. S. J. Brickner, D. K. Hutchinson, M. R. Barbachyn, P. R. anninen, D. A. Ulanowicz, S. A. Garmon, K. C. Grega, S. K. Hendges, D. S. Toops, C. W. Ford and G. E. Zurenko, *J. Med. Chem.*, 1996, **39**, 673.
9. S. J. Brickner, P. R. Manninen, D. A. Ulanowicz, K. D. Lovasz and D. C. Rohrer, 'Abstracts of Papers', 206th National Meeting of the American Chemical Society, Chicago, IL, American Chemical Society, Washington, D.C., August 22-27, 1993, ORGN 089.
10. K. L. Kirk, *J. Heterocyclic Chem.*, 1976, **13**, 1253.
11. M. Vaultier, N. Knouzi and R. Carrié, *Tetrahedron Lett.*, 1983, **24**, 763.
12. M. R. Barbachyn, D. K. Hutchinson, S. J. Brickner, M. H. Cynamon, J. O. Kilburn, S. P. Klemens, S. E. Glickman, K. C. Grega, S. K. Hendges, D. S. Toops, C. W. Ford and G. E. Zurenko, *J. Med. Chem.*, 1996, **39**, 680.
13. N. J. Leonard and C. R. Johnson, *J. Org. Chem.*, 1962, **27**, 282.
14. S. W. Kaldor and M. Hammond, *Tetrahedron Lett.*, 1991, **32**, 5043.
15. M. R. Barbachyn, D. S. Toops, K. C. Grega, S. K. Hendges, C. W. Ford, G. E. Zurenko, J. C. Hamel, R. D. Schaadt, D. Stapert, B. H. Yagi, J. M. Buysse, W. F. Demyan, J. O. Kilburn and S. E. Glickman, *Bioorg. Med. Chem. Lett.*, 1996, **6**, 1009.
16. G. E. Zurenko, B. H. Yagi, R. D. Schaadt, J. W. Allison, J. O. Kilburn, S. E. Glickman, D. K. Hutchinson, M. R. Barbachyn and S. J. Brickner, *Antimicrob. Agents Chemother.*, 1996, **40**, 839.
17. C. W. Ford, J. C. Hamel, D. M. Wilson, J. K. Moerman, D. Stapert, R. J. Yancey, Jr., D. K. Hutchinson, M. R. Barbachyn and S. J. Brickner, *Antimicrob. Agents Chemother.*, 1996, **40**, 1508.

3
From the Michael Reaction to the Clinic

J. C. Barrière and J. M. Paris*

RHÔNE-POULENC RORER, CENTRE DE RECHERCHE DE VITRY-ALFORTVILLE, 13, QUAI JULES GUESDE, F 94403 VITRY SUR SEINE CEDEX, FRANCE

1. INTRODUCTION

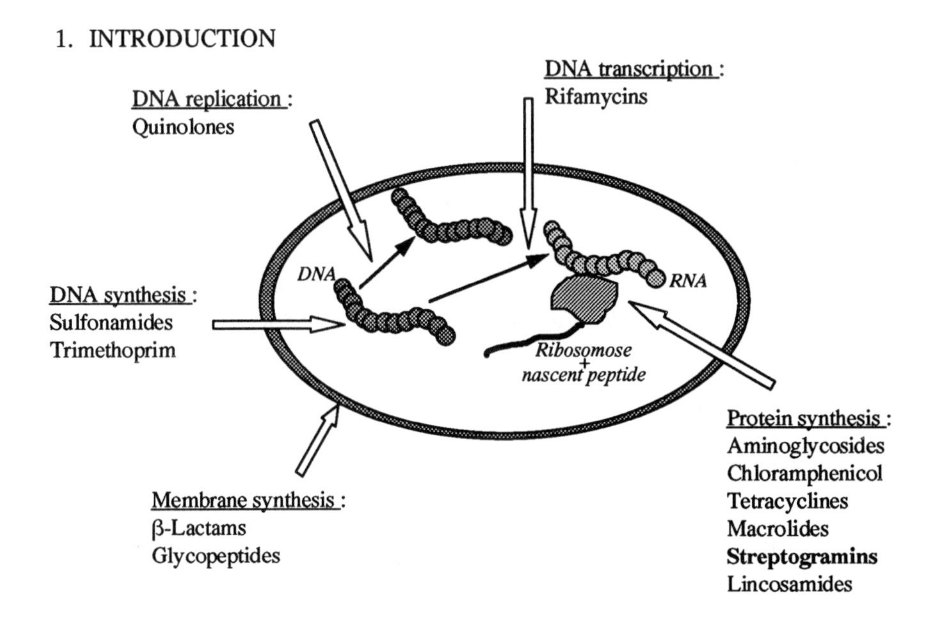

Figure 1 *Targets of the major classes of antibiotics*

The major classes of antibiotics exert their action on a limited number of targets within bacteria : cell wall synthesis, DNA synthesis, DNA replication , DNA transcription and protein synthesis (Figure 1). Among the many antibiotics that inhibit protein synthesis in prokaryotic cells, the streptogramins are unique in that they consist of two kinds of structurally unrelated molecules. The synergy between the two types of molecule confers real advantages over other classes of antibiotics, and the streptogramins may prove effective in the treatment of severe infections caused by multiresistant Gram-positive bacteria.[1,2]

2. PROPERTIES OF THE STREPTOGRAMINS

2.1 Structures

A great number of natural streptogramins has been isolated; these compounds can be considered a structurally homogeneous family.[3,4] They are produced by fermentation principally from *Streptomyces*. The two major streptogramins, pristinamycin and virginiamycin, are produced on the scale of several tens of tons a year. The streptogramins are constituted of two groups of synergistic components : 15 to 40% of group B streptogramins (pristinamycin I or PI) and 60 to 85% of group A streptogramins (pristinamycin II or PII).[5,6] The molecules of group A are polyunsaturated macrolactones, whereas the group B components are peptidic macrolactones (depsipeptides) (Figure 2).

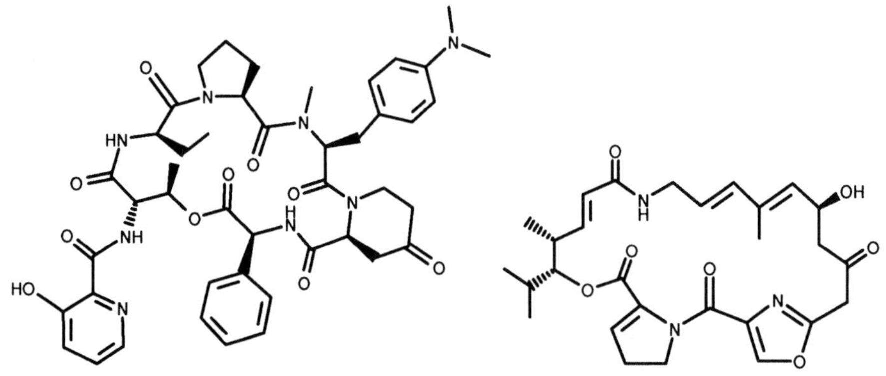

pristinamycin I_A pristinamycin II_A

Figure 2 *Structures of the main components of pristinamycin*

2.2 Biological activities

Streptogramins are active against Gram-positive bacteria (especially staphylococci and streptococci), some Gram-negative bacteria (generally those involved in respiratory tract infections, such as *Haemophilus*, *Neisseria* and *Legionella*) and mycoplasma. The components both inhibit protein synthesis by binding to the 50 S subunit of the ribosome.

As shown in Table 1, the association of the two components of pristinamycin is synergistic : the *in vitro* activity of the mixture is ten-fold greater than the average of the activities of the individual compounds. This phenomenon of synergy is of importance in the case of resistant strains (see entry 3 in Table 1), since in this example pristinamycin I is only slightly active *in vitro* and pristinamycin II is practically inactive whereas the association displays an excellent activity. This synergy can also be observed in an *in vivo* model of *S. aureus* infections, in which the activity (curative dose 50%) of the association is 50 mg/kg for individual values ranging from 75 to 300 mg/kg. The synergistic effect may involve a conformational change of the ribosomes induced by the pristinamycin II_A component.[7,8]

In addition, each component alone displays only a bacteriostatic activity while the synergistic association shows a strong bactericidal action against susceptible bacteria.

Table 1 *Comparison of antibacterial activity of separate components of pristinamycin and of the association*

	In vitro activity MIC (mg/l)			In vivo activity# CD50 (mg/kg)		
	PI	PII	PI+PII	PI	PII	PI+PII
Staphylococcus aureus	2 - 4	1 - 2	0.12	60-100	300	50
Streptococcus pyogenes	2	0.25	0.03			
Str. pneumoniae Ery R*	16	128	0.25	>150	>300	120
Haemophilus influenzae	64	4	1			
Mycoplasma pneumoniae	8	1	0.06			
Neisseria gonorrhoeae	4	0.25	0.06			

#Mouse septicemia, * Erythromycin resistant

The main advantage of the synergy is schematised in Figure 3 : although pristinamycin I is active only on 70 % of the *S. aureus* strains tested at about 2 mg/l and pristinamycin II on 80% of these strains at about 1 mg/l, the mixture of the 2 components is active on 97 to 98% of tested *S. aureus* strains at approximately 0.1 mg/l.

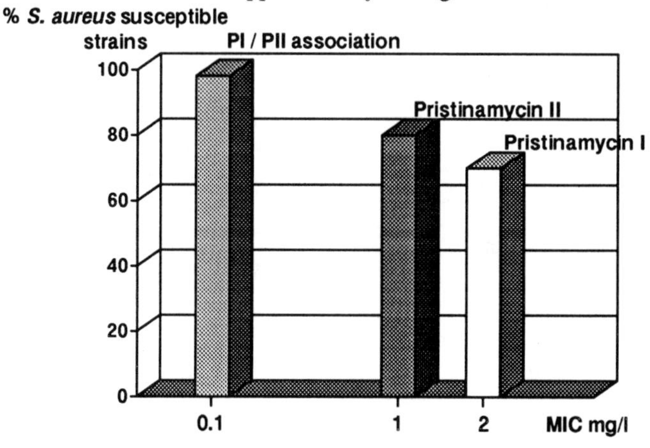

Figure 3 *Effect of synergy on the potency and the spectrum of pristinamycins*

In spite of their effectiveness, these naturally occurring antibiotics have not been extensively used : their poor water solubility means that they cannot be applied to treat severe bacterial infections. In order to solve this problem, an intensive semisynthetic study was initiated with the aim to synthesise water-soluble derivatives of the two groups.[9]

3. PRISTINAMYCIN I CHEMISTRY

3.1 Introduction

Examination of the pristinamycin I molecule suggests several sites for possible chemical modifications : the 3-hydroxy picolinic residue (O and N), the lactone function, the dimethylamino phenylalanine residue (N and aromatic reactivities) and the keto group and its α-positions (Figure 4).

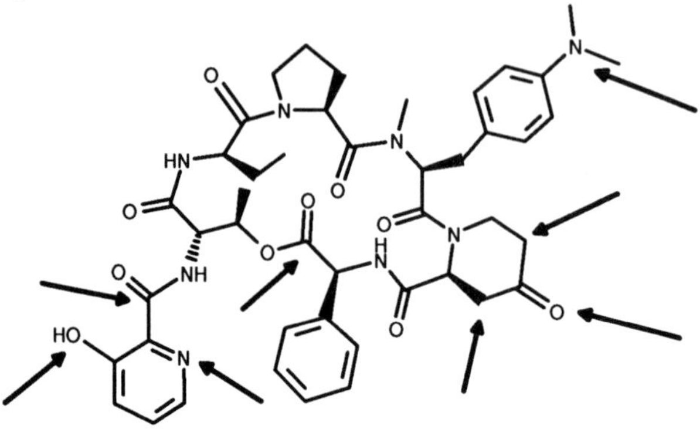

Figure 4 *Reactive sites of pristinamycin I*

It rapidly became apparent that the lactone function is required for the biological activity: the opening of the macrolactone ring results in completely inactive compounds. Hereafter are described the major transformations performed in order to identify water-soluble derivatives.

3.2 Reactions on 3-Hydroxy Picolinoyl Nucleus (Scheme 1)

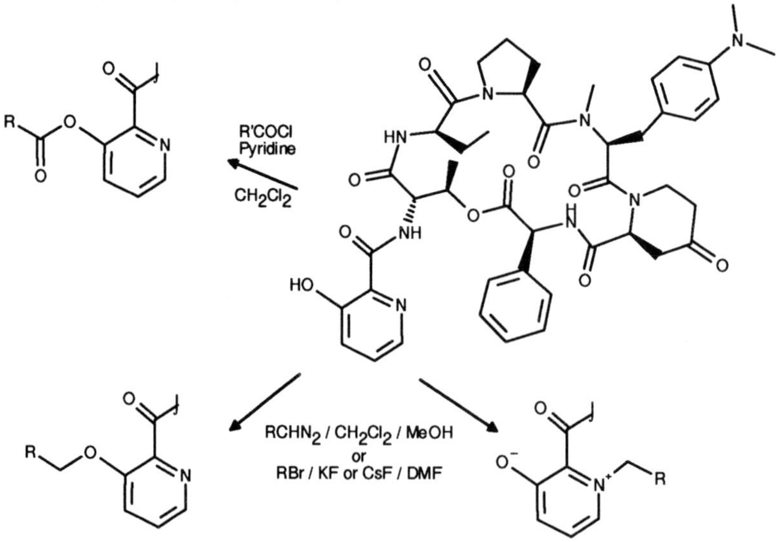

Scheme 1 *Reactions on 3-hydroxy picolinoyl residue*

Acyl derivatives of the "phenolic" function prepared by classical methods generally displayed less activity *in vitro* and *in vivo* than did pristinamycin I.

Alkylated derivatives of the hydroxy group were obtained, in mixture with betaines, by treatment of pristinamycin I with diazoalkanes. These compounds alone displayed no activity, which demonstrates the importance of this hydroxy function for the activity of group B streptogramins.[10] The betaines showed only a weak biological activity alone or in association with pristinamycin II. An improved method for preparing betaines of virginiamycin S has been published,[11] in which alkyl halides are used in the presence of potassium fluoride in DMF, instead of diazo reagents.

3.3 Reactions on 4-Dimethylamino Phenylalanine Residue

Reaction of pristinamycin I with alkyl halides yielded, as expected, the corresponding quaternary ammonium derivatives (Scheme 2),[12] which displayed no antibacterial activity of interest. The treatment of pristinamycin I_A with N-chloro succinimide or N-bromo succinimide led to derivatives of pristinamycin I_A having a chlorine or bromine atom in the position ortho to the dimethylamino group in 75% yields.

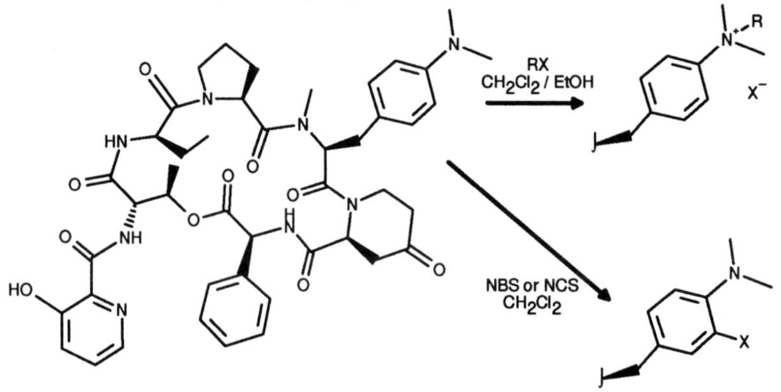

Scheme 2 *Reaction on the dimethylamino phenylalanine residue*

3.4 Modifications using the Reactivity of the Keto Group

The keto group was chosen, for its reactivity and for the induced reactivity in the α-position.

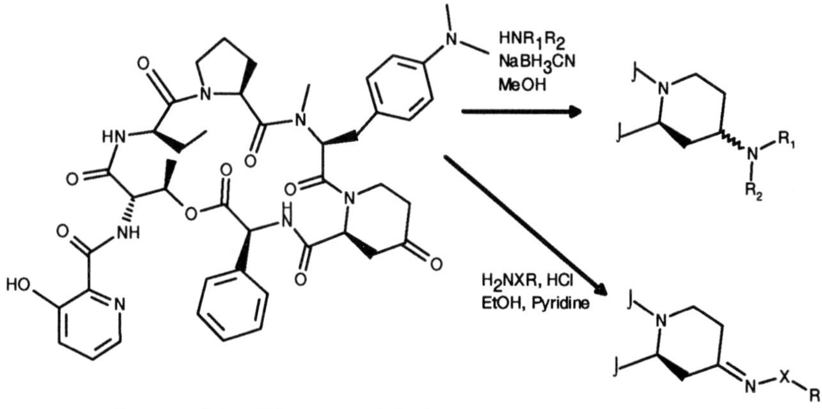

Scheme 3 *Modifications of the keto group of pristinamycin I*

3.4.1 Modification of the Keto Group. This function had already been modified by formation of substituted oximes and hydrazones under classical conditions, but the derivatives obtained were not sufficiently active to be considered for development (Scheme 3).

Reductive amination using the described conditions led to 5γ-deoxy 5γ-amino pristinamycin I_A derivatives in moderate yields (30-50%) (Scheme 3). This type of derivative alone displayed less activity than pristinamycin I; however, the synergy was retained, as the association of it with pristinamycin II was practically as active as the natural antibiotic (Table 2).[13,14] The acid salts of these derivatives were soluble in water.

3.4.2 Modifications in the Position α to the Keto Group. This position seemed attractive since it had never been explored before. The main problem associated with reactions in this position was to find reagents that would spare the lactone function and yet provide versatile intermediates. Among the various choices, the reaction of DMF-acetals on activated methylene groups appears interesting since it works in nearly neutral medium and yields enaminones, which possess numerous possibilities of reaction. A short study of the reactivity of DMF-acetals and analogues on pristinamycin I led to the selection of Bredereck's reagent :[15] at room temperature, the reaction of pristinamycin I with an excess of this reagent provided regioselectively the enaminone in 75% yield (Scheme 4).[16] This regioselectivity is a consequence of the conformation of the molecule: the dimethylamino-phenyl nucleus hides the 5β methylene, while the other methylene group (5δ) remains accessible.

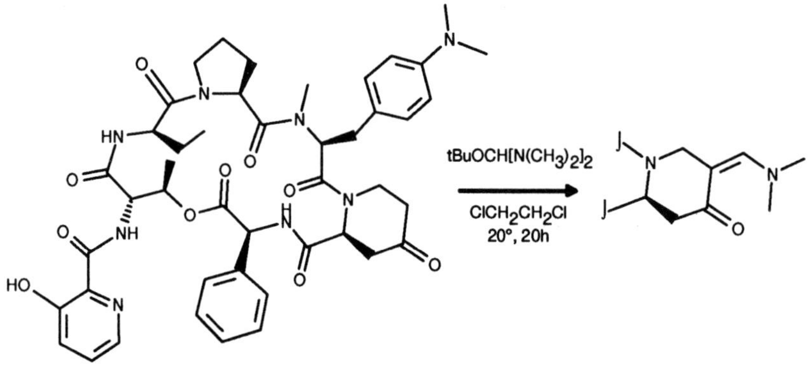

Scheme 4 *Enamination of pristinamycin I*

In scheme 5 are depicted the most interesting products obtained from the enaminone. Water-soluble derivatives were obtained via transenamination the enaminone with diamines in acidic medium; this type of enaminone is stable in water provided that the enaminic nitrogen bears a hydrogen, which forms a stabilising hydrogen bond with the keto group.

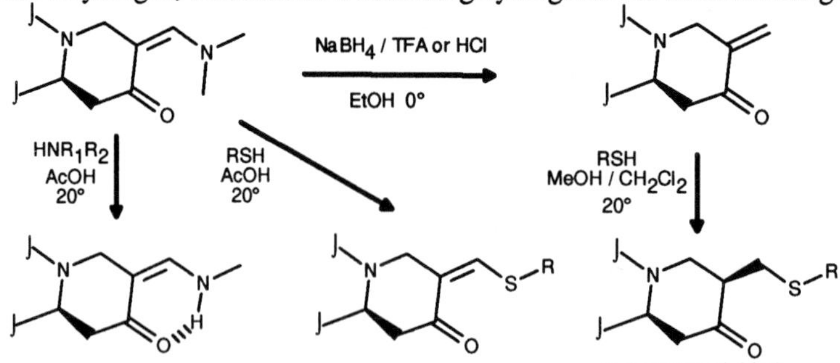

Scheme 5 *Reactions performed on the enaminone derivative of pristinamycin I*

A similar reaction with amino thiols provided thiovinylic derivatives of which the corresponding acid salts are soluble in water.

These 2 types of compounds alone displayed slightly less *in vitro* activities than did the natural product, but showed activity in association with pristinamycin II (Table 2).[17]

In order to eliminate the double bond in the position α to the keto group, the enaminone was reduced with sodium borohydride in acidic medium, yielding the methylene ketone. Then via a Michael-type reaction with aminothiols, this methylene ketone provided 5δ– thiomethyl derivatives. In this case again the conformation of the molecule induced a stereoselectivity : the equatorial isomer was generally formed in a proportion greater than 80%. This family of derivatives exhibited an *in vitro* antibacterial activity equivalent to that of pristinamycin I, both alone and in association with pristinamycin II (Table 2);[18] owing to its route of administration (s.c. or i.v.), however, it displayed better *in vivo* activity than the natural compound, which is necessarily given by the oral route.

Table 2 *Activity of water-soluble pristinamycin I derivatives*

	Water solubility	*In vitro* activity MIC (mg/l)	
		Alone	PII association
Pristinamycin I_A	0.1 %	2 - 4	0.12
	6 %	8	1
	1 %	1	0.25
	10 %	1	0.12
	10 %	1	0.25

The optimisation of this series led to the selection, on chemical, bacteriological and toxicological criteria, of the 3S-quinuclidinethiomethyl derivative (quinupristin) as the best compound for further development (see below).

3.5 Large-scale preparation of quinupristin

Since the pathway described above for the synthesis of the selected pristinamycin I derivative could not be easily scaled up for the preparation of large quantities, a new method was designed :

-the Mannich reaction was used to produce a mixture of the 2 diastereoisomers of a 5δ-aminomethyl derivative of pristinamycin I_A

-this intermediate was not isolated but was treated, in a 2-phase medium, with a sodium acetate/acetic acid buffered solution to give rise to the methylene derivative as a crystalline powder

-reaction of this compound with 3(S)-quinuclidinethiol in acetone at -20°, in order to favour the formation of the major isomer (5δR), led to quinupristin in a 75% yield as a crystalline powder (Scheme 6).[19]

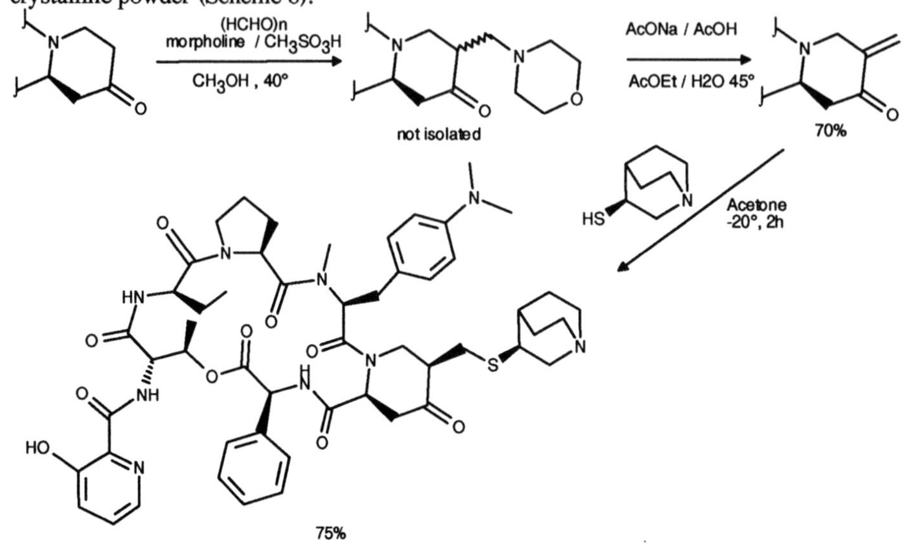

Scheme 6 *Large-scale preparation of quinupristin*

4. PRISTINAMYCIN II CHEMISTRY

4.1 Introduction

There are several obvious sites for chemical modifications of pristinamycin II_A :[6] the hydroxy group, the keto group and its α position, the dienic system and the double bond conjugated with the lactone function (Figure 5). Chemically modifying these components appeared to be much more difficult than doing so with the pristinamycin I components because the molecules were less stable. This instability is due to the following features :

- the macrolactone function is easily cleaved
- the hydroxy group (in the allylic position of a diene and beta to a ketone) in acidic medium is easily dehydrated, providing an inactive trienone, and in basic medium, it can induce a retro-aldolisation reaction, leading to mixtures.

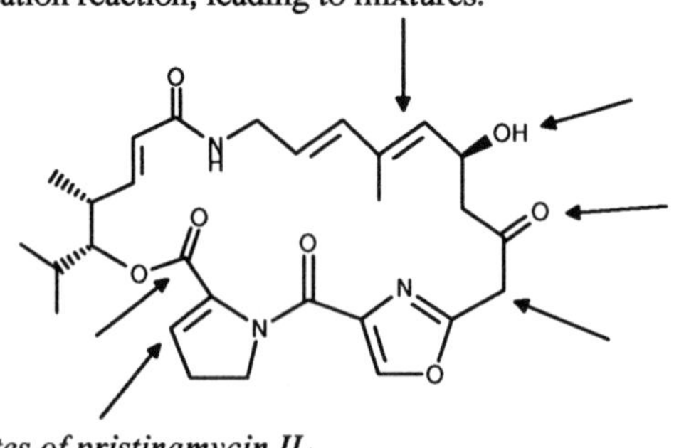

Figure 5 *Reactive sites of pristinamycin II_A*

4.2 Modifications of the hydroxy and keto functions of pristinamycin II_A

Chemical modifications of the hydroxy and keto functions were performed. Acyl derivatives (Scheme 6) alone did not display any biological activity *in vitro*. The keto group reacted with substituted hydroxylamines or hydrazines, but none of the corresponding derivatives had any interesting biological activity.

Scheme 6 *Acylation of the hydroxy group, transformations of the ketone and reactions at the position α to the keto group of pristinamycin II_A*

Derivatives substituted on the methylene activated by the keto group and the oxazole nucleus were prepared by reaction with DMF-acetals followed by a transenamination process (Scheme 6). The compounds obtained in these ways were devoid of antibacterial activity. These results suggested that the area of the molecule around the hydroxy and the keto groups might be of importance for the biological activity and prompted us to explore other parts of the molecule.

4.3 Michael-type addition reaction on pristinamycin II_A

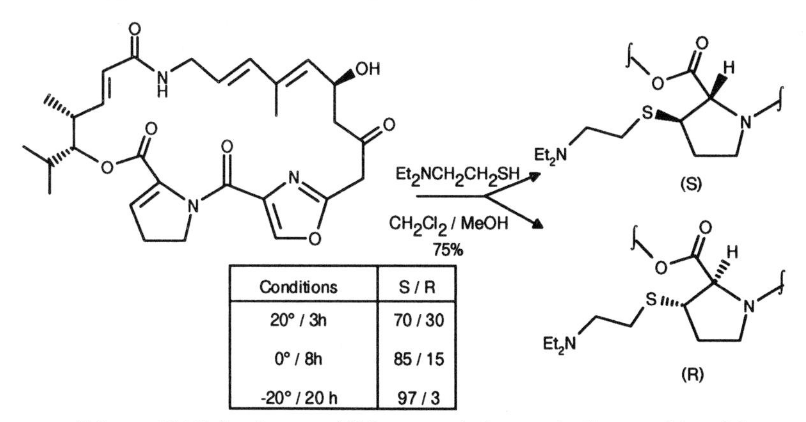

Conditions	S / R
20° / 3h	70 / 30
0° / 8h	85 / 15
-20° / 20 h	97 / 3

Scheme 7 *Michael-type addition on pristinamycin II at position 26*

The most interesting derivatives obtained to date are those prepared via a Michael-type addition of thiols to the double bond conjugated with the lactone function (Scheme 7).[20] Conjugated additions of alkylaminoalkyl thiols occurred readily in a mixture of chlorinated solvents and alcohols; the maximum of reactivity was observed when there were 2 or 3 carbon atoms between the sulfur and the nitrogen atoms. Decreasing the reaction temperature below -20° made it possible to control the stereochemistry of the addition. These 26-thio derivatives, of which the corresponding acid salts were soluble in water, were devoid of activity in the *in vitro* tests, both alone and in association with pristinamycin I, but displayed excellent *in vivo* activity in animal models (Table 3). This behaviour suggested that a transformation occurs *in vivo*, leading to active metabolites, and prompted us to synthesise oxidised derivatives.

4.4 Oxidation of the 26-thio derivatives

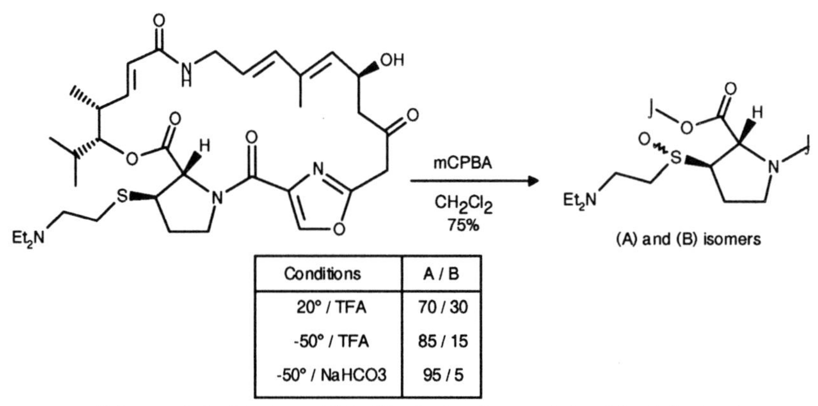

Conditions	A / B
20° / TFA	70 / 30
-50° / TFA	85 / 15
-50° / NaHCO3	95 / 5

Scheme 8 *Oxidation of 26-thio derivatives to 26-sulfinyl derivatives*

Oxidation of the sulfur atom occurred under classical conditions with mCPBA (in the presence of trifluoroacetic acid to protect the basic nitrogen atom from oxidation), providing a mixture of the 2 possible sulfoxides in a ratio depending on the temperature. The solution for obtaining one isomer in a ratio greater than 95% was found when the sulfide derivative was treated with mCPBA at -50° in the presence of an excess of sodium bicarbonate. All the compounds of this series displayed an activity both alone and in association with pristinamycin I equivalent to that of pristinamycin II.

In order to avoid the problem of asymmetry generated by the sulfoxide function, we decided to synthesise the corresponding sulfones. Due to the presence within the molecule of several functions sensitive to oxidation, this goal appeared challenging.

H_2O_2 / SeO_2

MeOH / CF_3CO_2H
20°

Scheme 9 *Oxidation of 26-thio derivatives to 26-sulfonyl derivatives*

After numerous attempts, the use of hydrogen peroxide associated with selenium dioxide provided the expected sulfones, but only in poor yields (<10%) (Scheme 9).[21]

A selection based on chemical, bacteriological and toxicological criteria was made in this series leading to the identification of dalfopristin for further development. This 26-sulfonyl pristinamycin II demonstrated biological activity equivalent to that of the natural compound (Table 3 entry 4).

Table 3 *Activity of water-soluble pristinamycin II derivatives*

	In vitro activity MIC (mg/l)		*In vivo* activity CD_{50}(mg/kg)*	
	Alone	PI association	Alone	PI association
pristinamycin IIA	2	0.12	300	50 (p.o.)
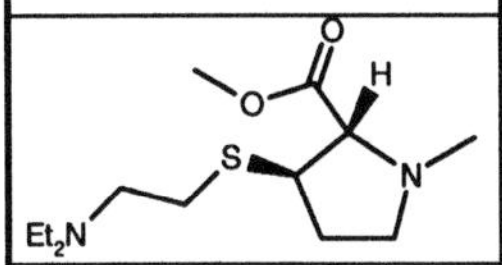 O, O, H, S, N, Et_2N	1000	4	>300	20 (s.c.)
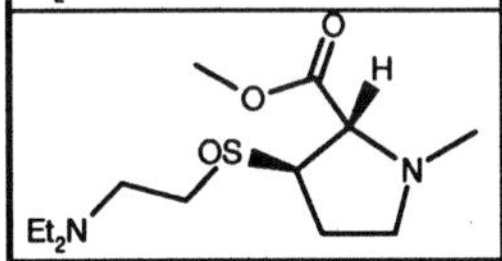 O, O, H, OS, N, Et_2N	4	0.25	220	20 - 26 (s.c.)
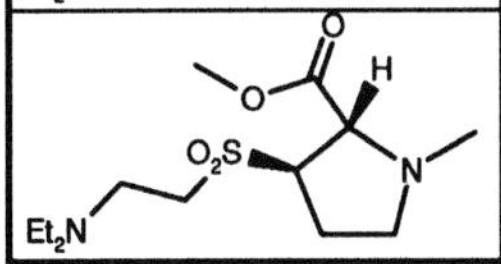 O, O, H, O_2S, N, Et_2N	2 - 4	0.25	120	10-20 (s.c.)

* *Staphylococcus aureus* mouse septicemia

4.5 Large-scale preparation of dalfopristin

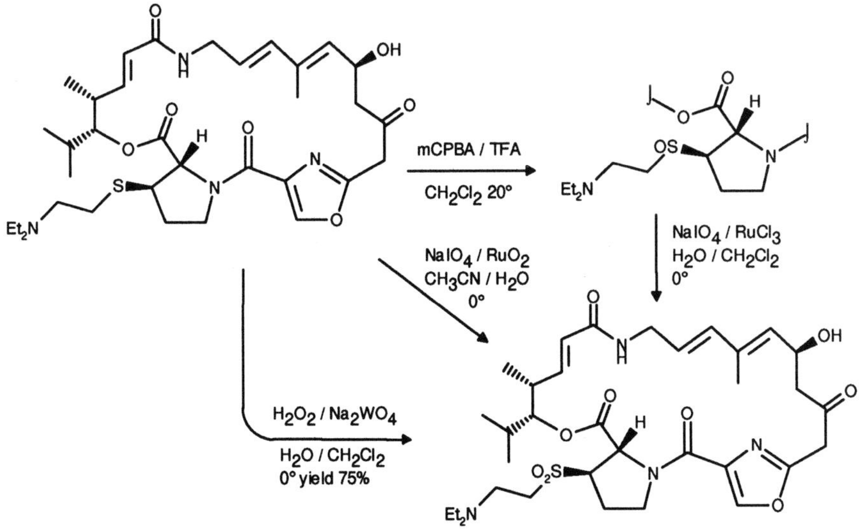

mCPBA / TFA

CH_2Cl_2 20°

$NaIO_4$ / $RuCl_3$

H_2O / CH_2Cl_2

0°

$NaIO_4$ / RuO_2

CH_3CN / H_2O

0°

H_2O_2 / Na_2WO_4

H_2O / CH_2Cl_2

0° yield 75%

Scheme 10 *Strategies for the large-scale production of dalfopristin*

The key issue for the preparation of dalfopristin was the oxidation step required for the transformation of the sulfur function into a sulfone function. A solution was found by using sodium periodate associated with ruthenium dioxide to directly oxidise the sulfur derivative in sulfone. An improvement, in terms of yield, was to use sodium periodate associated with ruthenium trichloride to oxidise the mixture of sulfoxide derivatives.[22] Finally, oxidation of the sulfide derivative by using hydrogen peroxide with sodium tungstate in a 2-phase medium gave the expected sulfone in good yields (Scheme 10).[23]

5. PROPERTIES OF THE QUINUPRISTIN/DALFOPRISTIN ASSOCIATION

5.1 Chemistry

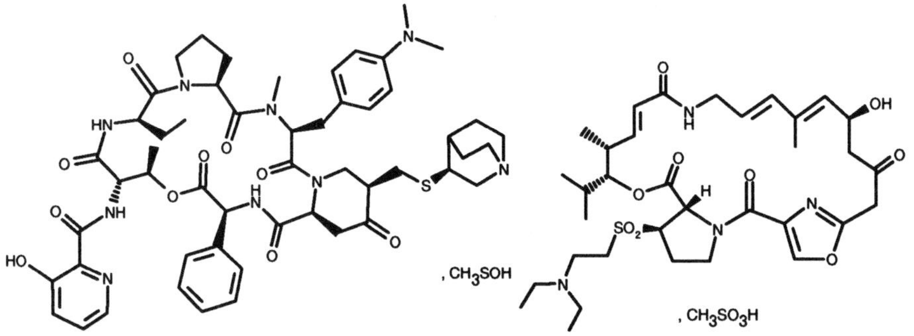

Quinupristin® 30% Dalfopristin® 70%

Figure 6. *Composition of the quinupristin/dalfopristin association*

The association is made up of the methanesulfonate salts of 5δR-[(3S)-quinuclidinyl] thiomethyl pristinamycin I (quinupristin) and of 26S-(diethylamino ethyl)sulfonyl pristinamycin II_B (dalfopristin) in a 30/70(w/w) ratio (Figure 6). This compound displays a 5% water solubility at pH of around 4.5. It is presented in a freeze-dried form to be dissolved in sterile isotonic solutions.

Table 4 *Comparison of the antibacterial activities of natural pristinamycin and the quinupristin/dalfopristin association*

	In vitro activity MIC (mg/l)		*In vivo* activity#
	Ery S*	Ery R*	CD_{50} (mg/kg)
Pristinamycin	0.12	1	50 (oral)
Quinupristin/dalfopristin	0.12	0.5	10 (s.c.)

*Erythromycin sensitive or resistant, #*Staphylococcus aureus* mouse septicemia

5.2 Biological properties

5.2.1 Antibacterial Spectrum. All the studies performed so far demonstrate clearly that this water-soluble association retains the *in vitro* properties of natural pristinamycin . It displays MIC_{50} ranging from 0.12 to 1 mg/l against *S. aureus* whatever the resistance phenotype (for example resistance to penicillin, methicillin, erythromycin and quinolones) (Table 4). Owing to its route of administration (s.c. or i.v.), however, it displays a better *in vivo* activity than the natural compound, which is necessarily given by the oral route (Table 4). The quinupristin/dalfopristin association is active against :

- Gram-positive cocci including multi-resistant isolates (methicillin-resistant *S. aureus* vancomycin-resistant *Enterococcus faecium* for example)
- a number of important Gram-negative respiratory pathogens
- intracellular pathogens

5.2.2 Bactericidal Effect. The bactericidal activity was demonstrated *in vivo* in a mouse model of *S. aureus* septicemia where the effect of the association at 120 mg/kg s.c. began 0.5 h after treatment and lasted up to 8 h after treatment. In comparison, vancomycin at the same dose had only a poor bactericidal activity which did not begin until 4 h after treatment, even though its maximum serum concentration, in this model, was more than 10 times higher than that of quinupristin/dalfopristin (Figure 6).[25]

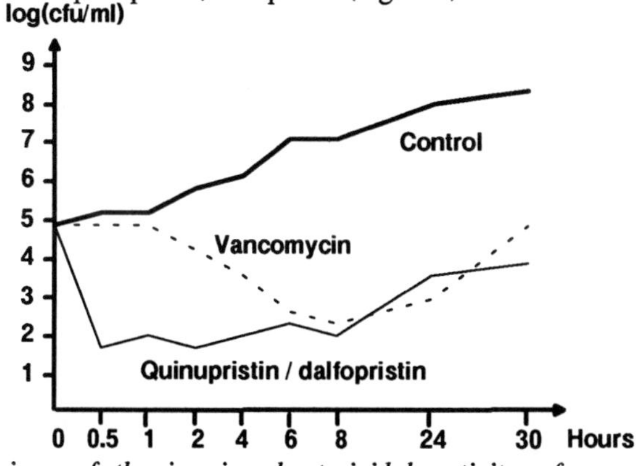

Figure 6 *Comparison of the* in vivo *bactericidal activity of vancomycin and of quinupristin/dalfopristin in a* S. aureus *septicemia in mouse*

5.2.3 Post Antibiotic Effect. The post-antibiotic effect (PAE) is the persistent suppression of bacterial growth after a short exposure of bacteria to antimicrobial agents. This phenomenon can be measured either by the time required for the count of C.F.U.(colony forming units/ml) to increase tenfold after drug removal or by the time required to restore maximal growth rate after drug removal. Quinupristin/dalfopristin had a long post-antibiotic effect on *S. aureus.*[26] For example, a 30-min. exposure to 8 x MIC of quinupristin/dalfopristin resulted in a PAE of 6.4 h under conditions in which vancomycin, erythromycin and oxacillin displayed, respectively, PAE of 2.2, 1.8 and 1.8 h.

5.2.4 Tissue concentration. The good penetration of quinupristin/dalfopristin into tissues, as demonstrated in an experimental endocarditis in rabbit (Figure 7), suggests that this compound may also be useful for the treatment of certain severe infections.[27]

6. CONCLUSION

Semi-synthetic modifications of the two major components of pristinamycin, pristinamycin I and pristinamycin II, yielded novel water-soluble pristinamycin derivatives that retain all the biological properties of the natural products, particularly their synergism, and that can be administered by the intravenous route. This first water-soluble streptogramin, quinupristin/dalfopristin, possesses the properties required for the treatment of severe infections caused by Gram-positive bacteria.

Quinupristin/dalfopristin is currently being assessed in phase III clinical trials for the treatment of severe infections caused by staphylococci, streptococci and vancomycin-resistant *E. faecium*.

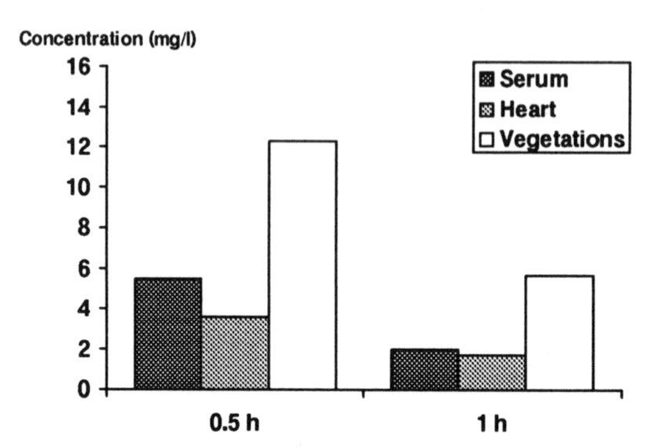

Figure 7 *Concentrations of quinupristin/dalfopristin in serum and different cardiac tissues during the treatment (20 mg/kg i.v.) of* S. aureus *experimental endocarditis in rabbit*

7. ACKNOWLEDGEMENTS

We acknowledge all the people from RHÔNE-POULENC RORER who have been involved in working on this topic. We thank D. Bouanchaud, J.F. Desnottes and N. Berthaud for helpful discussions and K. Pepper for her help with the preparation of this manuscript.

References

1. L.L. Silver and K.A. Bostian, *Antimicrob. Agents Chemother.*, 1993, **37** (3) 377.
2. J.F. Barrett, D.H. Klaubert, *Curr. Opin. Invest. Drugs.*, 1993, **2**(3): 245.
3. D. Vazquez, 'The streptogramin family of Antibiotics.' In : Antibiotics. Corcoran JN., Hahn FE. eds , Springer, Berlin, Heidelberg, New-York, 1975, Vol 3, p 521.
4. C. Cocito, *Microbiological Reviews*, 1979, **43**(2), 145.
5 (a) J. Preud'homme, A. Belloc, Y. Charpentié and P. Tarridec, *Compt. Rend. Acad. Sc. Paris,* 1965, **260**, 1309.
(b) J. Preud'homme, P.Tarridec, A. Belloc, *Bull. Soc. Chim. Fr.*, 1968, 585.
6. J.C. Barrière, D. Bouanchaud , J.F. Desnottes , J.M. Paris, *Expert Opin Invest. Drugs*,1994, **3**(2) , 115.
7. C.Cocito, ' Antibiotics' eds., J.N. Corcoran, F.E. Hahn, Springer, Berlin, Heidelberg, New-York 1983, Vol 6, p. 296.
8. D.Videau, *Path Biol*, 1982, **30**, 529.
9. J.M.Paris, J.C. Barrière, C. Smith and P.E. Bost, 'Recent Progress in the Chemical Synthesis of Antibiotics' (G.Lukacs, M. Ohno, eds), Springer Verlag, Berlin, Heidelberg, 1990, p 183.
10. J.M.Paris, J. François, C. Molherat, F. Albano, M. Robin, M.Vuilhorgne and J.C. Barrière, *J. of Antibiot.*, 1995, **48**(7), 676.
11. N.K. Sharma, and M.J.O. Anteunis, *Bull. Soc. Chim. Belges*, 1988, **97**, 365.
12. M. Bodansky and N.J. Princeton. (E.R. Squibb and Sons, Inc), 1969, U.S Patent: 3,420,816.

13. J.P. Corbet, C. Cotrel, D.Farge and J.M. Paris, 1983, French Patent: 2,549,062.
14. J.M. Paris, O. Rolin, J.P. Corbet, C. Cotrel and D. Bouanchaud, *Path. Biol.*,1985, **33**, 493.
15. R.F Abdulla and R.S. Brinkmeyer, *Tetrahedron*, 1979, **35**, 1675.
16. J.P Corbet, C. Cotrel, D. Farge and J.M. Paris, 1983, French Patent 2,549,064.
17. J.P. Corbet, C. Cotrel, D. Farge and J.M. Paris, 1983, French Patent 2,549,063.
18. J.C Barrière, D.H. Bouanchaud, J.M. Paris, O. Rolin, N.V. Harris and C. Smith, *J. Antimicrob. Chemother.*, 1992, **30** (supp A), 1-8.
19. J.P. Bastard, J.M. Paris and X. Radisson, 1992, Eur. Patent Appl., 043029.
20. J.P. Corbet, C. Cotrel, J.M. Farge and J.M. Paris, 1983, French Patent: 2,549,065.
21. J.C. Barrière, C. Cotrel and J.M. Paris, 1985 French Patent: 2,576,022.
22. D. Chaterjee, N.V. Harris, T. Parker, C. Smith and J.P. Warren, Eur.Patent Appl: 252,720.
23. X. Radisson. 1992, World Patent WO 92/ 01692-A.
24. J.C .Barrière and J.M. Paris, *Drugs of the future*, 1993, **18**(9), 833.
25. N. Berthaud, G. Montay, B.J. Conard and J.F. Desnottes, *Antimicrob. Chemother.*, 1995, **36** (2), 365.
26. W. Craig and S. Ebert, 33nd, ICAAC, New Orleans, 17-20/10/1993, AB : 470.
27. B. Fantin, R. Leclercq, M. Ottaviani, J.M. Vallois, B. Mazière, J. Duval, J.J. Pocidalo and C. Carbon, *Antimicrob. Chemother.*, 1994, **38** (3), 432.

4
The Synthesis of Salinomycin

Richard C. D. Brown and Philip Kocienski
DEPARTMENT OF CHEMISTRY, THE UNIVERSITY, SOUTHAMPTON SO17 1BJ, UK

1 INTRODUCTION

The first polyether antibiotics were isolated in 1951[1, 2] but it was not until the discovery that monensin[3] as a potent anticoccidial in poultry that an intensive search for new members of this class of compounds began. The ensuing search resulted in the isolation of a large number of new polyether antibiotics which included salinomycin **1**[4]. The salinomycin-producing organism is a *Streptomyces* designated as the strain number 80614, which was closely related to *Streptomyces albus* according to a taxonomical study[4]. The crude polyether antibiotic was obtained by extraction of the culture broth with organic solvents. Chromatography on alumina and silica gel followed by crystallisation allowed the isolation of pure salinomycin. Very closely related polyether antibiotics include: narasin A and B[5], deoxy-(*O*-8)-salinomycin, deoxy-(*O*-8)-epi-17-salinomycin[6], and A-28086D[7].

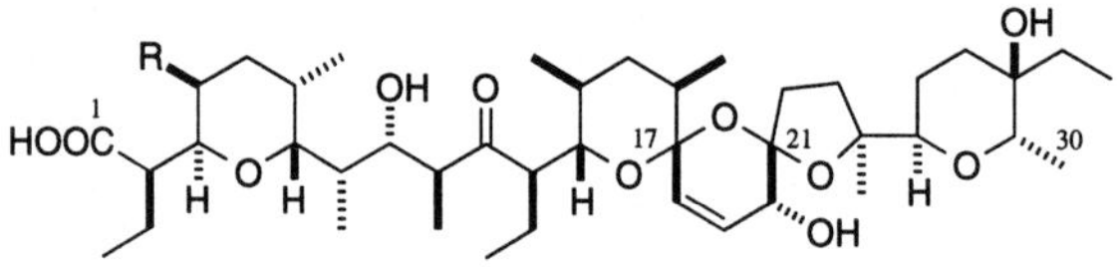

1 R = H **Salinomycin**
2 R = Me **Narasin A**

Salinomycin is a weakly acidic (pK_a = 6.4 in DMF) amorphous solid mp 112.5-113.5°C, $[\alpha]_D = -63$ (*c* 1.0, EtOH). Its structure was obtained from X-ray crystallographic analysis of a *p*-iodophenacyl ester[8]. The ^1H and ^{13}C NMR spectra of the free acid[9] as well as alkali metal salts have been assigned[10] and the mass spectrum of the free acid has also been disclosed[11]. Although the free acid of salinomycin can be prepared by acidification of an ethereal solution of its sodium salt with 0.1 N HCl, salinomycin gradually decomposes under acidic conditions, demonstrated by a loss of antimicrobial activity[4]. Decomposition of the natural product has been reported with formic acid[12] and under basic conditions[9].

2 BIOLOGICAL ACTIVITY AND MODE OF ACTION

Salinomycin is active against Gram-positive bacteria and some filamentous fungi but it exhibits no activity against Gram-negative bacteria or yeast[4, 13]. It is a commercially significant product for controlling coccidiosis in poultry and enhancement of feed utilisation in cattle[14] but its acute toxicity (LD_{50} in mice 18 mg/kg intraperitoneally and 50 mg/kg orally[4]) precludes its use as a general antibiotic.

Salinomycin renders membranes permeable to cations[15], particularly potassium, thereby disrupting the concentration gradients across the lipid bilayer. A model describing the transport of sodium and potassium cations across lipid bilayers has been presented by Riddell[14, 16]: at physiological pH the ionophore is ionised with its carboxylate group in the aqueous medium. The water of solvation of an approaching cation is successively replaced by the ligating groups of the ionophore to form a neutral ionophore-cation complex which diffuses across the membrane and then dissociates at the other interface. A cation of the same or another type or a proton may be transported in the reverse direction.

3 TOTAL SYNTHESES OF SALINOMYCIN AND NARASIN

Three total syntheses of salinomycin (**1**) have been reported. The first was published in 1981 by Kishi[17] in conjunction with the synthesis of the structurally similar polyether narasin **2**. The second synthesis was reported in 1985 by Yonemitsu[18], and has since been the subject of a number of publications[19, 20, 21, 22]. The third synthesis by Brown and Kocienski appeared in 1994[23, 24].

3.1 The Kishi Synthesis

Kishi's retrosynthetic analysis of narasin (**2**) is shown in Scheme 1. The feasibility of the stereoselective connection of the left (C1-C9) fragment **3** or **4** with the C10-C30 fragment **5** using a directed aldol reaction had been well demonstrated during the synthesis of the lasalocid system[25, 26]. Further analysis of the C10-C30 fragment lead back to three smaller fragments: δ-lactone **6**, acetylene **7** and γ-lactone **8**. Kishi assumed that narasin and salinomycin exist in their most stable configurations at both spiroacetal positions, which should allow the stereochemistry of the spiroacetals to be secured by equilibration under acid catalysis at the final stage of the synthesis.

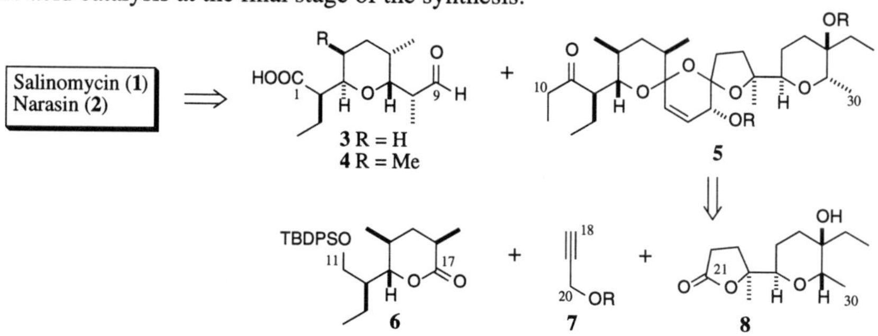

Salinomycin (1)
Narasin (2)
⟹
HOOC
R
1
9
H
3 R = H
4 R = Me
+
10
OR
OR
30
5
TBDPSO
11
17
6
+
18
20 OR
7
+
OH
21
30
8

Scheme 1

Substrate and reagent controlled asymmetric epoxidation of allylic alcohols combined with regio- and stereoselective oxirane cleavage by organocuprates were important features of Kishi's assembly of the acyclic precursor **18** to the left fragment (Scheme 2). The synthesis began with the enantiomerically enriched alcohol **9** which was converted to allylic alcohol **10** using conventional methods. Epoxidation proceeded with a very high level of stereoselectivity to afford oxirane **11**, whose subsequent cleavage with vinyl cuprate completed the introduction of the C2 and C3 stereocentres in 1,3-diol **12**. The protocol was repeated to secure the C5 and C6 substituents stereoselectively in **15**. Application of standard methodology allowed the synthesis of (*E*)-olefin **16** in three steps from 1,3-diol derivative **15** with a selectivity of 98:2.

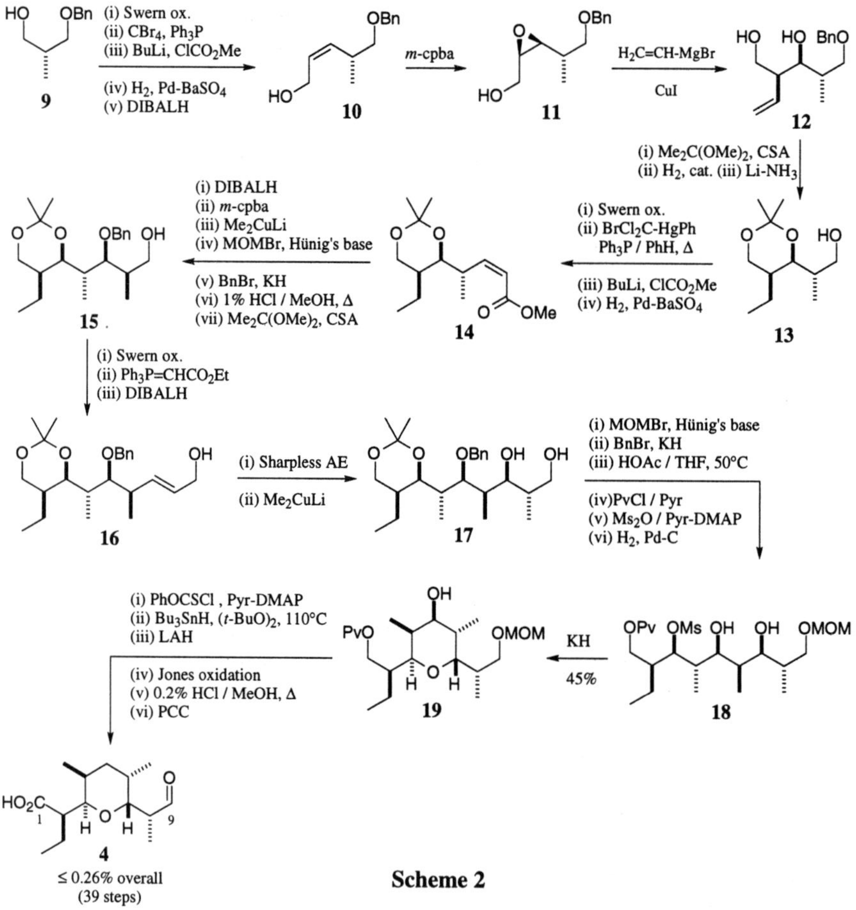

Scheme 2

Substrate controlled epoxidation of allylic alcohols such as **16** with *m*-CPBA usually affords only modest diastereoselectivity in the wrong sense required for the synthesis of salinomycin and narasin. However, reagent-controlled epoxidation according to the procedure of Sharpless[27, 28] gave the desired epoxy alcohol with less than 5% of other isomeric epoxides. Subsequent oxirane cleavage with lithium dimethylcuprate was only modestly regioselective, leading to a mixture of 1,2-diol and 1,3-diol **17** in a ratio of 1:4. Removal of the unwanted 1,2-diol was achieved using $NaIO_4$. The cyclisation precursor, hydroxymesylate **18**, was prepared from 1,3-diol **17** in six steps using conventional methods.

Cyclisation of the hydroxymesylate **18** (Scheme 2) was problematic, particularly in the narasin series, providing the substituted oxane ring **19** in only 45% yield. A major by-product was the olefin resulting from elimination of the mesylate **18**. An improved yield of 60% was observed for the salinomycin series, the increased efficiency of cyclisation being attributed to less steric congestion in the transition states for the salinomycin series. Examination of the two possible chair conformations for the cyclisation product **19a,b** (Scheme 3) in the narasin series reveals that both conformers suffer serious steric compression. The absence of the C4 methyl substituent in the salinomycin series removes an unfavorable 1,3-interaction from each conformer, which should be manifested in a lower energy transition state and increased yield.

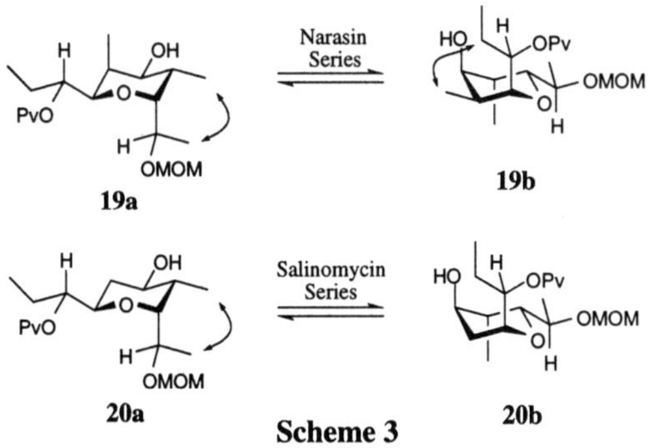

Scheme 3

Deoxygenation of secondary alcohol **19** (Scheme 2) also provided problems, but was finally achieved in neat tributyltin hydride at 110°C. Subsequent functional group manipulation provided the desired fragment **4**. In summary, the C1-C9 fragment of narasin was synthesised in less than 0.26% yield over thirty nine steps with the stereochemistry at the C2, C3 and C6 positions arising from substrate controlled diastereoselection directed by the C4 stereocentre present in the starting alcohol **9**. The stereochemistry at C7 and C8 was introduced by reagent controlled Sharpless epoxidation. Although Kishi also reported the synthesis of the left fragment **3** of salinomycin, no specific details have been disclosed.

More recently Kishi published an improved synthesis of the C1-C9 fragments of narasin and salinomycin[29], which avoided the problematic cyclisation of hydroxymesylate **18** by the installation of the C8-C9 unit onto the pre-formed oxane nucleus by means of a C-glycosylation reaction (Scheme 4). The attack of the nucleophile on the oxonium ion intermediate generated from **21** only resulted in the isolation of the stereoelectronically favoured products of axial attack. The level of stereocontrol at the C8 position was not so high with a ratio of 3.5:1 in favour of the desired product **22a** being observed. Fortunately, the minor isomer **22b** could be transformed into the desired alcohol **22a** by an elimination-hydroboration sequence. The synthesis of the starting oxane fragment **21** was not reported, but it was obtained by degradation of narasin. Furthermore, no detail was given for the synthesis of the central fragment **6** from the alcohol **9**.

The C21-C30 fragment **8** was synthesised by a protocol precedented in a synthesis of lasalocid A ketone[25, 26]; however, since details are scant, an outline of the likely approach is shown in Scheme 5. Epoxidation of the diene **24** followed by diastereoselective reduction of the ketone gave a mixture of epoxy alcohols (*ca* 9:1) whose acid-catalysed cyclisation gave racemic hydroxy oxolane **25** which was easily separated from its stereoisomer by column chromatography. Resolution of the hemiphthalate of alcohol **25** *via* its strychnine salt followed by protection and ozonolysis then yielded aldehyde **26**. Aldehyde **26** was probably transformed into lactone **28** *via* tertiary alcohol **27.** Finally, solvolytic ring expansion of the mesylate derived from deprotection of **28** gave the C21-C30 fragment **8**.

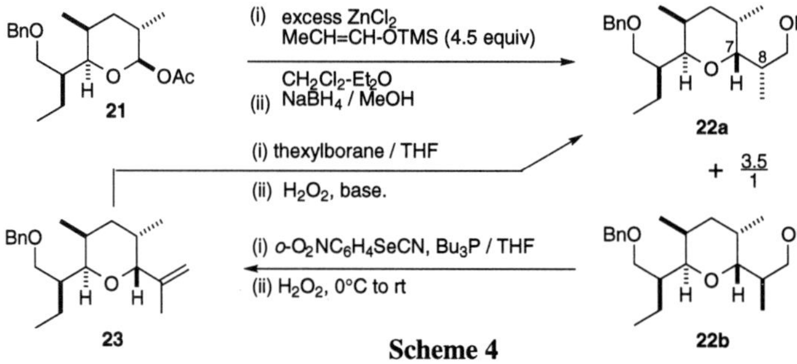

Scheme 4

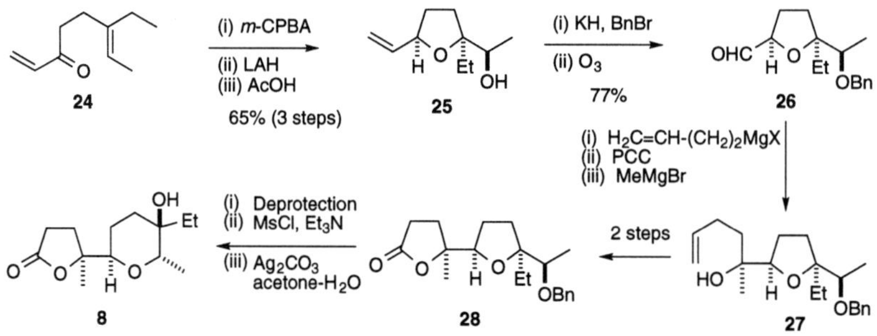

Scheme 5

Coupling of the central C11-C17 fragment **6** with the C21-C30 fragment **8** was achieved as shown in Scheme 6. The propargylic aldehyde **29** was condensed with the lithiated dithiane **30,** derived from fragment **8** in three steps (76% yield). Unfortunately, the desired propargylic alcohol **31** was the minor diastereoisomer produced from the coupling reaction of **29** and **30**. The major isomer from the coupling, which had the wrong stereochemistry at C20, could be converted to the desired diastereoisomer **31** by oxidation to the C20 ketone followed by reduction; however, the reduction was not selective so the operation needed to be repeated to maximise its efficiency. Dithiane cleavage, followed by treatment with acid lead to closure of the C24 tertiary alcohol onto the masked ketone to give a mixture of epimeric acetals.

To complete the synthesis of narasin (Scheme 7), partial reduction of the triple bond in **32** and submission of the resulting (*Z*)-olefin **33** to aqueous acetic acid afforded a single diastereoisomeric dispiroacetal **34** in reasonable yield. It should be noted that dispiroacetal

34 is epimeric to narasin and salinomycin at the C17 position. The synthesis of the C10-C30 fragment **5** was completed in six steps using standard methodology.

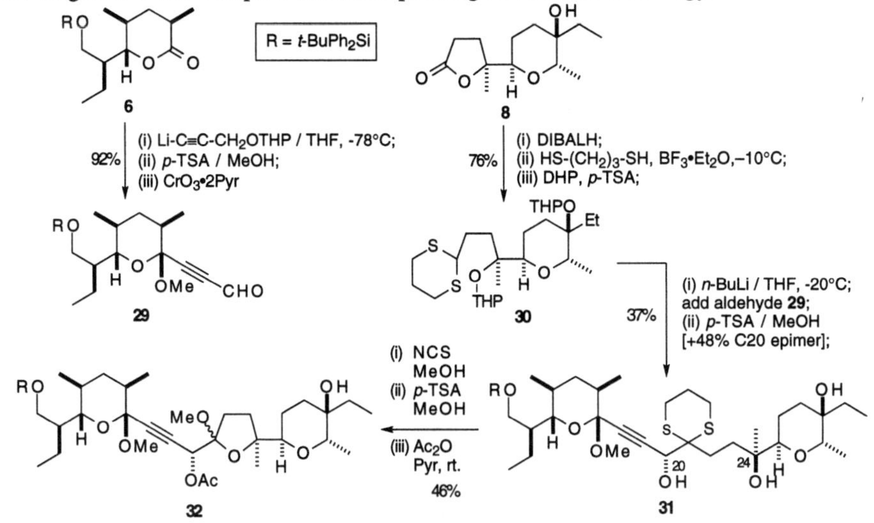

Scheme 6

Kishi investigated many conditions to effect the crossed aldol reaction of the C10-C30 fragment **5** with the left fragment **4**. The best results were obtained using dicyclohexylamidomagnesium bromide as the base, which led to the formation of a single aldol **36** in 58% yield after cleavage of the silyl ether. The product corresponded to 17-*epi*-narasin.

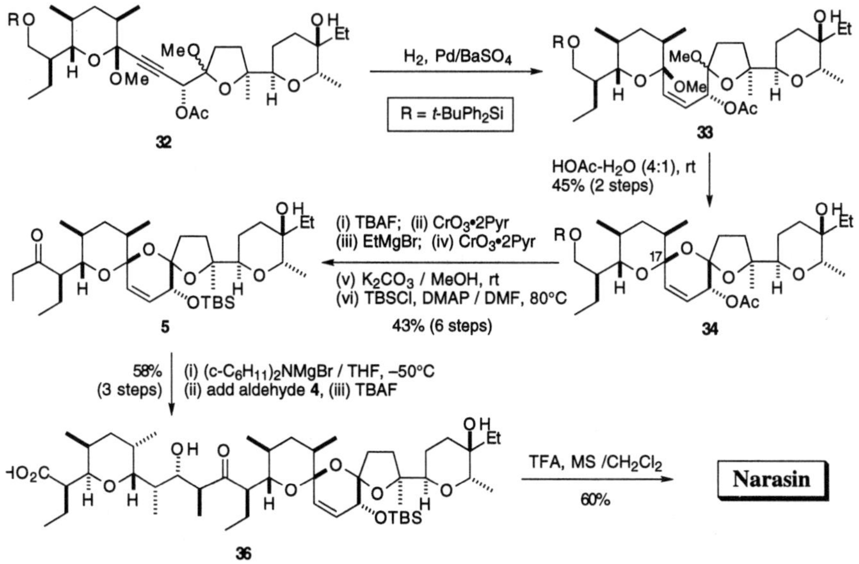

Scheme 7

At the outset, Kishi's synthetic plan was founded on the premise that diastereoisomers with the incorrect stereochemistry at either of the two spiroacetal centres would eventually undergo thermodynamically controlled epimerisation to give narasin or salinomycin with the correct configuration at C17 and C21. Fortunately, treatment of the 17-*epi*-narasin **36** with trifluoroacetic acid (TFA) in dichloromethane containing molecular sieves resulted in the desired epimerisation to the natural product **2** in 2% overall yield (23 steps) from the C21-C30 fragment **8**.

3.2 The Yonemitsu Synthesis

A distinguishing feature of Yonemitsu's synthesis of salinomycin is that most of the stereochemistry was ultimately derived from carbohydrate precursors. Yonemitsu[19] adopted Kishi's first disconnection of salinomycin (Scheme 8) to a C1-C9 fragment **3** and a C10-C30 fragment **5**. Further retrosynthetic analysis of the C10-C30 fragment **5** suggested a central C10-C17 fragment **37** and a C18-C30 fragment **38**.

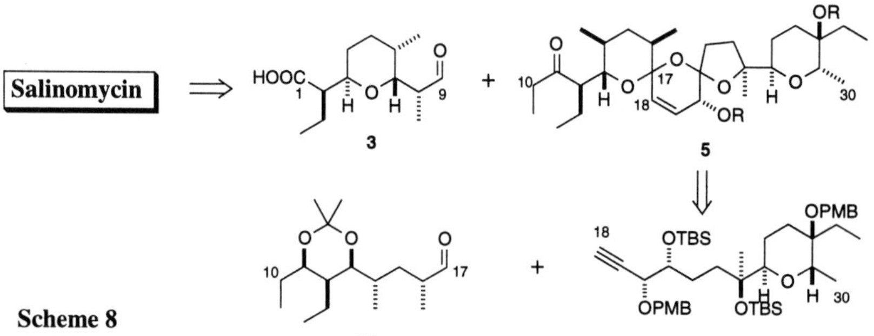

Scheme 8

The synthesis began (Scheme 9) with the conversion of the glucofuranose derivative **39** to the aldehyde **40**—a sequence which required a total of 11 steps (14.4% overall)[30, 31]. Standard functional group manipulation converted aldehyde **40** to aldehyde **41** which underwent Wadsworth-Emmons condensation with phosphonate **42** (itself carbohydrate-derived) followed by catalytic hydrogenation to give ketone **43**. Chelation-controlled addition of vinylmagnesium bromide occurred with good stereoselectivity (13:1) and subsequent functional group manipulation provided the epoxyalcohol **45**. Acid catalysed cyclisation of epoxyalcohol **45** followed by oxidation of the resulting primary alcohol afforded the desired oxane **46** in good yield. Unfortunately, decarbonylation of aldehyde **46** required forcing conditions and gave oxane **47** in only 28% yield. Selective removal of the *p*-methoxybenzyl protecting group in the presence of a benzyl group occurred cleanly on hydrogenation with Raney nickel to give the alcohol **48** from which the target C1-C9 fragment **3** was derived in three concluding steps. The synthesis of C1-C9 fragment **3** required 34 steps from **39** and proceeded in 0.8% overall yield.

The synthesis of the central C10-C17 fragment **37** (Scheme 10) began with displacement of the tosylate group from **49** with lithium dimethylcuprate to complete the C12 ethyl appendage. The remaining two carbon unit was added using the Wadsworth-Emmons reaction to afford an unsaturated lactone, which was transformed stereosectively to the acetal **51**. After reductive cleavage of the benzyl ether from olefin **51**, hydrogenation of the double bond using a rhodium catalyst allowed the formation of the C16 stereogenic centre with a stereoselectivity of 13:1. Oxidation of alcohol **52** to the aldehyde followed by addition of ethylmagnesium bromide gave the Cram adduct **53** as the sole product. Standard methodology was used to complete the synthesis of the central fragment **37** in 16.4% overall yield from tosylate **49** in 17 steps; however, the tosylate required an additional 7 steps to prepare[30, 31] from ketone **39** in an estimated yield of 17.8%. When the extra steps are considered the overall yield becomes 2.9% over 24 steps.

Synthesis of the right (C18-C30) fragment **38** was again based on the manipulation of carbohydrate precursors. Three sub-fragments were prepared; the first, a C9-C22 aldehyde **59** was synthesised from diacetone glucose **54** in 32.7% yield over 11 steps using well known methodology (Scheme 11).

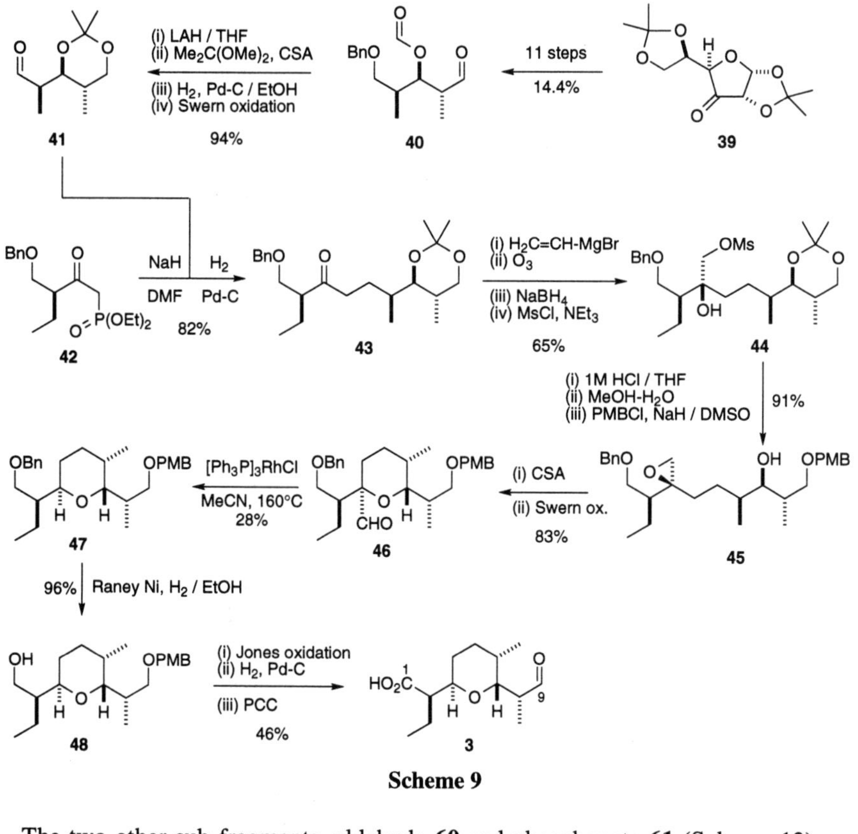

Scheme 9

The two other sub-fragments, aldehyde **60** and phosphonate **61** (Scheme 12), were coupled using the usual olefination-reduction protocol to afford ketone **62** in 92% yield. Chelation-controlled addition of ethylmagnesium bromide proceeded with excellent stereocontrol to provide tertiary alcohol **63**, which was transformed to the cyclisation precursor, tosylate **64**, using conventioal methods. Efficient base-induced cyclisation yielded the substituted oxane **65** *via* the intervention of an epoxide intermediate. A single carbon was then appended as the phosphonate **66**.

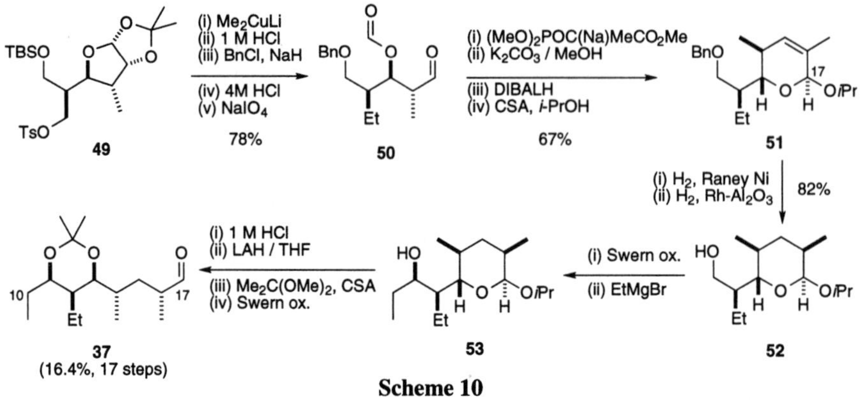

Scheme 10

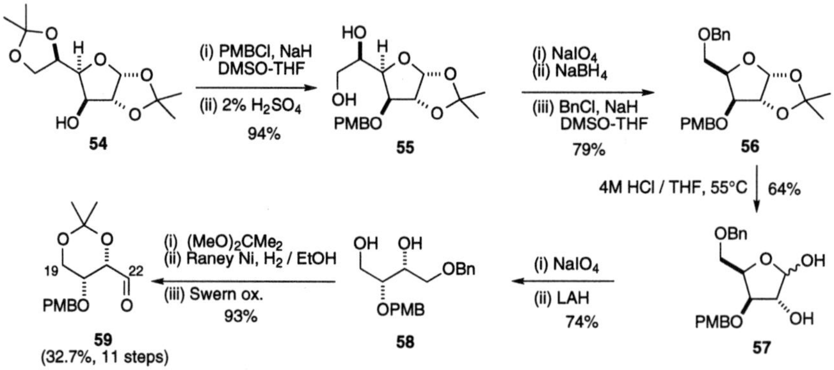

Scheme 11

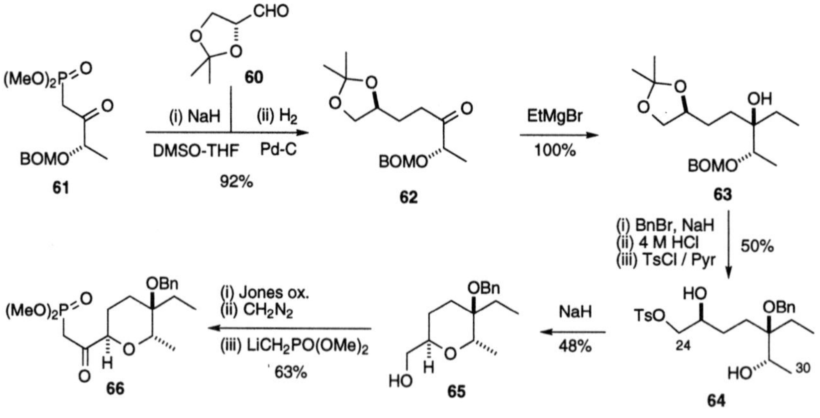

Scheme 12

To complete the synthesis of the C18-C30 fragment **38** (Scheme 13), the phosphonate **66** (Scheme 12) was joined to the C19-C22 aldehyde **59** (Scheme 11) *via* a Wadsworth-Emmons reaction and the resultant enone reduced by catalytic hydrogenation to give **67**. Ketone **67** was unreactive towards methylmagnesium bromide; however, chelation-controlled addition of methyllithium proceeded at –93°C giving the desired tertiary alcohol **68** in quantitative yield. Standard functional group interconversions allowed the completion of the synthesis of the C18-C30 fragment **38** in a total of 23 steps (4.7% overall) from aldehyde **60** (derived from mannitol) and phosphonate **61** (derived from ethyl L-lactate).

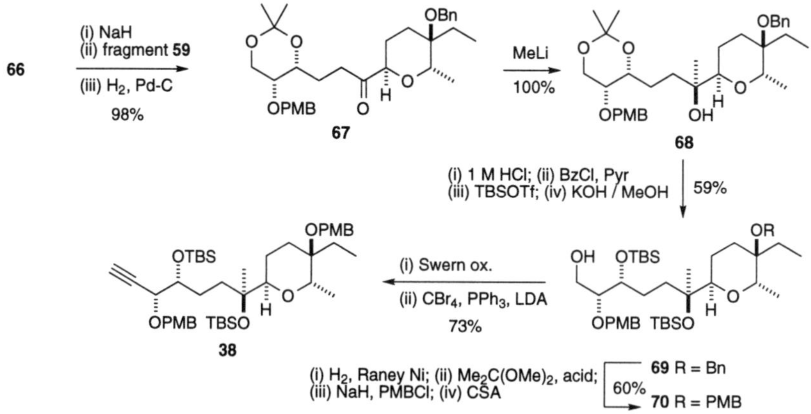

Scheme 13

The completion of Yonemitsu's synthesis of salinomycin is shown in Scheme 14. The key spirocyclisation reaction of (*Z*)-olefin **72** produced two diastereoisomeric dispiroacetals **5a** and **5b** in a ratio of 2:1. Neither of the dispiroacetals had the correct configuration at both C17 and C21. The crossed aldol reaction between the C10-C30 fragment **5b** and the C1-C9 fragment **3** was achieved under Kishi's conditions. Oxidative cleavage of the *p*-methoxybenzyl protecting groups from the aldol **73** released 17-*epi*-salinomycin which was epimerised to give the natural product **1**. Compound **5a** was also converted to salinomycin along similar lines.

3.3 The Southampton Synthesis

At the outset of our work, we benefited substantially from the prior art of both Kishi and Yonemitsu. For example, we knew that the C9-C10 bond could be created using a directed aldol reaction and this was adopted as a feature of our retrosynthetic analysis (Scheme 15). Secondly, we knew that acid-catalysed isomerisation of *late* intermediates harbouring the entire salinomycin skeleton led to the correct configuration of the spiroacetal centres at C17 and C21. Finally, we exploited the known degradation of salinomycin to provide a C1-C9 fragment in lieu of a product derived from total synthesis. With the foregoing considerations in mind, we focused our attention on the synthesis of the C10-C30 fragment **5** which was constructed from the acylfuran **75** serving as a latent enetrione intermediate **74**. In turn, acylfuran **75** was prepared from the 5-lithio-2-alkylfuran **76** and the γ-lactone **77**.

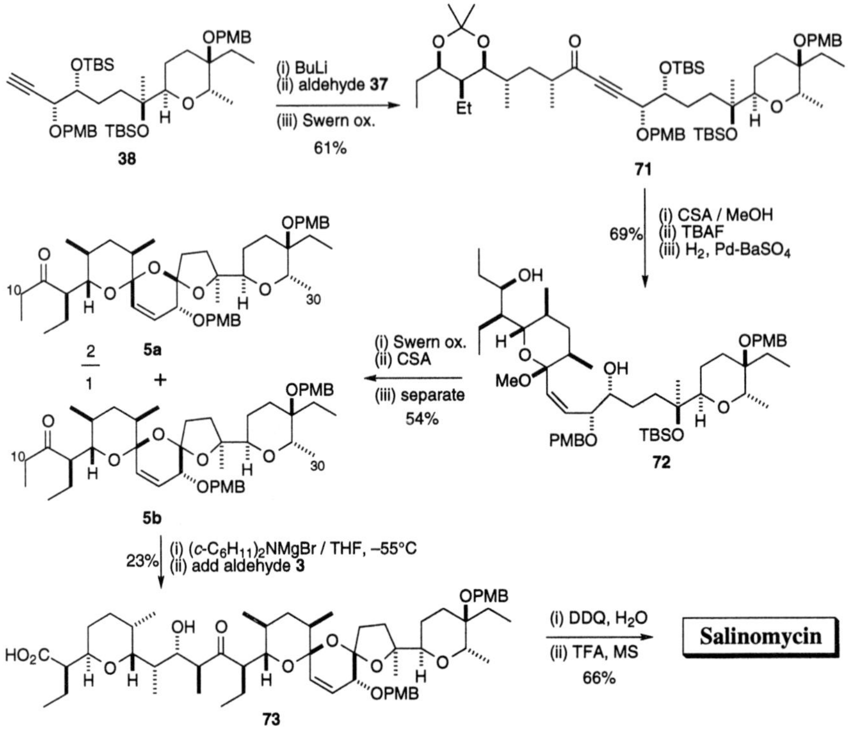

Scheme 14

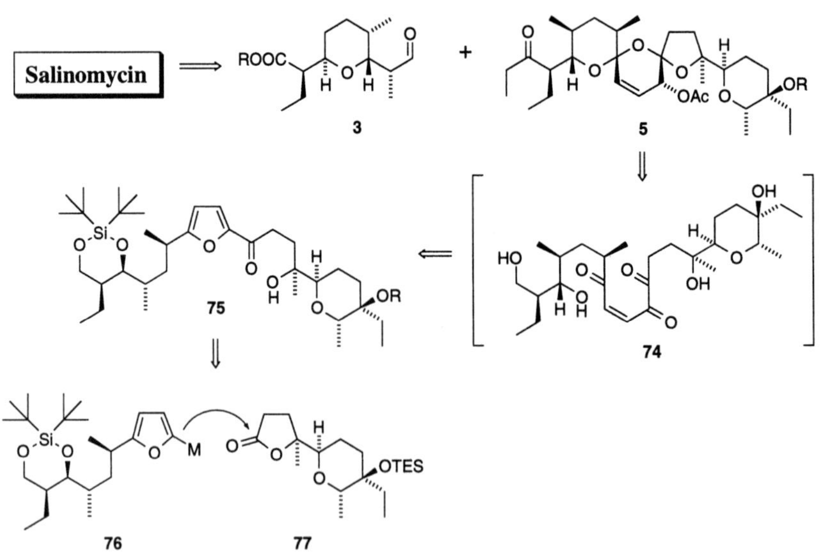

Scheme 15

3.3.1 Synthesis of the C1-C9 Fragment 3 by Degradation of Salinomycin (Scheme 16). Efficient reverse aldol reaction of salinomycin under base induced conditions is not possible due to base-promoted fragmentation of the central spiroacetal ring[9]. However, the thermal retro-aldol reaction of 20-*O*-acetyl salinomycin methyl ester proceeded as reported[22] furnishing aldehyde **78** and ketone **5** both as colourless solids in excellent yields. Unfortunately, methyl ester **78** could not be hydrolysed using conventional conditions. Even after many days at room temperature or several hours at reflux with sodium or lithium hydroxide in mixtures of methanol, water and THF, the acid was only released in low yield accompanied by epimerisation. The solution to the problem was circuitous: protection of aldehyde **78** as dioxolane[32] **79** followed by reduction of the methyl ester with $LiBHEt_3$ to give a primary alcohol which was then oxidised first to the aldehyde using the Dess-Martin periodinane, then to the carboxylic acid **80** with sodium chlorite. The aldehyde was released by acid catalysed hydrolysis of dioxolane group to provide the desired C1-C9 fragment **3**, which was used without purification.

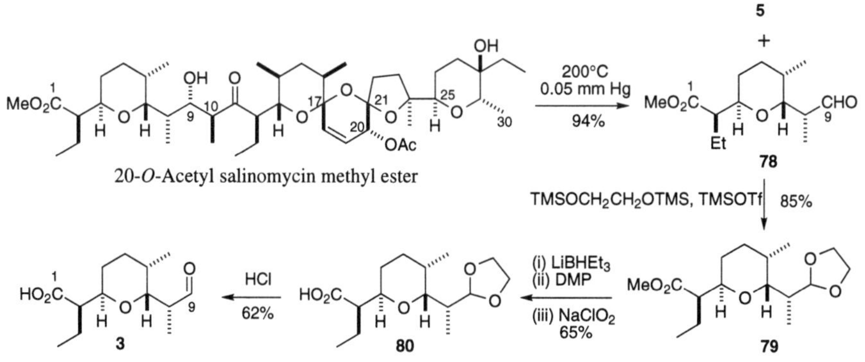

Scheme 16

3.3.2 Synthesis of Furan Fragment 76 (Scheme 17). The enantiomerically pure carboxylic acid **81** (Note 1) was selectively reduced to the alcohol ($BH_3 \cdot SMe_2$) which was protected as the benzyloxymethyl (BOM) ether to give **82**. The ester function in **82** was reduced to the aldehyde **83** which served as a substrate in a highly diastereoselective Sn(II)-catalysed aldol reaction with (*S*)-*N*-butanoyl-4-isopropyloxazolidine-2-thione according to the procedure of Nagao and co-workers[33, 34]. A commendable feature of the oxazolidinethione chiral auxiliary was its easy reductive cleavage with $NaBH_4$ to give a 1,3-diol intermediate which was protected as a di-*tert*-butylsilylene derivative **85**[35, 36]. The aldehyde **86**, derived from **85** in two routine steps chosen for their compatibility with the sensitive di-*tert*-butylsilyene group, was transformed to the furan **88** in three steps (27% overall from **81**). Metallation of **88** with *t*-BuLi gave the target lithio derivative **76**.

3.3.3 Synthesis of Lactone Fragment 77 (Scheme 18). Alkylation of the readily available allylic chloride **89** (6 steps from nerol) with propynyldilithium[37] produced a terminal alkyne which was further extended by metallation followed by methoxycarbonylation to produce the ynoate ester **90** in 62% overall yield for the three steps. The ynoate ester **90** underwent an efficient and stereoselective carbometallation on reaction with Et_2CuLi at –80°C to give the (*Z*)-enoate ester **91** in 96% yield. In order to accomplish asymmetric elaboration of ester **91**, the rudimentary stereochemical information ensconced in the two trisubstituted alkenes, required a chiral auxiliary. Best suited to our needs was Oppolzer's (2*S*)-bornane-10,2-sultam[38] which was introduced in 4 steps (65% overall yield). The sultam derivative **92** underwent asymmetric oxidation[39] on treatment with $KMnO_4$ to create 4 asymmetric centres and a tetrahydrofuran ring giving **93** as a 6:1 mixture of diastereoisomers. Construction of the lactone ring in **94** was easy but further elaboration to the target **77** required six steps and a heavy penalty in yield (34% overall). The sequence began with selective reduction of the *N*-acyl sultam to a

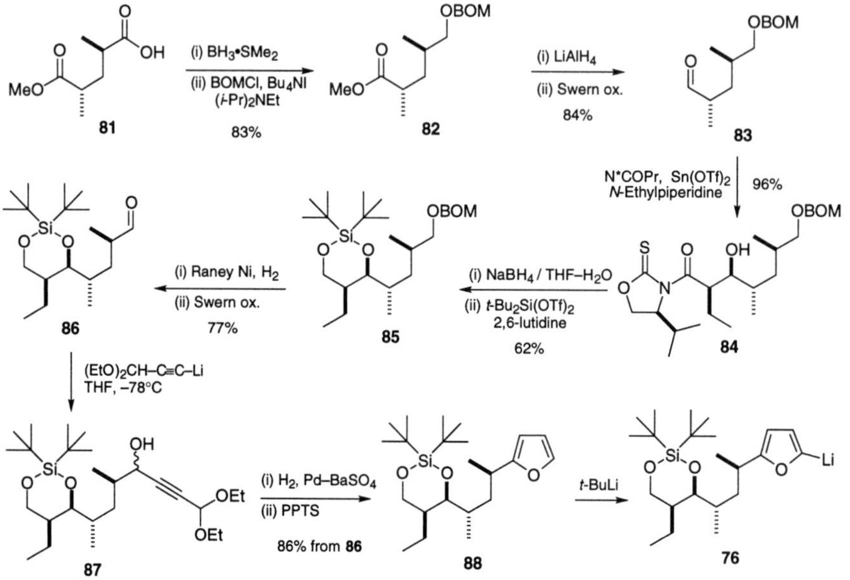

Scheme 17

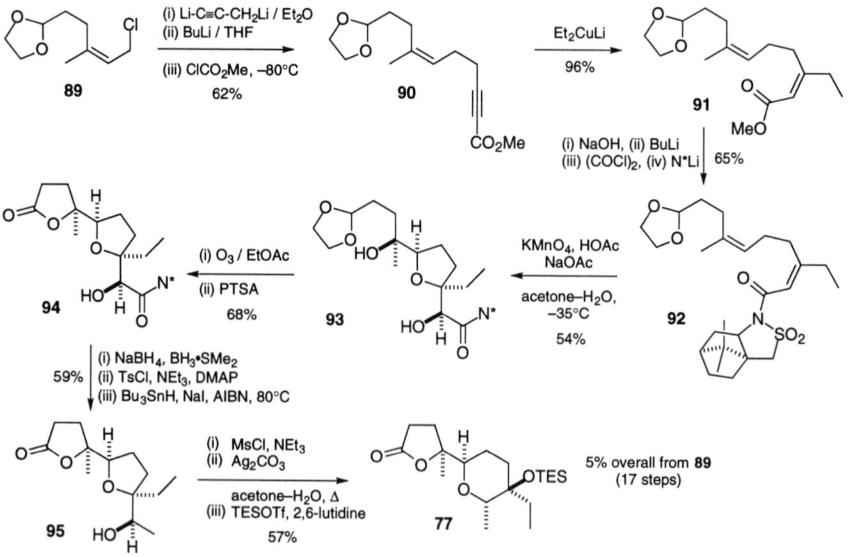

Scheme 18

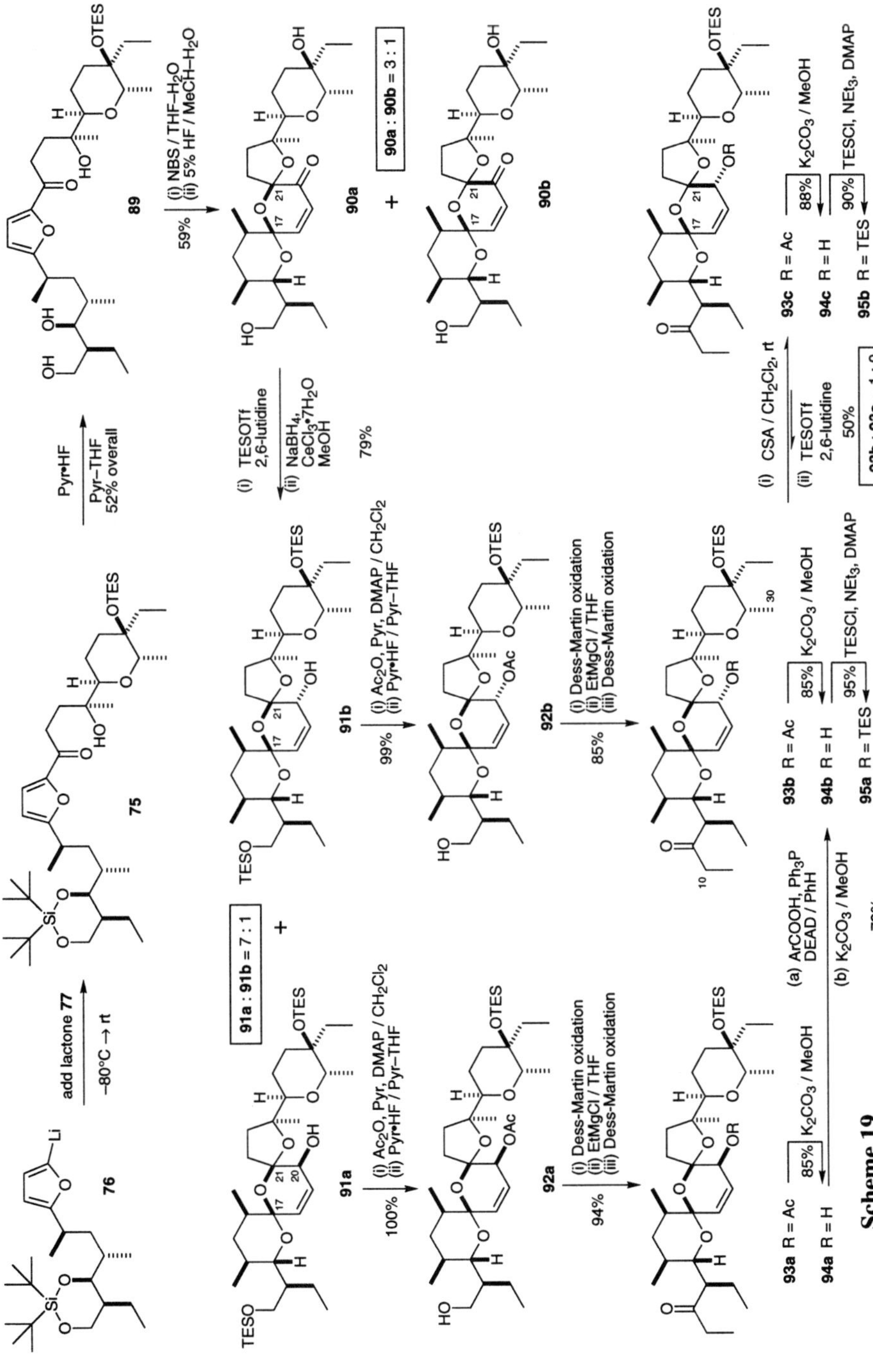

76
add lactone 77
−80°C → rt
75
Pyr•HF
Pyr–THF
52% overall
89
(i) NBS / THF–H2O
(ii) 5% HF / MeCH–H2O
59%
90a
90a : 90b = 3 : 1
90b
(i) TESOTf
2,6-lutidine
(ii) NaBH4,
CeCl3•7H2O
MeOH
79%
91a : 91b = 7 : 1
91a
91b
(i) Ac2O, Pyr, DMAP / CH2Cl2
(ii) Pyr•HF / Pyr–THF
100%
99%
92a
92b
(i) Dess-Martin oxidation
(ii) EtMgCl / THF
(iii) Dess-Martin oxidation
94%
85%
93a R = Ac
94a R = H
K2CO3 / MeOH
85%
(a) ArCOOH, Ph3P
DEAD / PhH
(b) K2CO3 / MeOH
73%
93b R = Ac
94b R = H
95a R = TES
95%
TESCl, NEt3, DMAP
(i) CSA / CH2Cl2, rt
(ii) TESOTf
2,6-lutidine
50%
93b : 93c = 1 : 9
93c R = Ac
94c R = H
95b R = TES
88%
90%

Scheme 19

diol[40] from which the primary hydroxyl was excised by reduction of the corresponding iodide[41]. The final task, expansion of the oxolane ring in **95** to the oxane ring in **77**, was accomplished by a protocol first reported by Kishi[25, 26]. Thus the mesylate derived from **95** was expelled with participation by the neighbouring oxygen to give an oxironium ion whose capture by water returned the desired oxane (59%, single diastereoisomer) along with 10% recovered **95**. A simple triethylsilylation completed the synthesis of **77**.

*3.3.4 Synthesis of C10-C30 Fragment **5** via Union of Furan **75** and Lactone **77** (Scheme 19).* The fulcrum of our synthetic plan involved the use of acylfuran **75** as a latent enetrione intermediate as depicted in Scheme 15. In particular, we intended to construct the 1,6,8-trioxadispiro[4.1.5.3]pentadec-13-ene core unit by an oxidative rearrangement of a 2-acyl furan precursor according to precedent established more than 30 years ago by Achmatowicz and co-workers[42, 43, 44].

Addition of lactone **77** to the metallated furan **76** produced the 2-acyl furan **75** which was deprotected with Pyr•HF to give the triol **89** in 52% overall yield. In the key step of the synthesis, the 2-acyl furan **89** was treated with 1.5 equiv. of NBS[45] in THF–H_2O (3:1) at –10°C for 30 minutes. The crude product was then treated immediately with 5% HF in MeCN–H_2O for 3 h to give a separable mixture of dispiroacetals (**90a** : **90b** = 3:1) in 59% yield. Unfortunately, neither of the products bore stereochemistry at the spiroacetal centres corresponding to salinomycin: the major isomer **90a** was epimeric at both C17 and C21 (salinomycin numbering) whereas the minor isomer **90b** was epimeric at C17.

Reduction of the *bis*-triethylsilyl ether of **90a** with $NaBH_4$ in the presence of $CeCl_3$•$7H_2O$[46] gave a separable mixture of 2 allylic alcohols (79% yield) predominating in the diastereoisomer **91a** having the incorrect stereochemistry at C20 (**91a** : **91b** = 7:1). After protection of the nascent hydroxyl as the acetate and selective deprotection of the primary TES ether, the alcohol **92a** was converted in three steps to the ketone **93a** harbouring the complete carbon skeleton from C10 to C30 albeit with the incorrect stereochemistry at *all three* positions in the central ring of the dispiroacetal core unit (C17, C20, and C21)! Fortunately, hydrolysis of the acetate in **93a** followed by Mitsunobu inversion at C20 gave the allylic alcohol **94b** after hydrolysis of the intermediate *p*-nitrobenzoate ester. Alcohol **94b** was also available from **91b** by a similar sequence of reactions used to convert **91a** to **94a**.

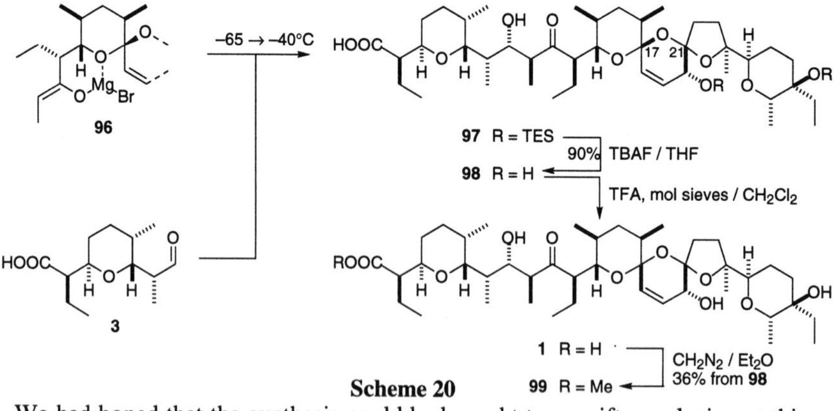

Scheme 20

We had hoped that the synthesis could be brought to a swift conclusion at this stage by an *anti*-selective aldol reaction of ketone **95a** with the C1-C9 aldehyde **3** akin to the transformations depicted in Scheme 20. Although the aldol reaction successfully conjoined the two fragments, attempts to isomerise the errant stereogenic centres at C17 and C21 with acid returned only traces of salinomycin. Assuming that decomposition was competing with isomerisation in the completed skeleton, we attempted the isomerisation of the dispiroacetal *before* the aldol reaction. Thus treatment of **93b** with camphorsulfonic acid caused isomerisation at C21 to give dispiroacetal **93c** (**93b** : **93c** = 1 : 9) now having only one incorrect stereogenic centre at C17.

3.3.5 Completion of the Synthesis (Scheme 20). Ketone **95b**, obtained on replacement of the acetate function in **93c** with a TES group, was converted to its magnesium enolate **96** according to precedent[17, 21] and condensed with aldehyde **3** to give a single major anti-adduct **97**. After removal of both TES groups with TBAF, the triol **98** isomerised on treatment with trifluoroacetic to give salinomycin (**1**) which was characterised by comparison of its methyl ester **99** with an authentic sample.

4 SYNTHESES OF 1,6,8-TRIOXADISPIRO[4.1.5.3]PENTADEC-13-ENE MODELS

The synthesis of the 1,6,8-trioxadispiro[4.1.5.3]pentadec-13-ene ring system has been the subject of a number of studies which have not yet culminated in application to salinomycin or any of its congeners. In a series of papers beginning in 1985 Baker[47, 48, 49] and Brimble[50, 51] described an approach based on the use of a Barton-type radical heteroannulation to generate the oxolane ring system. The most recent of these publications, summarised below, describes a synthesis of the dispiroacetal segment of deoxy-(*O*-8)-*epi*-17-salinomycin. Coupling of the lactone **6** with lithium acetylide **100** (Scheme 21) lead to a mixture of acetals **102** after treatment of the coupling product **101** with methanol under acid catalysis. Partial reduction of the triple bond by hydrogenation over Lindlar catalyst, followed by treatment with PPTS, resulted in the formation of two spiroacetals **103** (1:1 at C21, only one isomer shown) which were easily separated. Transformation of spiroacetal **103** to the hydroxy iodides **104** *via* Lewis acid catalysed opening of an oxirane intermediate and subsequent oxidative spirocyclisation lead to the formation of a mixture of dispiroacetals **105** and its C21 epimer in a ratio of 1.7:1 where the major isomer possessed the same relative stereochemistry as the natural product deoxy-(O-8)-epi-17-salinomycin. A similar sequence performed on the C21 epimer of **103** returned an identical mixture of dispiroacetals **105**.

In 1989 Kocienski and co-workers[52] exploited the stereoselective protonation of the methoxyallene intermediate **107** (Scheme 22) generated from carboxylation of lithiated methoxyallene **106**, to afford butenolide **108**. Cleavage of the silyl ether under acidic conditions allowed the direct formation of spirocyclic butenolide **109** in 82% yield overall. Reaction of the butenolide **109** with the 2-lithio 5,5-dimethyldihydrofuran, followed by treatment of the crude products with CSA gave a complex mixture from which the dispirocyclic enones **109a,b** were isolated in a disappointing combined yield of 15%. The stereochemistry of the dispirocyclic enones was assigned on the basis of NOE difference experiments.

In 1989 Kocienski[52] and Albizati[53] simultaneously reported syntheses of the dispirocyclic enones **109a,b** by an alternative route (Scheme 23) based on the oxidative spiroannulation of furan intermediates—a study which was to lay the foundation for the successful assault on salinomycin by the Southampton team described above.

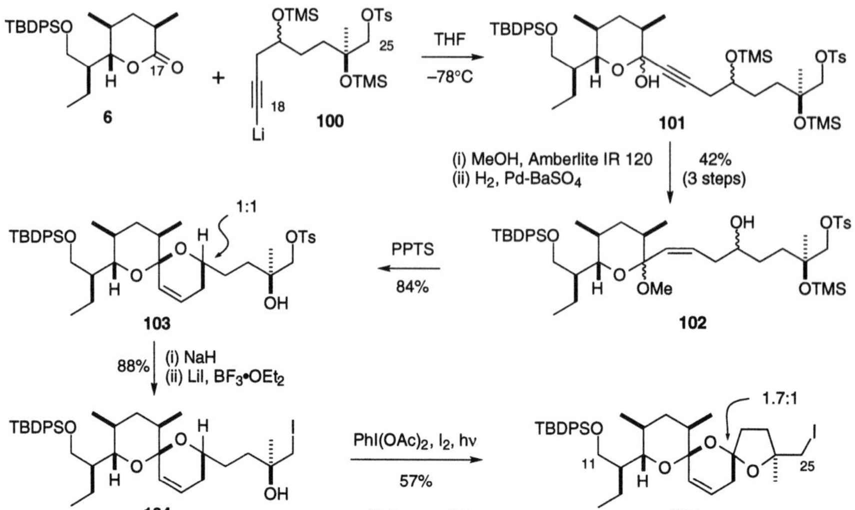

Scheme 21

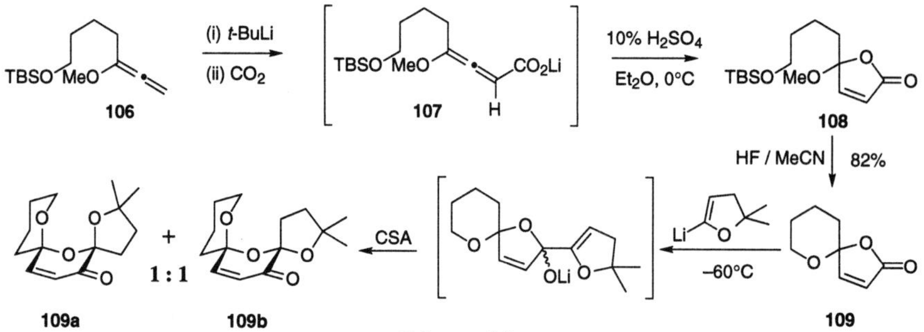

Scheme 22

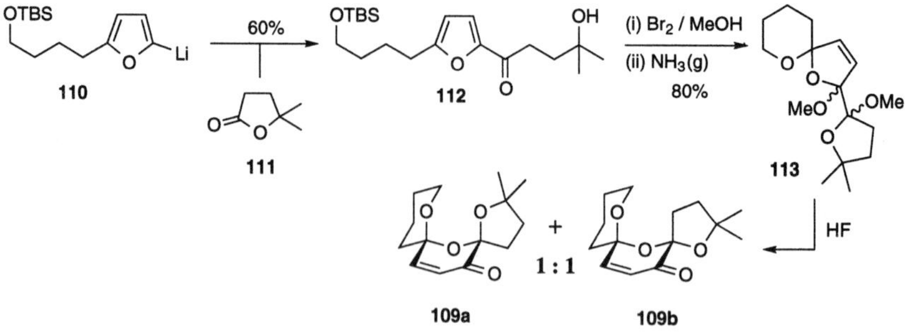

Scheme 23

1.4 Conclusions.

With its 18 stereogenic centres (262,144 diastereoisomers), salinomycin is one of the most complex polyether natural products conquered to date. Each of the 3 total syntheses described herein was based on a different strategy. Kishi's synthesis was an elegant demonstration of the power of hydroxyl-directed alkene epoxidation in conjuction with stereoselective oxirane cleavage as a route to complex polyketides. Yonemitsu's synthesis illustrated the advantages (and penalties) of using cheap, readily available chiral pool precursors for "off-the-shelf" stereogenicity. The oxidative cyclisation strategy used twice in the Southampton synthesis testifies to the value of furans as latent precursors to intermediates of high functional density and the rich stereochemical rewards attending substrate-controlled diene oxidation. All three syntheses indicate that long range non-bonded interactions can play a pivotal role in defining the stereochemistry at acetal centres in complex systems. The quantification of these long range effects and their deliberate manipulation for the purposes of stereocontrol remain a potential goal for the future.

Acknowledgements. We thank Merck, Sharp and Dohme and the EPSRC for generous financial support

References

1. R. L. Harned, P. H. Hidy, C. Y. Corum, and K. L. Jones, *Antibiot. Chemother.*, 1951, **1**, 594.
2. J. Berger, A. I. Rachlin, W. E. Scott, L. H. Sternbach, and M. W. Goldberg, *J. Am. Chem. soc.*, 1951, **73**, 5295.
3. A. Agtarap, J. W. Chamberlin, M. Pinkerton, and L. Steinrauf, *J. Am. Chem. Soc.*, 1967, **89**, 5737.
4. Y. Miyazaki, M. Shibuya, H. Sugawara, O. Kawaguchi, C. Hirose, and J. Nagatsu, *J. Antibiot.*, 1974, **27**, 814.
5. J. L. Occolowitz, D. H. Berg, M. Debono, and R. L. Hamill, *Biomed. Mass Spectrom.*, 1976, **3**, 272.
6. J. W. Westley, J. F. Blount, R. H. Evans, and C.-M. Liu, *J. Antibiot.*, 1977, **30**, 610.
7. H. Umezawa, *J. Antibiot.*, 1978, **31**, 78.
8. H. Kinashi, N. Otake, H. Yonehara, S. Sato, and Y. Saito, *Tetrahedron Lett.*, 1973, 4955.
9. H. Seto, Y. Miyazaki, K.-I. Fujita, and N. Otake, *Tetrahedron Lett.*, 1977, 2417.
10. F. G. Riddell and S. J. Tompsett, *Tetrahedron*, 1991, **47**, 10109.
11. H. Kinashi and N. Otake, *Agric. Biol. Chem.*, 1976, **40**, 1625.
12. J. L. Wells, J. Bordner, P. Bowles, and J. W. McFarland, *J. Med. Chem.*, 1988, **31**, 274.
13. C.-m. Liu, in *Polyether Antibiotics*, ed. J. W. Westley, New York and Basel, 1982.
14. M. D. Ruff, in *Polyether Antibiotics*, ed. J. W. Westley, New York and Basel, 1982.
15. R. W. Taylor, R. F. Kauffman, and D. R. Pfeiffer, in *Polyether Antibiotics*, ed. J. W. Westley, New York and Basel, 1982.
16. F. G. Riddell and S. J. Tompsett, *Biochim. et Biophys. Acta*, 1990, **1024**, 193.
17. Y. Kishi, S. Hatakeyama, and M. D. Lewis, in *Frontiers of Chemistry*, ed. K. J. Laidler; Pergamon: Oxford, 1982, p 287-304.
18. Y. Oikawa, K. Horita, I. Noda, and O. Yonemitsu, *Heterocycles*, 1985, **23**, 231.
19. K. Horita, Y. Oikawa, and O. Yonemitsu, *Chem. Pharm. Bull.*, 1989, **37**, 1698.

20. K. Horita, S. Nagato, Y. Oikawa, and O. Yonemitsu, *Chem. Pharm. Bull.*, 1989, **37**, 1705.
21. K. Horita, Y. Oikawa, S. Nagato, and O. Yonemitsu, *Chem. Pharm. Bull.*, 1989, **37**, 1717.
22. K. Horita, S. Nagato, Y. Oikawa, and O. Yonemitsu, *Chem. Pharm. Bull.*, 1989, **37**, 1726.
23. R. C. D. Brown and P. Kocienski, *Synlett*, 1994, 415.
24. R. C. D. Brown and P. Kocienski, *Synlett*, 1994, 417.
25. T. Nakata, G. Schmid, B. Vranesic, M. Okigawa, T. Smith-Palmer, and Y. Kishi, *J. Am. Chem. Soc.*, 1978, **100**, 2933.
26. T. Nakata and Y. Kishi, *Tetrahedron Lett.*, 1978, 2745.
27. T. Katsuki and K. B. Sharpless, *J. Am. Chem. Soc.*, 1980, **102**, 5974.
28. B. E. Rossiter, T. Katsuki, and K. B. Sharpless, *J. Am. Chem. Soc.*, 1981, **103**, 464.
29. J. A. Tino, M. D. Lewis, and Y. Kishi, *Heterocycles*, 1987, **25**, 97.
30. Y. Oikawa, T. Tanaka, K. Horita, I. Noda, N. Nakajima, N. Kakusawa, T. Hamada, and O. Yonemitsu, *Chem. Pharm. Bull.*, 1987, **35**, 2184.
31. A. Rosenthal and M. Sprinzl, *Can. J. Chem.*, 1969, **47**, 3941.
32. T. Tsunoda, M. Suzuki, and R. Noyori, *Tetrahedron Lett.*, 1980, **21**, 1357.
33. Y. Nagao, S. Yamada, T. Kumagai, M. Ochiai, and E. Fujita, *J. Chem. Soc., Chem. Commun.*, 1985, 1418.
34. Y. Nagao, T. Kumagai, S. Yamada, E. Fujita, Y. Inoue, Y. Nagase, S. Aoyagi, and T. Abe, *J. Chem. Soc., Perkin Trans. 1*, 1985, 2361.
35. B. M. Trost and C. G. Caldwell, *Tetrahedron Lett.*, 1981, **22**, 4999.
36. E. J. Corey and P. B. Hopkins, *Tetrahedron Lett.*, 1982, **23**, 4871.
37. L. Brandsma and H. D. Verkruijsse, 'Synthesis of Acetylenes, Allenes, and Cumulenes: a Laboratory Manual', Elsevier Scientific, 1981.
38. W. Oppolzer and J.-P. Barras, *Helv. Chim. Acta.*, 1987, **70**, 1666.
39. D. M. Walba, C. A. Przybyla, and C. B. Walker, Jr., *J. Am. Chem. Soc.*, 1990, **112**, 5624.
40. S. Saito, T. Hasegawa, M. Inaba, R. Nishida, T. Fujii, S. Nomizu, and T. Moriwaki, *Chem. Lett.*, 1984, 1389.
41. C. Bonini and R. D. Fabio, *Tetrahedron Lett.*, 1988, **29**, 815.
42. O. Achmatowicz, P. Bukowski, B. Szechner, Z. Zwierzchowska, and A. Zamojski, *Tetrahedron*, 1971, **27**, 1973.
43. O. Achmatowicz and P. Bukowski, *Can. J. Chem.*, 1975, **53**, 1975.
44. O. Achmatowicz and R. Bielski, *Carbohydr. Res.*, 1977, **55**, 165.
45. M. P. Georgiadis and E. A. Couladouros, *J. Org. Chem.*, 1986, **51**, 2725.
46. J. L. Luche, *J. Am. Chem. Soc.*, 1978, **100**, 2226.
47. R. Baker, M. A. Brimble, and J. A. Robinson, *Tetrahedron Lett.*, 1985, **26**, 2115.
48. R. Baker and M. A. Brimble, *J. Chem. Soc., Chem. Commun.*, 1985, 78.
49. M. A. Brimble, G. M. Williams, and R. Baker, *J. Chem. Soc., Perkin Trans. 1*, 1991, 2221.
50. M. A. Brimble, *Aust. J. Chem.*, 1990, **43**, 1035.
51. M. A. Brimble and G. M. Williams, *J. Org. Chem.*, 1992, **57**, 5818.
52. P. J. Kocienski, Y. Fall, and R. Whitby, *J. Chem. Soc., Perkin Trans.1*, 1989, 841.
53. F. Perron and K. F. Albizati, *J. Org. Chem.*, 1989, **54**, 2044.

5
Inhibition of Protein Export in Bacteria: The Signalling of a New Role for β-Lactams

A. E. Allsop, M. J. Ashby, G. Brooks, G. Bruton,* S. Coulton, P. D. Edwards, S. A. Elsmere, I. K. Hatton, A. C. Kaura, S. D. McLean, M. J. Pearson, N. D. Pearson, C. R. Perry, T. C. Smale and R. Southgate
SMITHKLINE BEECHAM PHARMACEUTICALS, BROCKHAM PARK, BETCHWORTH, SURREY RH3 7AJ, UK

The continuous use of antibiotics over the last 50 years has resulted in the selection and development of several bacterial resistant mechanisms: antibiotic inactivation *via* biochemical modification, mutation of target site, increased permeability barriers and antibiotic efflux pumps. In particular, the incidence of resistant infections has risen significantly over the last fifteen years. Vancomycin resistant enterococci have risen from 2.4% in 1975 to 7.9% in 1993, resistant pneumococci have increased from 6.5% in 1987 to 16% 1994 and infections due to resistant *Staphylococcus aureus* has risen from 2.4% in 1975 to 32.1% in 1992. In 1995 alone the incidence of MRSA infections in teaching hospitals was estimated to be as high as 40%. Infections due to drug resistant bacteria thus pose a significant and increasing threat for the foreseeable future, a threat that cannot be met by merely elaborating existing classes of antibiotics. Future advances in antimicrobial therapy will be made by the development and introduction of novel of antibiotics acting upon new molecular targets.

Signal peptidase enzymes represent a novel class of antibacterial target, they are involved in one of the pathways by which bacteria export proteins across their cytoplasmic membranes, Fig 1[1]. Proteins destined for this pathway are synthesised as preproteins with an extra domain, the signal sequence, attached to the amino terminus. This signal sequence targets the preprotein to the cytoplasmic membrane where it inserts to form a trans-membrane loop.

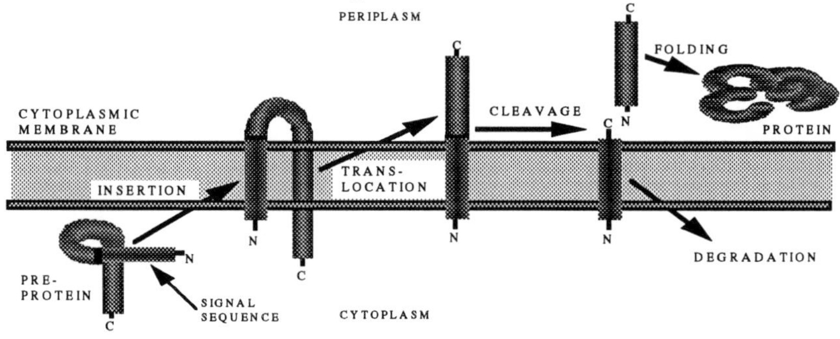

Fig 1.

A multi protein complex, the Sec proteins, then pumps the preprotein through the cytoplasm in an energy dependent process, but the preprotein remains in the periplasm tethered to the cytoplasmic membrane by the signal sequence. It is at this point that the signal peptidase enzyme plays its part by cleaving off the signal sequence and liberating the protein, which is then able to fold into its active conformation. By inhibiting signal peptidases, all proteins exported by this pathway will be left tethered to the cytoplasmic membrane. These proteins would remain functionally dormant and their accumulation would eventually compromises the integrity of the cytoplasmic membrane resulting in cell death.

Signal peptidases are atypical membrane bound serine proteases. Their catalytic centre contains a serine lysine diad, akin to β-lactamases, rather than the usual serine-histidine-aspartate triad. No protein structure currently exists to aid inhibitor design but analysis of the substrate, the signal sequence, has proved useful in the development of working models. A typical signal sequence (Fig. 2) consists of a positively charged amino terminus that remains in the cytoplasm, a helical hydrophobic trans-membrane domain, terminated by a helix breaker and a short carboxy terminus domain leading to the cleavage site. The cleavage site itself is characterised by the A-X-A motif with a near essential alanine at P1, the P3 alanine is less essential with natural substitutions by Val, Leu, Ser and Thr being known. Predictions of secondary structure indicate that two beta turns may be present close to the cleavage site. One of these occurs between 4-6 residues downstream from the cleavage site and the other was believed to be either just before or at the cleavage site[2].

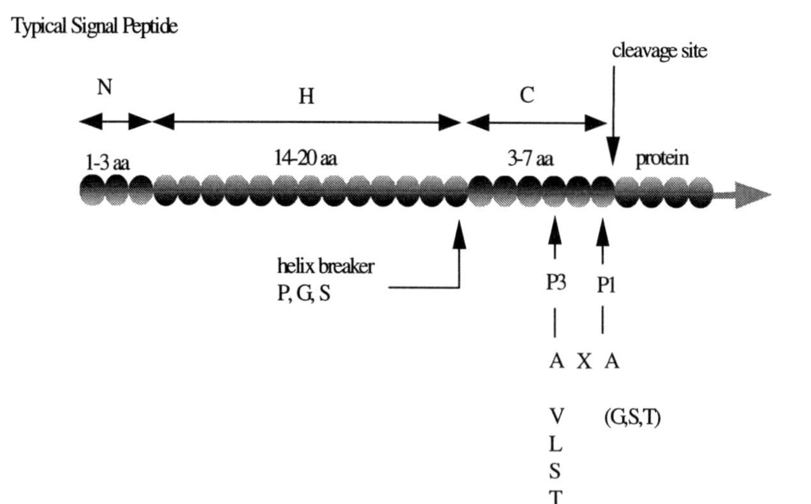

Fig 2.

A variety of known serine protease inhibitors were screened against the signal peptidase from *Escherichia coli*, LP1. The only activity of note was found in the racemic penem ester (1), subsequently it was found that all activity resided in the 5S diastereomer (2)[3]. This stereochemistry is the opposite to that required for inhibition of the bacterial penicillin binding proteins.

(1) (2)

Development of the SAR of the penem nucleus did not readily lend itself to a combinatorial approach, so it became necessary to develop a working model that would give direction to a program of chemical research. The first element in this working model was to be the "unnatural" 5S stereochemistry of the active penem (2), as it indicates the facial positioning of the active site serine in relation to the signal peptide substrate. The β -lactam carbonyl of a penem can potentially be attacked from each side of the plane of the β–lactam ring (Fig.3). The thiazolidine ring shields one face of the beta-lactam ring to some extent, but this steric impedance is not in itself the major controlling influence for facial selectivity towards nucleophilic attack at the β-lactam carbonyl.

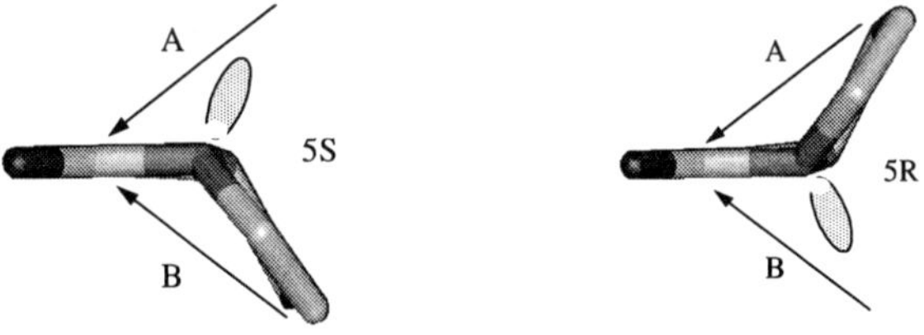

Fig 3.

5S and 5R penem ring systems viewed edge on to the β-lactam ring.

During serinolysis of a peptide bond a tetrahedral transition state is formed in which the electron density of the nitrogen lone pair finally resides, post inversion, on the same face of the peptide plane as the serine nucleophile. As the transition state breaks down, proton donation to the nitrogen lone pair occurs prior to loss of the P1' residue Thus both the incoming nucleophile and proton donation occur from the same face of the peptide bond. The position of the β-lactam lone pair (Fig.3) therefore directs the facial approach of the incoming nucleophile as it functions as the proton acceptor in the late stage transition state. The positioning of the lone pair is, of course, indirectly controlled by the stereochemistry at C5 and thus serinolysis of both the 5S and 5R penems will occur from the least hindered sides, vectors A and B respectively, but due to stereoelectronic rather than steric effects. The second element to the working model was the hypothesis that the cleavage site might lie on a β-turn as predicted by Chou and Fasman sequence analysis[2]. The residues from P5 to P3' around the cleavage site are all L-amino acids without the requirement for a glycine residue and this defines the turn as a Type I β -turn[4]. Each peptide bond of a Type I β -turn (Fig 4) represents a potential cleavage site which could be approached from either face of the peptide plane.

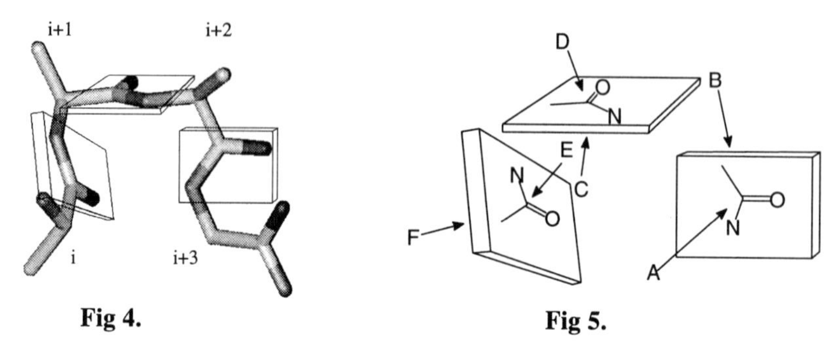

Fig 4. **Fig 5.**

These planes can be represented diagrammatically by (Fig 5). Initially we examined the steric impedance that would be experienced by an incoming nucleophile along each approach vector to a Type I β -turn composed of alanines, A-F (Fig 5). Vectors C and E could be ruled out due to severe steric clashes, vectors B, D and F were readily available to a nucleophile and vector A was only slightly impeded. In reality vectors F and A may experience some degree of steric impedance due to the incoming and outgoing protein chains. However, due to the variable nature of this aspect of β -turns and the lack of real structural data we were unable to use this information to help our analysis.

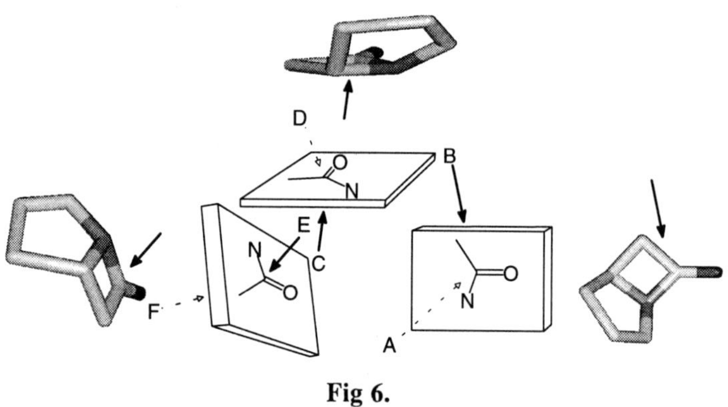

Fig 6.

At this point it was possible to use the 5S stereochemistry of our active penem (2) to finalise the working model. Each of the three potential cleavage sites could now be evaluated in terms of the nucleophilic approach vector required for serinolysis of a 5S penem sitting at each of these potential cleavage sites (Fig 6). The penem amide was overlaid upon the three β-turn peptide bonds in turn and the acceptable vectors of approach to the β -turn and the penem were compared, (Fig 6). In only one case was there a compatible match of approach vectors, namely vector B between the i+2 and i+3 positions of the β-turn. Our working model therefore became a Type I β-turn with the cleavage site between i+2 and i+3 (Fig 7).

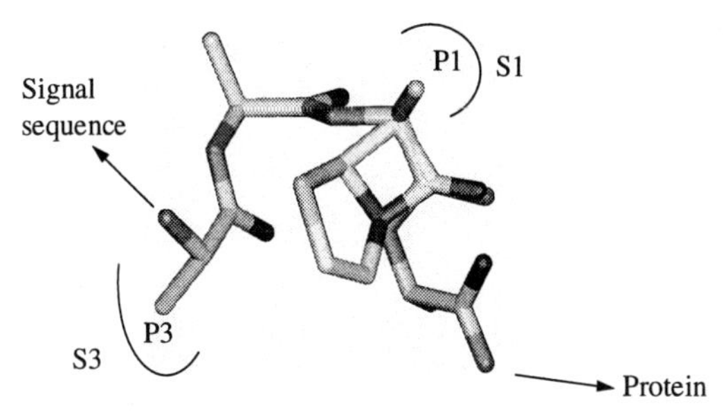

Fig 7.

From this model we were able to make predictions that would help drive the chemical effort with the positions of the β -turn (i, i+1, i+2 and i+3) equating to positions of the substrate (P3, P2, P1 and P1' respectively). The skeletal elements of the penem fell cleanly within the plane of the β-turn providing an ideal template to probe for further interactions with the enzyme. The penem ester equated to the P' side of the cleavage site, which is known to bear multiple ionised carboxylates in the substrate. Although the free acid of penem (2) was inactive, our model suggested that an acid functionality at this position should be tolerated. According to our model, accessibility to the S1 pocket (Fig 7) requires a *cis* arrangement across the top of the β -lactam ring system, while the C2 position of the penem appears to represent the only realistic position from which to access S3 pocket.

The P' side was explored through a plethora of penem esters and amides, representative examples are shown in Table 1. The enzyme assays did not lend themselves to full kinetic analysis, but IC_{50}s were attainable. For time dependent inhibitors, such as these penems, IC_{50}s can have little obvious comparative relevance due to their multi-component constitution. However, when comparing inhibitors based upon the same nucleus then their relative chemical reactivities could be regarded as equivalent. We also knew that these penems permanently inhibited LP1 with a partition ratio of 1, thus there was no turnover or off-rate and thus we were able to take these particular IC_{50}s as an approximate indication of the ranking order of K_i. Notably, both acids (Table 1) were inactive, otherwise this position was relatively tolerant to substitution. The activity profile of the amides (Table 1) indicates a degree of spatial restriction in the enzyme pocket that binds this C3 substituent. The planar *trans* amides have less conformational space available to them compared to the esters and were generally less well accommodated.

The ability of the enzyme to accommodate most of these esters and amides is not surprising as this area represents the P' side of the preprotein substrates, a highly variable region amongst the various preproteins substrates that are cleaved by LP1.
A significant improvement in binding would be expected if these penem inhibitors were to bind into the S1 and S3 recognition pockets of the enzyme. Our working model predicted that *cis* substitution at C6 would allow access to the former of these, the S1 pocket. The initial synthetic approach to these systems began with the generation of an azetidinone anion (Fig 8) which could be reacted with aldehydes or alkyl halides to introduce our probe for the S1 pocket.

Table 1

Esters

R	IC_{50} μM
OH	600
OMe	2
$OCH_2C{:}CH_2$	4
OCH_2Ph	3
OCH_2Ph-p-CO_2H	>100
OCH_2CCl_3	30
$O(CH_2)_4CH_3$	21
$O(CH_2)_4Ph$	12
$O(CH_2)_{15}Me$	IA

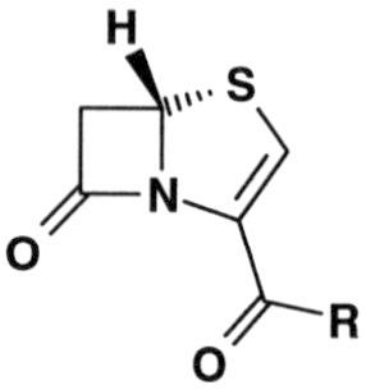

Amides

R	IC50 μM
NHPh	58
NHCH2Ph	>100
NMeCH2Ph	50
NEt2	58
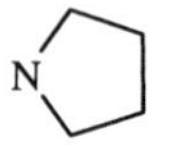 N	10
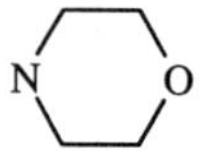 N O	115

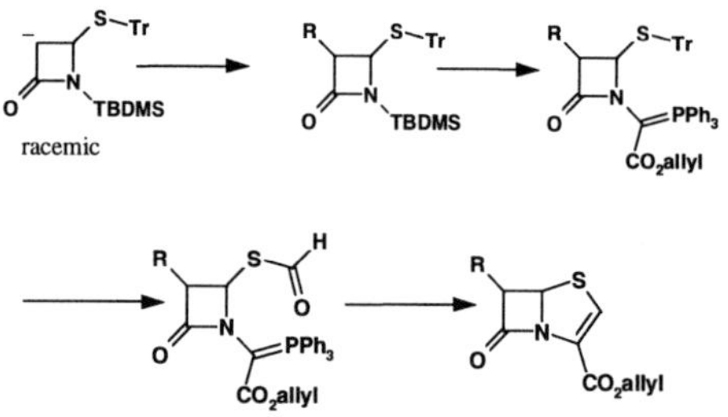

Fig 8.

This azetidinone was then elaborated and cyclised to the C6 substituted penem systems *via* the well known Wittig phosphorane route (Fig 8)[5]. This non-chiral route was useful as a starting point as we were able to generate all possible stereochemistries for initial evaluation, including *trans* substituted systems to challenge our model as well as the predicted active *cis* geometry. However, we were to also require chiral routes to 5S *cis* substituted penem systems and one of these began with the chiral 5R 6-bromo-penem (Fig 9). Zinc mediated aldol condensations gave *trans* substituted 5R penems which upon thermolysis equilibrated at C5 to give a *circa* 1:1 mixture of 5R and 5S penems *via* homolytic S1-C5 cleavage and recyclisation. Chromatographic isolation gave pure 5S-*cis*-substituted penems.

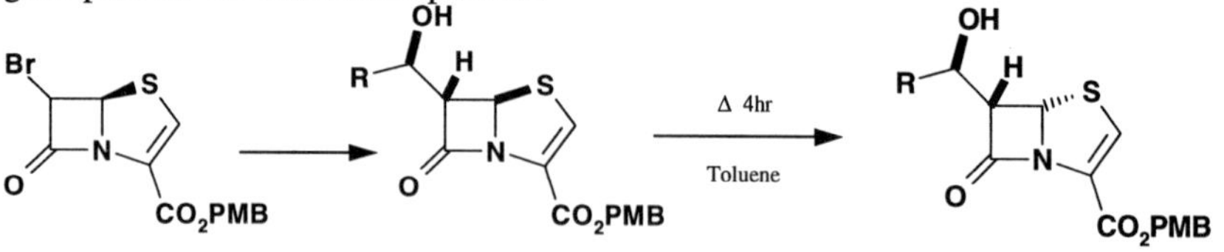

Fig 9.

A closely related and representative series of C6 substituted penems synthesised by these routes are shown in Table 2. All the *trans* substituted penems were devoid of activity as expected, but only one of the two possible *cis* alcohols formed from aldol condensations was active (Table 2a).

The specific stereochemical requirement for the C6 substituent, 6S 1'R, could be understood by reference to our working model (Fig 10) in which the active [5S, 6S, 1'R] 6-acetoxyethyl penem benzyl ester is fitted to the Type I β-turn. The C6 acetoxyethyl substituent is shown in its low energy conformation where it is stereochemically locked into occupying the S1 pocket with a perfect fit to the P1 alanyl methyl. Inversion of the acetoxyethyl stereochemistry to 1'S would place the unacceptable acetoxy group into the S1 pocket resulting in loss of activity.

The S1 pocket should only be able to accommodate an alanyl methyl group and we tested this assumption with a series of *cis* substituted 5S penems, Table 2b, which confirmed the steric limitations at this position. Notably, now that an S1 recognition element had been incorporated into the system we were able to maintain activity with a free carboxylate at C3 as expected, (5) and (6) Table 2a.

In order to both increase the activity of our inhibitors and to challenge our model, we set out to probe for the S3 pocket. The S3 pocket, as defined by the P3 substituent in Fig 10 is accessible from the C2 position of the penem but lies slightly behind the

Table 2

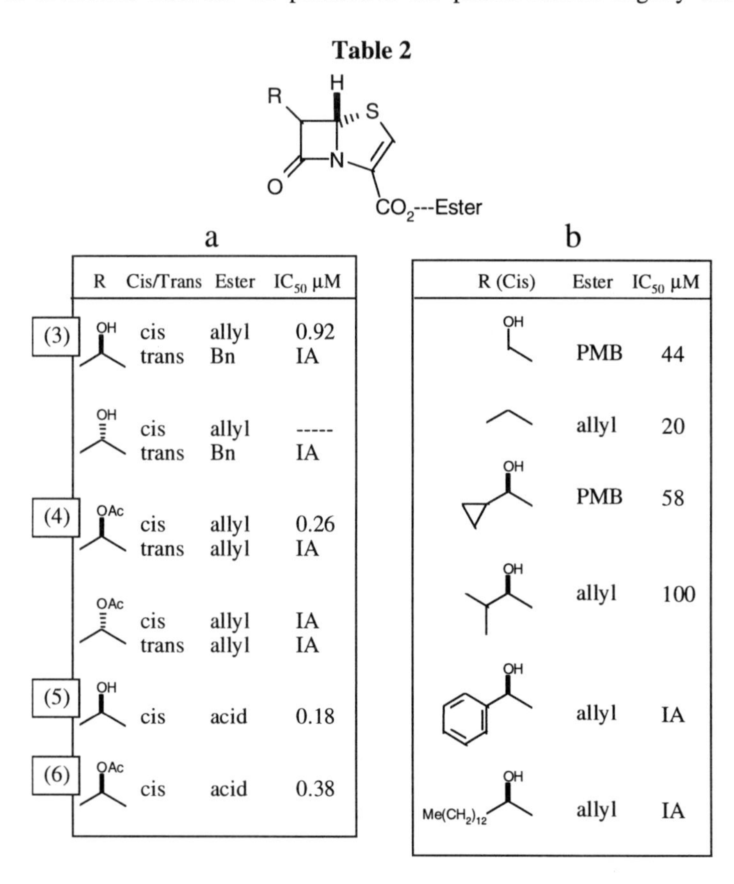

a

	R	Cis/Trans	Ester	IC_{50} μM
(3)	OH	cis	allyl	0.92
		trans	Bn	IA
	OH	cis	allyl	-----
		trans	Bn	IA
(4)	OAc	cis	allyl	0.26
		trans	allyl	IA
	OAc	cis	allyl	IA
		trans	allyl	IA
(5)	OH	cis	acid	0.18
(6)	OAc	cis	acid	0.38

b

R (Cis)	Ester	IC_{50} μM
OH	PMB	44
	allyl	20
OH	PMB	58
OH	allyl	100
OH	allyl	IA
$Me(CH_2)_{12}$ OH	allyl	IA

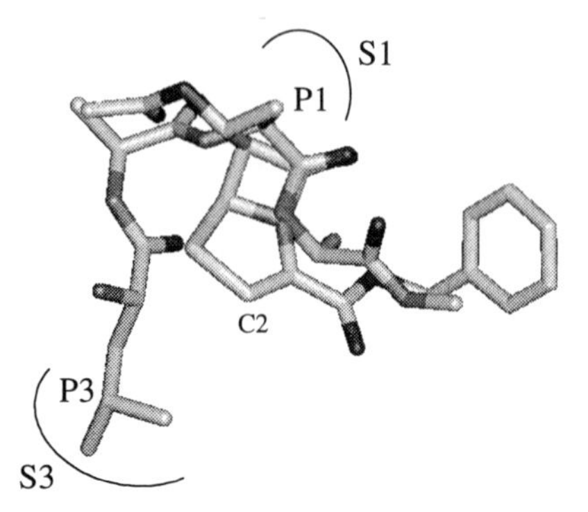

Fig 10.

Computational fit of an active C6-acetoxyethyl-penem-benzyl ester with a Type I β-turn consisting of Val-Ala-Ala-Ala.

plane of the thiazoline ring. To reduce the loss of entropic energy upon binding into the S3 pocket, thus maximising binding affinity, we wished to control the conformational space available to our C2 substituent and direct it towards the S3 binding pocket. Molecular modelling had shown that a tertiary sp3 carbon atom bound to C2 would act as a stereochemical lock (Fig 11). The R stereochemistry (Fig 11a) locks the oxygen atom on the convex face and the methyl group on the concave face of the penem.

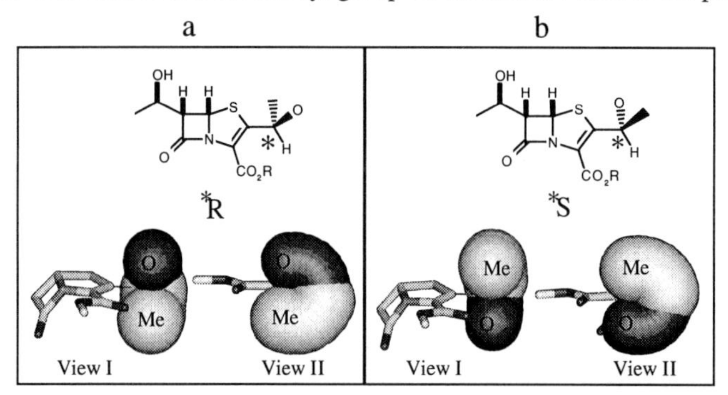

Two views of each penem showing the oxygen atoms and methyl carbon atoms of the C2 substituents displayed in Van der Waal's radii spheres for all allowed conformations at ambient temperature. View II is View I rotated around the Y axis by 90^0.

Fig 11.

The opposite is true for the S stereochemistry (Fig 11b) where the oxygen and carbon atoms are now on the concave and convex faces respectively. There were two approaches available for accessing the S3 pocket (Fig 10). In one approach we could derivatise a hydroxyethyl substituent (Fig 12a) where the R stereochemistry is expected to be active, and in the other we would use the hydroxyl as the stereochemical lock and derivatise the methyl group (Fig 12b).

Fig 12.

In this latter case the active geometry would have the S configuration. The C2 hydroxyethyl substituted penems (Fig 12a), could be synthesised in homochiral form by introduction of D-or L-thiolactic acid into the C4 position of the chiral azetidinone (12) (Fig 15) followed by elaboration and penem formation *via* the usual Wittig phosphorane route. The chiral azetidinone (12) is also the starting point for the carbon linked S3 probe (Fig 12b). In this case the azetidinone is elaborated to the chiral penem (7) (Fig 13) which on uv initiated homolytic epimerisation at C5 to gave a *circa* 2:1 mixture of the 5S and 5R penems[6].

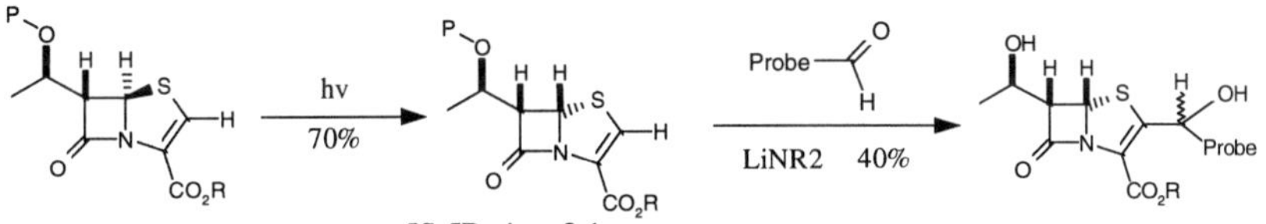

Fig 13.

After chromatographic separation of the desired 5S penem, the anion generated at C2 with lithium diisopropylamide was condensed with various aldehydes. The two isomers of these condensation products were readily separable by chromatography and a representative series is shown in Table 3. We mainly focused on an aromatic link to a probing methyl group as this gave us further spatial control which could be analysed through molecular modelling. In each case (Table 3) the less polar isomer was the most active and although the best S3 probes (8) and (9) were active, they were no more potent than the C2 unsubstituted penem (5) (Table 2). These results indicated that the less polar isomers was merely being tolerated in each case, while the more polar isomers had detrimental contributions to binding. In no case did we observe the jump in activity that would be associated with the introduction of an additional binding element.

It was conceivable that either the aromatic rings were not acceptable linkages, or that the methyl probe of our compounds was not adequately interacting with the S3 pocket, which is known to successfully accommodate valine residues as an alternative to the usual alanyl methyl. Our second series of S3 probes derived from D and L thiolactic acid (Fig 12a) were not open to such criticisms. We chose a urethane link to an isopropyl group as our probe, modelling showed that that this should be able to both access and fill the S3 pocket. The results (Table 4) show that although the urethanes were more active than their parent C2-hydroxyethyl parents, they were no more active than the C2 unsubstituted penem (5) (Table 2). The S isomer was also the more active which was contrary to our predictions for accessing the S3 pocket sited on the convex face of the penem nucleus.

Table 3

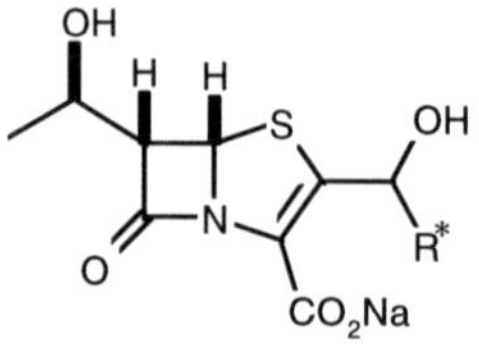

R	Isomer *	IC_{50} µM	R	Isomer *	IC_{50} µM
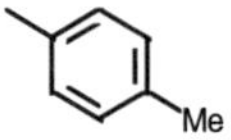 Me	1	2.3	(9) 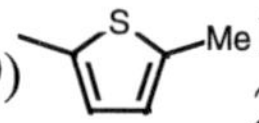 S Me	1	0.26
	2	20		2	1.5
(8) 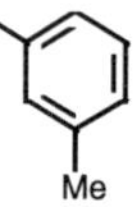 Me	1	1	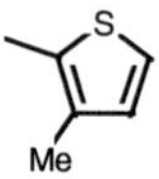 S Me	1	0.85
	2	6.7		2	6.6
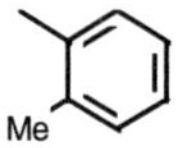 Me	1	0.9	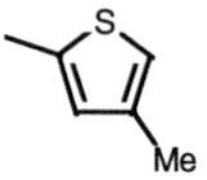 S Me	1	0.93
	2	4.6		2	3.8

Table 4

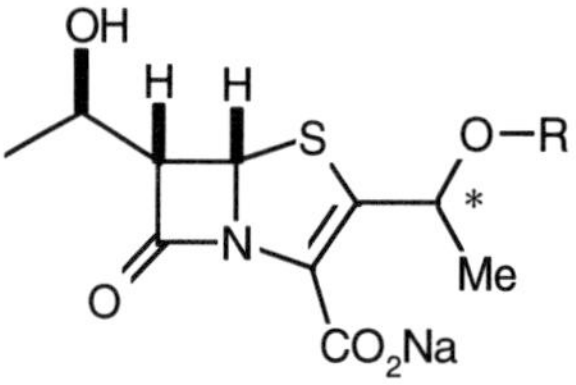

R	Isomer *	IC_{50} µM
$CONHCHMe_2$	S	1.7
H	S	12
$CONHCHMe_2$	R	3.8
H	R	24

The conclusions from the preceding two series was that the S3 pocket was not in the region of space as predicted by our β-turn model. This conclusion was further supported by an analogous result from a designed LP1 inhibitor. Using our β -turn model we designed two azetidinone inhibitors, (10) and (11). The 4-vinyl azetidinone (10) spans the β -turn directly from P1 to P3 (Fig 14) and showed an excellent 5 atom molecular fit of key atoms with an rms deviation of <0.2 A. This system would also be electronically activated towards a nucleophile *via* the enaminone functionality. Our second system was the sophisticated spiro-azetidinone (11) which, although unactivated, was able to mimic various hydrogen bonding elements of the β -turn backbone.

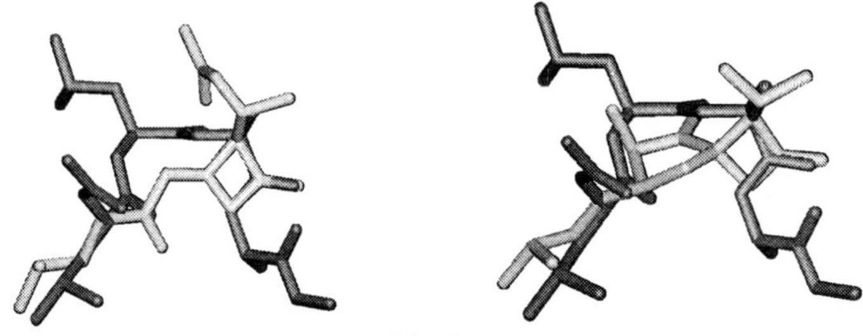

Fig 14.

A computational fit of the azetidinones (10) and (11) to a Type I β-turn [Val-Val-Ala-Ala].

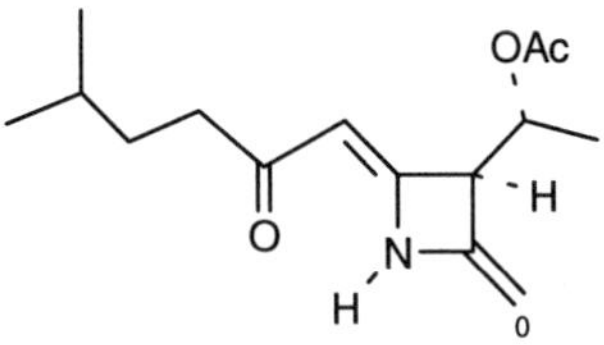

(10) 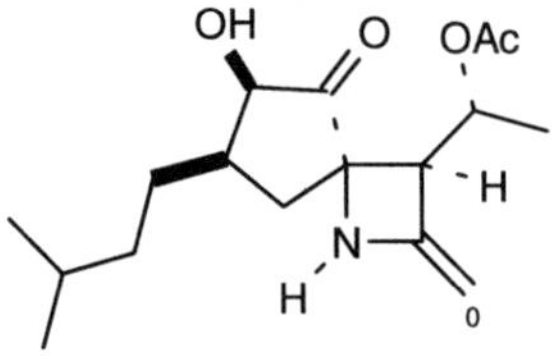 (11)

The stereochemical requirements for the spiro-azetidinone were not trivial, so we chose to synthesise the simpler but more activated 4-vinyl azetidinone in the first instance. The chiral azetidinone (12) was condensed with the stabilised anion (13) to introduce the S3 probe (Fig 15). After deprotection of the hydroxyethyl substituent it was possible to introduce the 4-vinyl moiety by thermal elimination of methylsulphinic acid.

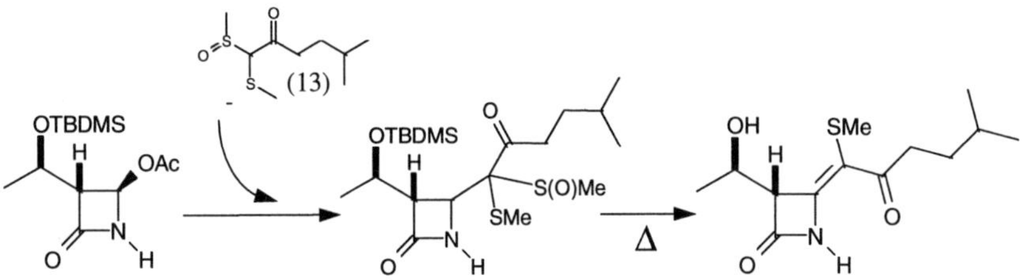

(13)

(12)

Fig 15

The E and Z geometrical isomers formed from this elimination were not separable by chromatography. When tested together they showed only 7% inhibition at 100 uM, despite the electronic activation of the azetidinone carbonyl which showed an I.R. absorption at 1810 cm^{-1}. These results confirmed those obtained from the penem inhibitors, and indicate that the position of the S3 pocket is not compatible with a β - turn conformation at the cleavage site. As the β -turn model is the only secondary structure that could place the S3 pocket adjacent to the C2 position of a penem, we were able to conclude that C2 substitution of a penem system would not be able to generate a significant increase in activity *via* binding into the S3 recognition pocket.

We tested several of our inhibitors for *in vivo* activity by monitoring the relative levels of β-lactamase preprotein and protein using a pulse-chase assay system (Table 5).

We could readily demonstrate complete inhibition of processing in the *E. coli* strain ESS, a strain with a 'leaky' outer membrane, but lost activity progressively with standard and clinical strains. This loss of activity could be restored, or partially restored, by testing in the presence of polymyxin B nonapeptide, a cell wall permeability enhancer.

Table 5

Activity of Four Penem Inhibitors - E.coli Pulse-chase

Penem	IC_{50} µM	(% inhibition β-lactamase processing at 100µM penem) ESSpBR3222	JM109pBR3222	JT4
(4)	0.07	100% (100%)	10% (100%)	0% (100%)
(3)	0.18	100% (100%)	50% (100%)	0% (50%)
(6)	0.69	100%	80%	40%
(5)	0.26	50%	30%	NT

( ) Pulse-chase in presence of 12ug/ml polymyxin B nonapeptide

In conclusion, we have demonstrated that penem esters, amides and salts inhibit signal peptidases *in vitro* and *in vivo* and that 5S stereochemistry is essential for this activity. Substituents at C6 equate to the P1 of the substrate and must be *cis* orientated with respect to the thiazoline ring. The β-turn predicted to be in the carboxy terminus of the signal peptide precedes the cleavage site and that the facial approach of the active site serine to the substrate is atypical for serine proteases.

References

1. a) R. E. Dalbey, *Mol. Micro*, 1991, **5**, 2855. b) M. Inouye and S. Halegoua, *CRC Crit. Rev. Biochem.*, 1980, **7**, 339.
2. a) B. M. Austen, *FEBS Let.*, 1979, **103**, 308. b) G. Von Heijne, *Eur. J. Biochem.*, 1983, **133**, 17. c) G. P. Vlasuk, S. Inouye and M. Inouye, *J. Biol. Chem.*, 1984, **259**, 6195.
3. A. E. Allsop *et al*, *Bio. Med. Chem. Lett.*, 1995, **5**, 443.
4. a) G. D. Rose, *Advan. Protein Chem.*, 1985, **37**, 1. b) C. M. Wilmot and J. M. Thornton, *J. Mol. Biol.*, 1988, **203**, 221.
5. 'Topics in Antibiotic Chemistry,' P. Sammes, Ellis Horwood, Chichester, 1980, Vol 4, Chapter 2, 100.
6. A. E. Allsop, G. Brooks, P. D. Edwards and A. C Kaura, *J. Antibiotics*, 1996, in press.

6

Synthetic Studies toward Complex Polyether Macrolides of Marine Origin

S. D. Burke,* J. R. Philips, K. J. Quinn, G. Zhang, K. W. Jung, J. L. Buchanan and R. E. Perri

DEPARTMENT OF CHEMISTRY, UNIVERSITY OF WISCONSIN-MADISON, 1101 UNIVERSITY AVENUE, MADISON, WI 53706-1396, USA

1 INTRODUCTION

Complex natural products of marine origin show great promise as chemotherapeutic agents against a variety of human health disorders, including microbial and viral infections, hypertension, and cancer, *inter alia*. This paper describes work in an ongoing program directed at the laboratory synthesis of the potent antimitotic agent halichondrin B [(1), Scheme 1]. This fascinating substance, with 32 asymmetric centers adorning a 54-carbon backbone, is both structurally and biologically interesting. Its structural complexity, scarcity, and powerful *in vivo* anticancer activity combine to make it an attractive target for chemical synthesis.

1.1 Isolation, Biological Activity, and Published Synthetic Efforts

Halichondrin B (NSC 609395) is the most potent of a class of polyether macrolides isolated in low yield (1.8 x 10^{-8} to 4.0 x 10^{-5}%) from four different sponge genera (*Halichondria*, *Axinella*, *Phakellia* and *Lissodendoryx*).[1] Although this distribution suggests exogenous production of microbial origins, the ten halichondrin congeners remain exceedingly scarce. National Cancer Institute (NCI) researchers have found that halichondrin B has a tubulin-based mechanism of action, with antimitotic activity analogous to vinblastine, maytansine, rhizoxin and dolastatin 10, among others.[2] Comparatively, the IC_{50} of halichondrin B for L1210 murine leukemia cells is 0.3 nM; other values include dolastatin 10 (0.5 nM), rhizoxin (1 nM), and vinblastine (20 nM). Halichondrin B appears to bind in the vinca domain of tubulin, with consequential inhibition of tubulin-dependent GTP hydrolysis.[2a] Halichondrin B and rhizoxin are nearly identical in their behavior as inhibitors of tubulin polymerization and GTP hydrolysis. NCI researchers E. Hamel and M. R. Boyd responded to our inquiries with data describing the potent *in vivo* activity of halichondrin B against chemoresistant human solid tumor xenografts in immune deficient mice.[3] Specifically, LOX melanoma, KM20L colon, FEMX melanoma and OVCAR-3 ovarian tumor xenografts responded. Based upon this,

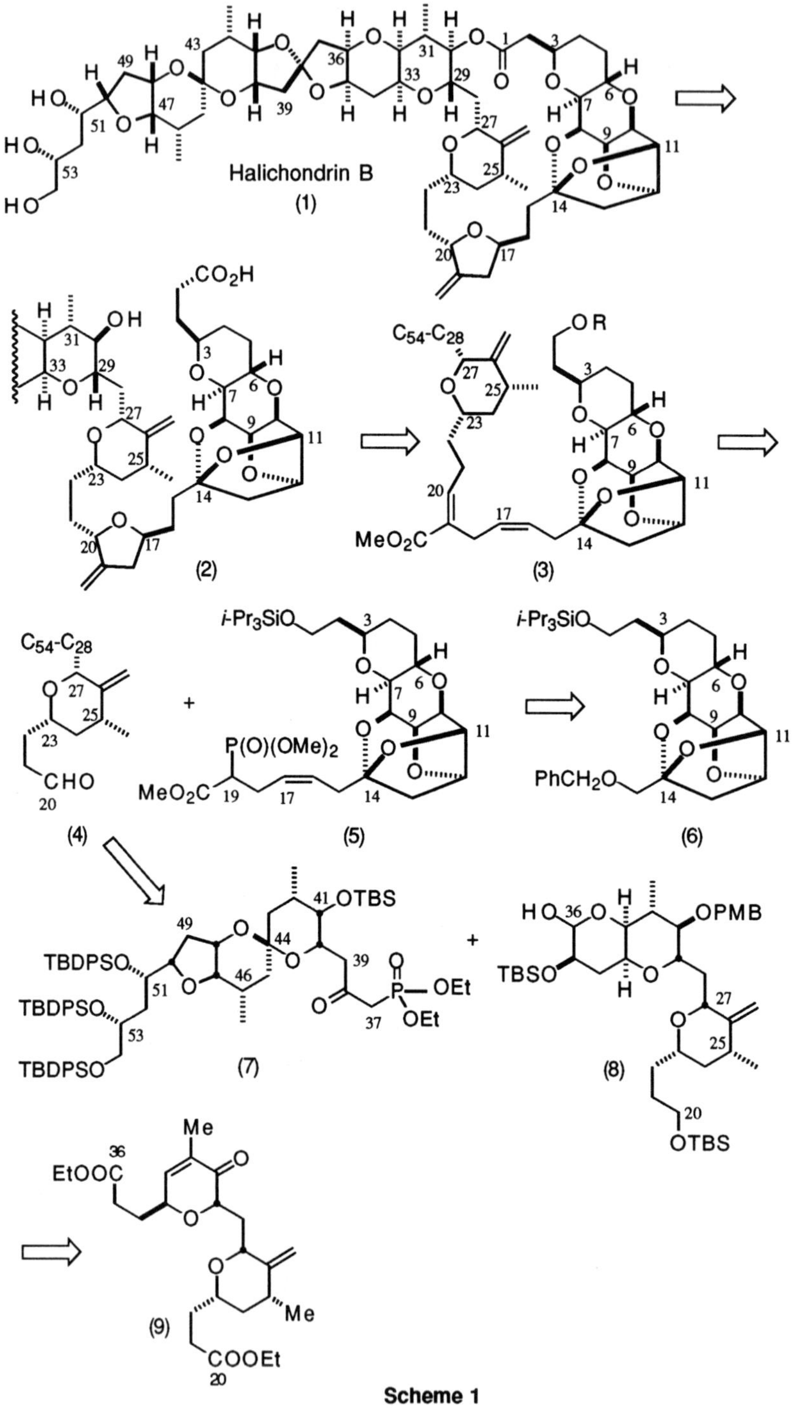

Halichondrin B (1)

(2) (3)

(4) (5) (6)

(7) (8)

(9)

Scheme 1

halichondrin B was recommended for stage A preclinical development.[3] Four research groups (Kishi,[4] Solomon,[5] Yonemitsu[6] and Burke[7]) have published synthetic studies directed at the halichondrins. Kishi's efforts have culminated in an impressive total synthesis of halichondrin B and norhalichondrin B. However, since this synthesis required approximately 150 steps, the issue of supply and access to analogs does not yet appear to be solved.

2 RETROSYNTHETIC ANALYSIS OF HALICHONDRIN B

A convergent strategy is obviously appropriate for a target molecule of such complexity. Our overall strategy along these lines is illustrated in Scheme 1. Formation of the 31-membered macrocyclic lactone in halichondrin B (1) is deferred until last; the seco acid (2) is to arise via a triply convergent approach in which C37-C54, C20-C36, and C1-C19 pieces are brought together via Horner-Wadsworth-Emmons couplings [see (4) + (5), (7) + (8)].[8] The primary focus of this paper is on the enantioselective preparations of the C1-C15 fragment (6) and the C20-C36 precursor (9).

2.1 Retrosynthesis of the C1-C19 Subunit

Simplification (Scheme 2) of the C1-C19 coupling subunit (5) to the propargyl phosphonoacetate (10) and the C1-C15 subunit (6) allows us to focus on the intriguing

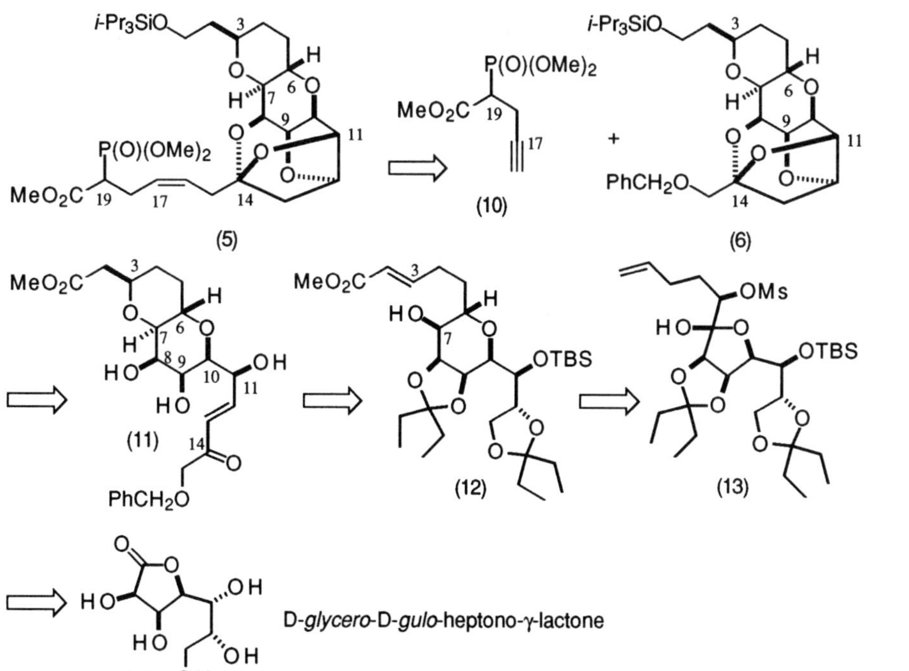

D-*glycero*-D-*gulo*-heptono-γ-lactone

Scheme 2

trans-dioxadecalin and C8-C14 trioxatricyclic cage moieties that constitute the latter. It was envisioned that the cage structure would ensue from enone triol (11) via a tetrahydrofuran-forming Michael addition of the C9 hydroxyl to the enone, followed by ketalization of the C14 carbonyl by the C11 and C8 hydroxyls. Control of the C3 stereocenter in (11) was to be accomplished via a reversible (thermodynamically controlled) Michael addition of the C7 hydroxyl in (12) to the C3 enoate. The tetrahydropyran in (12) was anticipated via a pinacol-type ring expansion[9] of hydroxy-mesylate (13), which was to arise from elaboration of the inexpensive ($0.07/g) D-*glycero*-D-*gulo*-heptono-γ-lactone (14), a chain extension product of D-glucose.

2.2 Retrosynthesis of the C20-C36 Subunit

We hoped to exploit an element of hidden symmetry in the C20-C36 region of halichondrin B (1) to simplify the synthesis. Specifically (Scheme 3), the methylene-linked bis(hydropyran) segment comprising C23-C33 was expected to arise from (9), which has been pursued via a two-directional chain synthesis/terminus differentiation strategy[10] via (15), in which the two hydropyrans are substantially differentiated. In bis(dihydropyran) (16), the expected precursor of (15), one can see the challenge of differentiating the two nearly identical rings. Were any of the four stereocenters in (16) inverted, the two hydropyrans would be rendered identical via either meso or C_2-symmetry. Formation of both of these subtly different dihydropyrans was planned to occur at once via a double [3,3] sigmatropic rearrangement of the bis(dioxanone) (17). A key element of our strategy for this subunit was the enantioselective desymmetrization of

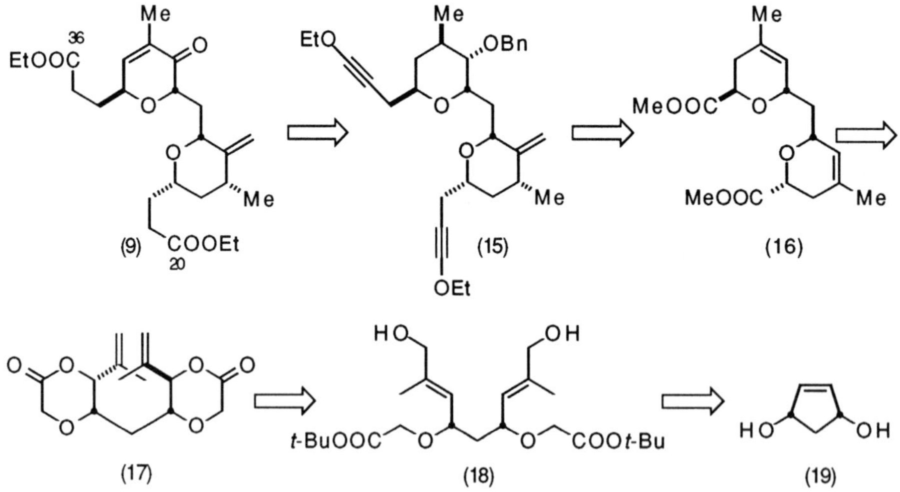

Me 36 EtOOC O O Me (9) COOEt 20

EtO Me OBn O Me (15) OEt

Me MeOOC O O MeOOC Me (16)

O O O O O O (17)

HO OH t-BuOOC O O COOt-Bu (18)

HO OH (19)

Scheme 3

the meso bis(allylic alcohol) (18), which was to arise from the achiral, meso cyclopentene-1,4-diol (19).

3 SYNTHESIS OF THE C1-C15 POLYCYCLIC CAGE KETAL

3.1 Choice of D-*glycero*-D-*gulo*-heptono-γ-lactone (14) as starting material

The choice of (14) (Scheme 2) as the starting material for the C1-C15 subunit was predicated upon several factors, including the aforementioned low-cost availability. Of structural pertinence were the α,β,γ-stereocenters, correct in functionality and absolute stereochemistry to serve as the C8, C9 and C10 carbons of the trioxatricyclic cage in (6). The terminal vicinal diol was expected to yield a C12 aldehyde as a precursor to (11) and the remaining carbinol in (14) would, upon inversion, serve as the C11 stereocenter. Of course, this necessitated a regiocontrolled differentiation of the hydroxyl groups in pentaol (14), preferably one in which the nascent C11 stereochemistry could be fixed. Bis(acetonide) formation from (14) is known to proceed with regioselectivity inappropriate for this task.[11] Bis(acetonide) formation leaving the C11 hydroxyl unmasked can be accomplished on the lactol (hemiacetal) derived from reduction of (14), but this protocol is extremely awkward and difficult to reproduce.[12] Fortunately, formation of the bis(pentylidene) acetal[13] from (14) directly provides the C11-unprotected derivative, greatly facilitating the synthesis.

3.2 Elaboration of D-*glycero*-D-*gulo*-heptono-γ-lactone (14) to the C1-C15 Subunit (6)

Regioselective formation of the bis(pentylidene) acetal from (14) gave (20) (Scheme 4) in which the incorrectly configured C11 carbinol center was left unprotected. Inversion was accomplished via oxidation,[14] followed by stereoselective reduction with $Zn(BH_4)_2$[15] to give (21) after silylation of the alcohol. Addition of the α-alkoxyorganolithium reagent derived from stannane (22)[16] to the lactone carbonyl in (21) gave the desired diol diastereomer, which was selectively mesylated to afford the pinacol rearrangement substrate (13). Regio- and stereospecific 1,2-migration[9] of the C-O bond afforded the C7 pyranone, which was stereoselectively reduced to (23), in which the C7 and C6 stereocenters were established. One-pot ozonolysis and Wittig homologation gave the *E*-enoate (12) in high yield. Establishment of the *trans*-fused dioxadecalin (24), with the C3 equatorial acetic ester side chain, was accomplished via a Michael addition effected under thermodynamic control.

Turning to the elaboration of the C8-C14 trioxatricyclic caged ketal, hydrolysis of the side-chain acetal was selectively accomplished, leaving the fused C8-C9 diol acetal intact.[17] Oxidative glycol cleavage with sodium periodate yielded aldehyde (25), which was reacted with Wittig reagent (26)[5b,18] to afford enone (27) in high yield. The final

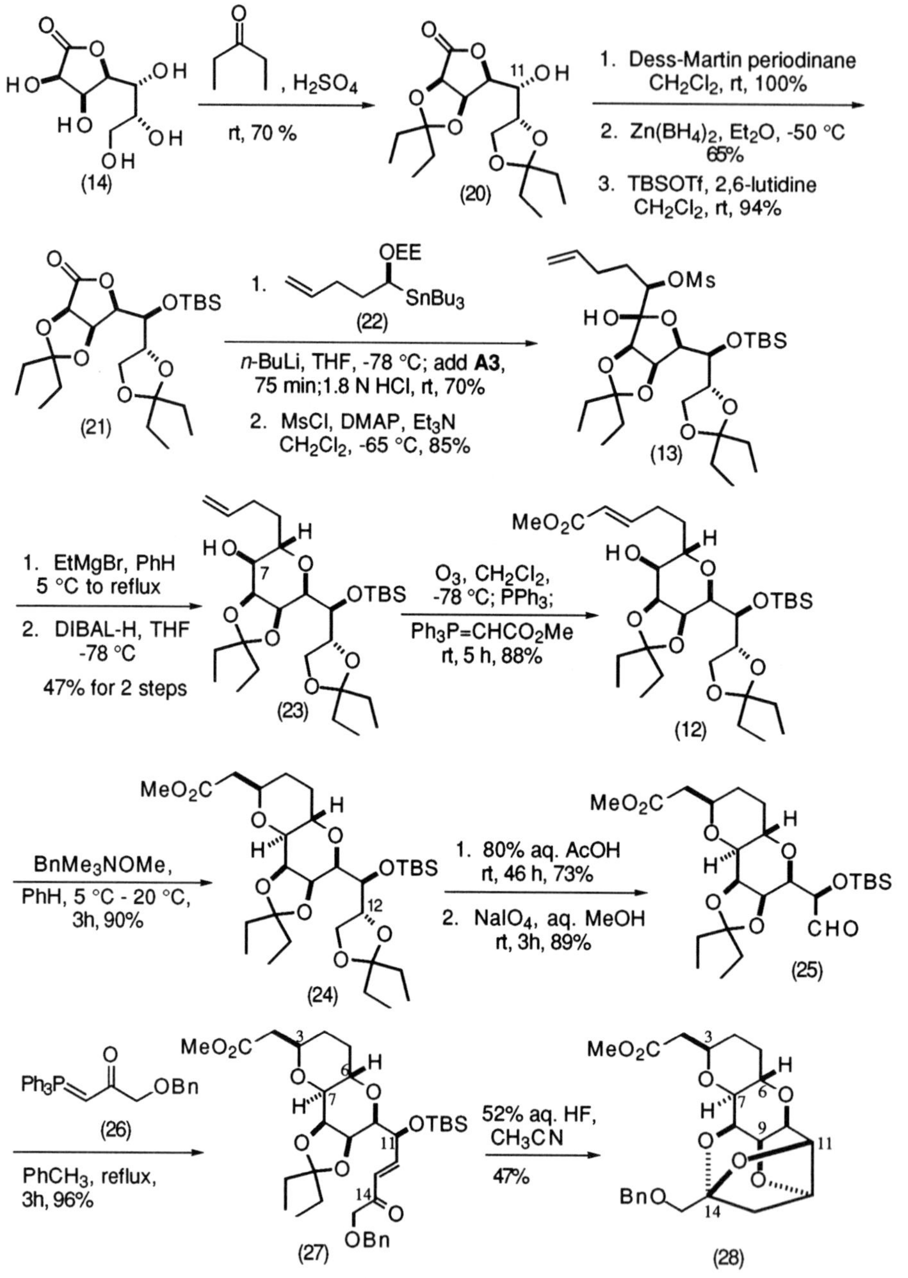

Scheme 4

step is actually a sequence of four reactions leading to the trioxatricyclic cage structure (28). *In situ* C11 silyl ether cleavage, pentylidene acetal hydrolysis, Michael addition of the C9 hydroxyl to the enone, and ketalization of the C14 ketone by the C8 and C11 hydroxyls effected this transformation as planned. The C1-C15 skeleton, with nine asymmetric centers, was thus assembled in fifteen steps. Ester reduction and primary alcohol protection completes the C1-C15 subunit (6).

4 SYNTHESIS OF THE C20-C36 BIS(HYDROPYRAN) SUBUNIT

As described in section 2.2 and Scheme 3, we sought to employ a two-directional chain synthesis/terminus differentiation strategy[10] for this halichondrin subunit. The construction of two polysubstituted hydropyran rings that were capable of being distinguished and individually functionalized was required, and we chose to apply our dioxanone-to-dihydropyran rearrangement[19] to this task.

4.1 Dioxanone-to-Dihydropyran and Double Dioxanone-to-bis(Dihydropyran) Rearrangements

Because of the ubiquity of polysubstituted hydropyran and hydrofuran rings as structural components of polyether natural products, we have developed versatile methods for their synthesis.[19,20] Among these methods is the dioxanone-to-dihydropyran Claisen rearrangement,[19] illustrated in Scheme 5. Readily available dioxanones of general structure (29), upon conversion to the corresponding silylketene acetals and thermolysis, afford the dihydropyran (31) via the bracketed transition structure (30). Similarly, the diastereomeric dioxanone (32) leads, via (33), to (34). As with all Claisen rearrangements,[21] these enjoy the following attributes: 1) sacrifice of a C-O bond for a new C-C bond; 2) conversion of remote sp^2-carbon geometry into vicinal (C2,C3) sp^3-carbon stereochemistry; and 3) suprafacial, suprafacial sigmatropic migration along the two π-systems. It is to be noted that this Claisen rearrangement variant requires that both the pre-existing heterocyclic and the six-atom pericyclic arrays adopt boat-like conformations. Nevertheless, this requirement is constant, and the rearrangements are predictably stereospecific.

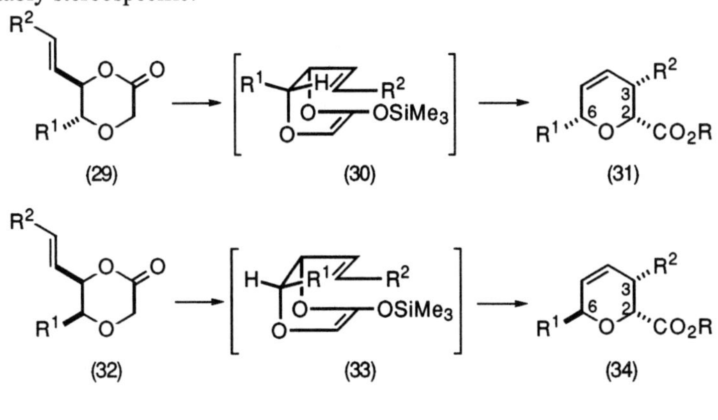

(29) (30) (31)

(32) (33) (34)

Scheme 5

An elaborate example of this method is shown in Scheme 6. In the context of an approach to erythronolide B using hydropyrans as templates for establishing the stereochemical and functional details of the seco acid backbone, a double dioxanone Claisen rearrangement was executed.[22] In this way, convergently assembled bis(dioxanone) (35) was converted, via the intermediates shown, to bis(dihydropyran) (36) in good yield. A similar transformation was planned as a key step in constructing the methylene-linked bis(hydropyran) C23-C33 segment of halichondrin B.

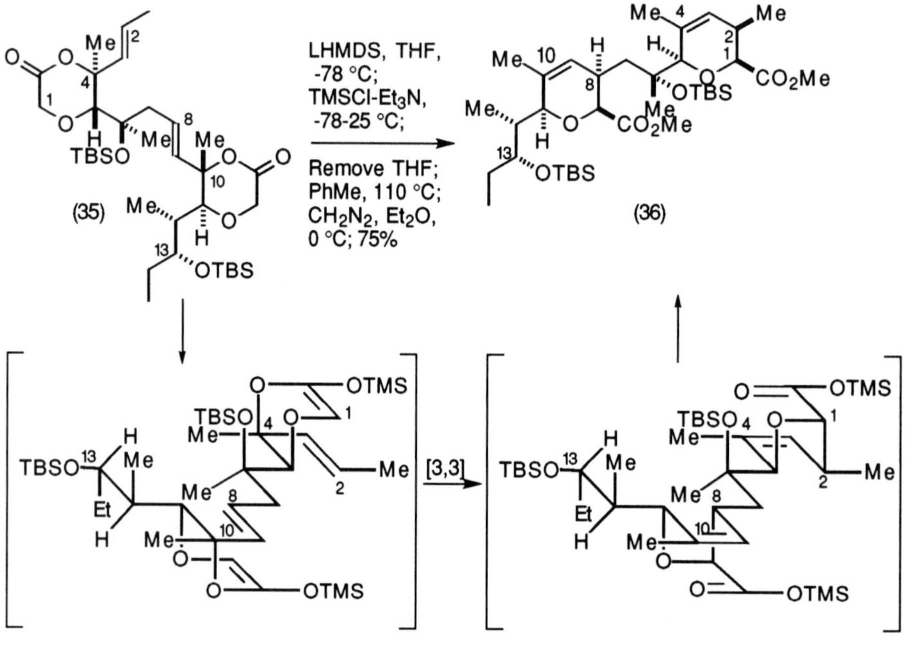

LHMDS, THF, -78 °C; TMSCl-Et_3N, -78-25 °C; Remove THF; PhMe, 110 °C; CH_2N_2, Et_2O, 0 °C; 75%

(35) (36) [3,3]

Scheme 6

4.2 Application of the Double Dioxanone Claisen Rearrangement for Two-Directional Chain Synthesis/Desymmetrization

Bis(*O*-alkylation) of meso cyclopentene-1,4-diol (19)[23] with *t*-butylbromoacetate under phase transfer conditions[24] gave [(37), Scheme 7]. Ozonolysis and two-directional Wittig homologation gave the bis(thioester) (38) as a single product,[25] which was readily reduced to the symmetric bis(allylic alcohol) (18).[26] Enantioselective desymmetrization of (18) was accomplished via the Sharpless asymmetric epoxidation,[27] yielding (39) in >99% enantiomeric excess. The high ee observed is a consequence of two asymmetric transformations occurring on a single substrate; minor stereochemical errors are effectively removed via diastereomer (rather than enantiomer) formation.[28]

HO OH (19)

$BrCH_2COOt$-Bu, 40 % NaOH, n-Bu_4NHSO_4, PhH, r. t., 93 %

t-BuOOC COOt-Bu (37)

O_3, CH_2Cl_2, -78 °C; PPh_3, -78-0 °C; $Ph_3P{=}C(CH_3)COSPh$, CH_2Cl_2, reflux, 74 %

PhSOC COSPh t-BuOOC COOt-Bu (38)

$NaBH_4$, EtOH, 0 °C to r. t., 88 %

HO OH t-BuOOC COOt-Bu (18)

L-(+)-DET, Ti(Oi-Pr)$_4$, 4 A MS, t-BuOOH, CH_2Cl_2, -40 °C, 90 %

HO OH t-BuOOC COOt-Bu (39)

1. CH_3SO_2Cl, Et_3N, CH_2Cl_2, -10 °C
2. NaI, 2-butanone, 95 % (two steps)

OH HO t-BuOOC COOt-Bu (40)

CF_3COOH, $PhCH_3$, 80 °C, quant.

(17)

LHMDS, THF, -78 °C; 1:1 TMSCl-Et_3N, -78 °C to r. t.; -THF, +$PhCH_3$, 110 °C; H^+, CH_2N_2, Et_2O; 71 %

H H H Me OTMS Me TMSO (41)

H H Me Me $TMSO_2C$ H OTMS H (42)

Me MeOOC Me MeOOC (16)

1. DIBAL-H, Et_2O, -78 °C to r. t.
2. TBSCl, imidazole, CH_2Cl_2, r. t. 95 % (two steps)

Me 31 30 29 28 27 26 25 Me TBSO TBSO (43)

BH_3-THF, THF, -78 °C to 0 °C; H_2O_2, OH^-, 62 %

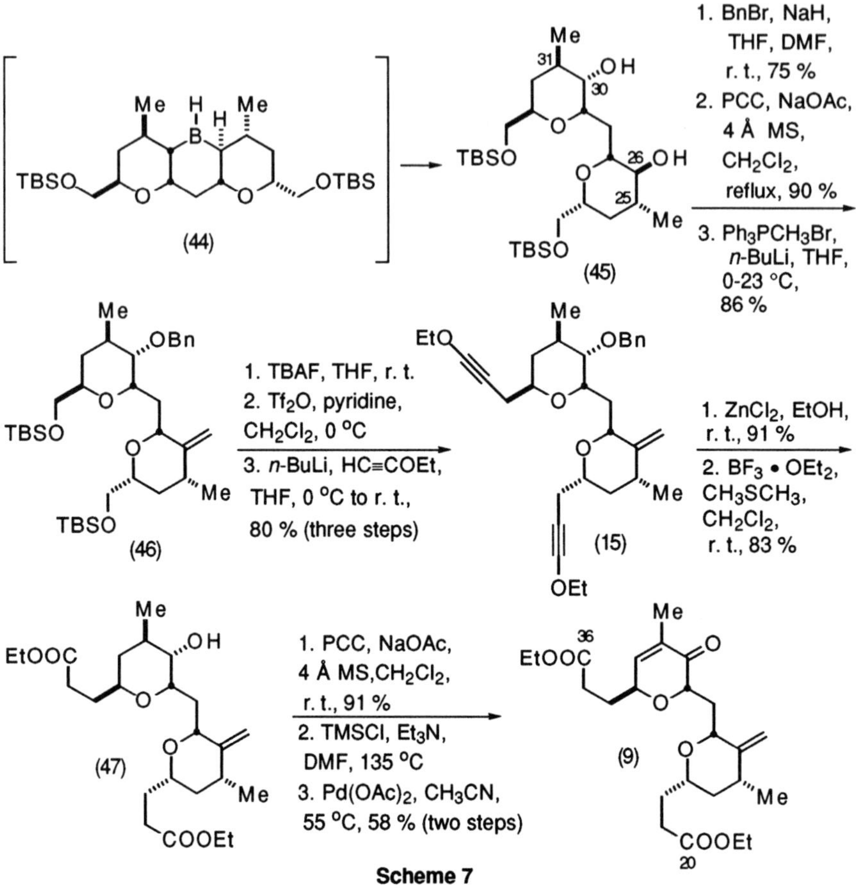

(44) → (45): 1. BnBr, NaH, THF, DMF, r. t., 75 %; 2. PCC, NaOAc, 4 Å MS, CH_2Cl_2, reflux, 90 %; 3. Ph_3PCH_3Br, *n*-BuLi, THF, 0-23 °C, 86 %

(46) → (15): 1. TBAF, THF, r. t.; 2. Tf_2O, pyridine, CH_2Cl_2, 0 °C; 3. *n*-BuLi, HC≡COEt, THF, 0 °C to r. t., 80 % (three steps)

(15) → (47): 1. $ZnCl_2$, EtOH, r. t., 91 %; 2. $BF_3 \bullet OEt_2$, CH_3SCH_3, CH_2Cl_2, r. t., 83 %

(47) → (9): 1. PCC, NaOAc, 4 Å MS, CH_2Cl_2, r. t., 91 %; 2. TMSCl, Et_3N, DMF, 135 °C; 3. $Pd(OAc)_2$, CH_3CN, 55 °C, 58 % (two steps)

Scheme 7

Reductive fragmentation of the bis(epoxy alcohol) (39) was accomplished via the dimesylate by treatment with excess sodium iodide in 2-butanone. The more traditional NaI/acetone/Δ conditions had allowed the mesylate-to-iodide conversion as a prelude to Zn reduction, but a small amount (~5%) of alkene-containing product was also observed. Increasing the solvent boiling point allowed the straightforward use of excess iodide as the reductant in this transformation, which directly gave (40) in high yield. Quantitative acid-induced closure to bis(dioxanone) (17) set up the key double Claisen rearrangement. Conversion of (17) to the bis(silylketene acetal) (41) and rearrangement via the indicated intermediates (note boat-like transition state requirements) gave bis(dihydropyran) (16) in good yield. Reduction of the esters and protection as the *t*-butyldimethylsilyl (TBS) ethers gave critical intermediate (43).

To reiterate, it was now necessary to distinguish the two hydropyran rings so that they could be elaborated into a suitable precursor to the fully elaborated C23-C33 region.

Again, the near-identity of the two hydropyrans in (43) is best realized by noting that epimerization at any one of the four asymmetric centers would render (43) symmetric in the C_2 or meso sense.

Our original intent was to subject (43) to a double hydroboration, with the expectation that both double bonds would be attacked from the β-face, as drawn, due to the C29 and C27 α-oriented substituents. We were surprised to find that the C25-C26 alkene was intrinsically much more reactive than the C30-C31 locus. As a result, an *intermolecular* β-face hydroboration at C25-C26 was followed by an *intramolecular* α-face hydroboration of the C30-C31 alkene, leading via (44) to (45). This set the C25 configuration correctly and gave appropriate functionality of C26 and C30. Although the stereocenters at C30 and C31 were inverted relative to those in halichondrin B, these centers were planned for revision via planarization (*vide infra*).

The ultimate distinction of the two hydropyran rings was dependent upon a second surprising reactivity differential; the C30 carbinol was much more reactive than the C26 carbinol. Acylation or oxidation could be accomplished at C30 in a completely regioselective fashion. Even benzyl ether formation occurred regioselectively at the C30 hydroxyl in (45); oxidation of the residual C26 carbinol and Wittig methylenation of the ketone afforded (46).

Simultaneous two-directional chain elongation [(46) → (15)] was accomplished via displacement with ethoxyacetylide[29] on the derived ditriflate. Mild, Lewis-acid induced hydration[30] of the alkyne moieties and benzyl ether cleavage[31] gave hydroxy diester (47). The final stage in distinguishing the two hydropyrans rested in the conversion of (47) to (9) via ketone formation and Saegusa oxidation[32] to afford the enone. This substrate is expected to facilitate the C32-C36-containing hydropyran ring formation and the establishment of the correct configurations of the C31 and C30 stereocenters.

4.3 Conclusions

In summary, direct and feasible routes to two major subunits (C1-C15 and C20-C36) of halichondrin B have been described. The first of these subunits (6), with nine asymmetric centers constituting C1-C15, is available from an inexpensive carbohydrate starting material in only 17 steps.[7b] The C20-C36 subunit (9) served as an excellent vehicle to demonstrate the two-directional chain synthesis/terminus differentiation strategy.[7a,c]

Acknowledgements: We are grateful to Pfizer, Inc., Merck, Inc., Wyeth-Ayerst Research, Bristol-Myers Squibb and the National Institutes of Health for direct and indirect support of this research.

References

1. (a) Hirata, Y.; Uemura, D. *Pure & Appl. Chem.* **1986**, *58*, 701. (b) Pettit, G. R.; Ichihara, Y.; Wurzel, G.; Williams, M. D.; Schmidt, J. M.; Chapuis, J.-C. *J. Chem. Soc., Chem. Commun.* **1995**, 383. (c) Pettit, G. R.; Herald, C. L.; Boyd, M. R.; Leet, J. E.; Dufresne, C.; Doubek, D. L.; Schmidt, J. M.; Cerny, R. L.; Hooper, J. N. A.; Rutzler, K. C. *J. Med. Chem.* **1991**, *34*, 3339. (d) Litaudon, M.; Hart, J. B.; Blunt, J. W.; Lake, R. J.; Munro, M. H. G. *Tetrahedron Lett.* **1994**, *35*, 9435.
2. (a) Hamel, E. *Pharmac. Ther.* **1992**, *55*, 31. (b) Bai, R.; Paull, K. D.; Herald, C. L.; Malspeis, L.; Pettit, G. R.; Hamel, E. *J. Biol. Chem.* **1991**, *266*, 15882. (c) Paull, K. D.; Lin, C. M.; Malspeis, L.; Hamel, E. *Cancer Res.* **1992**, *52*, 3892. (d) Luduena, R. F.; Roach, M. C.; Prasad, V.; Pettit, G. R. *Biochem. Pharmacol.* **1993**, *45*, 421.
3. (a) Personal correspondence with Ernest Hamel, M.D., Ph.D., Senior Investigator, Laboratory of Molecular Pharmacology, Developmental Therapeutics Program, Division of Cancer Treatment, National Cancer Institute, NIH. Dr. Hamel informed us that halichondrin B cures *in vivo* human tumors transplanted into immunodeficient nude mice. (b) Personal correspondence with Michael R. Boyd, M.D., Ph.D., Chief, Laboratory of Drug Discovery Research and Development, NCI Cancer Research Center, Frederick, MD. Dr. Boyd shared with us his data and slides that resulted in the NCI Decision Network Committee's selection of halichondrin B for drug development.
4. (a) Aicher, T. D.; Buszek, K. R.; Fang, F. G.; Forsyth, C. J.; Jung, S. H.; Kishi, Y.; Matelich, M. C.; Scola, P. M.; Spero, D. M.; Yoon, S. K. *J. Am. Chem. Soc.* **1992**, *114*, 3162. (b) Duan, J. J.-W.; Kishi, Y. *Tetrahedron Lett.* **1993**, *34*, 7541. (c) Aicher, T. D.; Buszek, K. R.; Fang, F. G.; Forsyth, C. J.; Jung, S. H.; Kishi, Y.; Scola, P. M. *Tetrahedron Lett.* **1992**, *33*, 1549. (d) Buszek, K. R.; Fang, F. G.; Forsyth, C. J.; Jung, S. H.; Kishi, Y.; Scola, P. M.; Yoon, S. K. *Tetrahedron Lett.* **1992**, *33*, 1553. (e) Fang, F. G.; Kishi, Y.; Matelich, M. C.; Scola, P. M. *Tetrahedron Lett.* **1992**, *33*, 1557. (f) Aicher, T. D.; Kishi, Y. *Tetrahedron Lett.* **1987**, *28*, 3463.
5. (a) Kim, S.; Salomon, R. G. *Tetrahedron Lett.* **1989**, *30*, 6279. (b) Cooper, A. J.; Salomon, R. G. *Tetrahedron Lett.* **1990**, *31*, 3813. (c) DiFranco, E.; Ravikumar, V. T.; Salomon, R. G. *Tetrahedron Lett.* **1993**, *34*, 3247. (d) Cooper, A. J.; Pan, W.; Salomon, R. G. *Tetrahedron Lett.*, **1993**, *34*, 8193.
6. (a) Horita, K.; Hachiya, S.; Nagasawa, M.; Hikota, M.; Yonemitsu, O. *Synlett* **1994**, 38. (b) Horita, K.; Nagasawa, M.; Hachiya, S.; Yonemitsu, O. *Synlett* **1994**, 40. (c) Horita, K.; Sakurai, Y.; Nagasawa, M.; Hachiya, S.; Yonemitsu, O. *Synlett* **1994**, 43. (d) Horita, K; Sakurai, Y.; Nagasawa, M.; Maeno, K.; Hachiya, S.; Yonemitsu, O. *Synlett* **1994**, 46.
7. (a) Burke, S. D.; Zhang, G.; Buchanan, J. L. *Tetrahedron Lett.* **1995**, *36*, 7023. (b) Burke, S. D.; Jung, K. W.; Phillips, J. R.; Perri, R. E. *Tetrahedron Lett.* **1994**, *35*, 703. (c) Burke, S. D.; Buchanan, J. L.; Rovin, J. D. *Tetrahedron Lett.* **1991**, *32*, 3961.
8. (a) Hammond, G. B.; Blagg Cox, M.; Wiemer, D. F. *J. Org. Chem.* **1990**, *55*, 128. (b) Marshall, J. A.; DeHoff, B. S.; Cleary, D. G. *J. Org. Chem.* **1986**, *51*, 1735.
9. (a) Gilchrist, T. L.; Stanford, J. E. *J. Chem. Soc. Perkin Trans. 1* **1987**, 225. (b) Schoenen, F. J.; Porco, J. A., Jr.; Schreiber, S. L.; VanDuyne, G. D.; Clardy, J. *Tetrahedron Lett.* **1989**, *30*, 3765. (c) Sisti, A. J.; Vitale, A. C. *J. Org. Chem.* **1972**, *37*, 4090. For a recent review on pinacol rearrangement, see: (d) Rickborn, B. In *Comprehensive Organic Synthesis*; Trost, B. M., Fleming, I., Eds.; Pergamon: London, 1991; Vol. 3, p 721.
10. For recent reviews, see: (a) Magnuson, S. R. *Tetrahedron* **1995**, *51*, 2167. (b) Poss, C. S.; Schreiber, S. L. *Acc. Chem. Res.* **1994**, *27*, 9.
11. (a) Shing, T. K. M.; Tsui, H.-c.; Zhou, Z.-h.; Mak, T. C. W. *J. Chem. Soc., Perkin Trans. I* **1992**, 887. (b) Shing, T. K. M.; Tsui, H.-c.; Zhou, Z.-h. *J. Chem. Soc., Chem. Commun.* **1992**, 810.

12. (a) Brimacombe, L. S.; Tucker, L. C. N. *J. Chem. Soc. (C)* **1968**, 567. (b) Stork, G.; Takahashi, T.; Kawamoto, I.; Suzuki, T. *J. Am. Chem. Soc.* **1978**, *100*, 8272. (c) Wolfrom, M. L.; Wood, H. B. *J. Am. Chem. Soc.* **1951**, *73*, 2934.
13. Masamune, S.; Ma, P.; Okumoto, H.; Ellingboe, J. W.; Ito, Y. *J. Org. Chem.* **1984**, *49*, 2834.
14. Czernecki, S.; Georgoulis, C.; Stevens, C. L.; Vijayakumaran, K. *Tetrahedron Lett.* **1985**, *26*, 1699.
15. Iida, H.; Yamazaki, N.; Kibayashi, C. *J. Org. Chem.* **1986**, *51*, 3769 and references cited therein.
16. (a) Still, W. C. *J. Am. Chem. Soc.* **1978**, *100*, 1481. (b) Chan, P. C.-M.; Chong, J. M. *J. Org. Chem.* **1988**, *53*, 5584. (c) Chong, J. M.; Mar, E. K. *Tetrahedron* **1989**, *45*, 7709. (d) Chong, J. M.; Mar, E. K. *Tetrahedron Lett.* **1990**, *31*, 1981. (e) Chan, P. C.-M.; Chong, J. M. *Tetrahedron Lett.* **1990**, *31*, 1985. (f) The preparation of stannane (22) from 4-pentenal is shown below:

CHO
1. *i*-PrMgCl, Bu_3SnH, Galvinoxyl, 56%
2. (*R*)-BINAL, THF, -78 °C, 48%
OH, $SnBu_3$ (86% ee)
Cl, O; Me_2NPh, CH_2Cl_2, 98%
O, O, $SnBu_3$ (22)

17. 4,5-Fused 1,3-dioxolane rings are more stable than those derived from acyclic vicinal diols. See: Haines, A. H. *Adv. Carbohydr. Chem. Biochem.* **1981**, *39*, 13.
18. Bestmann, H. J.; Arnason, B. *Chem. Ber.* **1962**, *95*, 1513.
19. (a) Burke, S. D.; Armistead, D. M.; Schoenen, F. J.; Fevig, J. M. *Tetrahedron* **1986**, *42*, 2787-2801. (b) Burke, S. D.; Armistead, D. M.; Schoenen, F. J. *J. Org. Chem.* **1984**, *49*, 4320. (c) Burke, S. D.; Lee, K. C.; Santafianos, D. *Tetrahedron Lett.* **1991**, *32*, 3957. (d) Burke, S. D.; Chandler, A. C., III; Nair, M. S.; Campopiano, O. *Tetrahetron Lett.* **1987**, *28*, 4147-4148. (e) Burke, S. D.; Schoenen, F. J.; Nair, M. S. *Tetrahedron Lett.* **1987**, *28*, 4143-4146. (f) Burke, S. D.; Schoenen, F. J.; Murtiashaw, C. W. *Tetrahedron Lett.* **1986**, *27*, 449-452.
20. (a) Burke, S. D.; Rancourt, J. *J. Am. Chem. Soc.* **1991**, *113*, 2335. (b) Burke, S. D.; Jung, K. W. *Tetrahedron Lett.* **1994**, *35*, 5837-5840. (c) Burke S. D.; Jung K. W.; Perri, R. E. *Tetrahedron Lett.* **1994**, *35*, 5841-5844.
21. (a) Ziegler, F. E. *Chem. Rev.* **1988**, *88*, 1423. (b) Wipf, P. in "*Comprehensive Organic Synthesis*," Trost, B. M., Ed.; Pergamon Press: Oxford, **1991**, Vol. 5, pp 827-873.
22. Unpublished work of J. Young, K. C. Lee and C.-S. Lee.
23. Kaneko, C.; Sugimoto, A.; Tanaka, S. *Synthesis* **1974**, 876.
24. (a) Pietraszkiewicz, M.; Jurczak, J. *Tetrahedron* **1984**, *40*, 2967. (b) Rabinovitz, M.; Cohan, Y.; Halpern, M. *Angew. Chem., Int. Ed. Engl.* **1986**, *25*, 960.
25. (a) Keck, G. E.; Boden, E. P.; Mabury, S. A. *J. Org. Chem.* **1985**, *50*, 709. (b) Keck, G. E.; Murry, J. A. *J. Org. Chem.* **1991**, *56*, 6606. (c) Emmer, G. *Tetrahedron* **1992**, *48*, 7165.
26. Liu, H.-J.; Bukownik, R. R.; Pednekar, P. R. *Synth. Commun.* **1981**, *11*, 599.
27. (a) Gao, Y.; Hanson, R. M.; Klunder, J. M.; Ko, S. Y.; Masamune, H.; Sharpless, K. B. *J. Am. Chem. Soc.* **1987**, *109*, 5765. (b) Rossiter, B. E. in *Asymmetric Synthesis* , Morrison, J. B., Ed., Vol. 5, p. 193, Academic Press (1985).
28. Schreiber, S. L.; Schreiber, T. S.; Smith, D. B. *J. Am. Chem. Soc.* **1987**, *109*, 1525.
29. Carling, R. W.; Holmes, A. B. *Tetrahedron Lett.* **1986**, *27*, 6133.
30. Broekema, R. J. *Rec. Trav. Chim.* **1974**, *94*, 209.
31. Fuji, K.; Kawabata, T.; Fujita, E. *Chem. Pharm. Bull.* **1980**, *28*, 3662.
32. Ito, Y.; Hirao, T.; Saegusa, T. *J. Org. Chem.* **1978**, *43*, 1011.

7
Trinems: Synthesis and Antibacterial Activity of a New Generation of Antibacterial β-Lactams

S. Biondi

GLAXOWELLCOME, MEDICINES RESEARCH CENTRE, VIA A. FLEMING 4, 37135 VERONA, ITALY

1 INTRODUCTION

β-lactam antibiotics are the agents of choice in the treatment of infectious diseases caused by bacteria.[1] Since the discovery of penicillin, a number of new classes have been found either from natural sources or by chemical synthesis.[2] Among them are: cephalosporins, carbacephems, oxacephems, monobactams, penems, carbapenems, and more recently trinems (formerly referred to as tribactams).[3]

Their extremely good safety profile and efficacy in fighting a wide range of bacterial strains have lead to a widespread use in healthcare units and common practice. However, their misuse and overuse has posed the favourable conditions for the onset of resistance, by selecting those bacteria that have the ability to survive in the presence of these antibiotics. Resistance to β-lactams is particularly diffused and causes serious concern among clinicians, as these effective drugs are losing efficacy.[4]

The last generation of β-lactams has targeted its spectrum of activity against resistant strains, by overcoming reduced permeability of membranes through active transport,[5] by increasing the hydrolytic stablity to chromosomic and plasmidic β-lactamases (TEM, OXA, SHV, K1, P99, PSE, Cph),[6] by enhancing their affinity to altered Penicillin Binding Proteins (PBP2a, Methicillin resistant *Staphylococcus aureus*, PBP2x, *Streptococcus pneumoniae* Penicillin resistant).[7]

In particular, trinems are a new class of β-lactams discovered in GlaxoWellcome laboratories at Verona, characterized by a broad spectrum of activity versus Gram positives and Gram negatives either aerobic and anaerobic and are stable to clinically relevant isolated β-lactamases and human dehydropeptidases (DHP-I).[8]

The structure of trinems (Fig.1) is made up of a tricyclic β-lactam **1** in which ring **C** could be a five-, six- or seven-membered ring, and a hydroxyethyl side chain at C10. Structure activity relationship studies showed that the best biological profile was observed in those compounds bearing a heteroatom attached at C4 and having absolute configuration (*4S,8S*), while the remaining three stereocenters were fixed.

An appropriate example is given by Sanfetrinem (GV104326) **2a**, which showed a particularly good biological profile, and its metabolically labile ester GV118819, **2b**, that are both currently undergoing phase II clinical studies.[9] This compound proved to be very active against Gram positives, Gram negatives, anaerobes, including β-lactamase producing strains, and is safe and very well tolerated in man.

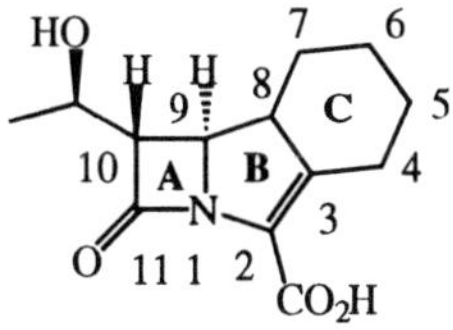

1

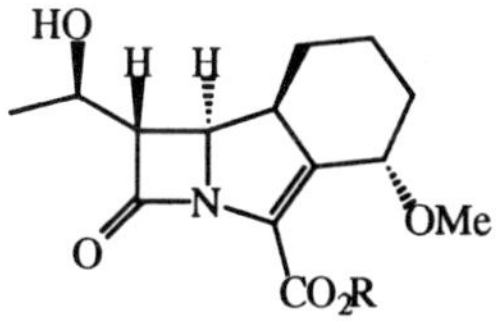

2a GV104326 R = Na
2b GV118819 R = $CH(Me)O_2COC_6H_{11}$

Fig.1 General structure of trinems **1** and Sanfetrinem **2.**

The research project has continued in our laboratories with the aim of exploring the biological properties of trinems.
This paper refers to the synthesis and microbiological activity of several templates featuring a side chain at C4 containing either an oxygen or a nitrogen atom directly linked to ring C (Fig.1), and their utilization for the production of several classes of compounds distinguishable by the functional group directly bound at such position. Figure 2 shows the general structure of trinem templates **A**, which contains the tricyclic core structure and a substituent X containing a heteroatom selected from O or N directly bound at C4.

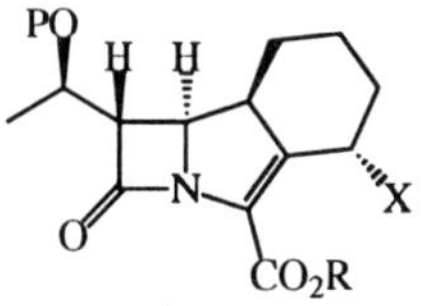

A

P = TBDMS, H

R = allyl, benzyl, H, Na

Fig. 2. General structure of trinem templates

2 GENERAL CHEMICAL METHODOLOGIES

The availability of an efficient and selective synthesis of advanced intermediates suitable for chemical derivatization to produce a large number of compounds in a relatively short period of time was considered a prerequisite for the objectives we posed in this project. The previously reported synthesis, successfully employed to obtain such derivatives, presented only a moderate diastereoselectivity in the first step, resulting in a loss of material.[10] In this view, an optimization study of several advanced intermediates has been performed allowing a highly selective multigram synthesis of the epoxide **6** (Scheme 1), in two steps from the commercially available 4-acetoxyazetidin-2-one **3**.[10]

Ring opening of epoxide **6** occurred regio and stereospecifically with inversion of configuration using a suitable nucleophile and a protic or Lewis acid catalyst, giving rise to intermediates **7**, **8** and **9** in satisfactory yields.

Another useful intermediate on the laboratory scale for the introduction of oxygen or nitrogen containing side chains is the epoxyphosphate **14**, obtainable in 5 steps from azetidinone **3** (Scheme 2), which can be easily converted into the intermediates **15** and **17**.[11]

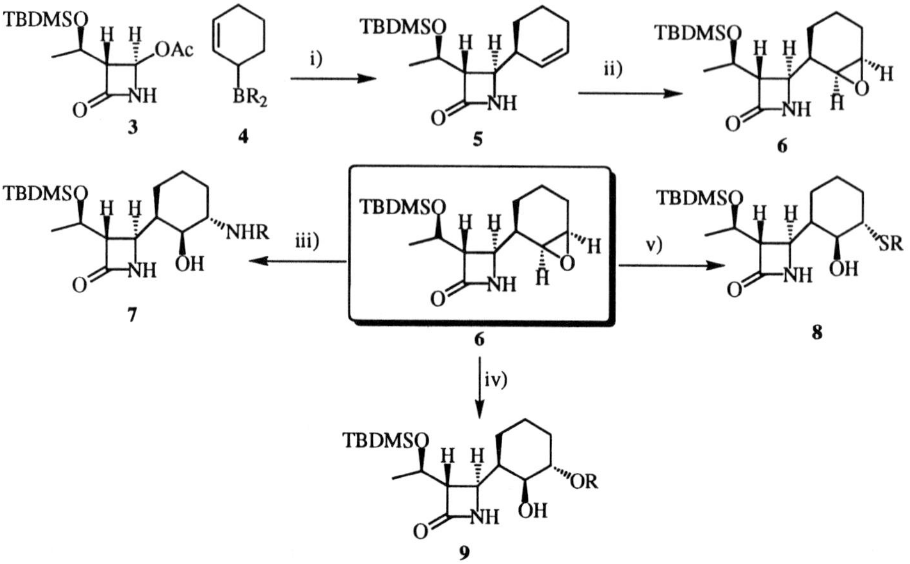

i) $ZnEt_2$, THF, ii) MMPTA, CH_2Cl_2, iii) NHR, $LiClO_4$, CH_3CN, iv) ROH, H^+ or CAN, v) RSH.

Scheme 1. Highly diastereoselective synthesis of key intermediates to trinems.

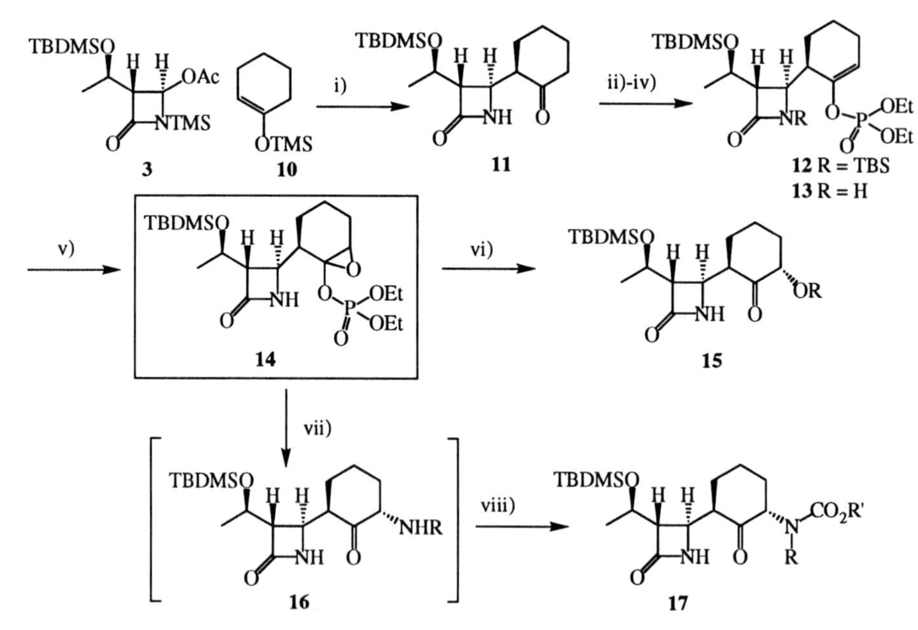

i) TMSOTf, CH_3CN, 0°C; ii) TBDMSCl, TEA, DMF; iii) a) LHMDSA, THF, -70°C, 1 hr; b) $ClP(O)(OEt)_2$, -70°C, 30'; iv) KF, MeOH; v) MCPBA, CH_2Cl_2, RT; vi) ROH; vii) NH_2R; viii) $ClCO_2R'$.

Scheme 2. Synthesis and utilization of epoxyphosphate **13**.

The ketoazetidinone **11**, readily available on large scale preparations, was temporarily protected at the nitrogen using *tert*-butyldimethylsilyl chloride and triethylamine in N,N-dimethylformamide at ambient temperature and the product so obtained was treated with lithium bis-trimethylsilylamide to generate the corresponding enolate, which was O-phosphorylated to give the enolphosphate **12**.

This intermediate can be isolated for characterization, but we usually removed the *tert*-butyldimethylsilyl group at the amidic nitrogen using KF in methanol and isolated the enolphosphate **13**.

Reaction of compound **13** with *m*-chloroperoxybenzoic acid in dichloromethane afforded the versatile intermediate **14** as a single isomer, that can be used to introduce a heterosubstituted side chain according to Scheme 2.

Azetidinone derivatives of type **15** and **17**, obtainable either from intermediates **7** and **9** (Scheme 1) by Swern oxidation of hydroxy function, or following the route outlined in Scheme 2, can be converted into trinems having general formula **20** or **21** as shown in Scheme 3. Compounds **15** or **17** were reacted with allyl oxalyl chloride and triethylamine to form an oxalimide intermediate that was treated with triethylphosphite in xylene to form the tricyclic ring system of compounds **18** and **19** respectively. Sequential removal of silyl and allyl protecting groups afforded alkoxy trinems **20** and alkylamino trinems **21**.

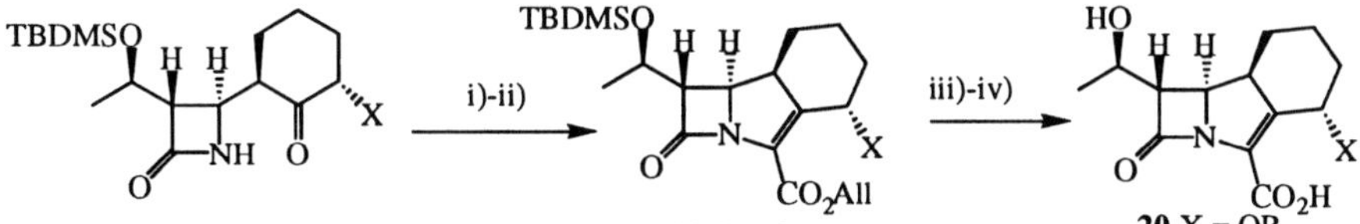

15 X = OR
17 X = NRCO2All

18 X = OR
19 X = NRCO2All

20 X = OR
21 X = NHR

i) $ClCOCO_2All$, TEA, CH_2Cl_2; ii) $P(OEt)_3$, xylene, 110-140°C; iii) TBAF, AcOH, THF, RT; iv) $Pd(Ph_3P)_4$, Na-2-ethylhexanoate or dimedone, THF.

Scheme 3. Conversion of azetidinones **15** and **17** into trinems.

3 ALKOXY TRINEMS

The antibacterial activity of several alkoxy trinems synthesized according to Scheme 3 is reported in Table 1.

Comparison of MIC data for compounds **20b, g, h** against *Escherichia coli* 851E (wild type) and the permeable, β-lactamase producer *Escherichia coli* 1919E, indicates how an increase in lipophilicity of the side chain results in a lower antibacterial activity against Gram negatives, probably due to permeability problems. Introduction of polar electron withdrawing substituents on the alkoxy side chain (compounds **20c, d, e**) improved the antibacterial activity against *Escherichia coli* 851E and other Gram negative strains, and when the hydrophilic hydroxy group (**20f**) was introduced, we observed a good and balanced spectrum of activity.

Compound **20i**, obtained according to Scheme 4 from iodide derivative **26e** by reaction with 4-mercaptopyridine in acetonitrile presented a moderate antibacterial activity against

Escherichia coli 851E (MIC =4). Formation of quaternary ammonium salts by alkylation of pyridine nitrogen did not improve their efficacy against Gram negative strains as indicated by compounds **20j, k.** None of these trinems showed an interesting antibacterial activity versus *Pseudomonas aeruginosa* 1911E.

Table 1. *In vitro antibacterial activity* (MIC, μg/ml)*of trinems **20** in comparison with Imipenem.*

20	R	S.a. 663	S.a. 853E	S.p. 3512	E.c. 851E	E.c. 1919E	P.aer. 1911E	P.aer. 2033E	C.per. 615E	B.frag. 2017E
a	CH_3	**0.25**	**0.25**	**0.06**	**0.5**	**0.5**	**>32**	**0.5**	**0.03**	**0.12**
b	CH_2CH_3	**0.5**	**0.5**	**n.t.**	**8**	**0.5**	**>32**	**n.t.**	**≤0.01**	**0.25**
c	$CH_2CH_2OCH_3$	**0.5**	**0.5**	**0.25**	**2**	**0.5**	**>32**	**n.t.**	**0.03**	**0.12**
d	CH_2CH_2CN	**0.2**	**0.2**	**0.03**	**2**	**0.5**	**>32**	**2**	**≤0.01**	**0.2**
e	CH_2CH_2F	**0.2**	**0.5**	**0.06**	**4**	**1**	**>32**	**8**	**0.03**	**0.2**
f	CH_2CH_2OH	**0.5**	**0.5**	**0.12**	**0.5**	**0.5**	**>32**	**n.t.**	**0.06**	**0.12**
g	CH_2–≡	**0.25**	**0.25**	**0.03**	**4**	**0.5**	**>32**	**n.t.**	**0.03**	**0.12**
h	Ph	**0.25**	**0.25**	**0.25**	**32**	**1**	**>32**	**n.t.**	**0.06**	**1**
i		**0.2**	**0.2**	**0.03**	**4**	**0.5**	**>32**	**2**	**0.03**	**0.5**
j		**0.2**	**0.2**	**0.06**	**8**	**0.5**	**>32**	**4**	**0.06**	**0.5**
k		**0.12**	**0.25**	**n.t.**	**8**	**0.5**	**>32**	**2**	**0.03**	**0.5**
	Imipenem	**0.06**	**0.12**	**0.03**	**0.5**	**0.5**	**4**	**1**	**0.03**	**0.12**

* Minimum Inhibitory Concentrations (MIC) determined in Mueller Hinton broth; Anaerobes Schadler broth Inoculum $5 \bullet 10^5$ CFU/ml. S.a. 663 = *Staphylococcus aureus* 663; S.a. 853E = *Staphylococcus aureus* 853E (β-lactamases producing strain); S.p. 3512 = *Streptococcus pneumoniae* 3512; E.c. 851E = *Escherichia coli* 851E; E.c. 1919E = *Escherichia coli* 1919E (permeable strain, TEM-1 β-lactamases producing strain); P.aer. 1911E = *Pseudomonas aeruginosa* 1911E; P.aer. 2033E = *Pseudomonas aeruginosa* 2033E (permeable strain); C.per. 615E = *Clostridium perfringens* 615E; B. frag. 2017E = *Bacteroides fragilis* 2017E.

In order to increase the efficacy against *Pseudomonas aeruginosa*, a number of derivatives containing a basic nitrogen or a quaternary ammonium salt on the substituent at C-4 were designed using compound **20f** as starting point.

Therefore, the epoxide ring opening of compound **6** (Scheme 4) with 1,2-dihydroxyethane in the presence of a catalytic amount of *p*-toluenesulphonic acid, was the first step towards the synthesis of the versatile intermediate **24**, which is the key point for the preparation of both trinem template **26** and **29** of Scheme 4.

Exhaustive deprotection of both primary and secondary hydroxy functional groups of compound **24** with tetrabutylammonium fluoride and acetic acid at room temperature led to intermediate **25**. We used this advanced trinem template to introduce amino groups with different basicity properties as shown in Scheme 4.

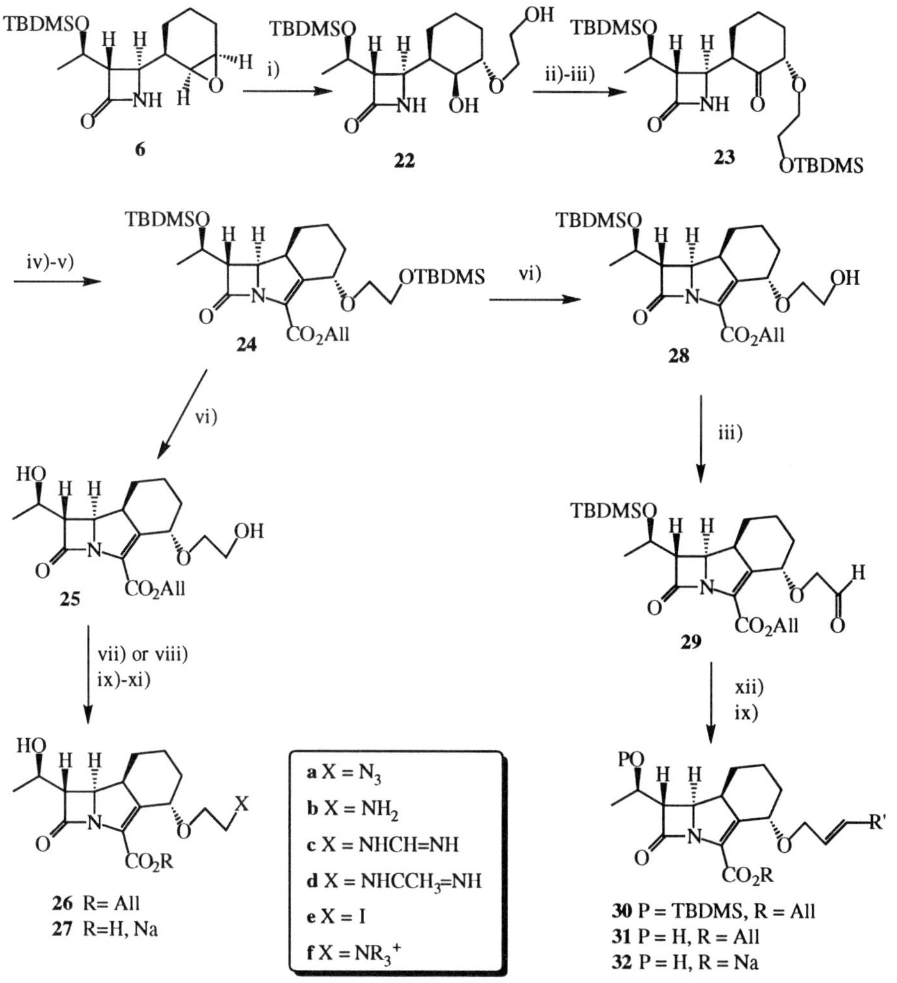
6
22
23
24
25
28
29
26 R= All
27 R=H, Na
a X = N3
b X = NH2
c X = NHCH=NH
d X = NHCCH3=NH
e X = I
f X = NR3+
30 P = TBDMS, R = All
31 P = H, R = All
32 P = H, R = Na
i)
ii)-iii)
iv)-v)
vi)
iii)
vii) or viii)
ix)-xi)
xii)
ix)

i) $HOCH_2CH_2OH$, CH_2Cl_2, TsOH, RT; ii) TBDMSCl, Imidazole, DMF; iii) $(COCl)_2$, DMSO, CH_2Cl_2; iv) $ClCOCO_2All$, TEA, CH_2Cl_2; v) $P(OEt)_3$, toluene, reflux; vi) TBAF, AcOH, THF; vii) PPh_3, CBr_4, NaN_3, DMF;viii) a: TsCl, TEA, CH_2Cl_2; b: KI, acetone; c: NR_3, CH_3CN; ix) $Pd(Ph_3P)_4$, Bu_3SnH, DMF or Na 2-ethylhexanoate, THF; x) H_2/Pd/$CaCO_3$, pH=7 phosphate buffer; xi) O-ethylformidate or O-ethylacetimidate, pH= 8 buffer; xii) Wittig reaction.

Scheme 4. Synthesis of oxygenated trinem templates.

Conversion of the primary hydroxy group using a saturated solution of sodium azide in DMF, triphenylphosphine and carbon tetrabromide at 0°C, gave the azido derivative **26a**.

Reduction of the azido group to the corresponding amine was thought to be problematic, both for the presence of the β-lactam ring, which is prone to undergo ring opening *via* nucleophilic attack to the carbonyl group, and for the presence of the Michael acceptor in ring B.

After several unsuccessful attempts to reduce the azido group of the allyl ester **26a**, we decided to investigate this reaction on the sodium salt **27a**. The hydrogenation reaction was performed in a 0.05M sodium phosphate buffered solution (pH=7), using 5% Pd/$CaCO_3$ and the product **27b** purified by preparative HPLC.

The amidine derivatives **27c** and **27d** were obtained by treating compound **27b** with benzylformimidate hydrochloride or ethylacetimidate hydrochloride in a 0.05M sodium phosphate buffer solution at pH=8.5 and isolating them by reverse phase HPLC.

The introduction of a quaternary amonium salt was resolved more easily by converting the primary hydroxy group of compound **25** into the iodide **26e** and treating it with a suitable tertiary amine to furnish **26f**. The removal of the allyl ester to yield the internal salt **27f** required the use of tributyltin hydride in DMF[12] followed by preparative HPLC purification.

Selective deprotection of the primary hydroxy group of compound **24** to give the monosilylated derivative **28**, could be obtained using tetrabutylammonium fluoride and acetic acid in THF by reducing the concentration of the solution and the reaction time.

Oxidation under Swern reaction conditions afforded the template **29**, that was reacted with several phosphoranes to give compounds of general formula **30**.

Removal of *tert-b*utyldimethylsilyl group gave compounds of general formula **31** that after deprotection of allyl ester afforded the corresponding trinems **32** either as sodium or internal salts.

Table 2. *In vitro antibacterial activity* (MIC, μg/ml)*of trinems 27 in comparison with Imipenem.*

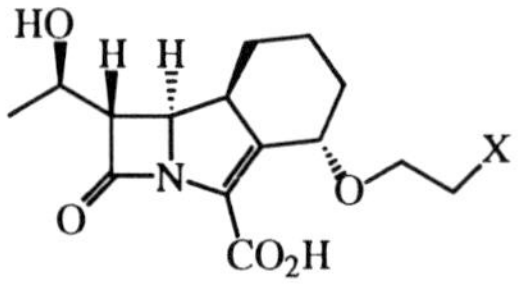

27	X	S.a. 663	S.a. 853E	S.p. 3512	E.c. 851E	E.c. 1919E	P.aer. 1911E	P.aer. 2033E	C.per. 615E	B.frag. 2017E
a	N_3	**0.25**	**0.5**	**0.25**	**8**	**1**	**>32**	**n.t.**	**0.03**	**0.06**
b	NH_2	**0.25**	**0.25**	**0.06**	**1**	**0.5**	**16**	**1**	**0.06**	**0.12**
c	NHC(Me)=NH	**1**	**2**	**n.t.**	**2**	**0.5**	**32**	**n.t.**	**0.12**	**0.5**
d	NHCH=NH	**0.25**	**0.5**	**0.03**	**2**	**1**	**32**	**n.t.**	**0.12**	**0.12**
f1	NMe_3^+	**0.2**	**0.5**	**0.03**	**1**	**1**	**8**	**n.t.**	**0.2**	**1**
f2	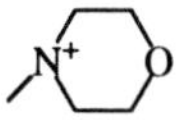 N-methylmorpholinium (N^+)	**0.5**	**0.5**	**0.1**	**2**	**1**	**8**	**2**	**0.2**	**1**
f3	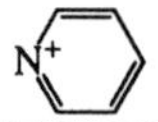 pyridinium (N^+)	**0.2**	**0.2**	**0.06**	**1**	**1**	**8**	**0.5**	**0.1**	**0.2**
	Imipenem	**0.06**	**0.12**	**0.03**	**0.5**	**0.5**	**4**	**1**	**0.03**	**0.12**

In table 2 the antibacterial activity of a series of compounds derived from compound **20f** is reported. With the exception of the azido trinem **27a**, which is also an intermediate in the synthesis of **27b,c,d**, all the compounds clearly showed a comparable antibacterial activity versus Gram positive and Gram negative strains. In particular, we found an interesting potency versus *Pseudomonas aeruginosa* 1911E.

The amino derivative **27b**, which showed activity against *Pseudomonas aeruginosa* 1911E, prompted us to undertake some chemical modification at the amino moiety. Unfortunately formation of acetamidino **27c** and formamidino **27d** failed to enhance the antibacterial activity.

Better results were obtained with ammonium salts having general formula **27f**, which proved to be comparable to Imipenem, supporting the hypothesis that a positively charged group facilitates the penetration of β-lactams into *Pseudomonas* spp.

Table 3. *In vitro antibacterial activity* (MIC, μg/ml)*of trinems 32 in comparison with Imipenem.*

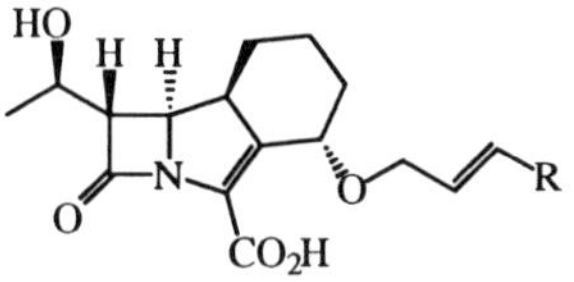

32	R	S.a. 663	S.a. 853E	S.p. 3512	E.c. 851E	E.c. 1919E	P.aer. 1911E	P.aer. 2033E	C.per. 615E	B.frag. 2017E
a	H	**0.2**	**0.2**	**≤0.01**	**2**	**0.2**	**>32**	**0.5**	**n.t.**	**n.t.**
b	$CONH_2$	**0.25**	**0.5**	**n.t.**	**2**	**0.5**	**>32**	**1**	**0.03**	**0.5**
c	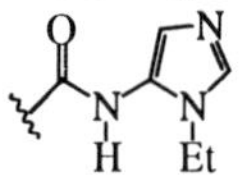	**0.5**	**0.5**	**≤0.01**	**32**	**2**	**>32**	**4**	**0.06**	**1**
d	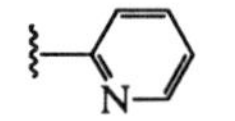	**0.5**	**0.5**	**0.03**	**32**	**2**	**>32**	**32**	**n.t.**	**0.5**
e	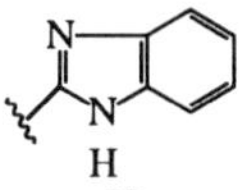	**0.25**	**0.25**	**≤0.01**	**32**	**1**	**>32**	**8**	**≤0.01**	**1**
f	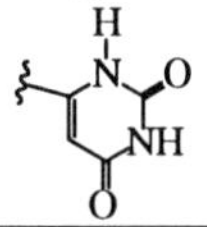	**0.25**	**0.5**	**≤0.01**	**16**	**1**	**>32**	**2**	**≤0.01**	**2**
	Imipenem	**0.06**	**0.12**	**0.03**	**0.5**	**0.5**	**4**	**1**	**0.03**	**0.12**

Another class of trinems containing an alkoxy side chain, readily obtainable from aldehyde **29** using Wittig type reactions is reported in Table 3. Although compounds **32a**, **b**, that bear small R groups showed a good potency, none of the compounds gave a satisfactory spectrum of activity against Gram negative strains possibly due to lack of permeability (compare MIC data for the wild type,*Escherichia coli* 851E, and the permeable strain *Escherichia coli* 1919E, and also the wild type *Pseudomonas aeruginosa* 1911E,with the permeable strain*Pseudomonas aeruginosae* 2033E).

4 AMINO TRINEMS

As the introduction of positively charged amino or ammonium group was found to be beneficial to the anti Pseudomonal activity of our compounds, we decided to explore a number of derivatives with a nitrogen atom directly linked at C-4 of the trinem ring system.[13]

The synthesis of trinems having an amino group in the side chains was accomplished according to the general method shown in Schemes 1-3. Epoxide **6** and epoxyphosphate **15** were converted into N-alkyl carbamates **17**, then following previously described procedures were transformed into the internal salts **21a-g** (Table4).

This class was particularly interesting possessing good potency and a broad spectrum of action; in particular compounds **21a-c** showed activity versus *Pseudomonas aeruginosa* 1911E comparable to Imipenem.

Benzyl derivatives **21d-g** showed an improved efficacy against Gram negative strains compared to compounds **21a-c** and a balanced spectrum of activity, independently of the substitution pattern on the aromatic ring.

Table 4. *In vitro antibacterial activity* (MIC, µg/ml)*of trinems **21** in comparison with Imipenem.*

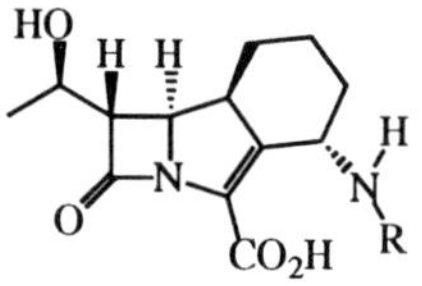

HO H H H N R O N CO_2H

21	R	S.a. 663	S.a. 853E	S.p. 3512	E.c. 851E	E.c. 1919E	P.aer. 1911E	P.aer. 2033E	C.per. 615E	B.frag. 2017E
a	H	**0.25**	**0.25**	**1**	**1**	**0.5**	**4**	**n.t.**	**0.25**	**0.25**
b	Me	**0.2**	**0.5**	**1**	**2**	**0.5**	**4**	**n.t.**	**0.5**	**2**
c	CH_2CH_2OH	**0.5**	**0.5**	**0.25**	**1**	**1**	**8**	**2**	**0.06**	**0.5**
d	CH_2Ph	**0.25**	**0.5**	**0.12**	**0.5**	**0.25**	**>32**	**1**	**≤0.01**	**n.t.**
e	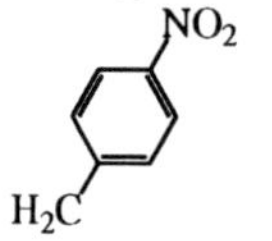 H_2C-C_6H_4-NO_2	**0.25**	**0.5**	**≤0.01**	**0.12**	**0.06**	**>32**	**16**	**0.03**	**0.5**
f	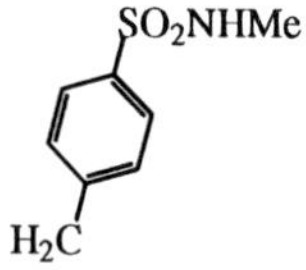 H_2C-C_6H_4-SO_2NHMe	**0.5**	**0.5**	**≤0.01**	**0.5**	**0.25**	**>32**	**0.5**	**0.03**	**0.5**
g	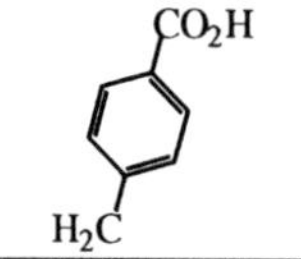 H_2C-C_6H_4-CO_2H	**1**	**1**	**0.25**	**0.25**	**0.25**	**32**	**1**	**0.03**	**0.5**
	Imipenem	**0.06**	**0.12**	**0.03**	**0.5**	**0.5**	**4**	**1**	**0.03**	**0.12**

The key trinem template **36** used for the synthesis of alkenylamino trinems **39**, was obtained from the epoxyphosphate **13** as outlined in Scheme 5.

Nucleophilic addition of O-*tert*-butyldimethylsilyl ethanolamine to the epoxyphosphate **13** afforded the ketoazetidinone **33**. In order to prevent side reactions during the next cyclization procedure, the amino group on the side chain of **33** was acylated by sequential addition of allylchloroformate and triethylamine. Intermediate **34** so obtained, was cyclized using triethylphosphite in refluxing xylene to give the fully protected trinem **35** in reasonably good yield.

The primary hydroxy group was then selectively deprotected and oxidized to the aldehyde **36** according to the same procedure as for compound **24** (Scheme 4).

The aldehyde moiety of this advanced intermediate was converted into an alkenyl group using Wittig reaction, giving rise to trinems of general structure **37**. Deprotection of the hydroxyethyl side chain gave trinems **38** that were converted into the internal salts **39** by reaction with $Pd(Ph_3P)_4$ and dimedone in THF.

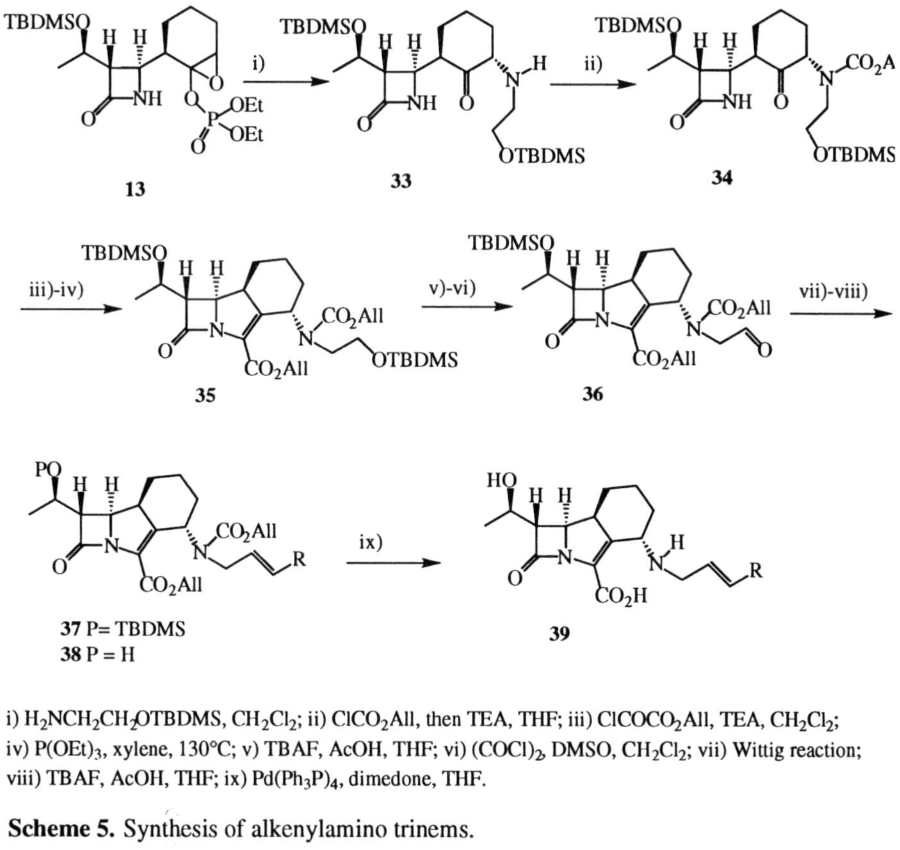

TBDMSO H H NH O O P OEt OEt O **13** i) TBDMSO H H NH O O N H OTBDMS **33** ii) TBDMSO H H NH O O N CO_2A OTBDMS **34**

iii)-iv) TBDMSO H H N O CO_2All N CO_2All OTBDMS **35** v)-vi) TBDMSO H H N O CO_2All N O CO_2All **36** vii)-viii)

PO H H N O CO_2All N R CO_2All **37** P= TBDMS **38** P = H ix) HO H H N O H N R CO_2H **39**

i) $H_2NCH_2CH_2OTBDMS$, CH_2Cl_2; ii) $ClCO_2All$, then TEA, THF; iii) $ClCOCO_2All$, TEA, CH_2Cl_2; iv) $P(OEt)_3$, xylene, 130°C; v) TBAF, AcOH, THF; vi) $(COCl)_2$, DMSO, CH_2Cl_2; vii) Wittig reaction; viii) TBAF, AcOH, THF; ix) $Pd(Ph_3P)_4$, dimedone, THF.

Scheme 5. Synthesis of alkenylamino trinems.

The antibacterial activity of compounds **39**, which are strictly related to compounds **32**, is reported in Table 5. The two classes of compounds posses a very similar spectrum of activity, but compounds having general formula **39** present higher potency against Gram negative strains, showing little or no difference among wild type and permeable strains.

In particular note that compound **39a** represents a rare case in which an acidic function in the side chain is compatible with high potency versus Gram negatives.

In Scheme 6, the transformation of trinems **21a**, **b** into amidino, guanidino and ureido derivatives is shown.

Amino acids **21a**, **b**, were dissolved in a phosphate buffer solution (pH = 8), then treated with the suitable reagent to give compounds **40a-d**, which were purified by preparative HPLC prior to microbiological evaluation (see Table 6).

Table 5. *In vitro antibacterial activity* (MIC, μg/ml)*of trinems 39 in comparison with Imipenem.*

39	R	S.a. 663	S.a. 853E	S.p. 3512	E.c. 851E	E.c. 1919E	P.aer. 1911E	P.aer. 2033E	C.per. 615E	B.frag. 2017E
a		**0.25**	**0.5**	**0.5**	**0.25**	**0.25**	**16**	**4**	**0.5**	**0.5**
b		**0.5**	**1**	**0.03**	**1**	**0.5**	**>32**	**1**	**0.03**	**2**
c		**0.25**	**0.5**	**0.12**	**1**	**0.5**	**32**	**2**	**n.t.**	**n.t.**
d		**0.5**	**0.5**	**n.t.**	**1**	**0.5**	**>32**	**4**	**0.06**	**2**
e		**0.25**	**0.5**	**0.25**	**0.25**	**0.25**	**>32**	**1**	**0.06**	**2**
	Imipenem	**0.06**	**0.12**	**0.03**	**0.5**	**0.5**	**4**	**1**	**0.03**	**0.12**

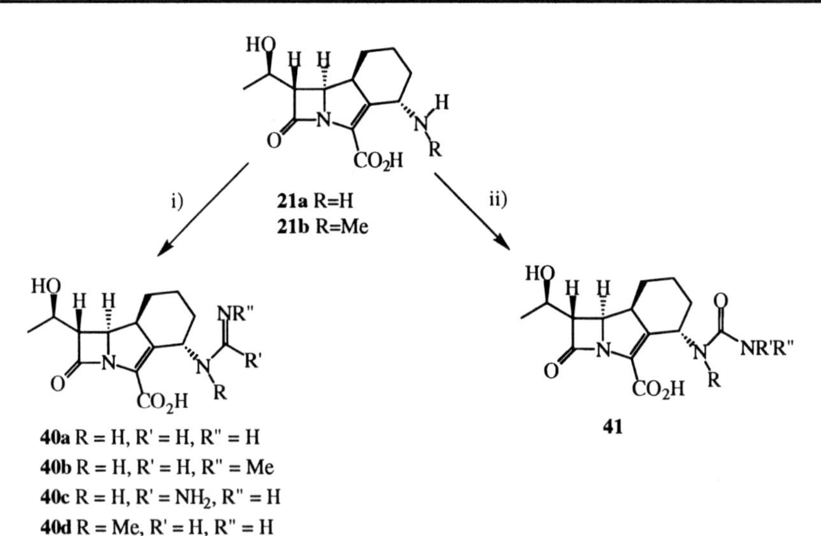

i) BnOCH=NH.HCl or EtOC(=NH)Me.HCl or $H_2NC(=NH)SO_3H$, pH=8 buffer;
ii) RN=C=O, acetone or CH_3CN.

Scheme 6. Synthesis of amidine trinems **40** and urea trinem **41**.

Formamidino derivatives **40a,b** and guanidino compound **40c** showed a good potency against Gram positives and moderate activity versus Gram negatives. The best results were obtained with compound **40d**, that gave good activity against *Pseudomonas aeruginosa* 1911E and an overall spectrum comparable to that shown by Imipenem.

The compound was also tested *in vivo* to determine the protective dose (ED_{50}) in a septicaemia model in mouse using Meropenem as reference compound. The formamidino trinem (GV129606) **40d**, was shown to be more effective against *Staphylococcus aureus* 853E ($ED_{50} \leq 0.01$ mg/Kg/dose) than Meropenem ($ED_{50} = 0.29$ mg/Kg/dose) and equivalent in challenging *Escherichia coli* 851E ($ED_{50} = 0.03$ mg/Kg/dose for both compounds). More interestingly the two compounds presented comparable efficacy against *Pseudomonas aeruginosa* 1911E ($ED_{50} = 0.48$ mg/Kg/dose for **40d** and $ED_{50} = 0.54$ mg/Kg/dose for Meropenem).

The extremely good biological profile, and the efficacy demonstrated in a series of *in vivo* mouse infection models led us to consider compound **40d** to be a lead for a β-lactam with extended spectrum.

Table 6. *In vitro antibacterial activity* (MIC, μg/ml)*of trinems 40in comparison with Imipenem.*

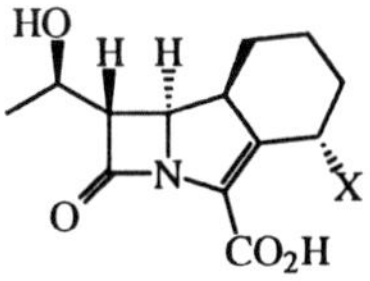

40	X	S.a. 663	S.a. 853E	S.p. 3512	E.c. 851E	E.c. 1919E	P.aer. 1911E	P.aer. 2033E	C.per. 615E	B.frag. 2017E
a	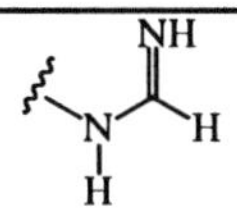 NH, N, H, H	0.2	0.5	0.06	4	8	16	n.t.	1	0.5
b	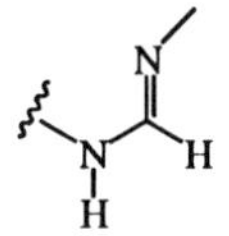 N, N, H, H	0.1	0.1	0.03	2	1	16	2	0.5	0.5
c	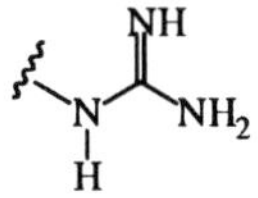 NH, N, NH_2, H	1	1	0.03	8	8	32	8	0.5	1
d	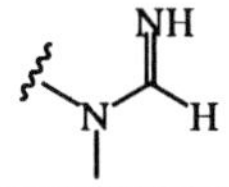 NH, N, H	0.1	0.1	0.06	0.5	0.2	4	0.5	0.03	0.1
	Imipenem	0.06	0.12	0.03	0.5	0.5	4	1	0.03	0.12

Reaction of trinems **21a,b** with suitable isocyanates afforded ureido derivatives having general formula **41** (see Table 7).

Trinem **41b** showed a markedly superior antibacterial activity in this series of ureido derivatives. In particular we would like to emphasize that a moderately good potency against *Pseudomonas aeruginosa* 1911E was also observed.

Unfortunately, the introduction of bulkier substituents (compounds **41c-f**) led to a decrease in efficacy versus Gram negative strains. As the class of ureido trinems showed

good chemical and enzymatic stability in blood serum, the pharmacokinetic profile of a series of ureido trinems was determined in mouse and *Cynomolgus monkey* .
In particular, compound **41c**, which bears a phenyl group attached at N', showed a prolonged serum half life either in rodents and primates.

Table 7. *In vitro antibacterial activity* (MIC, μg/ml)*of trinems **41** in comparison with Imipenem.*

HO H H N O X CO_2H

41	X	S.a. 663	S.a. 853E	S.p. 3512	E.c. 851E	E.c. 1919E	P.aer. 1911E	P.aer. 2033E	C.per. 615E	B.frag. 2017E
a	$NHC(O)NH_2$	**0.25**	**0.5**	**≤0.12**	**0.25**	**≤0.12**	**>16**	**0.5**	**≤0.12**	**0.25**
b	$N(CH_3)C(O)NH_2$	**≤0.12**	**≤0.12**	**≤0.12**	**0.25**	**≤0.12**	**8**	**0.5**	**≤0.12**	**≤0.12**
c	$NHC(O)NH{-}Ph$	**0.5**	**0.5**	**n.t.**	**1**	**0.25**	**>32**	**4**	**≤0.0**	**0.5**
d	$N(CH_3)C(O)NH{-}Ph$	**4**	**16**	**1**	**32**	**16**	**>32**	**32**	**1**	**32**
e	$NHC(O)NHCH_2Ph$	**0.5**	**1**	**0.25**	**1**	**0.25**	**>16**	**4**	**≤0.12**	**1**
f	$N(CH_3)C(O)NHCH_2Ph$	**0.25**	**0.5**	**≤0.12**	**4**	**0.5**	**>16**	**4**	**≤0.12**	**0.5**
	Imipenem	**0.06**	**0.12**	**0.03**	**0.5**	**0.5**	**4**	**1**	**0.03**	**0.12**

5 CONCLUSION

Research carried out in our laboratories in Verona has demonstrated that the trinems are a group of effective antibacterial agents.

In particular, chemical modification of the substituent at C-4 strongly influences the potency, spectrum of action and pharmacokinetic profile of compounds. Introduction of a nitrogen atom in the side chain in some cases resulted in an improvement of the antibacterial activity against *Pseudomonas* spp.

As a result of our studies, compound **40c** (GV129606) which showed a very good antibacterial activity *in vitro* over a wide spectrum of strains, high potency, valuable *in vivo* efficacy in septicaemia models in mouse, was selected as a lead compound.

Moreover we also found that the ureido trinem **41c**, which was chemically and enzymatically stable in blood serum demonstrated a prolonged half life in mouse and *Cynomolgus monkey*.

6 ACKNOWLEDGMENTS

I would like to thank all people involved in the synthesis analysis and biological evaluation of the compounds.
Contribution of the following people are particularly acknowledged:
G. Ageno, D. Andreotti, G.L. Araldi, G. Bonanomi, R. Broggio, D. Busetto, R. Carlesso, N, Case, S. Contini, O. Curcuruto, S. Davalli, E. De Magistris, R. Di Fabio, E. Di Modugno, D. Donati, I. Erbetti, L. Ferrari, G. Gaviraghi, S. Gehanne, C. Ghiron, M. Hamdam, S.M. Hammond, G. Kennedy, J. Lowther, M. Maffeis, C. Marchioro, A. Orlandi, A. Padova, D. Papini, M. Passarini, A. Pecunioso, G. Pentassuglia, A. Perboni, A. Pezzoli, L. Piccoli, E. Piga, D.A. Pizzi, S. Provera, E. Ratti, T. Rossi, A. Rottigni, D. Sabatini, G. Sbampato, G, Tarzia, R.J. Thomas, M.E. Tranquillini, A. Ursini, J.A. Winders, P. Zarantonello.

References

1. For recent reviews on β-lactam antibiotics see: a: F.C. Neuhaus, N.H. Georgopapadakou, in *Emerging Targets in Antibacterial and Antifungal Chemotherapy*, J. Sutcliffe, N.H. Georgopapadakou, Ed. Chapman and Hall, New York, 1992. b: N.H. Georgopapadakou, *Antimicrobial Agents Ann.*, 1988, **3**, 409. c: R. Southgate, S. Elson, in *Progress in the Chemistry of Organic Natural Products*, W. Herz, H. Grisebach, G.W. Kirby, C. Tamm, Springer Verlag, New York, 1985, p. 1. d: W. Durckheimer, J. Blumbach, R. Lattrell, K.H. Sheunemann, *Angew. Chem. Int. Ed. Engl.*, 1985, **24**, 180. e: *Recent Advances in the Chemistry of β-Lactam Antibiotics*, A.G. Brown, S.M. Roberts, The Royal Society of Chemistry, Cambridge, U.K., 1984.
2. a: O.H. Hartwig, *Pharm. Ztg.*, 1996, **141**, 11. J.F. Martin, S. Gutierrez, *Antoine van Leeuwenhoek*, 1995, **67**, 181. b: *The Organic Chemistry of β-lactams*, G.I. Georg, Ed., VCH Publisher Inc, 1993.
3. a: B. Tamburini, A. Perboni, T. Rossi, D. Donati, D. Andreotti, G. Gaviraghi, R. Carlesso, C. Bismara, *Eur. Pat. Appl.* EP0416953 A2, 1991, *Chem Abstr.* 1992, **116**, 235337t. b: A. Perboni, T. Rossi, G. Gaviraghi, A. Ursini, G. Tarzia, WO 9203437, 1992, *Chem. Abstr.* 1992, **117**, 7735m. c: A. Perboni, B. Tamburini, T. Rossi, D. Donati, G. Tarzia, G. Gaviraghi, in *Recent Advances in the Chemistry of anti-infective agents* P.H. Bentley, R. Ponsford, Ed. The Royal Society of Chemistry, Cambridge, 1993, p. 21.
4. a: H.C. Neu, *Science*, 1992, **257**, 1064. b: J. Davies, *Science*, 1994, **264**, 375.
5. a: H. Nikaido, *Science*, 1994, **264**, 382. b: H. Nikaido, *Molecular Microbiology*, 1992, **6**, 435. c: H. Nikaido, *J. Biol. Chem.*, 1994, **269**, 3905. d: H. Nikaido, *Antimicrob. Agents Chemother.*, 1989, **33**, 1831.
6. a: K. Bush, *Antimicrob. Agents Chemother.* 1989, **33**, 264. b: P. Ledent, X. Raquet, B. Joris, J. Van Beeumen, J.-M. Frère, *Biochem. J.*, 1993, **292**, 555. c: J.-M. Frère, *Molecular Microbiology* 1995, **16**, 385.
7. N.H. Georgopapadakou, *Antimicrob. Agents Chemother.* 1993, **37**, 2045.

8. G. Gaviraghi, *Eur. J. Med. Chem.*, 1995, Suppl. to Vol. **30**, 467S.

9. a: E. Di Modugno, I. Erbetti, L. Ferrari, G.L. Galassi, S.M. Hammond, L. Xerri, *Antimicrob. Agents Chemother.*, 1994, **38**, 2362. b: R. Wise, J.M. Andrews, N. Brenwald, *Antimicrob. Agents Chemother.*, 1996, **40**, 1248.

10. T. Rossi, S. Biondi, S. Contini, R.J. Thomas, C. Marchioro, *J. Am. Chem. Soc.*, 1995, **117**, 9604.

11. a: A. Perboni, *Eur. Pat. Appl.* EP0502488 A2, 1992. b: S. Biondi, G. Gaviraghi, T. Rossi, *Bioorg. Med. Chem. Lett.,* 1996, **6**, 525.

12. a: B.G. Christensen, T.N. Salzmann, S.M. Schmitt, *Eur. Pat. Appl.* 292191; *Chem Abstr.* 1989, **110**, 231331. b: M. Lang, E. Hungerbühler, P. Schneider, R. Scartazzini, W. Tosch, E.A. Konopka, O. Zak, *Helv. Chim. Acta* 1986, **69**, 1576.

13. M.E. tranquillini, G.L. Araldi, D.Donati, G. Pentassuglia, A. Pezzoli, A. Ursini, *Bioorg. Med. Chem. Lett.*, in press.

8
New Concepts of Inhibition of Penicillin Sensitive Enzymes

P. Pflieger,* P. Angehrn. M. Böhringer, K. Gubernator, I. Heinze-Krauss, Ch. Hubschwerlen, Ch. Oefner, M. G. P. Page, R. Then and F. Winkler
PRECLINICAL RESEARCH, F. HOFFMANN-LA ROCHE LTD., CH-4070 BASLE, SWITZERLAND

1 INTRODUCTION

Production of β-lactamases by bacteria is the most frequent form of resistance to β-lactam antibiotics. Among the various existing classes of β-lactamases, chromosomally encoded class C β-lactamases produced by Gram-negative bacteria, represent an increasing problem for third generation cephalosporins in clinical situation. No combination between a β-lactamase inhibitor and a β-lactam antibiotic that effectively covers bacteria producing class C β-lactamases has yet reached the market[1]. We describe in this paper our efforts to find new inhibitors of class C β-lactamases. These resulted in the discovery of the bridged carbacephems, which proved to be not only potent β-lactamase inhibitors, but also a new class of β-lactamase-stable antibiotics. The SAR within this new class of PSE inhibitors will be described.

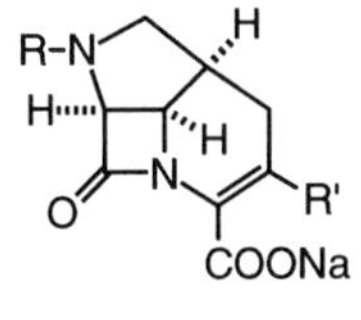

The solution of the X-ray structure of the class C β-lactamase from *Citrobacter freundii* gave us insight into the architecture of the active centre of this class of enzymes and their mechanism of action[2].
The observation that a rotation is necessary at the stage of the acyl-enzyme intermediate to allow deacylation in class C β-lactamases (Scheme 1)[3] led to the general concept of restricted rotation (Scheme 2).
The wide applicability of this concept was demonstrated by the synthesis of various potent class C β-lactamase inhibitors: the bridged monobactams, sulbactams, carbacephems, isooxacephems and isocephems (Scheme 3) [4, 5, 6].
In contrast to the bicyclic compounds (I), the tricyclic derivatives (II) displayed, beside their β-lactamase inhibitory properties, significant activities against the essential transpeptidases, resulting in antibacterial activity against a wide range of bacteria.

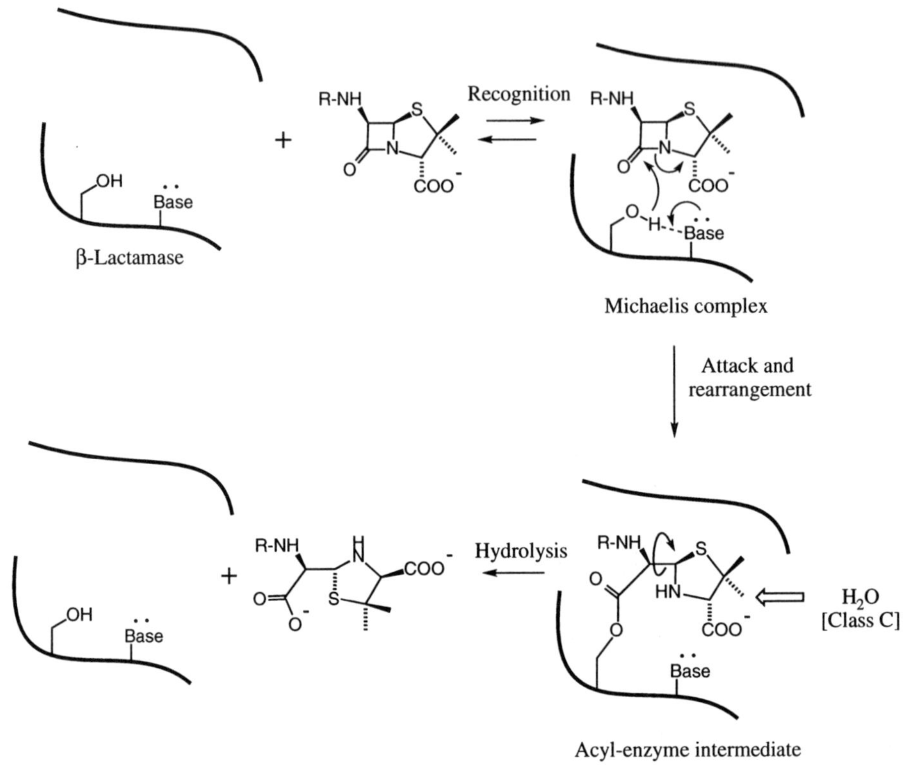

Recognition

Michaelis complex

Attack and rearrangement

Hydrolysis

H_2O [Class C]

Acyl-enzyme intermediate

β-Lactamase

Scheme 1 *Postulated mechanism of hydrolysis of Penicillins by class C β-lactamases*

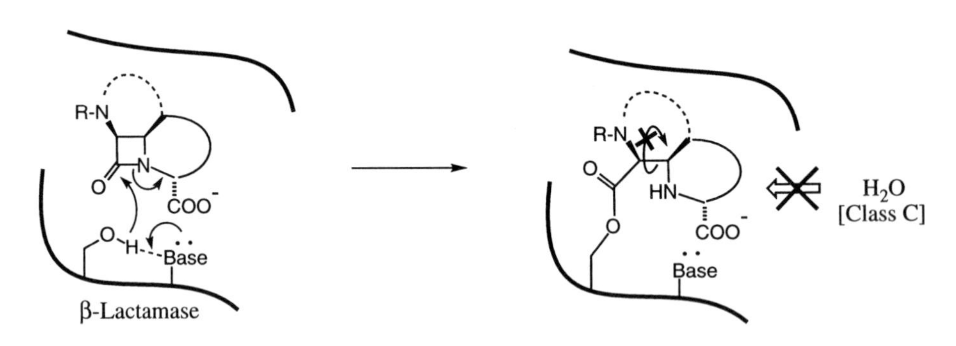

Scheme 2 *Concept of restricted rotation*

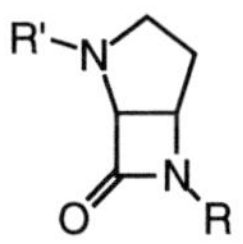

(I)

R = SO_3Na: bridged monobactams
= OSO_3Na: bridged sulbactams

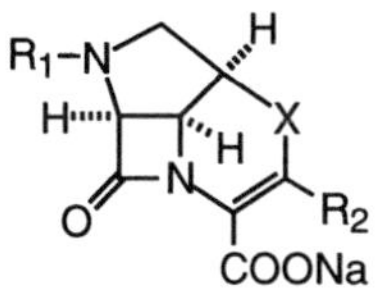

(II)

X = CH_2: bridged carbacephems
= O : bridged isooxacephems
= S : bridged isocephems

Scheme 3 *Lead structures*

2. CHEMISTRY

The bridged carbacephems can be divided in two major subclasses depending on their substitution in position 3 (Scheme 4): the "directly linked" compounds (3) and their "methylene homologues" (5). The "directly linked" compounds (3) were prepared in 3 steps from the tricyclic enol (2) whereas the "methylene homologues" (5) were easily obtained in 3 steps from the tricyclic allylic alcohol (4). Both key intermediates (2) and (4) were obtained from the bicyclic alcohol (1), which was prepared in large scale starting from vitamin C[6]. The key step in the synthesis of (2) was a Dieckmann-type cyclization[7] whereas the ring closure leading to (4) was based on a oxalimide cyclization[8].

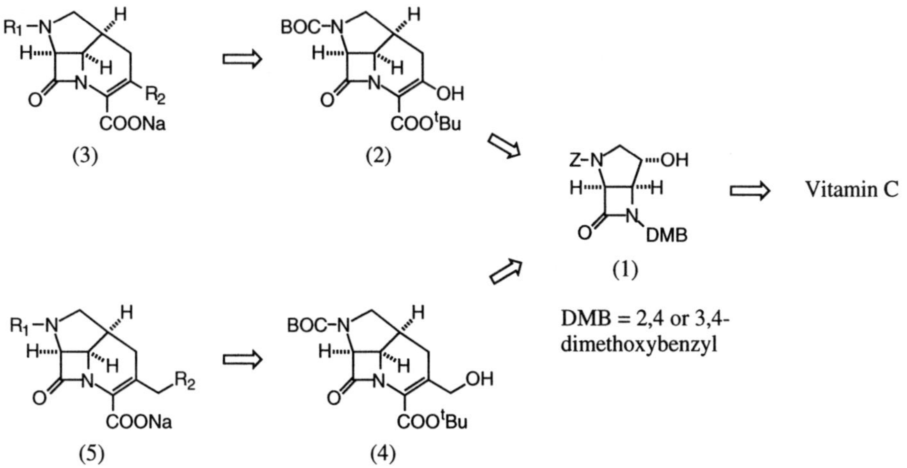

(3) (2) (1) Vitamin C

(5) (4)

DMB = 2,4 or 3,4-dimethoxybenzyl

Scheme 4 *Retrosynthesis*

The synthesis of key intermediate (2)[5] started with a Swern oxydation of (1) (Oxalyl chloride / DMSO) (Scheme 5), followed by a Wittig reaction with a stabilized phosphorane (Ph_3P=$CHCOOCH_2Ph$) leading to a mixture of Z and E olefins (6). Stereoselective hydrogenation of the double bond in the presence of BOC-anhydride led to the carboxylic acid (7). This acid was first reprotected ($PhCH_2OH$ / DCC) before removal of the DMB group using an oxidative procedure ($K_2S_2O_8$ in CH_3CN / H_2O). The β-lactam nitrogen was then alkylated with tert-butyl-bromoacetate using a sterically hindered strong base (lithium bis-trimethylsilyl amide at -78°C) and the benzyl ester was cleaved by catalytic hydrogenation, leading to (9).

In order to convert this intermediate into the tricyclic enol (2) by a Dieckmann-type cyclization, we developed a one pot procedure based on the conversion of the acid into an activated amide (Im_2CO in THF at rt), followed by deprotonation with lithium bis-trimethylsilyl amide at -78°C . The resulting anion reacted spontaneously with the activated amide, leading to the tricyclic enol (2).

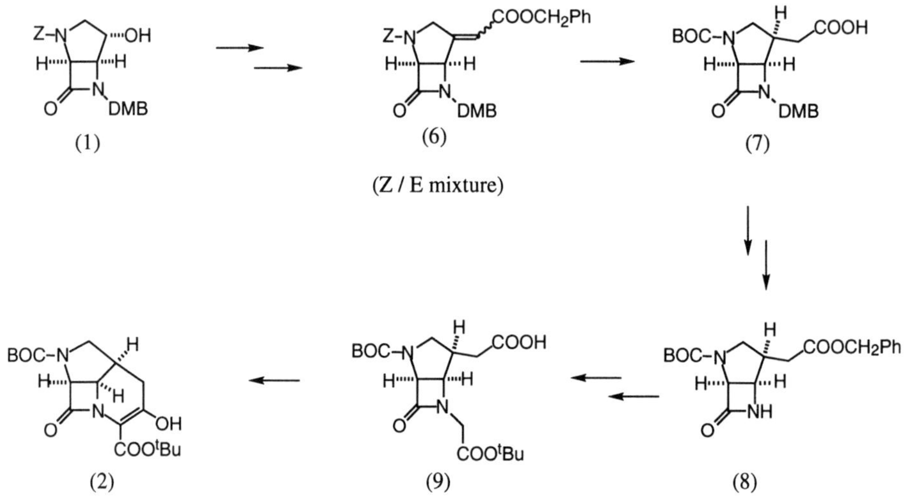

(1) (6) (7)

(Z / E mixture)

(2) (9) (8)

Scheme 5 *Synthesis of key intermediate (2)*

Conversion of (2) into bridged carbacephems (3a-i) is outlined in Scheme 6.
Formation of the triflate (10b) (Tf_2O / Hünig's Base at -78°C in CH_2Cl_2) was followed by substitution with various mercaptoheterocycles (deprotonated with sodium hydride in THF) affording intermediate (11). In some cases the triflate proved to be rather unreactive and extremely long reaction times (7 days at room temperature in the case of sodium 5-mercapto-1-methyltetrazole) were necessary to obtain complete conversion. The extreme stability of the triflate (10b) prompted us to consider the mesylate as a side chain. Reaction of the tricyclic alcohol (2) with mesyl chloride led to (10a) which was subsequently deprotected, leading to (12e). Deprotection of (10a) and (11) proved to be very sensitive reactions: direct treatment at room temperature with trifluoroacetic acid in dichloromethane led to complete decomposition of the starting material. Among many different reaction conditions investigated, best results were obtained by using a sequential treatment with trifluoroacetic acid in dichloromethane at -18°C, until the BOC group was totally removed, followed by few hours at room temperature, necessary to remove the tert-butyl ester group. Final acylation of (12a-e) with various reagents led to the bridged carbacephems (3a-i).

The synthesis of key intermediate (4) is outlined in Scheme 7. The alcohol (1) was first converted into the corresponding ketone (Swern oxidation: oxalyl chloride / DMSO), and submitted to a Wittig-Horner reaction ($(MeO)_2POCH_2COCH_2OTBDMS$ / LiOH) leading to a mixture of Z and E olefins (13). Stereoselective hydrogenation of the double-bond in parallel with direct exchange of the Z group by a BOC (H_2 / 10 % Pd/C / $(BOC)_2O$

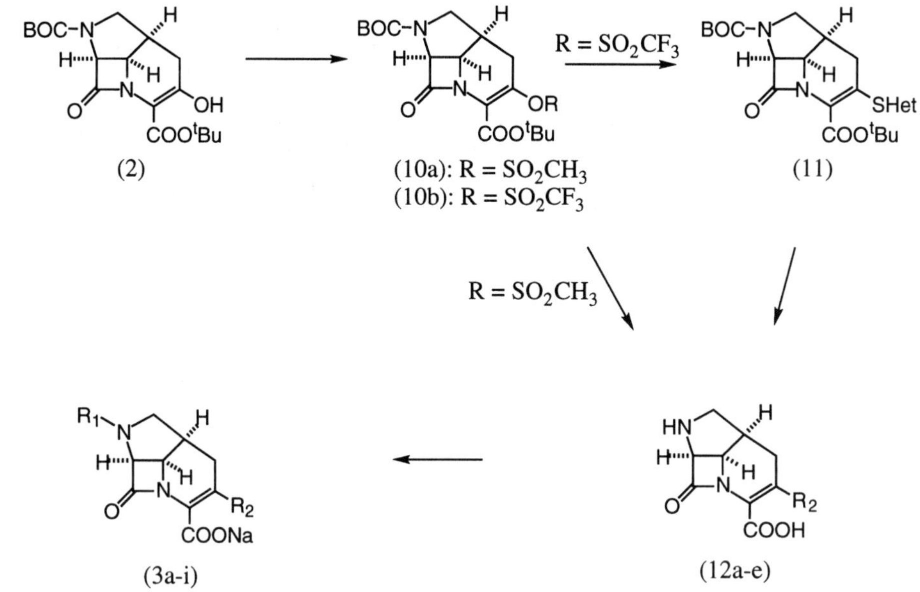

(2) → (10a): R = SO_2CH_3; (10b): R = SO_2CF_3 —R = SO_2CF_3→ (11)

R = SO_2CH_3 → (12a-e) ← (11)

(12a-e) → (3a-i)

For R_1 and R_2 see Tables 1, 2, 3, 5

Scheme 6 *Synthesis of "directly linked" bridged carbacephems (3a-i)*

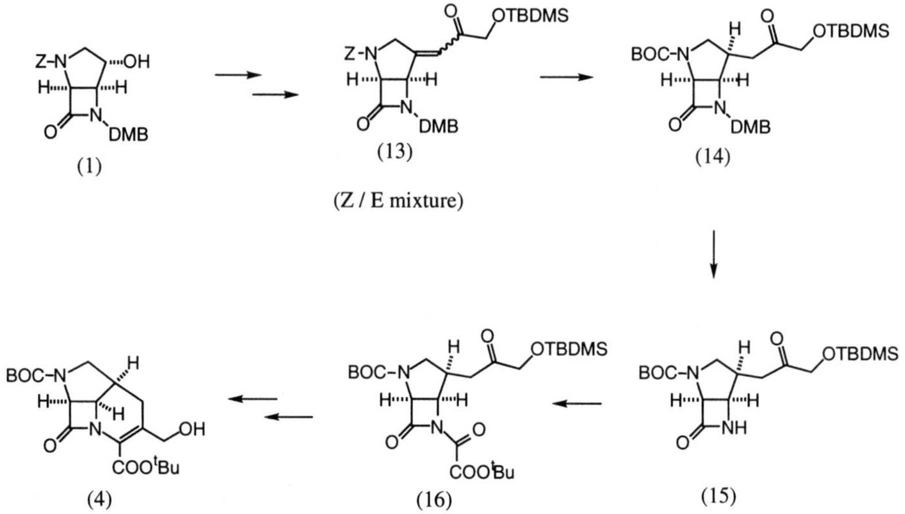

(1) → → (13) (Z / E mixture) → (14) → (15) → (16) → → (4)

Scheme 7 *Synthesis of key intermediate (4)*

in MeOH) led to (14). Removal of the DMB-group ($K_2S_2O_8$ in CH_3CN / H_2O) was followed by acylation of the β-lactam nitrogen with tert-butyl chloroglyoxylate in the presence of calcium carbonate and Hünig's Base. After isolation through acidic work-up, the oxalimide (16) was rapidly used in the cyclization step, to avoid decomposition occuring within hours at room temperature. Standard cyclization conditions using 2 equivalents of triethylphosphite in refluxing toluene, followed by removal of the TBDMS group under acidic conditions (HCl in THF / H_2O) led to the allylic alcohol (4).

Conversion of (4) into various substituted compounds (17) was mainly based on standard chemistry (Scheme 8). The mesylate formed in situ (MsCl / TEA / -40°C) from (4) was treated with various mercaptoheterocycles in the presence of triethylamine at 0°C. The pyridinium derivative was obtained by reacting (4) with MsCl in pyridine at room temperature. Finally carbamate groups were introduced either by treating (4) with various isocyanates, or by first forming a mixed carbonate (DSC / TEA) which was subsequently reacted with various amines[9].The deprotection and acylation steps leading to (5a-l) were similar to those reported for the preparation of (3a-i).

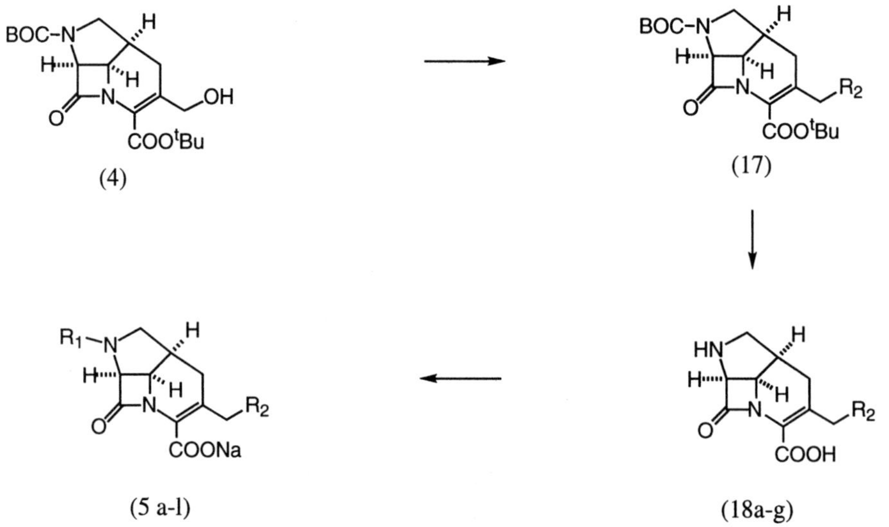

BOC-N, H, H, OH, O, N, COO^{t}Bu (4) → BOC-N, H, H, R_2, O, N, COO^{t}Bu (17) → HN, H, H, R_2, O, N, COOH (18a-g) → R_1-N, H, H, R_2, O, N, COONa (5 a-l)

For R_1 and R_2 see Tables 2,3,4,5

Scheme 8 *Synthesis of of "methylene homologues" (5a-l)*

3. BIOLOGICAL EVALUATION

1) β-lactamase inhibition

We first investigated the influence of both substituents (R_1 and R_2) of the bridged carbacephems on β-lactamase inhibitory properties. All compounds were tested against a series of β-lactamases (class C and class A) isolated from Gram-negative bacteria (members of the Enterobacteriaceae and *Pseudomonas aeruginosa*). Their synergistic action was evaluated by measuring the MIC of ceftriaxone (CRO), in combination with the inhibitor against the bacteria producing these β-lactamases.

In **Table 1**, the influence of R_1 on the β-lactamase inhibitory properties of a series of directly linked compounds (3a-e), is compared to one of the bridged monobactams that has so far given the greatest synergy with CRO (19):

Table 1: Influence of R_1 on ß-lactamase inhibition

N°	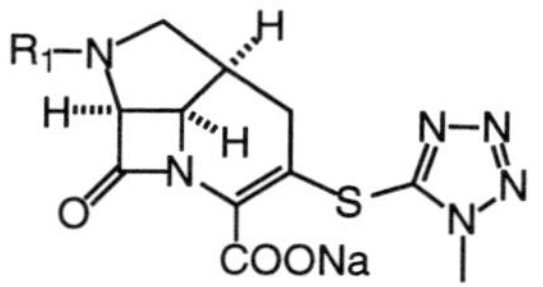 R_1	Inhibition of ß-Lactamases IC_{50} (nM)			MIC of combination with CRO* (µg/ml)		
		C. freundii 1982 (Class C)	*P. aerug* 18SH (Class C)	*E. coli* CF102 (Class A)	*C. freundii* 1982 (Class C)	*P. aerug* 18SH (Class C)	*E. coli* CF 102 (Class A)
					128	128	16
(12a)	H	12'700	120'000	Inactive**	8***	128	16
(3a)	BOC	9	24	1'800	4	16	4
(3b)	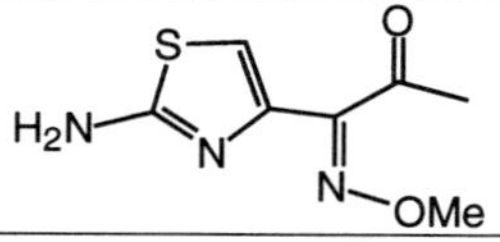	4'160	300'000	2'100	16	128	8
(3c)	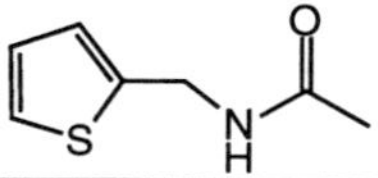	22	27	33'300	1	16	4
(3d)	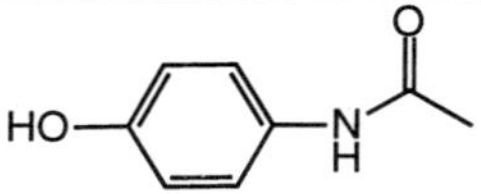	10	14	400'000	0.5	2	4
(3e)	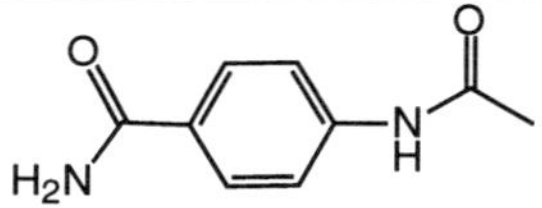	4	3	330'000	0.25	8	4
(19)	Bridged monobactam	53	217	Inactive**	0.5	2	16

* A combination of 4 equivalents of the inhibitor with 1 equivalent of Ceftriaxone (CRO) was used. The MIC refers to CRO.

** IC_{50} >10mM

*** Due to intrinsic antibacterial activity of the inhibitor (Table 3, MIC = 32µg/ml)

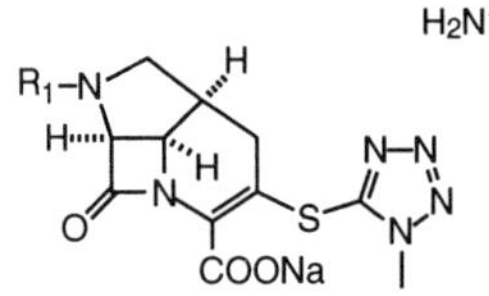

(3a-e)
(12a): R_1 = H

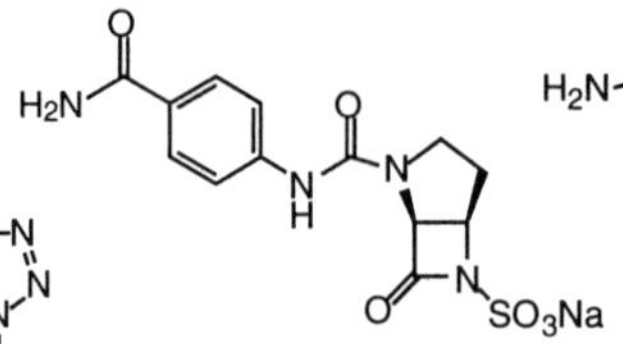

(19)

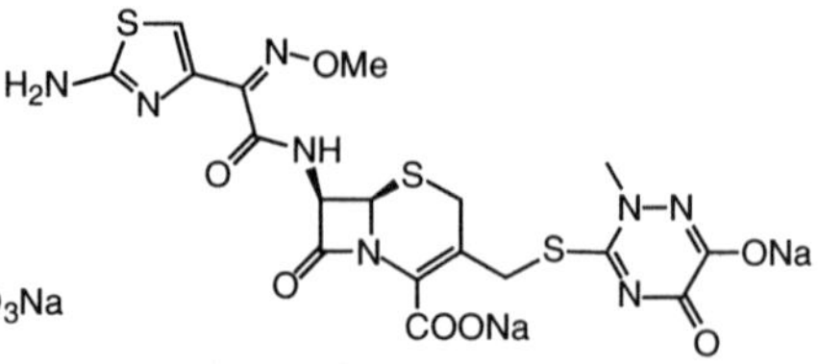

Ceftriaxone (CRO)

Against the bacterial strains included in this study, ceftriaxone alone was inactive (MIC of 128 µg/ml against *C. freundii* and *P. aeruginosa*) or practically inactive (MIC of 16 µg/ml against *E. coli*).

In the absence of any substituent on the pyrrolidine nitrogen (R_1 = H; (12a)), only weak activity against the class C β-lactamase from *Citrobacter freundii* 1982 was observed. In this case, the unexpected MIC value of the combination with CRO against *Citrobacter freundii* 1982 (8 µg/ml) was due to weak intrinsic antibacterial activity of the bridged carbacephem against this organism (see Table 3). These properties will be discussed later in this paper.

Introduction of a tert-butoxycarbonyl (BOC, (3a)) substituent resulted in an extremely potent inhibitor of class C β-lactamases, resulting in significant synergy with CRO against *C. freundii* (128 µg/ml -> 4 µg/ml). The weak synergy observed against *P. aeruginosa* is probably due to restricted penetration through the outer membrane.
Introduction of a third generation cephalosporin side chain (methoxyimino amino thiazolyl type side chain, (13b)) was detrimental to all activities, demonstrating that SAR derived from β-lactam antibiotics may not be useful in our case.

Among many different variations, compounds bearing urea type side chains showed the strongest inhibition of class C β-lactamases. Synergy against all members of the Enterobacteriaceae resistant by production of class C β-lactamase was observed (complete data not shown). A number of urea derivatives (3c, 3d, 3e) reached a level of activity comparable to the most potent member of the bridged monobactams (19). These derivatives combine extremely low IC_{50} values with improved penetration properties through the outer membrane of Gram-negative bacteria, resulting in improved synergy with CRO *in vitro* (compare (3a) to (3d). However the level of activity *in vitro* reached against *P. aeruginosa* is still below or at the borderline of our targeted value. Finally no significant activity against class A β-lactamases was observed, due to the different mechanism of deacylation (deacylating water attacking from the opposite side).

One major advantage of the bridged carbacephems over the bridged monobactams, is to possess a second substituent R_2, allowing to further enhance the binding into the active center of the enzymes, but also to optimize the penetration properties through the outer membrane of Gram-negative organisms.

Table 2 gives an idea about the influence of R_2 on the β-lactamase inhibitory properties of the "directly linked" (3) and "methylene homologues" (5) bridged carbacephems:

The "methylene homologues" reached a level of activity against class C β-lactamases similar to the "directly linked" compounds (compare (3e) to (5a) and (3f) to (5b)) but displayed

Table 2: Influence of R_2 on ß-lactamase inhibition

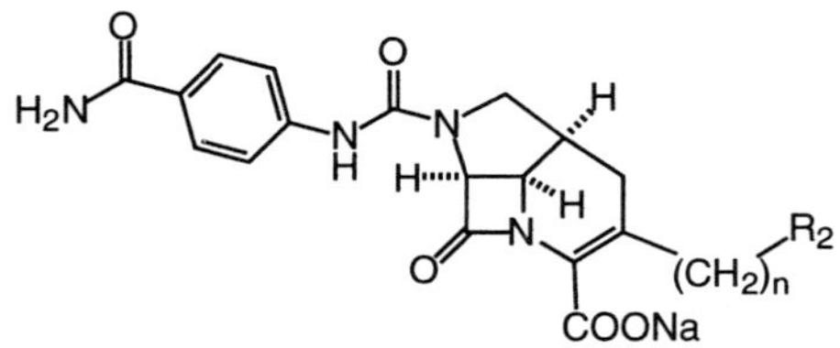

N°	n	R_2	Inhibition of ß-Lactamases IC_{50} (nM)			MIC of combination with CRO* (µg/ml)		
			C. freundii 1982 (Class C)	*P. aerug* 18SH (Class C)	*E. coli* CF102 (Class A)	*C. freundii* 1982 (Class C)	*P. aerug* 18SH (Class C)	*E. coli* CF 102 (Class A)
						128	128	16
(3e)	0	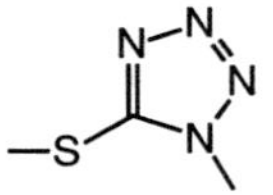	4	3	330'000	0.25	8	4
(3f)	0	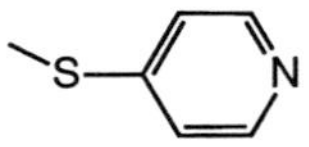	8	11	100'000	0.5	4	8
(3g)	0	$—OSO_2CH_3$	10	21	5'000'000	0.5	8	4
(5a)	1	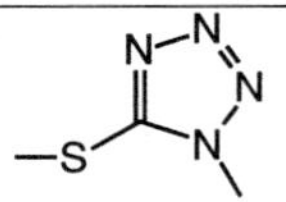	12	25	610	0.5	4	4
(5b)	1	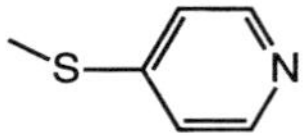	7	17	530	0.25	2	4
(5c)	1	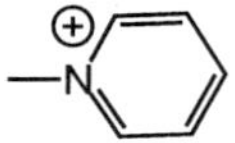	23	355	1'350	0.5	4	4
(5d)	1	$—OCONH_2$	6	24	4'030	≤ 0.06	4	8

* A combination of 4 equivalents of the inhibitor with 1 equivalent of Ceftriaxone (CRO) was used .The MIC refers to CRO.

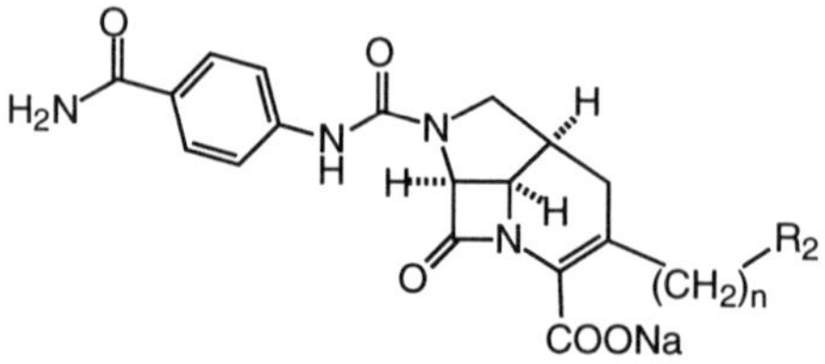

(3e-g): n = 0
(5a-d): n = 1

significantly improved IC_{50} values against class A β-lactamases. This may be attributed to a decreased deacylation rate. We assume that elimination of the mercaptoheterocycle occurs in the methylene homologues whereas this is impossible in the directly linked compounds. In the reaction of class A β-lactamases with cephalosporins, it has been shown that, if a good leaving group is present at the 3' position, this can be eliminated at the acyl-enzyme stage leaving a acyl moiety that is around one thousand fold less prone to hydrolysis[10]. However the β-lactamase is not irreversibly blocked because the acyl-enzyme complex can still be hydrolysed at a significant rate (deacylating water coming from the not hindered side). The low amount of enzyme still active is able to hydrolyse partially the incoming CRO, thus explaining the weak synergy with CRO *in vitro* against class A β-lactamase producing bacteria. Interestingly the mesylate derivative (3g) reached a level of activity comparable to inhibitors bearing more standard substituents.

In order to improve the synergy against *P. aeruginosa*, a positively charged substituent (compound (5c)) was introduced, however this tentative was unsuccessful. Finally, among many different side chains tested, introduction of small polar groups like carbamates, led to compound (5d), showing the best balance between low IC_{50} values and strong synergy *in vitro* against class C β-lactamase producing bacteria.

2) Intrinsic antibacterial activity

Among all the new componds synthesized so far, only few of them displayed, beside their β-lactamase inhibitory properties a marginal level of intrinsic antibacterial activity *in vitro*. Most of them were 7-unsubstituted compounds (R_1= H, eg (12a) in table 1).

Table 3 gives an idea about the inhibition of the essential transpeptidases from *E. coli* (PBP 1 to 3) and the resulting antibacterial activity *in vitro* of 7-unsubstituted compounds (R_1 = H):

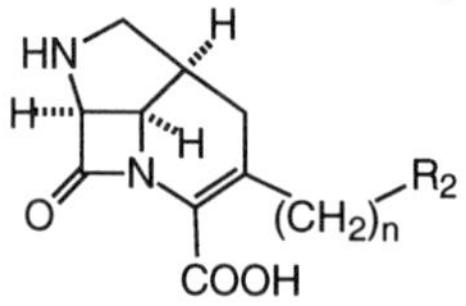

(12a-c): n = 0
(18c-g): n = 1

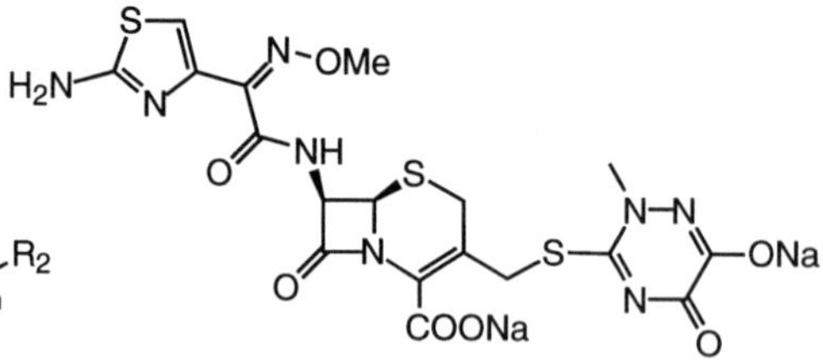

Ceftriaxone (CRO)

From this data set it is evident that the 7-unsubstituted compounds act by selective inhibition of PBP 2 from *E coli*. If one compares the antibacterial activities *in vitro* against *E. coli* CF-102, a strain producing class A β-lactamases (TEM-3 type), with the activities against *E.*

Table 3: Inhibition of essential PBP's and antibacterial activity: influence of R_2 (R_1=H)

N	n	R_2	Membrane bound PBP's of *E.coli* IC_{50} im µg/ml 1a	1b	2	3	MIC in µg/ml *E. coli* 5922 (No ß-Lact.)	*E. coli* CF102 (Class A)	*C. freundii* 1982 (Class C)
		HN, H, H, H, O, N, COOH, $(CH_2)_n$, R_2							
(12a)	0	—S, N–N, N, N, N	10	100	1	100	16	32	32
(12b)	0	—S, N–N, S, CH_3	10	100	1	100	8	32	8
(12c)	0	—S, N–N, S, NH_2	1	100	0.1	100	4	64	16
(18e)	1	—S, N–N, S, CH_3	100	100	0.1	10	16	16	16
(18f)	1	—S, N–N, S, NH_2	10	100	0.5	5	2	4	4
(18c)	1	—N⊕ (pyridinium)	100	100	5	100	32	32	32
(18d)	1	—$OCONH_2$	100	100	0.5	100	8	4	4
(18g)	1	—O, O, N–H, OH	10	50	0.5	5	4	8	8
		Ceftriaxone (CRO)	0.06	1.3	0.9	0.02	0.06	8	>32

coli 25922 (no β-lactamases), there is no significant difference within the methylene homologues (18c-g) whereas there is a clear lability of certain directly linked compounds (see (12b) and (12c)). This difference can be again attributed to a decreased deacylation rate, due to the presence of a leaving group at the 3' position (see also comments Table 2)[10]. Overall, the 7-unsubstituted compounds showed only weak antibacterial activities *in vitro*, due to an insufficient coverage of other essential transpeptidases (such as PBP 1a, 1b and 3).
We tried to improve these activities by further variations of R_1 and R_2.

In **Table 4** the influence of small acyl substituents R_1 on the inhibition of the essential transpeptidases and the resulting antibacterial activity was investigated:

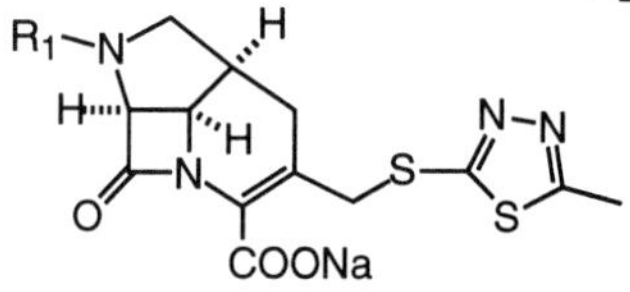

(5e-h)
(18c): R_1 = H

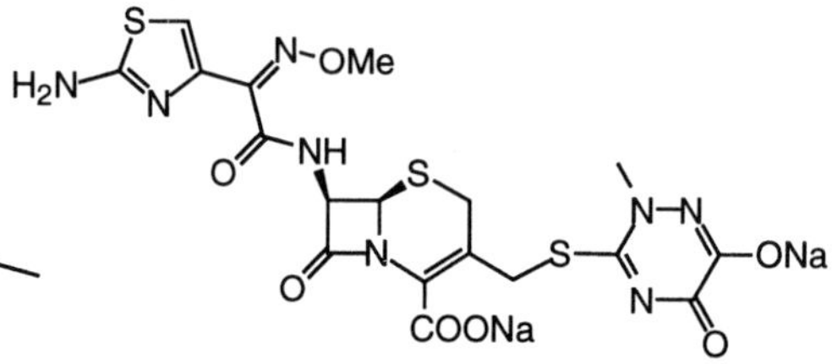

Ceftriaxone (CRO)

Introduction of small acyl substituents, especially of acetyl (5f) and formyl (5e), resulted in an improved coverage of all essential PBPs in *E. coli.* This improved the antibacterial activity *in vitro* against sensitive strains, but more importantly, improved potency against strains resistant either by production of class A or class C β-lactamases was also achieved. Against these strains the antibacterial activities are clearly superior to third generation cephalosporins like CRO, however there is still a lack of potency against sensitive strains when compared to CRO. Larger substituents are allowed if some flexibility is present (e.g. (5h)).

Finally, the influence of R_2 on the inhibition of the essential transpeptidases and the resulting antibacterial activity *in vitro* within the acetyl derivatives was investigated.
The results of this study are summarized in **Table 5**.

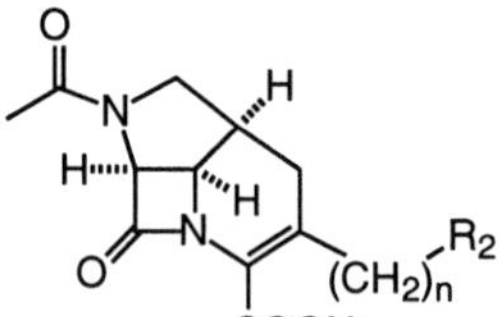

(3h-i): n = 0
(5f, i-l): n = 1

Directly linked compounds ((3h) and (3i)) were as active as the corresponding methylene homologues ((5f) and (5i)) against the essential transpeptidases, but the methylene homologues were overall more potent *in vitro*. Introduction of a pyridinium substituent resulted in a total loss of activity, whereas a carbamate group compensated partially the loss of enzymatic activity by improved penetration properties.

Table 4: Inhibition of essential PBPs and antibacterial activity: influence of R_1

	[structure: R1–N bicyclic cephem core, COONa, S-methylthiadiazole]	Membrane bound PBP's of *E. coli* IC_{50} in µg/ml				MIC in µg/ml *E. coli*		*C. freundii*
N°	R_1	1a	1b	2	3	25922 (No ß-Lact.)	CF102 (Class A)	1982 (Class C)
(18e)	H	100	100	0.1	10	16	16	16
(5e)	CHO	0.5	1	0.5	10	1	2	2
(5f)	CH_3CO	0.5	1	0.5	1	1	1	2
(5g)	CF_3CO	0.1	5	0.5	10	4	4	4
(5h)	[structure: H_2N-thiadiazolyl–S–CH_2–CO–]	0.1	5	1	50	4	2	4
	Ceftriaxone (CRO)	0.06	1.3	0.9	0.02	0.06	8	>32

Table 5: Inhibition of essential PBPs and antibacterial activity: influence of R_2 (R_1=Ac)

			Membrane bound PBP's of *E.coli* IC_{50} in μg/ml				MIC in μg/ml *E. coli*		*C. freundii*
N°	n	R_2	1a	1b	2	3	25922 (No ß-Lact.)	CF102 (Class A)	1982 (Class C)
(3h)	0	—S, N–N, S, CH_3	0.05	1	5	5	8	8	8
(3i)	0	—S, N–N, S, NH_2	0.05	5	0.5	1	2	4	4
(5f)	1	—S, N–N, S, CH_3	0.5	1	0.5	1	1	1	2
(5i)	1	—S, N–N, S, NH_2	0.1	50	1	5	1	1	4
(5j)	1	—S, N	0.5	50	1	5	4	1	8
(5k)	1	⊕ —N	1	100	50	100	32	32	32
(5l)	1	$—OCONH_2$	0.5	50	5	10	4	2	2
		Ceftriaxone (CRO)	0.06	1.3	0.9	0.02	0.06	8	>32

Finally, the best balanced spectrum of activity was obtained with some thioheterocyclic derivatives within the methylene homologues ((5f) and (5i)), which showed broad spectrum antibacterial activity against a wide range of Gram-negative bacteria (complete data not shown).

4. CONCLUSIONS

Based on the knowledge of the mechanism of hydrolysis of β-lactam antibiotics by class C β-lactamases, further exploitation of the concept of restricted rotation led to the synthesis of the bridged carbacephems. These compounds displayed potent inhibitory properties of class C β-lactamases, resulting in strong synergy with third generation cephalosporins against Enterobacteriaceae.

The activities could be extended to the essential transpeptidases through structural manipulation. This could be realized as intrinsic antibacterial activity against a wide range of bacteria. The strength of the bridged carbecephems lies in the broad coverage of members of the Enterobacteriaceae, resistant either by production of class C or class A β-lactamases. Further modifications are ongoing in order to extend the spectrum of antibacterial activity towards *P. aeruginosa* and MRSA. Finally we would like to thank E. Gisler and W. Müller for skilled technical assistance.

References

1. J. P. Rho, F. C. S. Takemoto, A. An and P. C. Norman, "β-Lactamase Inhibitors", *Infect. Dis. Ther.*, 1994, **9**, 151-167.
2. C. Oefner, A. D'Arcy, J. J. Daly, K. Gubernator, R. L. Charnas, I. Heinze, C. Hubschwerlen and F. K. Winkler, "Refined crystal structure of β-lactamase from *Citrobacter freundii* indicates a mechanism for β-lactam hydrolysis", *Nature*, 1990, **343**, 284-288.
3. I. Heinze-Krauss, P. Angehrn, R. L. Charnas, A. D'Arcy, K. Gubernator, C. Hubschwerlen, M. Kania, C. Oefner, M. G. P. Page, J. L. Specklin and F. Winkler, "Structure-based design of potent β-lactamase inhibitors", intended for *Science*.
4. C. Hubschwerlen, R. L. Charnas, K. Gubernator and I. Heinze, "Preparation of β-lactams as β-lactamase inhibitors" EP-A-0508234, 1991.
5. M. Böhringer, C. Hubschwerlen, P. Pflieger and J. L. Specklin, "New β-lactam compounds active against infective agents that produce β-lactamases", EP -A-0671401, 1995.
6. P. Angehrn, K. Gubernator, E. M. Gutknecht, I. Heinze-Krauss, C. Hubschwerlen, M. Kania, M. G. P. Page, J. L. Specklin and F. Winkler, "Bridged β-lactams as β-lactamase inhibitors: I. Synthesis and evaluation of bridged monobactams", intended for *J. Med. Chem.*
7. M. Hatanaka and T. Ishimaru, "A simple synthesis of carbacephem derivatives", *Tetrahedron Letters*, 1983, 4837-4838.
8. A. Afonso, F. Hon, J. Weinstein, A. K. Ganguly and A. T. McPhail, *J. Am. Chem. Soc.*, 1982, **104**, 6138-6139.
9. A. K. Ghosh, T. T. Duong, S. P. McKee and W. J. Thompson, "N,N'-Disuccinimidyl carbonate: a useful reagent for alkoxycarbonylation of amines", *Tetrahedron Letters*, 1992, 2781-2784.
10. W. S. Favari and R. F. Pratt, "Mechanism of inhibition of the PCI β-lactamase of *Staphylococcus aureus* by cephalosporins: importance of the 3'-leaving group", *Biochemistry*, 1985, **24**, 903-910.

9

Design, Syntheses and Studies of New Antibacterial, Antifungal and Antiviral Agents

Marvin J. Miller,* Ihab Darwish, Arun Ghosh, Manuka Ghosh, Jan-Gerd Hansel, Jingdan Hu, Chuansheng Niu, Allen Ritter, Karl Scheidt, Carsten Süling, Shunneng Sun, Deyi Zhang, Allen Budde,[§] Erik De Clercq,[¥] Sally Leong,[§] Francois Malouin,[‡] and Ute Moellmann[∇]
DEPARTMENT OF CHEMISTRY AND BIOCHEMISTRY, UNIVERSITY OF NOTRE DAME, NOTRE DAME, IN 46556, USA

1 INTRODUCTION

"Antibiotics: The End of The Miracle Drugs?" This was the headline on the cover of *Newsweek*, March 28, 1994. As described in the lead articles,[1] and in more scientific detail in several articles[2] in the April 15, 1994, issue of *Science* devoted to antibiotic resistance, the *Newsweek* headlines were not unfounded sensational journalism, but a real wake up call for an emerging health care crisis. D. E. Koshland's editorial, "The Biological Warfare of the Future", in the same issue of *Science*[3] ended with the warning "...the days of miracle drugs and universal vaccines are going. A long struggle with a premium on basic research to improve our strategems and applied research to develop new magic bullets is clearly the prognosis of the future." Indiscriminant use of antibiotics has been responsible for a considerable amount of the widespread development of antibiotic resistance. Microbes have altered their permeability barriers, drug target binding sites or even induced synthesis of enzymes to destroy the antibiotics (e.g., β-lactamases). The research in this paper describes methods for bypassing some of these microbial defense mechanisms by utilizing active transport nutrient (iron) assimilation processes to carry several known and some novel synthetic antimicrobial agents into microbial cells.

Iron has played an essential role in the evolution of nearly every form of life on earth. Although iron is one of the most abundant elements, its pivotal role depended on the development of effective methods for its assimilation. Ionic forms of iron, especially iron(III), its most common oxidation state, are very insoluble under physiological conditions. To circumvent the solubility problem, many microbial, plant and even higher organisms synthesize and utilize very specific low molecular weight iron chelators called siderophores.[4,5,6] Representative siderophore structures are shown below. Of the over 200 known siderophores, most contain hydroxamic acids, catechols, α-hydroxy carboxylic acids or combinations of these bidentate iron-binding ligands. Many siderophores contain three of these bidentate groups to most effectively chelate iron. When grown under iron deficient conditions, many microbes will synthesize and excrete siderophores in excess of their own dry cell weight to sequester and solubilize iron.[7] This extreme focus on the need for iron is reflected by its requirement for the proper function of the enzymes that facilitate electron transport, oxygen transport and other life-sustaining processes. In fact, competition for iron between a host and bacteria is one of the most important factors determining the course of a bacterial infection.[8] Because of their ubiquitous nature, microbial siderophores have been extensively studied, yet much still needs to be learned about their chemistry and biochemistry, as well as other important aspects of iron metabolism.

Representative Siderophores:

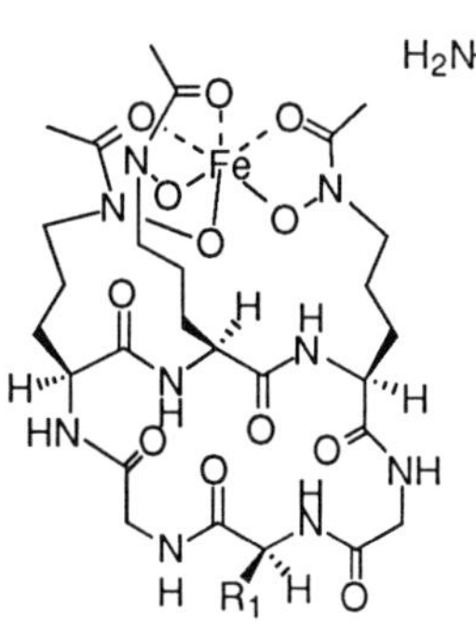

ferrichrome

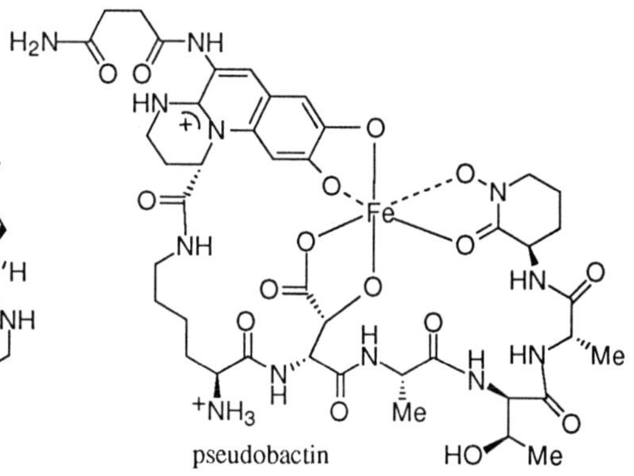

pseudobactin

mycobactins (shown without iron)

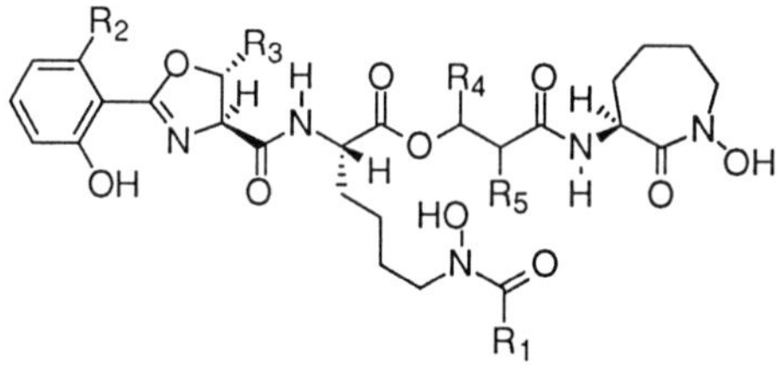

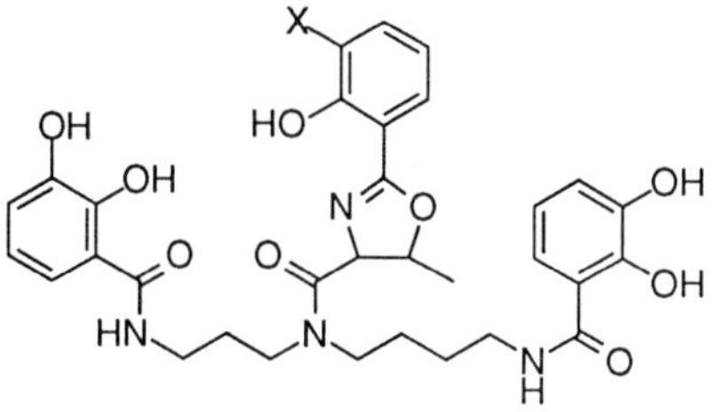

agrobactin, X=OH
parabactin, X=H
shown without iron

1.1 Therapeutic Potential of Siderophores

Siderophores and analogs have tremendous therapeutic potential,[9] yet few practical applications of siderophores have been realized. Suggested applications of siderophores to solve health related problems have included many areas, as described below.

1.1.1 Metal Ion Transport. a) Deferration. Disorders resulting in increased body iron are serious since extensive iron overload causes deposition of the metal in a number of organs causing tissue damage and early death. Primary hemochromatosis, which results from slow, continuous excess iron absorption, can be genetically or environmentally caused and is most easily treated by phlebotomy. Secondary hemochromatosis or transfusional siderosis is typified by sufferers of chronic aplastic anemia or β-thalassemia major, a genetic disease also called Cooley's anemia.[10,11] Cooley's anemia patients are born with the inability to produce adequate amounts of the β-chain of hemoglobin. In order to survive, these patients require frequent blood transfusions. Since the resulting excess iron in the body cannot be removed entirely by normal pathways, iron accumulates leading to organ malfunction and early death. The currently used deferration drug, Desferal®, is a salt of the natural siderophore, desferrioxamine. While relatively effective and nontoxic, neither it nor any of its reported derivatives are orally active. Hence, α-ketohydroxypyridines are being developed as alternative deferration agents.[12] Use of other natural siderophores for deferration also has the potential disadvantage of promoting microbial growth and consequent severe infections or septicemia. This might be avoided by the use of siderophore antibiotic conjugates described in this paper. b) Removal of other toxic metals. Siderophores and analogs have been used to bind and mobilize several different metals including aluminum and plutonium.[13]

1.1.2 Inhibition of Iron Induced Reactions (Antioxidant Properties, Inhibition of Brain Lipid Peroxidation). Reactions dependent on iron can include induction of protein hydrolysis by metal bound water or hydroxide, exacerbation of the toxicity of superoxide and hydrogen peroxide (free radical generation by the Fenton reaction) and initiation of DNA strand scission. These reactions may be detrimental to microbial growth, but they also result in mammalian toxicity. Evidence now indicates that iron chelators, such as desferrioxamine[14] and our synthetic siderophore analogs spermexatins and spermexatols,[15] can inhibit some of these detrimental reactions in mammals and may be especially useful in limiting tissue damage (i.e., ischemic heart and brain damage) by oxygen derived free radicals generated during reperfusion following resuscitation.[16] Iron requiring enzymes (e.g., lipoxygenases[17]) also can be inhibited by siderophores or analogs.

1.1.3 Design of New Antitumor Agents. The significance of iron assimilation in promoting or inhibiting tumor growth is just beginning to be explored.[18]

1.1.4 Plant Growth Regulation. During the last decade significant advances have been made in determining the influence of microbial siderophores on iron assimilation and metabolism by plants.[4,6]

1.1.5 Antimalarial Agents. Desferrioxamine has been shown to have antimalarial properties, including activity against chloroquine resistant strains of *Plasmodium falciparum*, the causative parasite.[19]

1.1.6 AIDS. Desferal® has retarded the progression of AIDS and iron chelation has been postulated as a key component in the treatment of immunocompromised patients whose natural iron withholding defense systems are being overwhelmed.[20]

1.1.7 Magnetic Resonance Imaging (MRI) Contrast Agents. Recent results not described in this paper indicate great potential for the use of our synthetic nontoxic siderophores as organ selective MRI contrast enhancing agents.

1.1.8 Antibiosis. Siderophores can possibly be used to assist antimicrobial therapy in several ways including: a) Deprivation of iron. Very effective chelators may deprive pathogenic microbes of iron essential for growth. This could be accomplished by competitive chelation of iron or by blocking the iron-siderophore receptor site with a nonfunctional siderophore analog or the natural siderophore substrate bound to another metabolically less useful metal.[21] For example, the growth of *Candida* is significantly inhibited by the presence of the catechol containing siderophore, parabactin,[22] derived from *Paracoccus denitrificans*.[23] Parabactin and related synthetic analogs apparently inhibit fungal growth by competing with the natural fungal siderophores for the physiologically available iron. b) Exploiting iron toxicity. Siderophore bound iron might be used to induce intracellular free radical damage for selective inhibition of microbial growth [also see *1.1.2* above]. c) Illicit drug transport. Direct attachment of antimicrobial agents to siderophores or analogs may allow drugs to be actively carried into cells by the iron transport system. The result may be the development of very effective and species selective antimicrobial conjugates represented by the generalized structure:

Further demonstration of the potential of iron transport-mediated drug delivery will be the main focus of this paper.

2 PRECEDENT FOR SIDEROPHORE-MEDIATED DRUG DELIVERY IN BACTERIA

A number of studies have demonstrated that the use of siderophore-mediated drug delivery is feasible. E-0702 is a semisynthetic iron chelating antipseudomonal cephalosporin derivative which has been demonstrated to be incorporated into *E. coli* cells by the *ton-B*-dependent iron transport system.[24] Albomycin[25] and the ferrioxamine B derivative, ferrimycin A_1,[26] are natural examples of the use of iron transport systems to deliver toxic substances to bacteria. Evidence now indicates that albomycin is actively carried into Gram positive and Gram negative bacterial cells by normal iron transport processes, and, once in the cells, the toxic thioribosyl moiety is enzymatically released, perhaps by a peptidase mediated cleavage.[27]

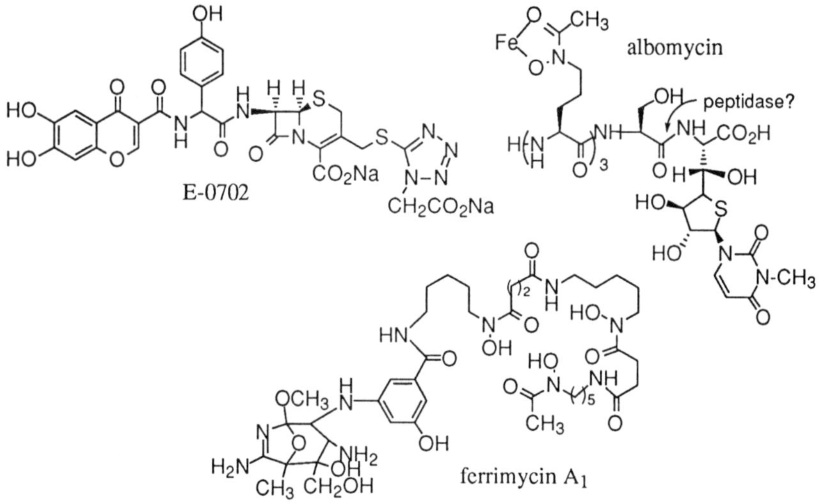

E-0702

albomycin

peptidase?

ferrimycin A_1

Recent studies in our laboratories indicate that the rational design and syntheses of antibacterially effective siderophore conjugates of antibacterial agents is possible.[15,28] For example, at the 1992 Cambridge Meeting,[28b] we reported the synthesis and preliminary antimicrobial activity of conjugates (**1** and **2**) of carbacephalosporins with separate hydroxamic acid-based and catechol-based siderophore components.[29] As expected, detailed biological assays revealed that the hydroxamate- and catechol-containing conjugates utilized different outer membrane receptor proteins to initiate cellular entry (Fhu and Cir, respectively).[30] The conjugates also were found to directly interact with penicillin binding proteins (PBPs) and inhibit growth of the parent organism. *In vitro* studies suggested that strains lacking appropriate outer membrane receptor proteins (as determined by outer membrane protein analysis) for **1** or **2** were rapidly selected. While these strains were resistant to the individual conjugates, they were still susceptible to the alternate conjugates, and subsequent studies revealed that the selected mutants were greatly impaired in their ability to survive in mammals, presumably since the mutant lacked a full complement of iron assimilation mechanisms and is at a growth disadvantage in physiologically iron-restricted conditions. Additionally, the combined use of hydroxamate and catechol-carbacephalosporin conjugates **1** and **2** resulted in more effective inhibition of microbial growth than the individual conjugates alone.

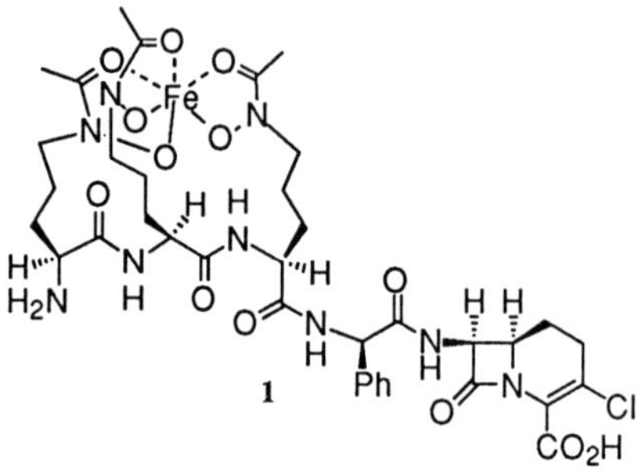

1 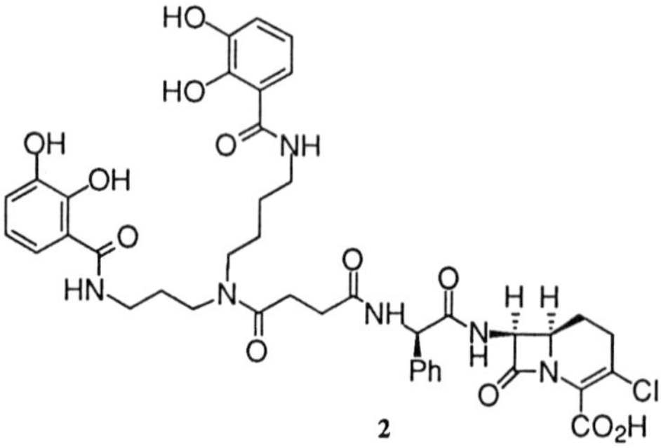 **2**

2.1 New Mixed-Ligand Siderophore-β-Lactam Conjugates

Based on these interesting observations, we postulated that combining catechol and hydroxamate components into a single conjugate structure might promote recognition and transport by multiple siderophore assimilation processes and minimize the development of resistance by selection of mutants defective in one type of siderophore recognition. Any multiply resistant strains also would be especially prone to iron starvation. Thus, we recently synthesized and performed preliminary biological studies of the first mixed hydroxamate and catechol-containing siderophore-like conjugates (**3** and **4**) of carbacephalosporins. Indeed, as described below, some forms of **3** and **4** can use multiple transport processes in *E. coli* and are not only effective against the parent strains, but also the individual mutants selected from prior incubation with conjugates **1** and **2**.

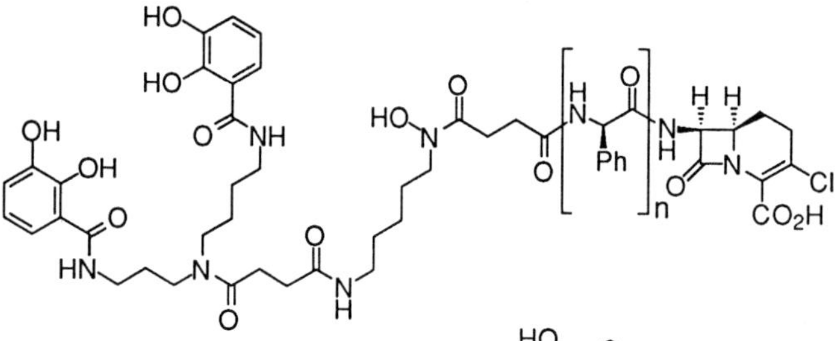

3a, n=0
3b, n=1

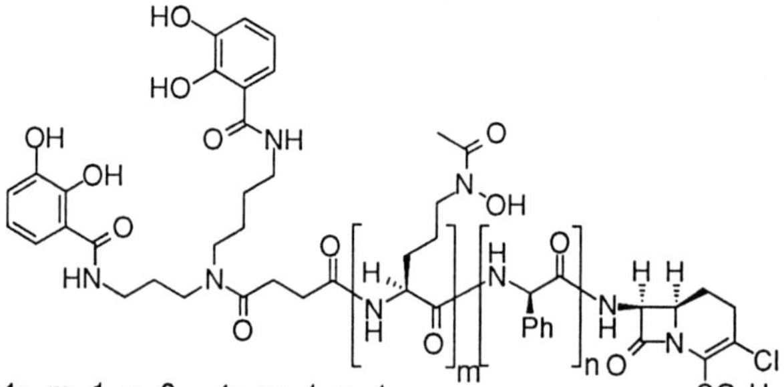

4a, m=1, n=0 **4c**, m=1, n=1
4b, m=3, n=0 **4d**, m=3, n=1

The syntheses of representative forms of **3** and **4** proceeded as expected. Thus, separate coupling of protected hydroxamate derivatives (**5** and **6**)[31,32] of both the parent carbacephem nucleus and Lorabid®[33] with protected *bis* catechol-containing spermidine derivative **7** afforded **3a,b**. Since the phenylglycyl side chain imparts better biological activity to carbacephalosporins, **4c** and **4d**, which include that side chain, also were prepared by coupling of protected forms of δ-*N*-hydroxyornithyl carbacephalosporin **8a** or the corresponding tripeptide derivative **8b** with **7**.

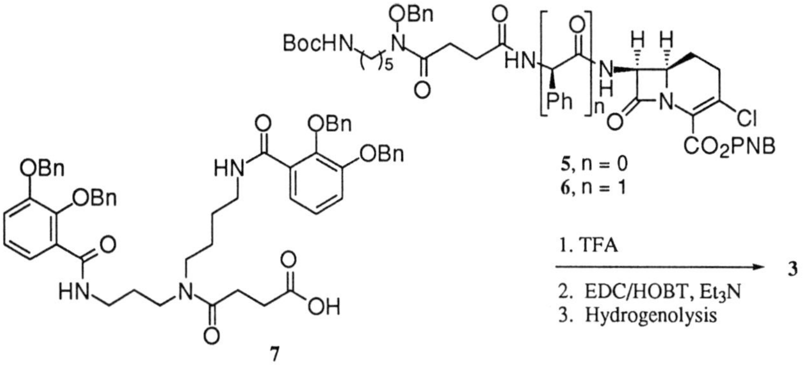

5, n = 0
6, n = 1

7

1. TFA
2. EDC/HOBT, Et_3N
3. Hydrogenolysis

→ **3**

8a, m=1, n=1
8b, m=3, n=1

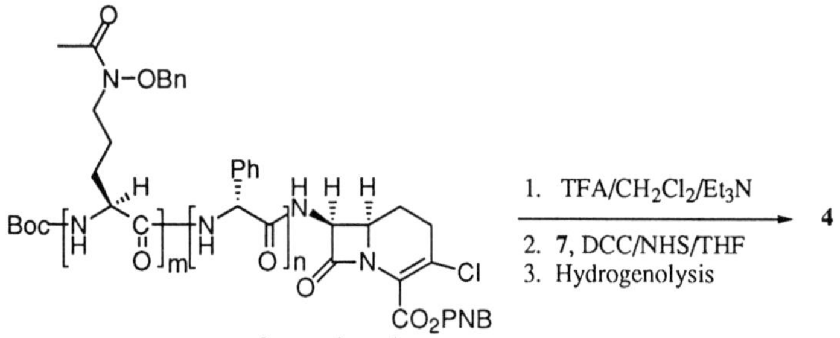

1. $TFA/CH_2Cl_2/Et_3N$
2. **7**, DCC/NHS/THF
3. Hydrogenolysis

→ **4**

Several standard and more detailed bacterial inhibition assays of conjugates **3a**,**b** and **4c**,**d** were performed. Agar diffusion assays of **3a**,**b** with select microbes suggested that **3b**, containing the important phenylglycyl side chain of Lorabid®, was a more effective growth inhibitor than **3a**. Not surprisingly, in antibiotic susceptibility tests performed by the standard agar dilution method,[34] compound **3a** displayed only moderate *apparent* activity against some Gram positive species and no activity against Gram negative bacteria. However, compound **3b** displayed mild antibacterial activity against some Gram positive and against a series of Gram negative species, including outstanding activity against *Acinetobacter* (MIC=0.03 μg/mL, whereas Lorabid® had MIC=64). Standard MIC data determined by the standard agar dilution method[34] for compounds **4c**,**d** also appeared disappointing for most organisms tested. Since, as described earlier,[30b,c] the standard agar or broth dilution assays do not differentiate growth of the parent strain from selection of mutants lacking specific siderophore receptors (thus, giving misleading high MIC values), more detailed growth kinetic studies in broth were performed with one organism. Incubation of conjugates **3a**,**b** and **4c**,**d** with β-lactam hypersensitive *E. coli* X580[35] induced significant inhibitory effects as shown by the delay in observed microbial growth compared to the control. The bacteria that eventually did grow were separately incubated again in the presence of each of the test compounds. No delay of growth relative to the control was observed in this reinoculation experiment. As determined with other siderophore-antibiotic conjugates,[30b,c] these results suggest that during the first incubation growth of the wild-type organism was completely inhibited until selection of resistant mutants from the parent *E. coli* strain occurred. These mutants may be deficient in some specific form of iron transport since growth thereafter was observed in the presence of the same conjugates that inhibit growth of the original organism, but not in the presence of other conjugates.

Iron transport deficient mutants selected from previous exposure to siderophore-antibiotic conjugates such as **1** or **2** were found to be greatly impaired in their ability to grow in mammals because of their decreased ability to assimilate physiologically present iron, even though they were deficient in only one of several siderophore outer membrane recognition proteins.[30c] Thus, it also was important to determine the mode of transport of the new mixed-ligand conjugates as they might utilize more than one receptor, indicating that mutants unable to recognize the conjugates might be missing more than one siderophore receptor and be especially prone to iron starvation under physiological conditions. Separate incubation of **3a**,**b** (with or without preforming the iron complex) with mutants previously isolated from the exposure of *E. coli* X580 to catechol conjugate **2**, and shown to be missing the Cir protein,[30c] resulted in delay of growth relative to the control similar to that observed from incubation with the original *E. coli* X580. This suggested that **3a**,**b** do not require just the Cir protein for transport. The same experiments with **4c**,**d** resulted in little change relative to growth of the control, perhaps indicating that this conjugate requires the Cir protein for transport. Incubation of **3b** and **4c** (with or without preforming the iron complex) with mutants previously isolated from the exposure of *E. coli* X580 to hydroxamate conjugate **1**, and shown to be missing the outer membrane triornithylhydroxamate receptor protein (FhuA),[30c] resulted in significant inhibition of growth, again suggesting that these conjugates do not require just FhuA for transport. The effect of **4d** on this mutant was less dramatic. Repetition of the growth inhibition/delay studies with *E. coli* X580 in the presence of 10 μM **3a**,**b** or **4c**,**d** and EDDA [ethylenediamine *bis*(*o*-hydroxyphenyl)acetic acid] to simulate a free iron deficient medium similar to mammalian serum, resulted in extended growth delay caused by **3a**, and complete inhibition of growth in the presence of **3b**, **4c** or **4d**. These results suggested that conjugates **3b** and **4c** may use an alternate receptor and/or transport system or more than one receptor/transport system compared to the original conjugates **1** and **2**, while **4d** might not be as versatile. Indeed, further studies with previously selected and characterized mutants indicated a dependence of the activity of **3b** on iron recognition/transport proteins TonB, Cir and Fiu.[36]

Thus, the new mixed ligand siderophore conjugates **3a**,**b** and **4c**,**d** appear to be able to utilize a variety of active transport processes to deliver antibiotics to and inhibit the growth of parent strains of pathogenic bacteria. Subsequently selected iron transport deficient mutants are iron starved and, based on precedent,[30b,c] might not be as virulent. Although we have not yet determined if the drug-component of these new siderophore-drug conjugates is or must be released from the siderophore to be effective, these results suggest that it is possible to design, synthesize and utilize drug conjugates of siderophores and analogs capable of active transport into microbial cells by multiple pathways. Proper choice of siderophore, linker and drug may allow selective targeting and destruction of pathogenic organisms, thus, providing significant therapeutic advantage during the development of new antimicrobial agents. By facilitating active transport, siderophore conjugation also may rejuvenate drugs to which resistance has developed by microbial alteration of membrane permeability.

2.2 Vancomycin-Siderophore Conjugates

As an extension of this concept we have begun to prepare siderophore conjugates with a number of drugs of varying activity and site of action. Subsequent studies will focus on the nature of the siderophore to drug linker with the intent of controlling drug release by chemical and/or enzymatic processes. Here we will describe preliminary studies of vancomycin-siderophore conjugates **12a**,**b**. Details of the syntheses have been published recently.[37] The position of attachment of the siderophore component to vancomycin was determined by fast atom bombardment (FAB) mass spectral analysis of **12a** and **12b** and comparison with related analyses of modified vancomycins in the literature.[38,39]

Conjugates **12a** and **12b** were screened for their antibacterial activity by using the broth-microdilution method. Conjugates were tested in Mueller Hinton broth, under iron-depleted as well as iron-sufficient conditions against a broad spectrum of Gram negative and Gram positive strains. Representative Gram negative strains included wild-type *E. coli* (EC14), wild-type *Pseudomonas aeruginosa* (X620), and an antibiotic hypersensitive strain (X621) of *P. aeruginosa*. The Gram positive strains studied included a penicillin-sensitive and a methicillin-resistant *Staphylococcus aureus*, and a *Micrococcus luteus*, a nonpathogenic test strain. Biological studies indicated that siderophore-modified vancomycins lost some activity (4 to 16 fold) against Gram positive bacteria relative to vancomycin itself, and were generally similar to vancomycin in activity against Gram negative bacteria under iron sufficient conditions. In this respect, the siderophore conjugates appear to be similar to other acylated vancomycins.[38] However, under iron depleted conditions which mimic human serum, mixed ligand conjugate **12b** displayed enhanced (> eight fold) antibacterial activity against an antibiotic hypersensitive strain of *Pseudomonas aeruginosa* compared to vancomycin itself. Details related to the potential need to include a more readily cleavable linker between the siderophore and vancomycin or whether the vancomycin conjugates even have the same specific mode of activity as the parent vancomycin have yet to be determined.

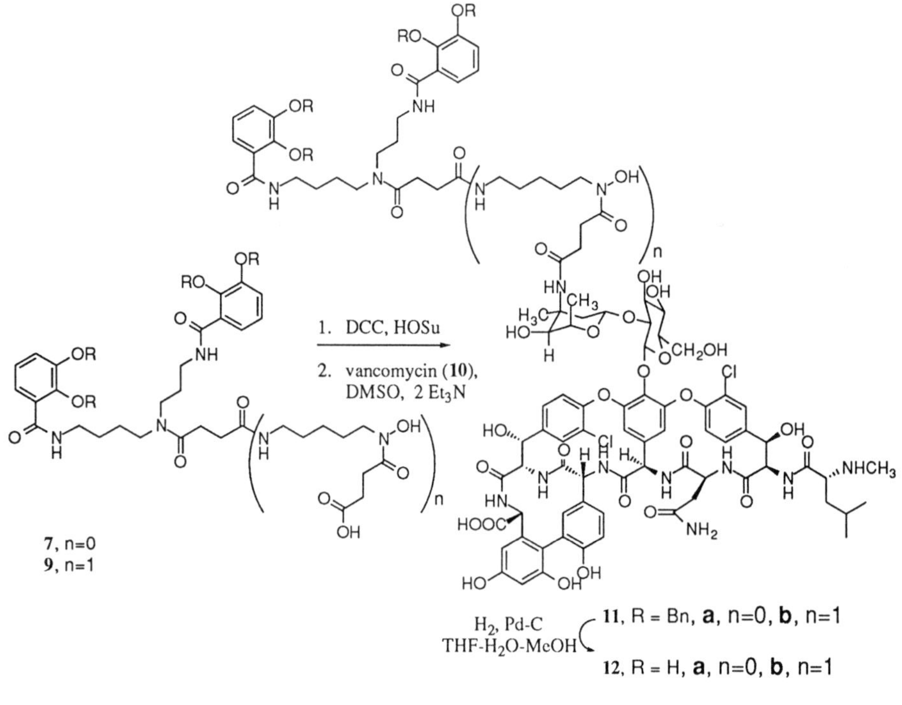

1. DCC, HOSu
2. vancomycin (**10**), DMSO, 2 Et_3N

7, n=0
9, n=1

H_2, Pd-C
THF-H_2O-MeOH

11, R = Bn, **a**, n=0, **b**, n=1
12, R = H, **a**, n=0, **b**, n=1

3 FUNGAL SIDEROPHORES AND DRUG DELIVERY

Having demonstrated the feasibility of iron transport (siderophore)-mediated drug delivery in bacteria, the focus shifted to determine if siderophore-mediated drug delivery could be effective in fungi. Although a number of pathogenic fungi are known to require siderophores for growth, structures of many of the siderophores have not yet been determined. Thus, we decided to develop a broad program related to the development of new antifungal agents and drug delivery systems based on siderophore chemistry and biochemistry.

3.1 Siderophore-Drug Conjugates as Antifungal Agents: Goals

Specific aims of this program include: 1) Utilize methodology developed in our laboratory to prepare a library of siderophores and components to determine which siderophores can be recognized and utilized by opportunistic pathogens with initial emphasis on *Candida albicans, Cryptococcus neoformans,* and *Aspergillus fumigatus* (though many others may be included in broader screening). 2) Synthesize and study siderophore-antifungal agent conjugates to determine if the siderophore can actively transport the antifungal agent (drug) into the cell or anchor the siderophore-drug conjugate in the cell membrane and, in either case, exert a lethal effect. Known and novel antifungal agents with various modes of action will be conjugated to the siderophores to determine a) optimal microbial selectivity and b) if mammalian toxicity of some drugs can be reduced by conjugation and/or siderophore-mediated targeting of the drug to fungi. 3) Study the influence and importance of linkers between the siderophore and antifungal agent and determine if drug release is necessary or desired. 4) Develop efficient and practical syntheses of novel antifungal agents (and siderophore conjugates). Here we describe our preliminary studies in these areas.

3.2 Determination of Specific Siderophore Utilization by Opportunistic Fungal Pathogens.

The production of siderophores by bacteria, nonpathogenic yeasts and some fungi has been well documented. By contrast, verification of the production and use of siderophores by opportunistic and pathogenic fungi which can infect humans has been limited.[40] Interestingly, in early studies, only hydroxamate based siderophores were suspected to be produced by fungi. For example, *Histoplasma capsulatum* has been shown to produce and use the trihydroxamate siderophore coprogen B. Ferrichrome is assimilated by strains of *Aspergillus. Cryptococcus melibiosum* utilizes the trihydroxamate ferrichrome C. More recent studies indicate that some fungi and yeasts, such as *Candida albicans*, produce both hydroxamic acid and catechol-based siderophores,[41] even though some catechol based siderophores have been demonstrated to inhibit the growth of *Candida.* In addition, *C. albicans* has an absolute requirement of iron for growth and appears to have an iron-uptake system that can obtain iron by using candidal and non-candidal siderophores.[42] (Therefore, it was anticipated to be susceptible to siderophore-drug conjugates.) Aside from the early isolation of pulcherriminic acid from *Candida*,[43] none of the structures of the candidal siderophores have been determined. Even the structure of the relatively simple pulcherriminic acid has been controversial and no definitive studies have been performed to demonstrate if this iron binding natural product exhibits siderophore activity for producer strains of *Candida.* Most importantly, no real structure-activity study had been performed on the siderophores that can be utilized by any of the fungi or yeasts responsible for opportunistic infections. Thus, determination of the types of siderophores or siderophore components that could be utilized by strains of opportunistic fungi was an important aspect of the project and initial emphasis was placed on development of a reliable assay for screening siderophore use by *Candida albicans.*

A number of chemical and biochemical methods are now used routinely for the detection of microbial siderophores. The methodology has been thoroughly reviewed by Neilands[10] and extended by many others.[44,45,46,47] Bioassays using *Candida albicans* in an iron-deficient agar have been developed, extensively utilized in our laboratory and are now published.[48] We sought a suitable liquid culture medium which would, by adding known iron chelators such as EDDA and 2,2'-dipyridyl, create an iron-deficient environment that might stimulate *Candida albicans* to begin synthesizing siderophores. Adding EDDA up to concentrations of 500 mg/mL to YM broth, Sabouraud dextrose broth, and potato dextrose broth had absolutely no effect on growth of *Candida*, though we know from our earlier studies that these are conditions which will limit the growth of *E. coli*. Realizing the problems of using complex media, which often contain yeast extract and thus possibly yeast siderophores, we turned our attention to chemically defined Lee's medium.[49] In order to determine how *Candida albicans* responds to the presence of iron chelators in this medium, and to find an optimal concentration of chelator which would limit growth, increasing amounts of EDDA (10 μg/mL up to 200 μg/mL) and 2,2'-dipyridyl (20 μM up to 200 μM) were added to separate flasks of Lee's medium. After monitoring the growth in these flasks at 600 nm, we found from dilutions into fresh cultures, that dipyridyl at high concentrations appears to be toxic to *Candida albicans*, although growth is possible at low concentrations. However, in this medium, EDDA at 20 μg/mL limits growth, but is not toxic. This result and the earlier analyses, which showed that *Candida albicans* can tolerate high concentrations of EDDA in complex media led us to an experiment in which EDDA was added to Lee's agar (1.5% agar) to final concentrations of 10 μg/mL, 25 μg/mL, 50 μg/mL, and 100 μg/mL. After storage at 4 °C for 48 hours, a 1:5000 dilution of *Candida albicans*, in the yeast form, from an overnight culture was added to the melted but cool agar, which was poured into sterile petri dishes. We wished to establish whether natural siderophores, such as ferrichrome, or any of our synthetic siderophore components and analogs would promote the growth of *Candida albicans* under these iron-deficient conditions (i.e., act as siderophores). Aliquots of 5 μL of various solutions were added to sterile filter paper disks placed on the surfaces of the solidified agar. Incubation upside down at 37 °C for 24 hours clearly revealed the presence or absence of zones of stimulation around filter paper disks containing known amounts of ferrichrome (as the iron complex), Desferal® (**13**, with and without iron), our synthetic tri δ-*N*-hydroxy ornithinyl peptide **21** (n=3), synthetic hexapeptide **22**, and the related zwitterionic peptide (**23**, without the *N*-terminal acetyl group of **22**), and the corresponding preformed Fe(III) complexes of the synthetic compounds. The results are summarized below. Zones of stimulation were measured after 36 hours. The zones of stimulation observed contained some white creamy surface growth, characteristic of *Candida albicans*. Careful examination in all cases confirmed that it was *Candida* and it was in the yeast form.

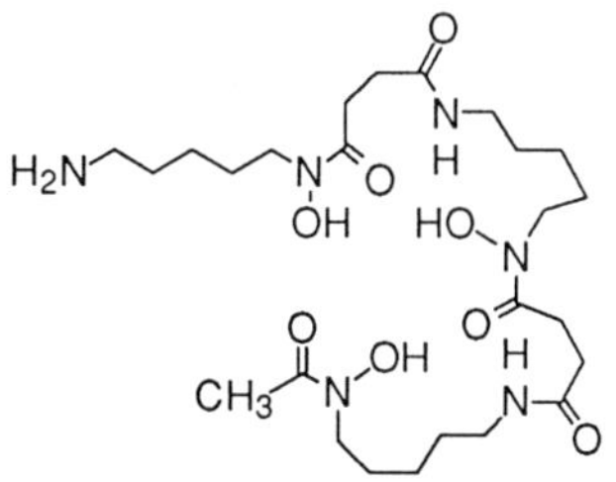

Desferal®, **13** 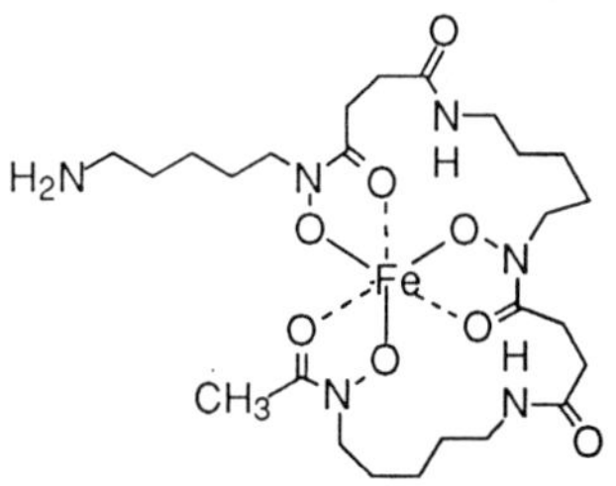 Desferal®, **13** with iron

This data indicated that we now had developed a convenient and very sensitive bioassay to determine siderophore activity for *Candida albicans*. Furthermore, we demonstrated that the natural siderophore ferrichrome is a potent growth factor for *Candida*, while Desferal® (desferrioxamine, another natural siderophore) is not. The significant activity of several of our synthetic siderophores, especially as their corresponding iron complexes, suggested that the rational design of siderophore-drug conjugates might be feasible and prompted intense study with a focus on *Candida* but, as will be described later, also found to be applicable to studies with *Cryptococcus*.

glutamic acid **14** → **15** ($NaBH_4$) → **16** (trocNHOCH$_2$Ph, DEAD / Ph_3P) → **17** (Zn/Ac_2O, HOAc) → **18** (TFA) → **19** → **20**

19 + **20** → **21** → **22**, **23**

Effect of Various Siderophores and Synthetic Analogs on the Growth of *Candida albicans* (ATCC 48130) in Lee's Agar Supplemented with Deferrated EDDA (Zones of Stimulation Given in mm)

	Fc[a]	desferal[b]	**21**(n=3)		**21**(n=3)+Fe(III)		**22**	**22**+Fe(III)	**23**	**23**+Fe(III)
	5[c]	5-25[c]	5[c]	25[c]	5[c]	25[c]	5[c]	5[c]	5[c]	5[c]
10[d]	>90	0	55	54	52	65	38	e	e	e
25[d]	56	0	29	9	36	47	10	28	-	27
50[d]	46	0				14	-	10		

[a]Fc = ferrichrome. [b]with or without Fe(III). [c]amount in nanomoles. [d]concentration of EDDA in μg/mL. [e]heavy background growth.

3.3 Structure-Activity Relationships (SAR) of Siderophores and Analogs

Although much of our effort emphasized development of practical methods for the syntheses of siderophores and components, important and insightful structure-activity relationships of siderophore recognition and transport were rapidly determined since our research group had several natural siderophores and had already prepared a number of siderophore analogs. Using our growth promotion assay we found a significant preference for trihydroxamate siderophore utilization by *Candida albicans*. Thus, monopodal (mono hydroxamates and catechols) and dipodal hydroxamates [the dipeptide of δ-*N*-hydroxy-δ-*N*-acetyl-L-ornithine **21**(n=2)], rhodotorulic acid, and spermidine-based *bis* catechols (such as the siderophore component of **2**) and related synthetic tricatechols were ineffective siderophores for *C. albicans*. However, as already indicated, *tri*-δ-*N*-hydroxy-δ-*N*-acetyl-L-ornithine (**21**, n=3) and extended peptides were excellent siderophores for *C. albicans*. Furthermore, totally synthetic trihydroxamates tethered to cyanurate (as in **26** and **27**) also readily substitute for natural siderophores.

1. NaH
2. Br–$(CH_2)_n$–N(troc)OBn

24

1. Zn/Ac_2O
2. H_2/Pd-C

25 n=4

Zn, succinic anhydride

26 (n=4), Siderophore analog, effective iron chelator and microbial iron transport agent. *J. Med. Chem.* **1985**, *28*, 323

27

The successful demonstration of the use of tripeptide **21** (n=3) based on δ-*N*-hydroxy-δ-*N*-acetyl-L-ornithine as a siderophore by *Candida albicans* prompted an intense focus on the synthesis of related drug conjugates to determine if siderophore-mediated drug delivery to fungi were feasible. Thus, the remaining decisions were related to the choices of drugs and linkers to the siderophore components. The type of linker might determine whether the attached drug would be chemically or enzymatically releasable, even though it was not yet known if this would be necessary (recall that in our earlier studies with bacteria, the intact conjugates of β-lactam antibiotics and siderophores were effective). A variety of drugs with differing modes of action were considered.

3.4 Synthesis and Study of Antifungal Agents and Siderophore-Antifungal Agent Conjugates

Although we had already demonstrated the ability to synthesize amino acid-based hydroxamate ligand components by several methods, including that summarized by the sequence shown earlier (**14** to **19** or **20**), we sought a more direct route to facilitate the synthesis of siderophore components. Studies indicate that siderophore hydroxamates are biosynthesized by oxidation of primary amines to hydroxylamines first and then acylation.[50] A related biomimetic synthesis of amino acid-based hydroxamates was anticipated to significantly simplify the preparation of important siderophores and components. While direct chemical oxidation of primary amines to hydroxylamines or their equivalent is difficult without inducing over oxidation, a direct method using dimethyldioxirane-induced oxidation chemistry has been realized in our laboratories and is demonstrated with a new short synthesis of ε-*N*-hydroxy-ε-*N*-acetyl-L-lysine (**31**) from lysine itself.[51] The chemistry has been repeated successfully on ornithine to give a practical route to δ-*N*-hydroxy-δ-*N*-acetyl-L-ornithine, the most common amino acid-based siderophore hydroxamate.

New direct oxidation method

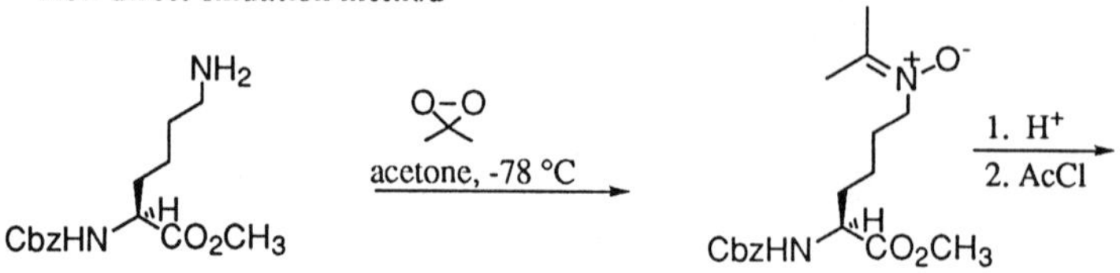

28, protected L-lysine → (dimethyldioxirane, acetone, -78 °C) → 29 → (1. H^+; 2. AcCl) → 30 → (1. KOH, 0 °C (75%); 2. H_2 / Pd-C, CH_3OH (94%)) → 31

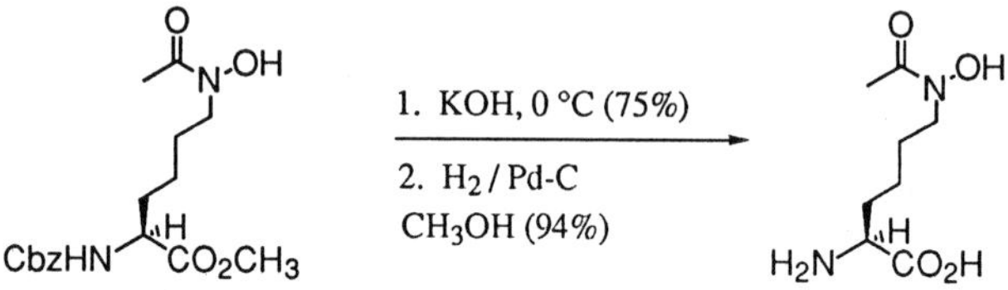

Representatives of each of several detailed structures illustrated below were synthesized and their antifungal activity determined.

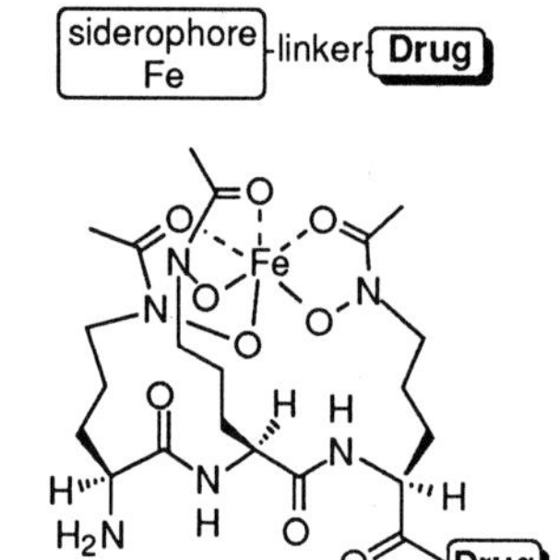

siderophore Fe -linker- Drug

32 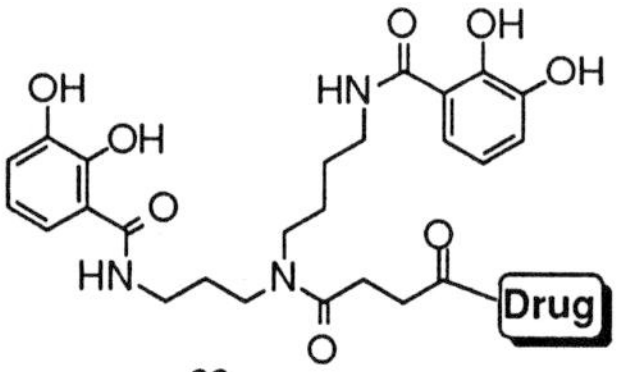 33 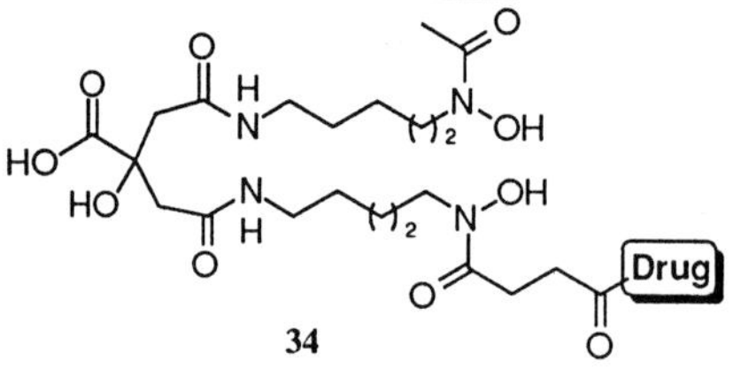 34 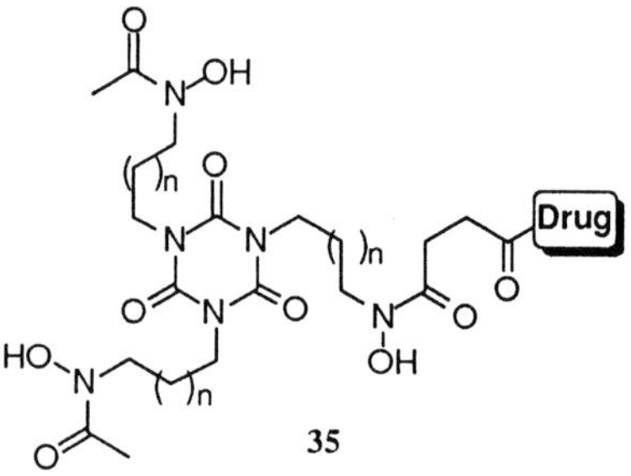 35

Drug = a, b, c, d

R=H, Ph or iPr

(n = 5, 6)

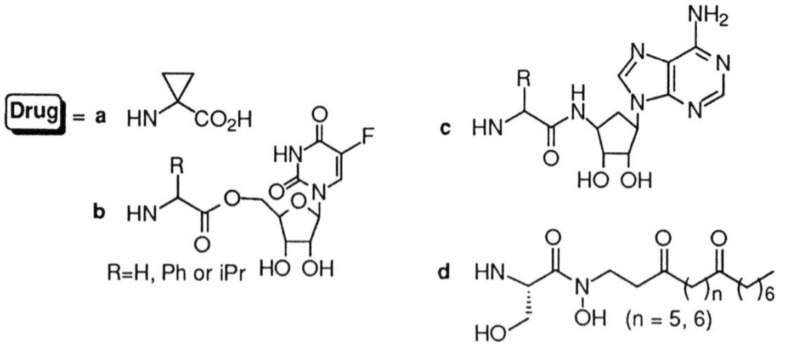

*3.4.1 Conjugates of ACCA ("Drug" **a** in the Structure List).*[52] The synthesis of drug conjugates was initiated by a simple extension of our peptide syntheses. Since 1-amino-1-cyclopropyanelcarboxylic acid (ACCA, drug "**a**") is an effective amino acid-based enzyme inhibitor, we prepared the corresponding benzyl ester and coupled it to the active ester of the protected parent tripeptide chelator. Deprotection provided the tetrapeptide conjugate **32a**. Interestingly, incubation with *Candida albicans* promoted, rather than inhibited growth. Thus, it was reasonably clear that the peptide was an effective iron chelator and transport agent, but that the ACCA constituent apparently was not made available for intercellular damage.

*3.4.2 Conjugates of 5-FU ("Drug" **b** in the Structure List).*[28b] The preliminary study with ACCA prompted a more ambitious program to conjugate more effective drugs to the siderophore component. Our next choice for the antifungal agent was 5-fluorocytosine. Even though 5-FC does contain a site for attachment to a siderophore, we chose instead to use one of its metabolites which contains the ribose moiety. 5-Fluorouridine was chosen as the 5-FC equivalent. We also thought that the conjugation should include a more labile linkage so that the drug might be hydrolytically released once carried into the cell. The 5' hydroxyl constituent was a convenient attachment site to the tripeptide, and this metabolite was anticipated to avoid some of the potential resistance mechanisms attributed to problems with the therapeutic use of 5-fluorocytosine. The amino acids glycine, phenylalanine and valine were used to link the tripeptide siderophore to the antifungal agent. The spacer amino acids were anticipated to help control the release of the drug once inside the cell. Our initial routes to these conjugates involved selective protection of the vicinal diol of 5-FU, several attempts at coupling to the protected amino acid spacers by the usual carboxyl activation methods, deprotection of the amino acid part, coupling to protected siderophore tripeptide, and final deprotection. However, as shown below, this procedure has been greatly simplified by direct coupling of the amino acid spacer to 5-FU using Mitsunobu conditions (DIAD/PPh_3). The synthetic targets are represented by structure **39**. In addition to the 5-fluorouridine conjugate, analog **40** containing uridine, and the corresponding tetrapeptides without either uridine or 5-FU were synthesized to use as controls in the biological testing.

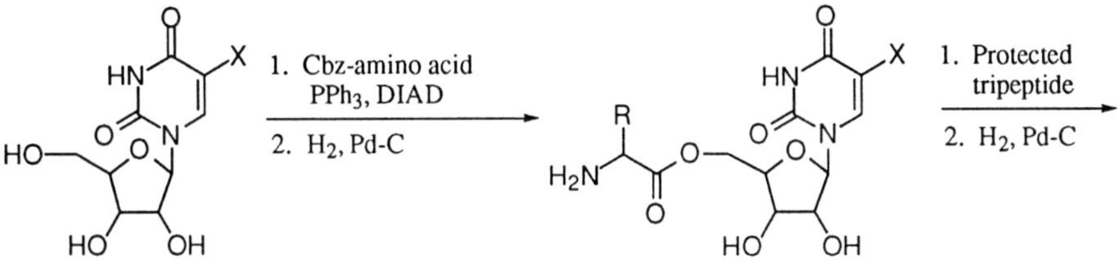

1. Cbz-amino acid PPh_3, DIAD
2. H_2, Pd-C

36, 5-FU, X = F
37, uridine, X = H

38

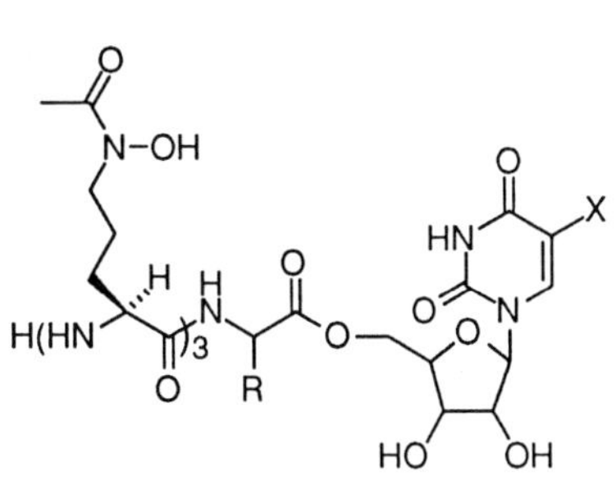

1. Protected tripeptide
2. H_2, Pd-C

39a, X = F, R = H $t_{1/2}$ = 13 hr
C. a. inhib. zone 33 mm
39b, X = F, R = Ph $t_{1/2}$ = 10 hr
C. a. inhib. zone 36 mm
39c, X = F, R = iPr $t_{1/2}$ = 241 hr
C. a. inhib. zone 35 mm
40, X = H, R = H

The lability of the ester linkage (to the 5-FU) in the biological test medium (Lee's) at 37 °C was determined by fluorine NMR studies which monitored the formation of free 5-FU by hydrolysis of the ester linkage. As noted above, half-lives varied and depended on the stereoelectronic factors of the amino acid spacer, thereby allowing variable release times, if in fact drug release was necessary. Growth promotion and inhibition studies using 5 or 25 nanomole of test compound were first carried out with *Candida albicans* in Lee's agar. In media containing EDDA to simulate iron deficient conditions, all of the peptides and the peptide uridine conjugate (**40**, X=H) stimulated growth, indicating that, as expected, they were effective siderophores for *C. albicans*. In the absence of EDDA, 5-FU and all of the 5-FU conjugates inhibited growth with zones of inhibition of 33, 36, and 35 mm with 25 nmol of **39a-c**, respectively (24, 24, 26 mm with 5 nmole) after 24 h. 5-FU also inhibited growth (48 mm zone). Preformation of the iron complexes did not alter the inhibitory effects. Incubation for longer times gave smaller zones with apparent selection of mutants similar to the observations with our earlier antibacterial β-lactam conjugates (**1** and **2**). The effects of the variation of the test media on the activity of conjugates **39a** and **39b** (glycine and phenylglycine spacers) also were examined. As shown in the table below, the lowest MIC values for both were obtained in RPMI & MOPS[3-(*N*-morpholino)propanesulfonic acid], and YNB (yeast nitrogen base) and dextrose. RPMI contains no iron compounds, YNB contains 200 μg/L of $FeCl_3$. The enhanced activity of the conjugates in low iron media suggests uptake by iron transport systems. The diminished activity of the conjugates in YNB may be due to exogenous siderophores in yeast extract. Although neither compound was more active than the 5-FC control for *C. albicans*, phenylglycyl conjugate **39b**, was more active in the RPMI and MOPS broth than was 5-FC.

In vitro Antifungal Evaluation (MIC in μg/mL) of 5-FC and 5-FU Conjugates

Compound	*Candida albicans* A26		*Candida parapsilosis* CP18	
	RPMI & MOPS	YNB & Dextrose	RPMI & MOPS	YNB & Dextrose
5-FC	0.078	0.078	0.625	0.078
39a	5.0	20	0.625	5.0
39b	2.5	10	0.312	5.0

The first two peptide-5-FU conjugates (**39a,b**) prepared also were submitted to antibacterial assays at Eli Lilly and Co. The conjugates were moderately active against several strains of Gram positive bacteria. The tripeptide conjugates were more active than the 5-FC control against strains of *Staphylococcus* (including clinically important methicillin resistant strains X-400, and S 13E, MICs=0.25 and 1 μg/mL, respectively) and had some activity against *Streptococcus* (MICs=2-16 μg/mL). Both the conjugates and 5-FC showed poor activity against strains of Gram negative bacteria. Interestingly, **38c** (R=iPr) displayed notable activity against a number of organisms, including MRSA (methicillin resistant *Staph. aureus*).

Encouraged by results with the 5-FU conjugates of hydroxamate-containing peptides, we prepared 5-FU conjugates (with the same amino acid spacers/linkers) of spermidine-based *bis* catechols, tricatechols, and cyanurate trihydroxamate **35b**. Consistent with siderophore assays described earlier, none of the catechol conjugates displayed anticandidal activity. However, the 5-FU cyanurate trihydroxamate conjugates had essentially the same activity as the peptide-based conjugates![53] This exciting result suggested that our totally synthetic, readily accessible (especially since it contains no asymmetric centers), artificial cyanurate-based trihydroxamate **35** could serve as a building block for anticandidal drug conjugates.

3.4.3 Conjugates of Novel Synthetic Carbocyclic Nucleosides ("Drug" **c** *in the Structure List)* The interesting activity of the 5-FU conjugates prompted efforts to prepare siderophore conjugates of carbocyclic nucleosides. Carbocyclic analogs of nucleosides are of considerable interest as antiviral, antitumor and antifungal agents and

Dr E. De Clercq and others have shown that isosteric replacement of the oxygen of furanose by a methylene group results in enzymatic and hydrolytic resistance and often a decrease in mammalian toxicity.[54] Replacement of the "normal" 5'-hydroxymethyl group of nucleosides with an amine would allow direct conjugation to the siderophores using amide linkages for comparison to the ester links described above. However, no 5'-azacarbocyclic nucleosides have been reported. The development of a novel and efficient asymmetric synthesis of 5'-azacarbocyclic nucleoside analogs is summarized below. The chemistry utilized asymmetric amino acid or sugar-based nitroso Diels-Alder reactions to assemble the carbocyclic nucleoside framework. Thus, oxidation of L-amino acid hydroxamates (**41**) to the corresponding nonisolatable acylnitroso compounds (**43**) in the presence of cyclopentadiene produced cycloadducts **44a**. The diastereoselectivity depended on the amino acid used (phe, 60%; ala, 75%; val, 80%, *t*-leu, 85%: the major diastereomer is shown below). The diastereomers were easily separated and we have demonstrated that each is optically pure.[55] Use of the D-amino acids gave the alternative diastereomers **46b**, corresponding to the enantiomeric bicycles **47b** after removal of the chiral auxiliary by Edman degradation. Hetero Diels-Alder reactions with sugar-derived chloronitroso compounds (i.e., **42**) and cyclohexadiene had been reported to proceed with high selectivity and had the advantage that the chiral auxiliary was released upon workup. However, the reaction with cyclopentadiene had not been reported and racemic earlier syntheses of forms of **45a** (R=H) indicated that they were unstable. However, we have found that the reactions of the isolatable (bright blue crystals) chloronitroso derivatives of D-mannose and D-ribose do provide the desired cycloadducts with each producing forms of the different enantiomers shown. Both optical isomers of amino acid bicycle derivatives (**44a**, **46b**) and the parent or protected bicycles (**45a**, **47b**) are now readily available. We have found that $Mo(CO)_6$-mediated reduction followed by acetylation produces amino acetates **48** cleanly. Direct Pd(0)-mediated reaction with the sodium salt of adenine gave the first 5'-azacarbocyclic adenosines **49**, with traces of separable *N*-3 and/or *N*-7 alkylated products. The incorporation of the adenine base into the carbocycle proceeds with net retention as expected from the Pd(0) reaction. Dihydroxylation of **49** with OsO_4/NMO produced the desired *anti* diol **50** as the major product (with contamination by the separable *syn* diol). Removal of the Boc protecting group from the amino acid constituent (alanine in the first case studied to date) gave free amine **51** which was coupled to our protected trihydroxamate tripeptide, deprotected and treated with ferric ion to produce the first siderophore conjugate of an azacarbocyclic nucleoside **52** (R'=Me, **32c**, from the general structure list shown earlier). The carbocyclic adenosine was chosen as the first target because of the ease of direct incorporation of the adenine base into the framework by the Pd chemistry and because of the product's similarity to sinefungin, an antifungal agent of considerable interest, aristeromycin and noraristeromycin, two potent carbocyclic antiviral agents.

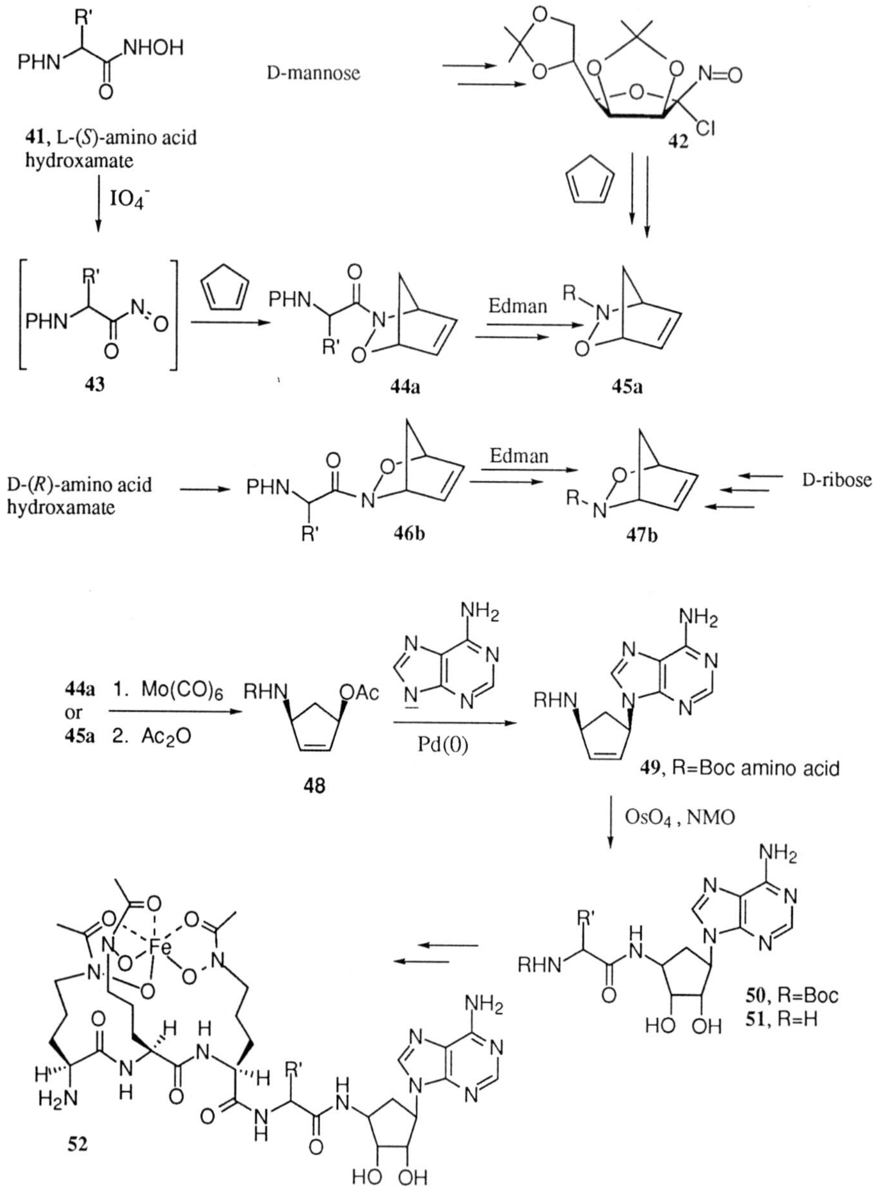

Antifungal tests with Boc-protected **50**, the corresponding free amine (**51**), and conjugate **52** have been initiated. Preliminary studies with *C. albicans* in our labs indicated no anticandidal activity. Broader antimicrobial screening has not yet been performed. Meanwhile, Dr. E. De Clercq's laboratory, has screened the Boc derivative, free amine and conjugate for antiviral and anticancer activity. While only slight activity was found against HIV and a number of other viruses, as shown below, notable selective antiviral and anticancer activity was found for these novel carbocyclic nucleosides.[56]

Compound	Minimum cytotoxic concentration[a] (μg/ml)	IC_{50} (μg/ml)	
		Parainfluenza-3 virus	Reovirus-1
50	> 400	> 400	>400
51	> 200	10	1
52	> 200	7	0.2
DHPA[b]	> 400	40	4
Ribavirin	> 400	150	7
C-c[3]Ado[c]	> 400	4	0.4

[a]50% inhibitory concentration
[b](S)-9-(2,3-dihydroxypropyl)adenine
[c]Carbocyclic 3-deazaadenosine

Inhibitory Effects on the Proliferation of Murine Leukemia Cells (L1210/0), Murine Mammary Carcinoma Cells (FM3A) and Human T-lymphocyte Cells (Molt4/C8, CEM/0)

		IC_{50} (μg/ml)		
Compound	L1210/0	FM3A/0	Molt4/C8	CEM/0
50, R=Boc-ala	> 100	> 100	> 100	> 100
51	21.5 ≥ 5.7	16.5 ≥ 12.6	27.7 ≥ 24.9	77.3 ≥ 21.2
52	9.71 ≥ 2.15	19.2 ≥ 16.2	43.5 ≥ 6.7	25.9 ≥ 12.1

The activities of these first samples of amino acid-containing azanoraristeromycins have encouraged us to prepare other amino acid derivatives to determine the scope of the activity. To date, besides the alanine derivatives, protected forms of phenylalanine, valine, *t*-leucine, proline, serine and aspartic acid have been prepared. Upon deprotection, the compounds will be submitted to antifungal, antiviral and anticancer screening. Intermediates, especially those containing the C=C (such as **49**) before dihydroxylation, also will be tested.

The results described, indicate that we will not only be able to prepare conjugates to siderophores by direct peptide coupling of the amino group of the nucleoside analogs to the carboxy terminus of the siderophore, but the chemistry provides access to a whole new class of nucleoside analogs. Early test results indicate diverse biological potential, including anticancer and antiviral activity, for these novel carbocyclic nucleosides, and the focus in the future will be development of analogs with antifungal activity.

3.4.4 Conjugates of antifungal neoenactins and analogs ("drug" ***d*** *in the structure list).* One of the new antifungal agents we wanted to conjugate to siderophores was neoenactin.[57] The neoenactins are a novel class of antifungal agents that, based on their structural similarity to lipoxamycins[58] and sphingofungins,[59] are believed to act on the fungal cell membrane by inhibition of serine palmitoyl transferase, the enzyme which condenses serine with a fatty acyl-CoA to form ketodihydrosphingosine in the first committed step of sphingolipid biosynthesis. We thought that it would be especially interesting to attach a membrane active antifungal agent to siderophores which might help initially target the "drug" to the fungal membrane by fungal recognition of the siderophore. As described below, the results have been exciting. The neoenactins also potentiate the activity of the polyene antifungal agents. Since we were unable to obtain an authentic sample of neoenactin, we decided to develop a short practical synthesis of neoenactin which would also allow the preparation of novel analogs. As shown below, we have completed the first total synthesis of neoenactin **58** (n=6), an analog norneoenactin **58** (n=5) and several derivatives (**59-63**) of each with variations of the amino acid component.[60] Biological assays in our lab and at Eli Lilly and Company are summarized below. Our first analog norneoenactin (**58**, n=5) is a very effective anti-candidal agent (MIC <0.5 μg/mL against *Candida albicans*) and also active against *Crytococcus neoformans* (M1-106) with an MIC of 0.625 μg/mL. Dr. Sally Leong's laboratory at the University of Wisconsin, Madison, has determined that norneoenactin is also very effective against a number of plant deleterious fungi (*Ophiostoma ulmi, Aphanomyces euteiches, Monilinia fructicola, Magnaporthe grisea* and *Fusarium solani).* Most interestingly, the tripeptide-neoenactin conjugate **64** and norneoenactin conjugate **65** showed diminished activity against *Candida* but enhanced activity against *Cryptococcus*, again demonstrating the potential of siderophore-mediated drug delivery for the development of species selective antimicrobial agents. Further structure-activity studies of this interesting new class of antifungal agents and conjugates were carried out. Removal of either or both of the hydroxyl groups from norneoenactin or neoenactin drastically reduced biological activity (see **61**, **62**). Removal of one of the oxo groups in **58**, significantly enhances antifungal activity.[61] A corresponding cycloserine analog (**59**), also has decreased activity. Dipeptide conjugates **66** and **67**, were synthesized since norneoenactin and neoenactin already contain a hydroxamate which might be able to participate in iron chelation, potentially obviating the need for the trihydroxamate tripeptide. However, bioassays revealed that while the dipeptide conjugate of neoenactin retained slight activity, that of norneoenactin was nearly devoid of activity. A synthetic spermidine-based catechol-containing siderophore was also coupled to norneoenactin to give **68**. The resulting conjugate was devoid of activity against the three organisms tested to date. However, conjugation of neoenactins to our totally synthetic artificial cyanurate siderophore to give **35d** restored significant activity (MIC<1) against *C. albicans* .[51]

•1/2 H_2SO_4

53, X=CH_2OH, P=Bn
54, X=CH_3, P=Bn
55, X, P=CH_2Ph

56, n=5
57, n=6

58, X=CH_2OH, Y=OH
59,**60**, X,Y=CH_2O (D-,L-)
61, X=CH_3, Y=OH
62, X=CH_2OH, Y=H
63, X=CH_3, Y=H

Compound	MIC, μg/mL *C. albicans*	*C. neoformans*	*A. fumigatus*
58 (n=6) (*S*)	0.625	0.156	20
58 (n=5) (*S*)	<0.5	0.625	ND
(*R*)	>4.25	ND	ND
62 (n=5)	20	2.5	>80
63	80	>80	>80
64 (**32d**) from tripep(**21**n=3) & **58**(n=6)	80	0.19	>80
65 (**32d**) from tripep(**21**n=3) & **58**(n=5)	80	0.32	>80
66 from dipep(**21**n=2) & **58**(n=6)	10	5	>80
67 from dipep(**21**n=2) & **58**(n=5)	80	40	>80
68 (**33d**) from catechol (**33**) & **58**(n=5)	80	80	>80

(ND - not determined)

4 CONCLUSIONS

Although only a limited number of the tremendous variety of possible siderophore-drug conjugates have been prepared and studied, the associated biological studies suggest considerable potential for the development of novel therapeutic agents based on siderophore-mediated drug delivery. To more completely determine the scope and limitations of this mode of drug development, considerable effort is needed on multiple fronts. Determination of the structures of natural siderophores used by pathogens would significantly facilitate the design of appropriate drug carriers. Fermentation processes and biosynthetic studies might provide efficient methods for the production of the quantities of siderophores or components needed for large scale preparation of conjugates. Still, continued attention needs to be given to the synthesis and semisynthesis of siderophores and components to provide important analogs for detailed structure-activity relationship (SAR) studies. Attention should be given to the mode of attachment of drugs to the siderophores by diversifying the linkers and determining if drug release is necessary. Variation of the drugs and incorporation of novel potential drugs also will expand the SAR, and it remains possible that conjugation of toxic drugs will decrease toxicity while still exerting targeted antimicrobial activity. Siderophore conjugation also may allow renewed use of drugs to which resistance has developed by effecting active iron-mediated transport and bypassing resistance mechanisms. Overall, the area has considerable opportunity for focused, yet interdisciplinary research.

5 Acknowledgments

MJM gratefully acknowledges the NIH for support of this research. Significant thanks are due to Eli Lilly and Company for providing the carbacephalosporin, other antimicrobial agents, and general antimicrobial assays. Early studies by E. K. Dolence, J. A. Mckee(Dolence), A. A. Minnick in MJM's laboratory are included in the references, but must be emphasized as being critical for the development of the project related siderophore-mediated drug delivery. The early assistance of A. Brochu and N. Brochu in F. M.'s laboratory with the biological evaluation of the conjugates is sincerely appreciated. Helpful discussions with Dr. Thalia Nicas, Jan Turner, Robert Gordee, Bill Turner (Lilly) and Professor J. B. Neilands (Berkeley) are appreciated. Desferal® was a gift from the late Professor Tom Emery. The patient assistance of Maureen Wickham in preparing the manuscript is sincerely appreciated.

References

§ USDA-ARS; Plant Disease Resistance Research Unit; Department of Plant Pathology; University of Wisconsin; Madison, WI 53706

¥ Rega Institute for Medical Research; Katholieke Universiteit Leuven; Minderbroedersstraat 10; B-3000 Leuven; Belgium

‡ Service d'Infectiologie; Centre Hospitalier de l'Univ. Laval; 2705 boul Laurier, bureau 9500; Quebec, Canada G1V 4G2 (present address: Versicor, 270 East Grand Avenue; So. San Francisco, CA 94080)

∇ Hans-Knoell-Institut fuer Naturstoff-Forschung; Bereich Mikrobiologie; Beutenbergstrasse 11; D-07745 JENA; Germany

1. a) Begley, S. "The End of Antibiotics," *Newsweek*, March 28, 1994, 46. b) Cowley, G. "Too Much of a Good Thing," *Newsweek*, March 28, 1994, 50. c) Adler, J. "The Age Before Miracles," *Newsweek*, March 28, 1994, 52.
2. a) Travis, J. "Reviving the Antibotic Miracle?," *Science,* 1994, **264**, 360. b) Davies, J., "Inactivation of Antibiotics and the Dissemination of Resistance Genes," *Science*, 1994, **264**, 375. c) Nikaido, H., "Prevention of Drug Access to Bacterial Targets: Permeability Barriers and Active Efflux," *Science*, 1994, **264**, 382. d) Spratt, B. G., "Resistance to Antibiotics Mediated by Target Alterations," *Science*, 1994, **264**, 388.
3. Koshland, D. E. "The Biological Warfare of the Future," *Science*, 1994, **264**, 327 (Editorial).

4. Winkelmann, G.; van der Helm, D.; Neilands, J. B., Eds, "*Iron Transport in Microbes, Plants and Animals*;" VCH: Weinheim, FRG; New York, 1987 and references therein.
5. Hider, R. C., "*Siderophore Mediated Absorption of Iron*," 1984, **58**, 25 and references therein.
6. Swinburne, T. R.; Ed. "*Iron, Siderophores, and Plant Diseases*," Nato ASI series; Seris A: Life Sciences Vol 117; Plenum, N.Y. 1986 and references therein.
7. Neilands, J. B., *Bacteriol. Rev.*, 1957, **21**, 101.
8. a) Payne, S. M. "*CRC Critical Reviews in Microbiology*," 1988, 16, 81. b) Bullen, J. J.; Griffiths, E., Eds. "*Iron and Infection*," Wiley: New York, 1987.
9. Neilands, J. B., "*Structure and Bonding*," Springer-Verlag: Berlin, 1984, 58, 1.
10. May, P. M.; Williams, D. R., "*Metal Ions in Biological Systems*," Sigel, H., Ed.; Marcel Dekker: New York, 1978, chapter 2.
11. a) "Tailor-Made Drugs Treat Genetic Blood Disease," *Chem. Eng. News*, 1977, **55**, (18) 24. b) "Iron Chelators May Aid in Anemia Treatment," *Chem. Eng. News*, 1980, **58**, (39), 42-45.
12. Sheppard, L. N.; Kontoghiorghes, G. J., "Synthesis and Metabolism of L1 and Other Novel Alpha-ketohydroxypyridine Iron Chelators and Their Metal Complexes," *Drugs of Today*, 1992, **28** *(Suppl. A)*, 3.
13. Kappel, M. J.; Raymond, K. N., *Inorg. Chem.*, 1982, **21**, 3437.
14. Gutteridge, J. M.; Richmond, R.; Halliwell, B., *Biochem. J.*, 1979, **184**, 469.
15. Miller, M. J.; Malouin, F., "*The Development of Iron Chelators for Clinical Use*" Bergeron, R. J.; Brittenham, G. M., Eds. CRC Press, Boca Raton, 1994, Chapter 13.
16. Halliwell, B., "*The Development of Iron Chelators for Clinical Use,*" Bergeron, R. J.; Brittenham, G. M., Eds. CRC Press, Boca Raton, 1994, Chapter 2.
17. Oheinerg, K. A.; Nguyen, V. N.; Schewender, C. F.; Singer, M.; Steber, M.; Ansell, J.; Hageman, W., *BioMed. Chem.*, 1994, **2**, 187.
18. a) Trowbridge, I. S.; Omary, M. B., *Proc. Natl. Acad. Sci.*, 1981, **78**, 3039. b) Finkel, T.; Cooper, G. M., *Cell*, 1984, **36**, 1115. c) Kameyama, T.; Takahashi, A.; Kurasawa, S.; Ishizuka, M.; Okami, Y.; Takeuchi, T.; Umezawa, H., *J. Antibiotics*, 1987, **40**, 1664. d) Bergeron, R. J.; McManis, J. S., *Tetrahedron,* 1989, **45**, 4939.
19. a) Mastrandea, S.; Carvajal, J. L.; Kaeda, J. S.; Kontoghiorghes, G. ; Luzzatto, L., "Growth Inhibition of *Pasmodium Falciparum* by Orally Active Iron Chelators," *Drugs of Today*, 1992, **28** *(Suppl. A,)*, 25. b) Gordeuk, V. R., "Iron Chelation Therapy for Malaria," Conference on Iron and Microbial Iron Chelators, Brugge, Belgium, November 5-6, 1993. c) Loyevsky, M.; Lytton, S. D.; Mester, B.; Libman, J.; Shanzer, A.; Cabantchik, Z. I., *J. Clinical Investigation*, 1993, **91**, 218.
20. Weinberg, E. D. "Consequences of Altered Iron Metabolism in HIV-Infection; Iron Chelation as Possible Strategy," Conference on Iron and Microbial Iron Chelators, Brugge, Belgium, November 5-6, 1993.
21. Plaha, D. S.; Rogers, H. J., *Biochim. Biophys. Acta*, 1983, **760**, 246.
22. Bergeron, R. J. in Winkelmann, G.; van der Helm, D.; Neilands, J. B., Eds, "*Iron Transport in Microbes, Plants and Animals*"; VCH: Weinheim, FRG; New York, 1987, Chap 16.
23. a) Peterson, T.; Falk, K.-E.; Leong, S. A.; Klein, M. P.; Neilands, J. B., *J. Am. Chem. Soc.*, 1980, **102**, 7715. b) Tait, G. H., *Biochem. J.*, 1975, **146**, 191.
24. a) Watanabe, N.-A.; Nagasu, T.; Katsu, K.; Kitoh, K., *Antimicrob. Agents Chemother.*, 1987, **31**, 497. b) Katsu, K.; Kitoh, K.; Inoue, M.; Mitsuhashi, S., *Antimicrob. Agents Chemother.*, 1982, **22**, 181.
25. a) Benz, G.; Born, L.; Brieden, M.; Grosser, R.; Kurz, J.; Paulsen, H.; Sinnwell, V.; Weber, B., *Liebigs Ann. Chem.*, 1984, 1408. b) Benz, G.;

Schröder, T.; Kurz, J.; Wünsche, C.; Karl, W.; Steffens, G.; Pfitzner, J.; Schmidt, D., *Angew. Chem. Suppl.*, 1982, 1322. c) Benz, G., *Liebigs Ann. Chem.*, 1984, 1399. d) Benz, G.; Schmidt, D., *Liebigs Ann. Chem.*, 1984, 1434. e) Paulsen, H.; Brieden, M.; Benz, G., *Liebigs Ann. Chem.*, 1987, 565.

26. a) Neilands, J. B.; Valenta, J. R., "*Metal Ions in Biological Systems*", Sigel, H., Ed. Marcel Dekker, N. Y.; Vol 19, Chapter 11. b) Rogers H. J. "*Iron Transport in Microbes, Plants and Animals*," Winkelmann, G.; van der Helm, D.; Neilands, J. B., Eds. VCH: Weinheim, FRG; New York, 1987, Chapter 13. c) Bickel, H.; Mertens, P.; Prelog, V.; Seibl, J.; Walser, A., *Tetrahedron*, 1966, Suppl. **8**, 171.
27. Braun, V.; Günthner, K.; Hantke, K.; Zimmermann, L., *J. Bacteriol.*, 1983, **156**, 308.
28. a) Miller, M. J.; Malouin, F., *Accts. Chem. Res.*, 1993, **26**, 241. b) Miller, M. J.; Malouin, F.; Dolence, E. K.; Gasparski, C. M.; Ghosh, M.; Guzzo, P. R.; Lotz, B. T.; McKee, J. A.; Minnick, A. A.; Teng, M., "*Recent Advances in the Chemistry of Antiinfective Agents,*" Bently, P. H.; Ponsford, R., Eds. Royal Society of Chemistry, Special Publication No. 119, Royal Society of Chemistry, Cambridge, 1993, Chap 10.
29. a) Dolence, E. K.; Minnick, A. A.; Miller, M. J., *J. Med. Chem.*, 1990, **33**, 464. (b) Dolence, E. K.; Minnick, A. A.; Lin, C.-E.; Miller, M. J., *J. Med. Chem.*, 1991, **34**, 956. (c) McKee, J. A.; Sharma, S. K.; Miller, M. J., *Bioconjugate Chem.*, 1991, **2**, 281. (d) Miller, M. J., *Chem. Rev.*, 1989, **89**, 1563.
30. (a) Dolence, E. K.; Minnick, A. A.; Lin, C.-A.; Miller, M. J., *J. Med. Chem.*, 1991, **34**, 968. (b) Minnick, A. A.; McKee, J. A.; Dolence, E. K.; Miller, M. J., *Antimicrob. Agents Chemother.*, 1992, **36**, 840. (c) Brochu, A; Brochu, N.; Nicas, T. I.; Parr, T. R., Jr.; Miller, M. J.; Malouin, F., *Antimicrob. Agents Chemother.*, 1992, **36**, 2166.
31. (a) Lee, B. H.; Miller, M. J., *J. Org. Chem.*, 1983, **48**, 24. (b) Milewska, M. J.; Chimiak, A.; Glowacki, Z., *J. Prakt. Chem.*, 1987, **329**, 447.
32. Ghosh, A.; Miller, M. J., *J. Org. Chem.*, 1993, **58**, 7652.
33. (a) Jones, R. N.; Barry, A. L., *J. Antimicrob. Ther.*, 1988, **22**, 315. (b) Jorgensen, J. H.; Redding, J. S.; Maher, L. A., *Antimicrob. Agents Chemother.*, 1988, **32**, 1477. (c) Howard, A. J.; Dunkin, K. T., *J. Antimicrob. Ther.*, 1988, **22**, 445. (d) Bodurow, C. C., *et. al.* & Vladuchick, W. C., *Tetrahedron Lett.*, 1989, **30**, 2321.
34. Kirst, H. A.; Wild, G. M.; Baltz, R. H.; Hammil, R. L.; Ott, J. L.; Counter, F. T.; Ose, E. E. *J. Antibiot.* 1982, **35**, 1675.
35. A gift from Eli Lilly and Company, Indianapolis, IN.
36. Moellmann, U.; Reissbrodt, R.; Miller, M. J. Detailed results to be published separately.
37. Ghosh, M.; Miller, M. J., *Bioorg. Med. Chem.*, 1996, **4**, 43.
38. Nagarajan, R.; Schabel, A. A.; Occolowicz, J. L.; Countea, F. T.; Ott, J. L., *J. Antibiotics*, 1988, **41**, 1430.
39. Roberts, G. D.; Carr, S. A.; Rottschaefer, S.; Jeffs, P. W., *J. Antibiotics,* 1985, **38**, 713.
40. Holzberg, M.; Artis, W. M., *Infect. Immun.*, 1983, **40**, 1134.
41. Ismail, A.; Bedell, G. W.; Lupan, D. M., *Biochem. Biophys. Res. Commun.*, 1985, **130**, 885.
42. Ismail, A.; Lupan, D. M., *Mycopathologia*, 1986, **96**, 109.
43. MacDonald, J. C., *Can. J. Chem.*, 1963, **41**, 165.
44. Holzberg, M.; Antis, W. M., *Infection Immun.*, 1983, **40**, 1134.
45. Arnow, L. E., *J. Biol. Chem.*, 1937, **118**, 531.
46. Young, I. G.; Cox, G. B.; Gibson, F., *Biochim. Biophys. Acta,* 1967, **141**, 329.

47. Emery, T.; Neilands, J. B., *J. Am. Chem. Soc.*, 1960, **82**, 4903.
47. Minnick, A. A.; Eizember, L. E.; McKee, J. A.; Dolence, E. K.; Miller, M. J., *Anal. Biochem.*, 1991, **194**, 223.
49. Lee, K. L.; Buckley, H. R.; Campell, C. C., *Sabouraudia,* 1975, **13**, 148.
50. a) Leong, S. A.; Mei, B.; Voisard, C.; Budde, A.; McEvoy, J.; Wang, J.; Xu, P.; Saville, B., "*The Development of Iron Chelators for Clinical Use,*" Bergeron, R. J.; Brittenham, G. M., Eds. CRC Press, Boca Raton, 1994, Chapter 10. b) Thariath, A. M.; Valvano, M. A.; Viswanatha, T., "*The Development of Iron Chelators for Clinical Use,*" Bergeron, R. J.; Brittenham, G. M., Eds. CRC Press, Boca Raton, 1994, Chapter 19.
51. Hu, J.; Miller, M. J., *J. Org. Chem.*, 1994, **59**, 4858.
52. McKee, J. A., Ph.D. Thesis, University of Notre Dame, 1991.
53. Ghosh, M.; Miller, M. J., *Bioorg. Med. Chem.*, 1995, **3**, 1519.
54. De Clercq, E., *Nucleosides and Nucleotides*, 1994, **13**, 1271 and references therein.
55. Ritter, A. R.; Miller, M. J., *J. Org. Chem.*, 1994, **59**, 4602.
56. Full synthetic details and biological studies will be described in a separate publication.
57. Okada, H.; Yamamoto, K.: Tsutano, S.; Inouye,Y.; Nakamura, S.; Furukawa, J., *J. Antibiotics,* 1989, **42**, 276.
58. Mandala, S. M.; Frommer, B. R.; Thornton, R. A.; Kurtz, M. B.; Young, N. M.; Cabello, M. A.; Genilloud, O.; Liesch, J. M.; Smith, J. L.; Horn, W. S., *J. Antibiotics*, 1994, **47**, 376.
59. a) VanMiddlesworth, F.; Giacobbe, R. A.; Lopez, M.; Garrity, G.; Bland, J. A.; Bartizal, K.; Fromtling, R. A.; Polishook, J.; Zweerink, M.; Edison, A. M.; Rozdilsky, W.; Wilson, K. E.; Monaghan, R. L., *J. Antibiotics,* 1992, **45**, 861. b) Horn, W. S.; Smith, J. L.; Bills, G. F.; Raghoobar, S. L.; Helms, G. L.; Kurtz, M. B.; Marrinan, J. A.; Frommer, B. R.; Thornton, R. A.; Mandala, S. M., *J. Antibiotics,* 1992, **45**, 1692.
60. a) Darwish, I. S.; Patel, C., *J. Org. Chem.,* 1994, **59**, 451.
61. Niu, C.; Sun, S.; Miller, M. J. to be submitted.

Advances in Antifungals

10

The Chemistry of Azole Antifungals: Fluconazole and Beyond

S. D. A. Street

DEPARTMENT OF DISCOVERY CHEMISTRY, PFIZER CENTRAL RESEARCH, SANDWICH, KENT CT13 9NJ, UK

In the first International symposium on the Recent Advances in the Chemistry of Anti-Infective Agents, held in 1992, Dr Ken Richardson of the Discovery Chemistry Department at Pfizer talked about the discovery and development of fluconazole[1]. This paper will continue the theme of antifungal research at Pfizer and will discuss: The Chemistry of Azole Antifungals, Fluconazole and Beyond.

This paper includes a brief introduction to fungi and fungal infections in man, indicating why the search for new and improved agents for the treatment of such infections is a focus of effort at Pfizer. The profile of fluconazole is reviewed, highlighting the advantages, and leading to a definition of the target profile for the next generation antifungal agent. The main section of the presentation describes the work, starting from fluconazole, that ultimately led to the identification of voriconazole, UK-109,496. Finally, some thoughts on the directions in which antifungal chemotherapy will move over the next few years are presented.

Fungal infections in man cover a wide spectrum, both in terms of severity and incidence. Thus, at the lower end, superficial infections such as athlete's foot are very common, but in most cases are seldom more than a minor problem frequently treated with 'over the counter' remedies. At the other end of the spectrum, infections by the opportunistic fungal pathogens, such as *Candida*, *Aspergillus* and *Cryptococcus,* especially in the immuno-compromised host, lead to significant morbidity and can be life-threatening. Moreover, in the last few years the increase in the number of immune compromised patients (AIDS patients, transplant recipients and patients undergoing cancer chemotherapy) has led to a dramatic increase in the incidence of serious fungal infections. At Pfizer, interest is in the identification of new agents to treat serious systemic fungal infections caused by these pathogens.

Against this background, a programme was started at Pfizer a number of years ago looking to discover an orally bioavailable azole antifungal. This was the work that ultimately led to the discovery and development of fluconazole. Fluconazole is now the drug of choice for treating infections caused by *Candida* and *Cryptococcus* in man[2,3]. As shown in **Figure** 1, fluconazole has good activity *in vitro* (MIC is defined as the minimum concentration of drug required to inhibit the replication of the pathogen *in vitro)* against *Candida albicans* and *Cryptococcus*. When combined with its excellent oral bioavailability and pharmacokinetics, this translates to very good clinical efficacy against *C. albicans* and *Cryptococcus*.

fluconazole

Pathogen	*in vitro* MIC (μg/ml)
C. albicans	1.0
Aspergillus	>100
Cryptococcus	9.6

Figure 1, *Structure of Fluconazole and in vitro MIC Data*

Since its introduction in 1988, well over 50 million patients have been treated with fluconazole world-wide. This not only reflects fluconazole's pharmacokinetic profile, but also its excellent safety and toleration. These properties also enable fluconazole to be used for suppression of relapse of cryptococcal meningitis in patients with AIDS. Moreover, the excellent pharmacokinetics, high water solubility and good stability allow for easy formulation for i.v. delivery. This route of administration is particularly important in patients with life threatening infections.

The mechanism of action of fluconazole has been extensively investigated and is well understood. Like other azoles, fluconazole prevents the conversion of lanosterol to ergosterol through inhibition of the fungal cytochrome P450 enzyme, lanosterol 14α–demethylase. Fungi have a critical dependence on a viable cell membrane and sterols, such as ergosterol, are the major constituents of this key organelle. Thus, by inhibiting lanosterol 14α–demethylase, fluconazole decreases the levels of ergosterol in the fungal cell membrane, ultimately stopping fungal growth.

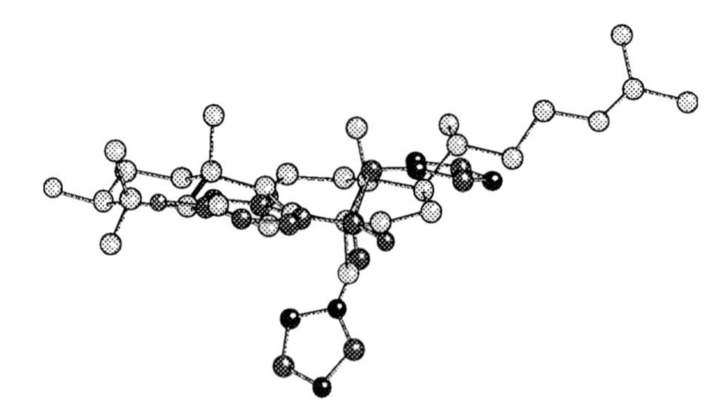

Figure 2, *Computer Generated Overlap of Lanosterol with Fluconazole (Dark Shaded)*

The lack of any X-ray data on lanosterol 14α–demethylase precludes detailed study of how fluconazole binds to, and inhibits its target enzyme. However, we have generated a model (**Figure 2**) by considering the single crystal X-ray structure of fluconazole in comparison to that of the naturally occurring substrate, lanosterol. The key aspects of the proposed overlap are that the 2,4-difluoro-phenyl group overlaps with the sterol B ring, one of the triazole groups stretches out along the sterol side-chain, and the second triazole ring occupies the same region of space as does the 14-α–methyl group of the lanosterol. This overlap allows one of the triazole rings of fluconazole to bind directly to the haem present in the cytochrome P450, thus preventing the necessary molecule of oxygen from binding to, and being activated by, the haem iron. Hence the enzyme is prevented from carrying out its normal function.

It is important to consider the overall spectrum of activity of fluconazole (**Table 1**) in order to define the profile required in a follow-on agent. As already discussed, fluconazole has very good activity against *Candida albicans*, the most common fungal pathogen in humans. In addition, its very good efficacy profile against *Cryptococcus* may be contributing to a gradual decrease in the incidence of cryptococcosis. However, fluconazole appears to be less active against the two emerging *Candida* species, *C. glabrata* and *C. krusei*. Infection with *Aspergillus*, although not common, is frequently life-threatening, and fluconazole is not used in the treatment of *Aspergillus*.

Table 1, *Activity Profile of Fluconazole against Main Fungal Pathogens in vitro*

Pathogen	Activity	Status
C. albicans	+	Common
C. glabrata	±	Emerging
C. krusei	±	Emerging
Aspergillus	-	Static
Cryptococcus	+	Decreasing

Therefore, the next generation of azole antifungals should feature a broader-spectrum of antifungal activity including *C. glabrata*, *C. krusei* and *Aspergillus spp*. In addition, a good safety profile combined with the ready availability of oral and i.v. formulations is very important. With the target profile defined, our strategy was to use fluconazole as a starting point, and to focus our effort on improving activity against *Aspergillus*. All the available data suggested to us that as activity against *Aspergillus* improves, so will that against the other fungi. Thus, we set ourselves the laboratory objective of an *Aspergillus* MIC <1μg/ml, based on our experience that reaching this level of *in vitro* activity provided the best chance of seeing good *in vivo* activity.

Our efforts started with an analysis of all the analogues of fluconazole that were available in our files. One of the first compounds that attracted our interest was the α–methyl derivative (UK-54,315), which was prepared using a similar route (**Scheme 1**) to that developed for fluconazole. Friedel-Crafts acylation of difluorobenzene (1), using either chloroacetyl chloride or 2-chloropropionyl chloride, provided the intermediate α–chloroketone derivatives, (3) R = H, and (4) R = Me. In turn (4) was reacted with sodium triazole to give the very useful intermediate triazole-ketone (6). Activation of (6) using the Corey ylide (7) yielded the epoxide (9), which, on treating with sodium triazole, readily opened to give the targets. In the case of the α–methyl derivative (11), the product was isolated as a mixture of four diastereoisomers that were not separated.

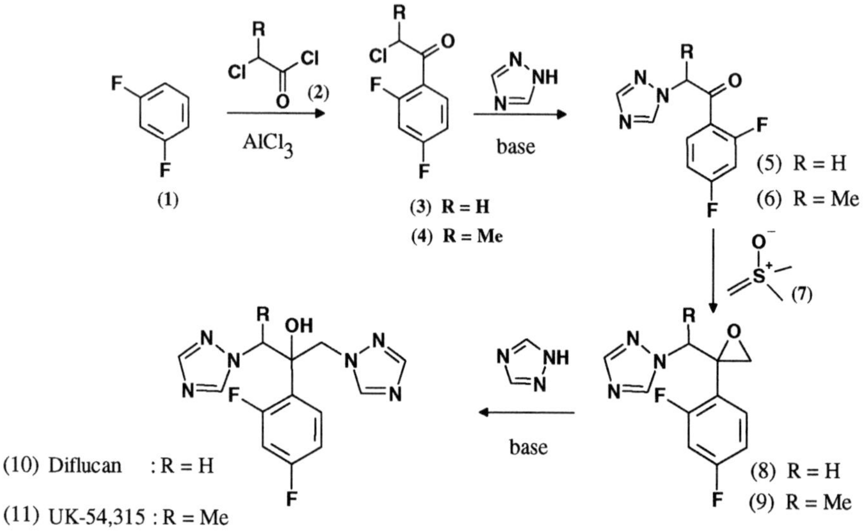

(1)

(2)

$AlCl_3$

(3) R = H

(4) R = Me

base

(5) R = H

(6) R = Me

(7)

(8) R = H

(9) R = Me

base

(10) Diflucan : R = H

(11) UK-54,315 : R = Me

Scheme 1

As shown in **Figure 3**, introduction of the α–methyl group improved *Aspergillus* activity. Although this level of activity is not outstanding, it is a marked improvement to that of fluconazole, which is essentially inactive against *Aspergillus* (MIC > 100µg/ml). Subsequent work in our laboratories confirmed that the α–methyl compound had indeed achieved the improved activity against *Aspergillus* through increased inhibition of ergosterol biosynthesis. In addition, we confirmed that as the activity improved against *Aspergillus*, so it did against the other pathogenic fungi. Thus, for the sake of clarity, only the *Aspergillus* activity data is described for the remainder of the paper. Unfortunately, although the *in vitro Aspergillus* potency was very encouraging, the α–methyl compound did not show good activity against *Aspergillus in vivo*, and was poorly tolerated at the necessary dose levels.

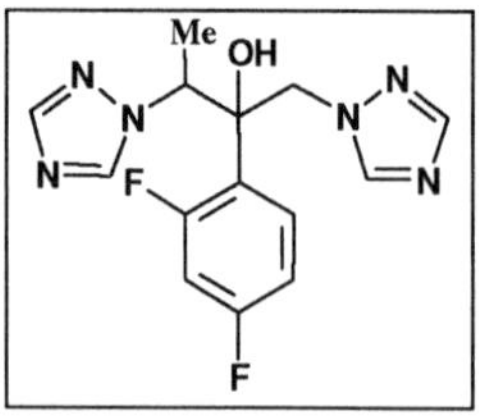

***In vitro Aspergillus* Activity**

α-methyl	**MIC 12.5µg/ml**
fluconazole	**MIC >100µg/ml**

Figure 3, *Structure of α-Methyl-diflucan and in vitro Aspergillus Activity Data*

Despite this disappointing *in vivo* profile, we continued to examine close analogues of fluconazole, and turned our attention to replacements for one of the triazole heterocycles. The proposed mechanism of inhibition led us to assume that it is important to keep at least one of the triazoles in order to bind effectively to Cytochrome P450. The 4-pyridyl analogue (13), shown in **Scheme 2**, was one of a series of pyridyl derivatives prepared by straightforward condensation of 4-methylpyridine (12) with triazole-ketone (5) described in **Scheme 1**.

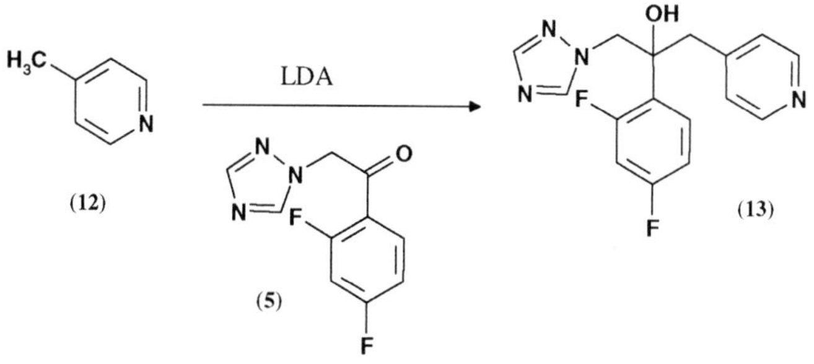

LDA

(12) (5) (13)

Scheme 2

We were delighted to see that replacement of one of the triazoles with pyridine also increased potency *in vitro* against *Aspergillus*, with MIC improving to 3.15μg/ml. Unfortunately, although this compound was better tolerated, we again observed no efficacy *in vivo*. Our hypothesis to explain this result was that we still had not reached our target *Aspergillus* MIC of <1μg/ml.

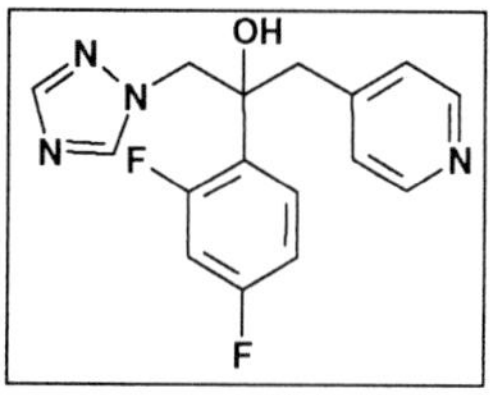

In vitro Aspergillus **Activity**	
4-pyridyl	**MIC 3.15μg/ml**
fluconazole	**MIC >100μg/ml**

Figure 4, *Structure of 4-Pyridyl Analogue (13) and in vitro Aspergillus Activity Data*

At this stage we decided to investigate a series of compounds incorporating both an α–methyl and a triazole replacement. The synthesis of one of the first targets, (±)-UK-82,137 (**Scheme 3**) used essentially the same chemistry as before, using 4-ethylpyridine (14), but with α–chloro-ketone (3). Depending on the precise reaction conditions in this sequence, we were able to isolate either chloro-alchohol (15) or epoxide formed *in situ*, by simple ring closure. In either case, reaction with sodium triazole gave us the target compound (16).

Once again, the product had two chiral centres and was isolated as a mixture of all four diastereoisomers. However, unlike the bis-triazole system, it proved possible to separate these diastereoisomers readily into *cis-* and *trans-* pairs by straightforward flash chromatography. (±)-UK 82,137, and all future compounds, are shown with *trans*-orientation of the methyl and hydroxyl groups, on the assumption that this is the active configuration. X-ray data will be described later that confirm this structure, but all the data shown, unless otherwise stated, are on racemic mixtures.

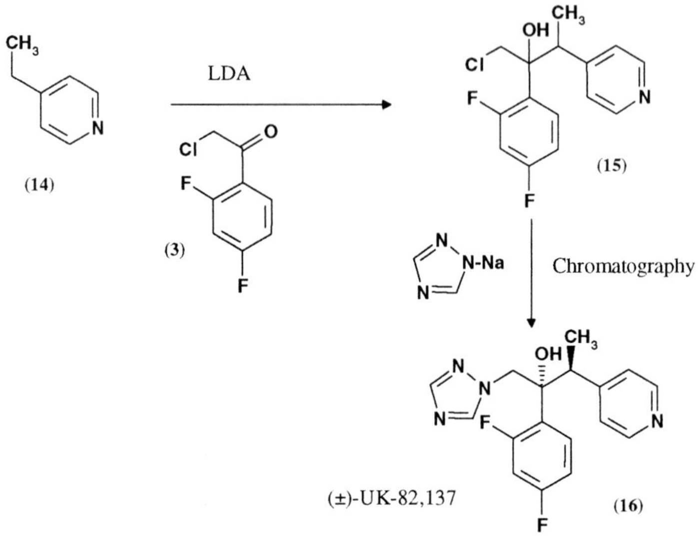

LDA

(14) (3) (15)

Chromatography

(±)-UK-82,137 (16)

Scheme 3

As shown in **Figure 5**, (±)-UK 82,137 was the first compound to reach our *in vitro* target for *Aspergillus* activity. In addition, we were intrigued to note that (±)-UK 82,137 also showed a fungicidal mode of action against *Aspergillus in vitro.*

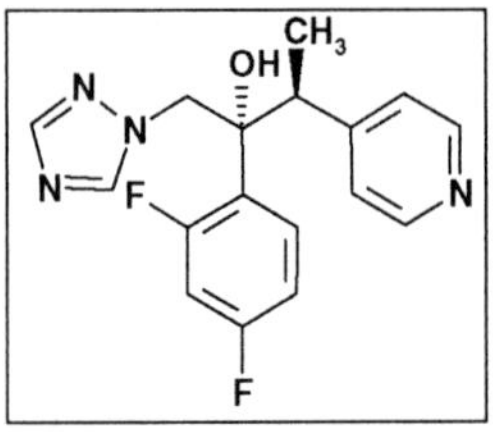

***Aspergillus* Activity**

In vitro	MIC 0.19μg/ml
Mouse	4/5 survivors
*in vivo**	0/5 cures

* 20mg/kg i.v. bid for 5 days

Figure 5, *Structure of (±)-UK 82,137 and Aspergillus Activity Data*

In vivo profiling was carried out in a mouse lethality model of systemic aspergillosis. Mice are infected with *Aspergillus* and are subsequently treated under a standard regimen of 20mg/kg i.v. for five days. There are two distinct endpoints, firstly the number of mice that survive, and secondly the number of animals cured, *i.e. Aspergillus* infection completely cleared. The cure rate indicates the level of fungicidality achieved *in vivo.*

Although (±)-UK 82,137 showed good *in vivo* activity in terms of survival, there was a disappointing lack of cures. Clearly, we expected a compound with this level of *in vitro* potency to show better *in vivo* activity.

At this stage (±)-UK 82,137 was examined in detail by our Drug Metabolism Department who observed that the compound was being rapidly metabolised to the pyridine N-oxide *in vivo*. Consequently, our main objective became to decrease the propensity of the (±)-UK-82,137 system to N-oxidation. One of the hypotheses considered was that this N-oxidation was related to heterocycle basicity, and thus we decided to investigate a series of compounds where the pyridine is replaced by less basic heterocycles such as a pyrimidine.

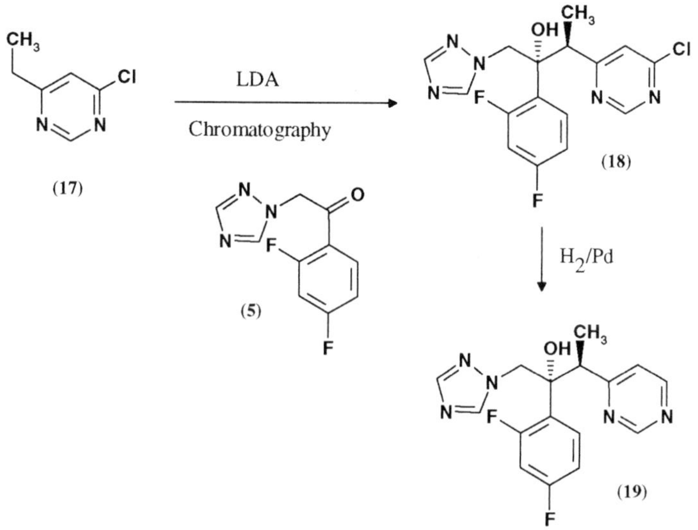

LDA

Chromatography

(17) (5) (18)

H_2/Pd

(19)

Scheme 4

The synthesis of (±)-UK-83,610 (19) (**Scheme 4**), the pyrimidine analogue of (±)-UK-82,137, uses chloro-substituted pyrimidine (17) which reacts under standard LDA conditions with triazole-ketone intermediate (5). Chromatography followed by removal of the chlorine *via* hydrogenation yields the product (19) as a racemic mixture.

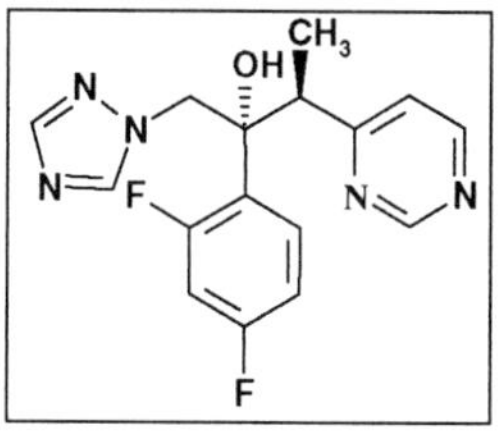

***Aspergillus* Activity**

In vitro	MIC 0.39μg/ml
Mouse	3/5 survivors
*in vivo**	0/5 cures

* 20mg/kg i.v. bid for 5 days

Figure 6, *Structure of (±)-UK-83,610 and Aspergillus Activity Data*

(±)-UK-83,610 shows comparable activity to the 4-pyridyl analogue *in vitro*, but again shows a disappointing cure rate *in vivo*, **Figure 6**. When the pharmacokinetics of (±)-UK-83,610 were studied in detail, they confirmed that N-oxidation was not taking place, rather, the major metabolite *in vivo* was ketone (5), presumably produced *in vivo* by a retro-aldol reaction. At the same time, detailed examination of the (±)-UK-83,610 synthesis showed that retro-aldol reaction, already identified as the major problem *in vivo,* was also a significant problem *in vitro.* Clearly we required a more stable target and our hypothesis was that the propensity for retro-aldol was also related to heterocycle basicity with pyrimidine protonation a likely first step in the retro-aldol degradation. Decreasing the pyrimidine basicity should decrease the likelihood of retro-aldol. Thus, our immediate targets became halo-substituted pyrimidines.

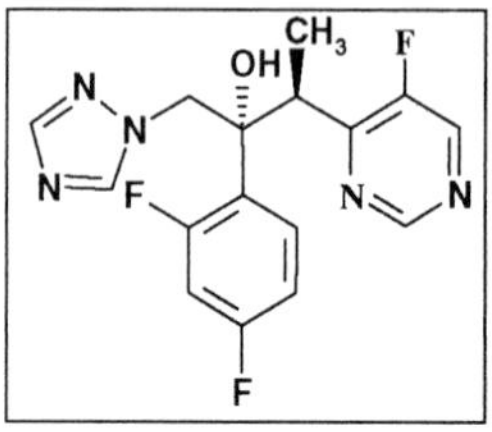

***Aspergillus* Activity**

In vitro	MIC 0.39μg/ml
Mouse	5/5 survivors
*in vivo**	4/5 cures

* 20mg/kg i.v. bid for 5 days

Figure 7, *Structure of (±)-UK-103,449 and Aspergillus Activity Data*

One of the first compounds prepared was the fluoro-substituted derivative, (±)-UK-103,449, which maintained excellent *in vitro Aspergillus* activity (**Figure 7**) To our delight, when the compound was studied in a mouse lethality model, not only did we see 5/5 survivors but, also very good cure rates indicating that excellent *in vitro* potency had translated to outstanding *in vivo* efficacy. Pharmacokinetic data on (±)-UK-103,449 subsequently confirmed that this improved *in vitro* to *in vivo* translation was, at least in part, due to the improved stability profile.

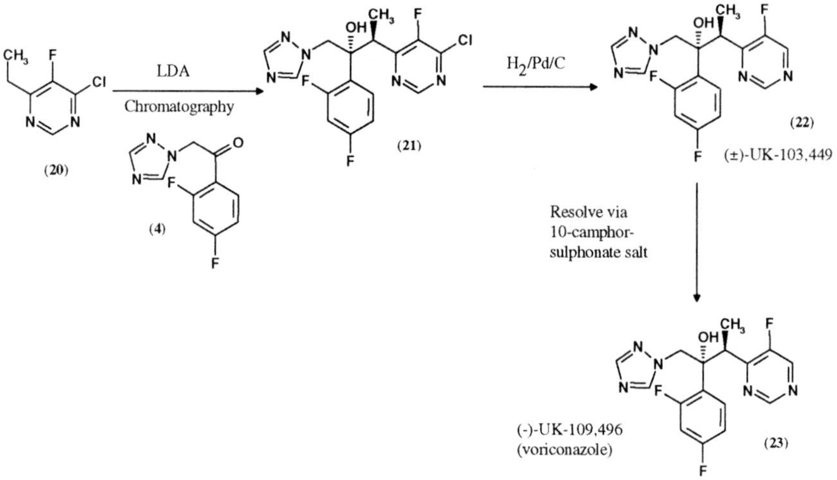

(20) LDA Chromatography (4) (21) H_2/Pd/C (22) (±)-UK-103,449 Resolve via 10-camphor-sulphonate salt (-)-UK-109,496 (voriconazole) (23)

Scheme 5

Having achieved this outstanding activity with the racemic mixture, we were very keen to investigate the profile of the individual enantiomers. As shown in **Scheme 5**, (±)-UK-103,449 was prepared using a comparable route to (±)-UK-83,610, the unsubstituted pyrimidine, and after some experimentation, (±)-UK-103,449 was eventually resolved by classical resolution using camphor sulphonic acid. Moreover, we were able to use the highly crystalline salt to obtain a single crystal X-ray of the active single enantiomer, UK-109,496, since named voriconazole, and confirm the assignment of its structure.

The X-ray[4] is shown in **Figure 8**, and it clearly confirms the *trans*-orientation of the two key substituents on the back-bone. Also shown in **Figure 8** is the computer generated overlap of voriconazole with lanosterol.

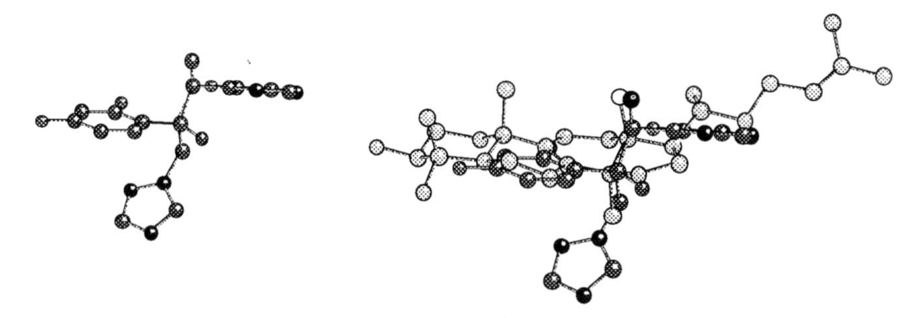

Figure 8, *(Left) Single Crystal X-Ray Structure of Voriconazole and (Right) Computer Generated Overlap of Lanosterol (Grey Shaded) and Voriconazole (Dark Shaded)*

There is a great deal of similarity to the earlier overlap of fluconazole with lanosterol. The triazole is positioned to be able to bind to the haem, the 2,4-difluoro-phenyl ring overlaps with the A and B rings of the steroid, and the pyrimidine points out along the sterol side-chain. The most significant observation is the overlap shown between the α–methyl introduced into the voriconazole system (filled black sphere) with one of the methyl residues in lanosterol (open sphere). We believe that the improved activity shown by voriconazole is due to additional binding to the target enzyme achieved through the interaction of this methyl group with the active site.

The activity profile on voriconazole, the active single enantiomer derived from (±)-UK-103,449, is summarised in **Figure 9**, and confirms that voriconazole has excellent *in vitro* activity against a wide range of fungal pathogens.

Pathogen	MIC (μg/ml)
C. albicans	0.03
C. glabrata	0.19
C. krusei	0.24
Aspergillus	0.09
Cryptococcus	0.39

Figure 9, *Structure of Voriconazole and in vitro MIC Data*

In pre-clinical studies against *Candida*, *Cryptococcus* and *Aspergillus*, voriconazole showed very good oral efficacy in animals. Thus, in models of systemic infection with either *C. albicans*, *C. krusei*, or *C. glabrata*, the fungal load in the target organ is significantly reduced (≥ fluconazole). Voriconazole is curative in animal models of aspergillosis, a profile consistent with the compound showing a fungicidal mode of action *in vitro* against *Aspergillus*. Finally, voriconazole is highly efficacious *in vivo* against *Cryptococcus* infections resulting in a significant reduction in fungal tissue load, whether the infection was introduced into the lung or brain.

In Phase I studies, voriconazole was well tolerated and achieved sustained blood and tissue levels after oral and i.v. dosing. Consistent with these data, Phase II trials have shown voriconazole to have fluconazole-like activity against some candidoses, and exciting efficacy against life-threatening aspergilloses. Voriconazole is currently in Phase III trials against a wide range of life-threatening fungal infections, in comparison with standard agents, including fluconazole.

The future direction of antifungal research is a subject that we have considered at length at Pfizer in an effort to identify where the next major breakthrough may occur. We were very interested in, and very excited by, the fungicidal mode of action that voriconazole shows against *Aspergillus* both *in vitro* and *in vivo*. These observations have helped lead us to the conclusion that the target profile for the next generation of compounds, after voriconazole, would be a broad-spectrum fungicidal agent, *i.e. Aspergillus*, *Candida* and *Cryptococcus*. Such a compound should lead to shorter treatment times and should produce cures *in vivo* irrespective of the immune status of the patient being treated. Taken together, both of these attributes should result in a reduced risk of the development of resistance to specific agents, or relapse when therapy is withdrawn. At Pfizer, we are convinced that novel target mechanisms will be required to identify a compound with this profile, and therefore, we have entered into a collaboration with Chemgenics, based in Boston, USA. In this collaboration we are using molecular techniques to identify novel fungicidal targets and screens that we believe will ultimately lead to novel mechanisms and inhibitors.

In conclusion, fluconazole has an outstanding profile and is clearly a breakthrough drug in the treatment of fungal infections. We have identified voriconazole as the next generation azole antifungal agent and this compound is currently being developed as therapy for life-threatening fungal infections such as aspergilloses. Voriconazole is active against all key target fungi, oral and i.v. formulations are available, and the compound is performing well in the clinic.

It is impossible to name everybody who has worked on the antifungal project since the discovery of fluconazole. However, I would like to acknowledge Roger Dickinson and Ken Richardson - two of the principal chemists who worked in the programme which led to the discovery of voriconazole[5]. In Biology, Chris Hitchcock and Peter Troke were in charge of the laboratory that did all the pre-clinical evaluation of the compounds described. I would also like to acknowledge the valuable contribution made to the project by Serge Jezequel and Dennis Smith in Drug Metabolism.

References

[1] K. Richardson, 'Recent Advances in the Chemistry of Anti-Infective Agents', P.H. Bentley and R. Ponsford, The Royal Society of Chemistry, Cambridge, 1992, Chapter 12, p. 182.

[2] N. C. Karyotakis and E. J. Anaissie, *Curr. Opin. Inf. Dis.*, 1994, **7**, 658.

[3] M. Zervos and F. Meunier, *Int. J. Antimicrob. Agents*, 1993, **3**, 147.

[4] Detailed X-ray crystallographic data for voriconazole have been deposited at the Cambridge Crystallographic Data Centre.

[5] R. P. Dickinson, A. S. Bell, C. A. Hitchcock, S. Narayanaswami, S. J. Ray, K. Richardson and P. F. Troke, *Biorg. Chem. Med. Lett.*, **1996**, *16*, 2031.

11
Tubulins: A Target for Antifungal Agents

D. W. Holloman,[1,*] J. A. Butters[1] and H. Barker[2]

[1]IACR-LONG ASHTON RESEARCH STATION, DEPARTMENT OF AGRICULTURAL SCIENCES, UNIVERSITY OF BRISTOL, LONG ASHTON, BRISTOL BS18 9AF, UK

[2]DEPARTMENT OF BIOCHEMISTRY, SCHOOL OF MEDICAL SCIENCES, UNIVERSITY OF BRISTOL, UNIVERSITY WALK, BRISTOL BS8 1TD, UK

INTRODUCTION

Microtubules are a major component of all eukaryotic cells. Since the first description of microtubules some 35 years ago, studies involving biochemistry, microscopy, immunology, genetics and molecular biology have created a wealth of information about their structure and function. The current status of this research was summarised recently in "*Microtubules*" (Hyams and Lloyd, 1994), which reveals the key role microtubules play in controlling vital cell functions including shape, mobility, division and the "trafficking" of organelles within cells. It is not surprising then, that disruption of tubulin proteins, which form the core of microtubules, has been successfully exploited as a target for antifungal agents. In this paper we explore the interaction of existing antifungal agents with tubulin, and expose some of the limitations of current understanding about tubulin structure. Much of this information relates to one of the best studied examples, the benzimidazoles, which are used in both agriculture and veterinary medicine to control fungal diseases and parasitic flatworms.

Tubulin structure and function

α, β and γ tubulins comprise a distinct protein family which share around 36-42% amino acid similarity with each other, but have little homology with other microtubule proteins. Two of these tubulins, α and β, are condensed non-covalently as heterodimers into tubulin elements (M_r 100,000) which form the core of microtubules. Although tubulin proteins have the propensity to self-aggregate, microtubule assembly requires the binding of 2 molecules of Guanosine Triphosphate (GTP) to each tubulin heterodimer. One GTP molecule binds to α tubulin; the other binds to an 'exchangeable' site on β-tubulin and, unlike α-tubulin GTP, is hydrolysed. Microtubules are dynamic structures, and the balance between GTP dependent assembly of the heterodimers and GTP hydrolysis determines whether microtubules grow or not. But β-tubulin is an atypical GTP-ase (Sage *et al.*, 1995), and few amino acid motifs shown to be involved in GTP binding in other GTPases occur in β-tubulin. Nevertheless, ultraviolet cross linking of GTP to β-tubulin, and antibody inhibition of GTP binding, has defined regions which interact with GTP, and are involved in the release of phosphate (Figure 1). One of these regions at amino acid codons 180→186 contains the longest stretch of identical amino acids (Val-Val-Glu-Pro-Tyr-Asn) found in both α and β subunits, and which has been implicated in the binding of the ribose moeity of GTP (Mandelkow *et al.*, 1986). A little closer to the N terminal end, around amino acid 144 on both α and β subunits, is a cluster of glycines (Gly-Gly-Gly-Thr-Gly) which are also involved in nucleotide binding (Sternberg and Taylor, 1984).

Figure 1. **Structure of the β-tubulin gene: GTP binding sites and benzimidazole resistance.**

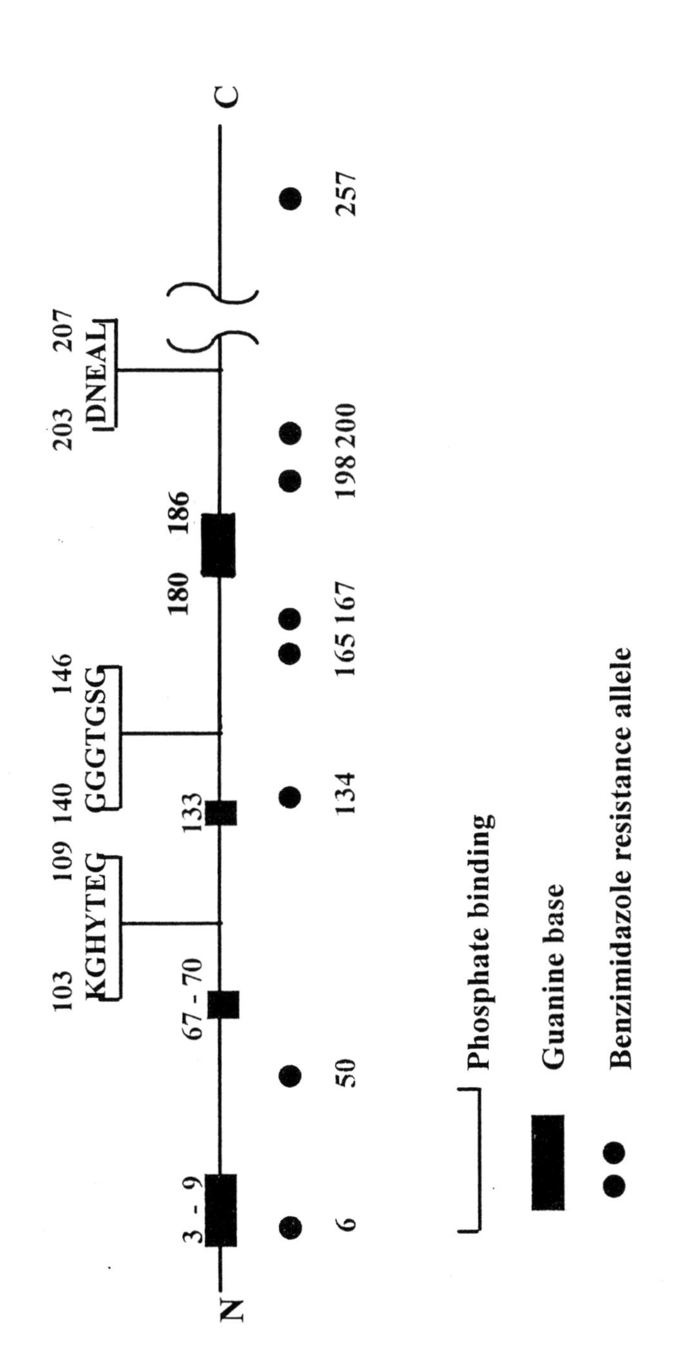

γ tubulin is the least abundant of the three tubulin proteins, and is localized in the microtubule organising centre (centrosome). This plays a part in initiating the formation of microtubules and establishing their polarity, probably through interaction with the β-tubulin subunit of the heterodimer (Oakley, 1994). Disruption of the γ-tubulin gene in *Aspergillus nidulans* blocks nuclear division showing that the protein is indeed essential for viability. Mutations in either α or β-tubulin can also affect growth and viability whilst overproduction of β-tubulin alone can be toxic to cells (Cooley, personal communication, 1992).

Tubulin proteins also interact with Microtubule Accessory Proteins (MAPs) forming a complex which regulates the dynamic properties of microtubules, and at the same time acts as links with cell organelles allowing them to be transported along "tracks" provided by the microtubule (Figure 2). Much of the information about these MAPs, however, is derived from studies on mammalian microtubules, and the role of similar proteins interacting with tubulin in filamentous fungi remains uncertain.

Over the last 10 years some 200 sequences of α, β, and γ-tubulin genes have been published, revealing a family of closely related tubulin proteins. Significant heterogeneity only exists in the C-terminal peptide which appears to provide the binding site for most MAPs (Little and Seehaus, 1988). When these highly divergent acidic carboxy-termini are excluded from sequence alignments, amino acid homologies between α and β-tubulin increase to 63% for α and β, to 51% for α and γ and 59% for β and γ (Burns, 1991), indicating that the three tubulin proteins have very similar tertiary structures. Homogeneity is even greater within each tubulin subunit, and amino acid identities in excess of 95% are not uncommon at least for β-tubulin sequences (Table 1).

Table 1 Fungal β-tubulin: sequence comparisons with *Neurospora crassa*

Fungi		Amino acid identity %	Amino acid codons 196-201
Neurospora crassa		100	SDETFC
Venturia inaequalis		97.5	SDETFC
Erysiphe graminis		96.6	SDETFC
Septoria nodorum		96.8	SDETFC
Rhynchosporium secalis		95.5	SDETFC
Aspergillus nidulans	Ben A	93.3	SDETFC
	Tub C	83.4	SDETFC
Saccharomyces cerevisiae		75.2	SDETFC

From Butters *et al.*, 1995

Table 2 Benzimidazole fungicides

Common name	Chemical name	Manufacturer trade name	Structure
Benomyl	Methyl 1-(butylcarbamoyl)-2-benzimidazole-carbamate	Du Pont Benlate® Chinoin Fundazol®	CO-NH-C_4H_9; NH-$COOCH_3$
Carbendazim	Methyl benzimidazole-2-yl carbamate (MBC)	Du Pont Delsene® BASF Bavistin® Hoechst Derosal®	NH-$COOCH_3$
Fuberidazole	2-(-2′-furyl)-1H-benzimidazole	Bayer Neo-Voronit®	
Thiabendazole	2-(4′-thiazolyl)-benzimidazole	Merck Mertect®	
Thiophanatemethyl	Dimethyl 4,4′-*o*-phenylene bis (3-thioallophanate)	Nippon Soda Cercobin-M® Topsin-M®	NH-C(=S)-NH-$COOCH_3$; NH-C(=S)-NH-$COOCH_3$

Figure 2. Structure of microtubules

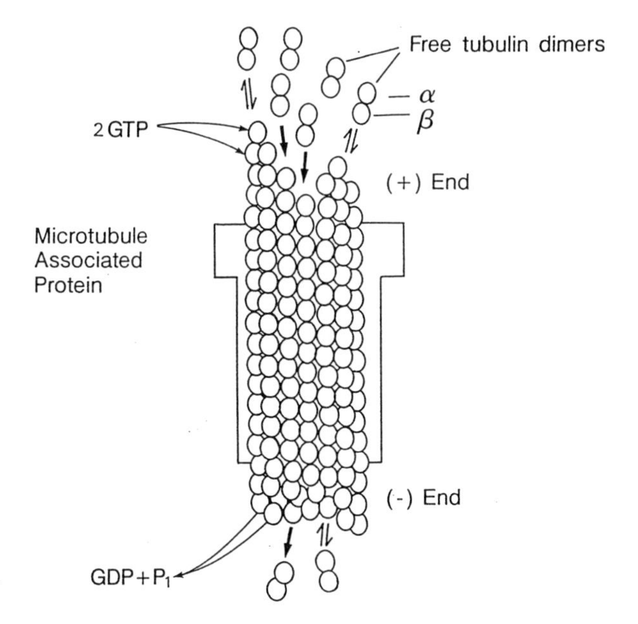

In many organisms multigene families exist for each tubulin subunit, and although in fungi there are seldom more than two isoforms, in the plant *Arabidopsis thaliana* there are seven α- and nine β-tubulin genes despite its small genome size (Kopczak *et al.*, 1992; Snustad *et al.*, 1992). It has been suggested (Raff, 1944) that each tubulin isoform carries out a separate function within the cell, but in general tubulins appear to be interchangeable. For instance, the β-tubulin gene normally involved in conidiation in *A. nidulans* can be replaced by the one normally expressed during vegetative growth (Weatherbee *et al.*, 1985). Despite the fact that tubulins are highly conserved across all organisms, sufficient diversity exists to provide scope for the selectivity needed for successful development of anti-tubulin agents either as fungicides, herbicides, antihelminthics, or anti-tumour agents. Indeed, a single amino acid difference may be all that is needed to create the required selectivity (or resistance). Although care must be taken in assigning particular amino acid changes to direct, and not indirect effects on binding, the presence of a phenylalanine at amino acid codon 200 in β-tubulin from fungi (Butters *et al.*, 1995) and free-living nematodes (Kwa *et* al., 1995) seems critical for fungicidal and chemotherapeutic activity. Unfortunately, it has not been possible so far to determine the three dimensional structure of any tubulin protein by crystallography or NMR, and although tubulin was recently resolved by electron crystallography to a resolution of 6.5 Å, this does not provide sufficient detail of the amino acid backbone to explain the molecular basis of selectivity (Nogales *et al.*, 1995).

Chemistries that interfere with tubulin function

A range of divergent chemistries, many of which are derived from natural products, interfere with microtubule assembly. Benzimidazole fungicides (Table 2), N-phenylcarbamates (diethofencarb; Figure 3) and griseofulvin are some that have been used commerically in agriculture and veterinary medicine. Others such as rhizoxin, the diterpene alkaloid taxol, and synthetic N-phenylformamidoxines (DCPF; Ishii and Takeda, 1982), N-phenylalanines (Ishii *et al.*, 1995) and benzamides (A0001, Young, 1991), have some antifungal activity but so far have only been exploited as research tools. In addition to the well known natural products colchicine and vinblastine (Wilson and Jordan, 1994) which have helped explore microtubule function in plants, synthetic chemistry has added other anti-tubulin agents such as nitroaniline and phosphoric amide herbicides (Morejohn and Foskett, 1991).

Some common themes have emerged from the many studies that have attempted to explain the mode of action of these many divergent chemistries. Often compounds show considerable phylogenetic specificity acting against only fungi, plants or mammals. Mutational studies with fungi, aimed at generating resistant mutants to many of these compounds point to an interaction with just β-tubulin, whereas biochemistry generally identifies an involvement with the tubulin heterodimer, possibly at the interface between α- and β-tubulin.

The majority of the natural products share common, or overlapping, binding sites towards the C-terminal end of β-tubulin (Burns and Surridge, 1994), and which are shared with various MAPs. Just how they inhibit tubulin polymerisation at the "plus"

Figure 3. Chemistry of fungicidal anti-tubulin agents.

N-(3,5-dichlorophenyl) carbamate = MDPC
Other structures are mentioned in the text.

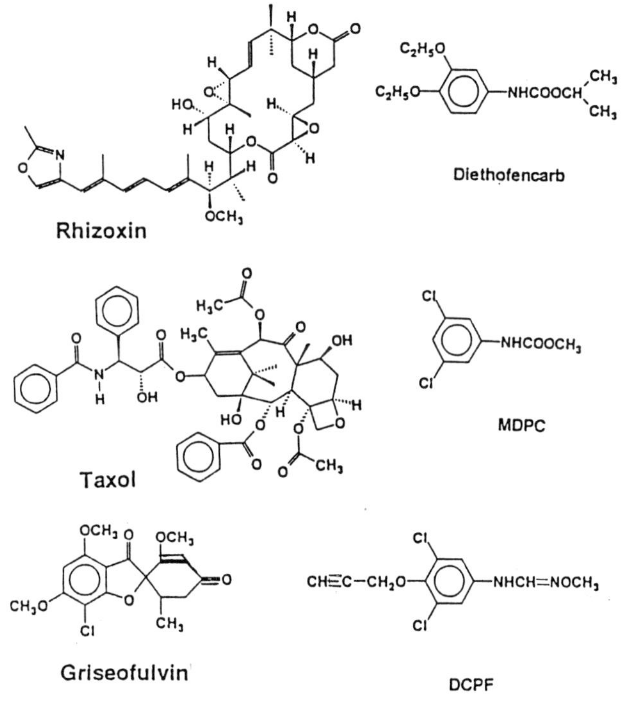

Figure 4. Possible interaction between carbendazim and β-tubulin

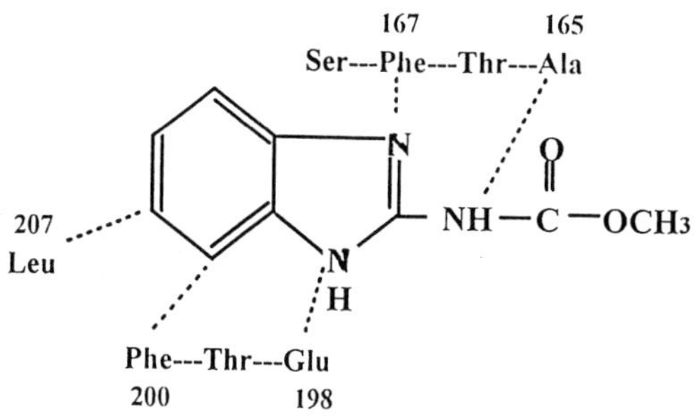

(assembly) end of microtubules is not clear, but it may be achieved by conformational changes which maintain the bound guanosine nucleotide in the GTP or GDP-P state. Unlike other anti-tubulin agents, Taxol (Paclitaxel) stablilizes microtubules rather than causing depolymerisation. Electron crystallography shows clearly that Taxol also binds to the N-terminal peptides (Nogales *et al.*, 1995) and probably bridges across adjacent subunits. As a result Taxol arrests mitosis through reorganisation of the mitotic spindle.

In the absence of an adequate three dimensional structure of tubulin, the most complete analysis of the binding of anti-tubulin agents is derived from mutational studies generating benzimidazole and N-phenylcarbamate resistance coupled with structure activity studies. Although point mutations at nine different sites within the β-tubulin protein confer benzimidazole resistance (Davidse and Ishii, 1995) it seems likely that only three regions, amino acid codons 165-167, 198-200, and 207, interact with the benzimidazole nucleus (Figure 4). At none of these points does benzimidazole binding overlap GTP binding sites (Figure 1). Substitution of Glu_{198} with smaller amino acids such as glycine or alanine destroys benzimidazole binding, but is accompanied by increased sensitivity to N-phenylcarbamates (Fujimura *et* al., 1992). Changes from either Glu_{198} to Lys_{198} or Val_{198}, or Phe_{200} to Tyr_{200} destroy the binding of both fungicide groups. The mutation Phe_{167} to Tyr_{167} in *Saccharomyces cerevisiae* confers resistance to carbendazim but actually increased sensitivity to benomyl, suggesting that aminoacid residue 167 of β-tubulin interacts with benzimidazoles in the vicinity of the 1-position (Li *et al.* 1996).

Molecular biology of benzimidazole binding to β-tubulin

Although genetic evidence strongly suggests that benzimidazole fungicides bind just to β-tubulin (M_r 52,000), biochemical studies do not confirm this. Where binding assays use gel filtration to separate bound and unbound radiolabelled fungicide, as in the case of our work with *R. secalis* (Figure 5, Kendall *et al.*, 1994), binding is usually associated with large molecular weight proteins similar in size to the tubulin heterodimer, as well as smaller ones of around 15kd. Little, if any, binding is associated with the β-tubulin monomer. In an attempt to resolve this, and establish whether benzimidazoles and N-phenylcarbamates bind to just β-tubulin, we have sought to express the *R. secalis* β-tubulin gene in *Escherichia coli*. Since bacteria lack microtubules, this approach should ensure that purified β-tubulin is free from MAPs. It is also one way to obtain sufficient β-tubulin from filamentous fungi, since tubulin levels within fungal cells are generally insufficient to allow purification by co-polymerisation. If successful, these experiments should help explain the molecular basis of the negative cross-resistance between benzimidazole and N-phenylcarbamate fungicides, and identify how to use these two fungicide groups in a durable anti-resistance strategy.

β-tubulin cDNA was prepared using a Marathon cDNA Kit (Clontech, Palo Alto, California), which involves using PCR to first produce overlapping 3' and 5' RACE fragments, based on already published sequence data of the genomic β-tubulin gene

Figure 5. Binding of ^{14}C carbendazim to tubulin in *Rhynchosporium secalis*.
A: Extracts from whole cells. B: Fusion protein

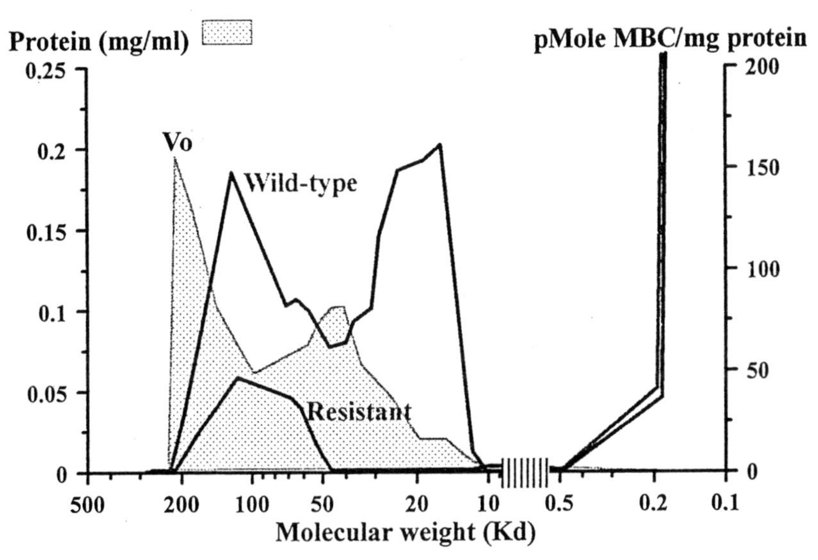

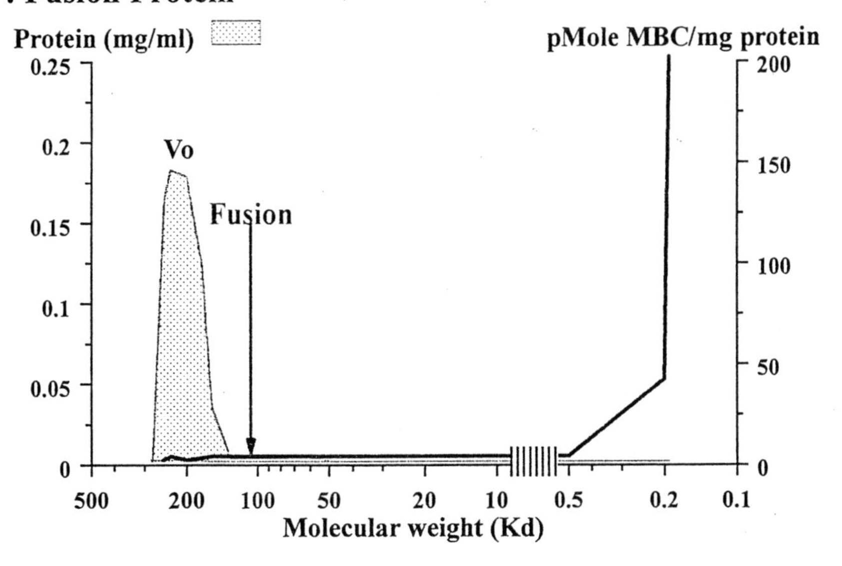

(Wheeler *et al.*, 1995). PCR primers were then designed to add EcoR1 and Hind III restriction sites immediately adjacent to the ATG start and TAA stop codons respectively. Using these two restriction sites, the β-tubulin cDNA was inserted in frame into the appropriate pMalE expression vector (Pharmacia, St Albans, UK) and transformed into *E. coli* strain XLI-Blue. 2 hours after induction with IPTG, considerable amounts of the soluble maltose binding protein - β-tubulin fusion (M_r 94,000) were produced. Expression was much better with the vector pMalE-C2, which accumulates protein in the cytoplasm, than with pMalE-P2 which secretes protein into the periplasm. Purification of the fusion protein on an amylose affinity column, and elution with maltose, yield preparations with concentrations in excess of 1 ml mg^{-1} protein.

Unfortunately, the Factor X cleavage site seems to be sufficiently buried in the recombinant protein that even long incubations with Factor X have not cleaved the fusion allowing purification of just β-tubulin. Consequently, fungicide binding studies were carried out on the fusion protein. Extracts (1 ml) were incubated with ^{14}C carbendazim (0.06μM, 529 mBq $mmole^{-1}$) for 60 min at 4° in 0.05 M sodium phosphate buffer, pH 6.8 containing 0.005 M $MgCl_2$, 0.1M KCl and 15% glycerol, and then separated on a Sephacryl HR300 column. GTP concentrations were varied between 0.2 and 1.0 mM, but unlike in cell extracts there was no evidence of carbendazim binding to protein (Figure 5). Instead, the fusion protein appeared to aggregate rapidly, and was eluted from the column well in advance of any protein with a M_r of 94,000. So far, we have been unable to identify conditions that prevent self-aggregation.

Concluding remarks

Tubulins are essential components of normally growing cells, and it is perhaps not surprising that commercially successful antifungal agents exploit this target. Almost all these products interact with β-tubulin, suggesting that target sites on other microtubule proteins remain to be exposed and developed. Despite the conserved nature of tubulin proteins, which almost certainly have similar, but not identical, tertiary structures, a lack of selectivity has not been a serious problem during development of existing fungicides. As DNA sequence analysis shows, selectivity for benzimidazole fungicides can be achieved by targetting regions of β-tubulin where a single amino acid difference between organisms is functionally significant. One consequence of this specificity is that a single amino acid change can cause resistance, but the functional importance of tubulin proteins ensures it seems that most amino acid changes carry a fitness penalty. For example, the number of different point mutations conferring benzimidazole resistance in field populations of many plant pathogens is very restricted, when compared with the many different resistant mutants generated in the laboratory where fitness constraints on artificial media are much less (Butters *et al.*, 1995).

Conformational changes conferring resistance to anti-tubulin agents often create novel binding sites for a second fungicide, but to properly exploit this negative cross-resistance, detailed knowledge of the interaction between fungicides and tubulins is required. This information is unlikely to become available simply from genetic studies coupled with structure activity studies. Although a great deal of research has been carried out, a useful three dimensional structure of tubulin proteins remains elusive. Their high molecular weight and propensity to self-aggregate ensures that analysis of tubulins remains beyond the scope of current crystallographic and NMR techniques. An ability to purify significant amounts of tubulin from filamentous fungi has also been a problem, but this can largely be overcome by cloning and expressing tubulin genes in *E. coli* or yeast. Electron crystallography has provided a structure of tubulin, albeit at too low resolution, but further developments in this imaging technology may provide sufficient resolution to enable individual amino acids to be identified within the protein chain. If a clearer picture of the interaction between antifungal agents and tubulin can be achieved it should allow development of the new tools, including novel chemistries, needed to exploit negative cross resistance as a durable anti-resistance strategy.

References

Burns RG (1991) α-, β- and γ-tubulins: sequence comparisons and structural constraints. *Cell Motil. Cytoskel* **20**: 181-189.

Burns RG, Surridge CD. (1994) Tubulin: conservation and structure. In: *Microtubules* (Hyams JS, Lloyd CW eds). Wiley-Liss, New York: 1-31.

Butters JA, Kendall SJ, Wheeler IE, Hollomon DW. (1995) Tubulins: lessons from existing products that can be applied to target new antifungals. In: *Antifungal Agents: Discovery and Mode of Action* (Dixon GK, Copping LG, Hollomon DW, eds) Bios Scientific, Oxford: 131-142.

Davidse LG, Ishii H. (1995) Biochemical and molecular aspects of the mechanisms of action of benzimidazoles, *N*-phenylcarbamates and *N*-phenylformamadoxines and the mechanisms of resistance to these compounds in fungi. In: *Modern Selective Fungicides.* 2nd edn. (Lyr H, ed). Gustav Fischer, Jena: 305-322.

Fujimura M, Oeda K, Inoue H, Kato T. (1992) A single amino-acid substitution in the beta-tubulin gene of *Neurospora* confers both carbendazim resistance and diethofencarb sensitivity. *Curr. Genet.* **21**: 399-404.

Hyams JS, Lloyd CW. (1994) Microtubules. Wiley-Liss, New York: pp.439.

Ishii H, Takeda H. (1982) Differential binding of a *N*-phenylformamidoxine compound in cell free extracts of benzimidazole-resistant and -sensitive isolates of *Venturia nashicola, Botrytis cinerea* and *Gibberella fujikuroi. Neth. J. Plant Pathol.* **95** (Suppl. 1): 99-108.

Ishii H, Josepovits J, Gasztonyi M, Miura T. (1995) Further studies on increased sensitivity to *N*-phenylalanines in benzimidazole-resistant strains of *Botrytis cinerea* and *Venturia nashicola. Pestic. Sci.* **43**: 189-193.

Kendall SJ, Hollomon DW, Ishii H, Heaney SP. (1994) Characterisation of benzimidazole-resistant strains of *Rhynchosporium secalis. Pestic. Sci.* **40**: 175-181.

Koenraadt H, Somerville SC, Jones AL. 1992) Characterization of mutations in the beta-tubulin gene of benomyl-resistant field strains of *Venturia inaequalis* and other plant pathogenic fungi. *Phytopathology* **82**: 1348-1354.

Kopczak SD, Haas NA, Hussey PJ, Silflow CD, Snustad DP. (1992) The small genome of Arabidopsis contains at least six expressed α-tubulin genes. *Plant Cell* **4**: 539-547.

Kwa MSG, Veenstra JG, Van Dijk M, Roos MH. (1995) β-tubulin genes from the parasitic nematode *Haemonchus contortus* modulate drug resistance in *Caenorhabditis elegans*. *J. Mol. Biol.* **246**: 500-510.

Li J, Katiyar SK, Edlind TD. (1996) Site-directed mutagenesis of *Saccharomyces cerevisiae* β-tubulin: interaction between residue 167 and benzimidazole compounds. *FEBS Lett.* **385**: 7-10.

Little M, Seehaus T. (1988) Comparative analysis of tubulin sequences. *Comp. Biochem. Physiol.* **90B**: 655-670.

Mandelkow EM, Schultheiss R, Rapp R, Müller M, Mandelkow E. (1986) On the surface lattice of microtubules *Helix* starts protofilament number seam and handedness. *J. Cell Biol.* **102**: 1067-1073.

Morejohn LC, Foskett DE. (1991) The biochemistry of compounds with anti-microtubule activity in plant cells. *Pharmacol. Ther.* ***51***: 217-230.

Nogales E, Wolf SG, Khan IA, Ludnena RF, Dowing KH. (1995) Structure of tubulin at 6.5 Å and location of the taxol binding site. *Nature* **375**: 424-427.

Oakley BR. (1994) γ Tubulin. In: *Microtubules* (Hyams JS, Lloyd CW, eds). Wiley-Liss, New York: 33-45.

Raff EC. (1994) The role of multiple tubulin isoforms in cellular microtubule function. In: *Microtubules* (Hyams JS, Lloyd CW, eds). Wiley-Liss, New York: 85-109.

Sage CR, Dougherty CA, Davis AS, Burns RG, Wilson L, Farrell KW. (1995) Site-directed mutagenesis of putative GTP-finding sites of yeast β-tubulin: evidence that α, β and γ-tubulins are atypical GTPases. *Biochemistry* **34**: 7409-7419.

Snustad DP, Haas NA, Kopczak SD, Silflow CD. (1992) The small genome of Arabidopsis contains at least nine expressed β-tubulin genes. *Plant Cell* **4**: 549-556.

Sterberg MJE, Taylor WR. (1984) Modelling the ATP-binding site of oncogene products, the epidermal growth factor receptor and related proteins. *FEBS Lett.* **175**: 387-392.

Weatherbee JA, May GS, Gambino J, Morris NR. (1985) Involvement of a particular species of β-tubulin (β3) in conidial development in *Aspergillus nidulans*. *J. Cell. Biol.* **101**: 706-711.

Wheeler IE, Kendall SJ, Butters J, Hollomon DW, Hall L. (1995) Using allele-specific oligonucleotide probes to characterise benzimidazole resistance in *Rhynchosporium secalis*. *Pestic. Sci.* **43**: 201-209.

Wilson L, Jordan MA. (1994) Pharmacological probes of microtubule function. In: **Microtubules** (Hyams JS, Lloyd CW, eds). Wiley-Liss, New York: 59-83.

Young DH. (1991) Effects of Zarilamide on microtubules and nuclear division of *Phytophthora capsici* and tobacco suspension-cultured cells. *Pestic. Biochem. Physiol.* **40**: 149-161.

12
Cell-Wall Active Antifungals and Emerging Targets

N. H. Georgopapadakou
DEPARTMENT OF MOLECULAR BIOLOGY, PRINCETON UNIVERSITY, PRINCETON, NEW JERSEY, USA

1 INTRODUCTION

The fungal cell wall is a multilayered, complex structure external to the cell membrane with a high polysaccharide content (1-6). It is a structure essential for cell viability: it determines the organism's shape, integrity, and, in pathogenic fungi, adhesion to the mammalian host and antigenicity (7). Importantly, it has no mammalian counterpart and thus presents an attractive target for new antifungals which are potentially fungicidal, a significant feature for the typically immunocompromised host.

The composition of the cell wall varies among species, but its major components are chitin, variously linked glucans (β-1,3-, α-1,3- and β-1,6) and mannoproteins (Figure 1). Chitin (chitosan in some fungi) and glucan fibrils form the scaffolding of the cell wall, while embeded mannoproteins form the matrix responsible for the wall's porosity and immunogenicity. Extensive hydrogen bonding between carbohydrate fibrils gives the fungal cell wall its mechanical strength, similar to plant cell walls where fibrils of cellulose, a 1,4-β-glucan homopolymer, form the scaffolding and unlike bacterial cell walls where a single, net-like peptide/sugar macromolecule (peptidoglycan) is responsible for the wall's strength. Although fungal cell walls are rigid enough to resist internal, osmosis-generated hydrostatic pressure, they are also flexible enough to expand with the growing cell and change composition in response to external and internal signals. For example, mannan fibril length of surface mannoproteins is temperature regulated in *C. albicans* and both chitin and glucan synthesis appear to be under cell cycle control.

This review will focus on the cell wall of two fungi: the non-pathogenic but extensively studied *Saccharomyces cerevisiae* and the pathogenic but less well studied *Candida albicans*. The latter organism, a commensal of the gastrointestinal tract in healthy individuals, is the major opportunistic fungal pathogen in neonates, diabetics and immunocompromised individuals such as transplant, AIDS and cancer patients (8,9). The review will highlight chitin and β-1,3-glucan synthase inhibitors and mannoprotein-interacting antifungals in clinical development as well as recent progress towards identifying and exploiting new cell wall targets.

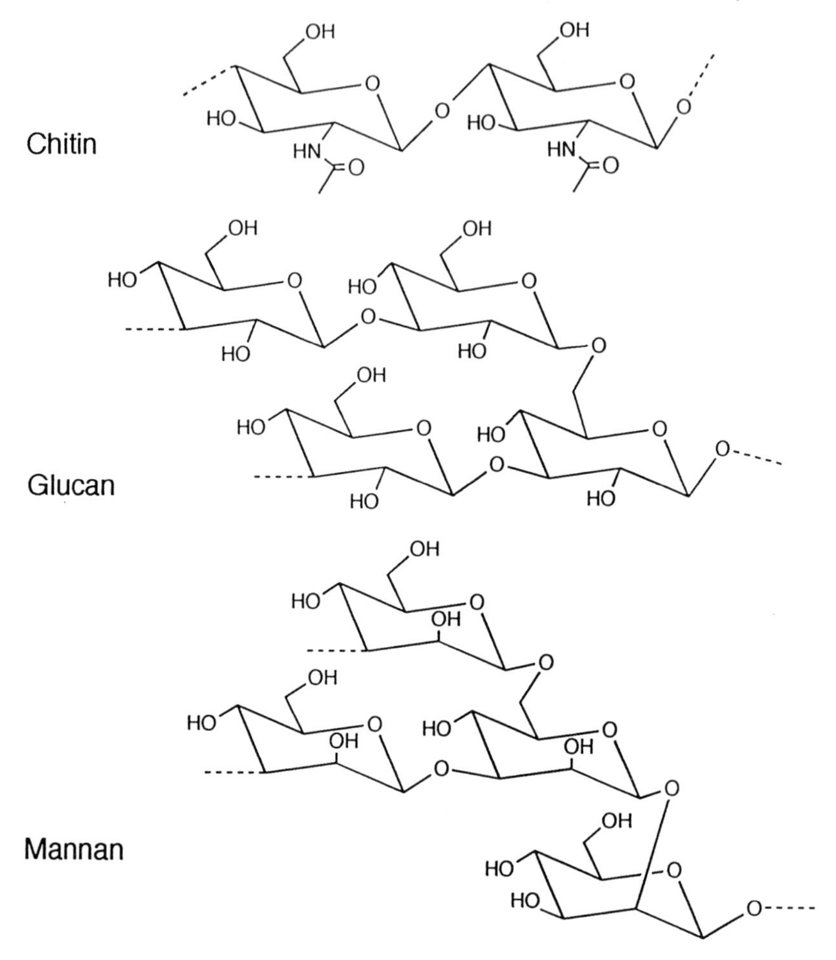

Figure 1 *Cell wall carbohydrate polymers: chitin, glucan and mannan. Shown are: for chitin, the sterochemical repeating unit; for glucan, 1,3- and 1,6- linkages; for mannan, 1,2-, 1,3- and 1,6-linkages*

2 CHITIN

Chitin is a linear β-1,4-linked N-acetylglucosamine (GlcNAc) homopolymer whose proportion in the cell wall varies among species and different growth phases in the same species (3, 10, 11). In *S. cerevisiae* and the yeast form of *C. albicans*, chitin is localized primarily in the septal region and the bud scars, with much lower amounts uniformly dispersed in the cell wall where it is covalently linked to β-glucan. Chitin is conveniently visualized in cells by staining with Calcofluor White M2R, which interacts specifically with β-1,4-linked D-glucopyranose polymers, such as chitin and cellulose, giving intense blue fluorescence (12, 13).

2.1 Chitin Synthesis and its Inhibition

Chitin synthesis is an important process in the fungal cell cycle and is under both spatial and temporal control. In *S. cerevisiae*, a burst of chitin synthesis before budding lays down a chitin ring at the presumptive bud-emergence site. A second burst of chitin synthesis at the end of mitosis deposits the primary septum on the mother-cell side within the chitin ring and becomes the bud scar after cell separation. An additional, "basal" level of chitin synthesis throughout the cell cycle deposits small amounts of chitin in the lateral cell wall. Chitin is overproduced over the entire cell surface in some *S. cerevisiae* cell cycle (*cdc*) mutants arrested in G1 phase (14) and in wild-type cells arrested in G1 phase by exposure to ergosterol synthesis inhibitors (15).

Chitin synthesis is vectorial: GlcNAc is polymerized on the cytoplasmic surface of the plasma membrane with concommitant translocation of the linear polymer through the membrane to the cell surface.

$$n \text{ UDP-GlcNAc} \text{ --------->} [\text{GlcNAc-}\beta\text{-1,4-GlcNAc}]_{n/2} + n \text{ UDP}$$

Chitin chains (average length, 100 units) subsequently crystallize as α-chitin through extensive hydrogen bonding outside the cell. Some chitin chains are covalently linked to β-glucan in the cell wall. The polymerization reaction is catalyzed by multiple, genetically distinct, membrane enzymes, three in *S. cerevisiae* and *C. albicans* (16-18), which may form transmembrane channels through which chitin is translocated as it is being formed. However, evidence for this is lacking; chitin synthases have been studied primarily in microsomal preparations of fungal cells. Very little is known about the biochemistry of the polymerization reaction other than that the substrate is UDP-GlcNAc, is stimulated by GlcNAc, requires divalent cations and, at least *in vitro*, does not require a primer (10).

The major *in-vitro* activity, Chs1 in *S. cerevisiae* (Chs2 in *C. albicans* (19)), is a nonessential repair enzyme since the only phenotypic defect of *chs1* mutants is partial bud lysis (17). Of the other two activities, Chs2 in *S. cerevisiae* (Chs1 in *C. albicans*) is involved in septum formation, since *chs2* mutants lack defined primary septum (17). Chs3 in both organisms is involved in cell wall maturation and bud ring formation since *chs3* mutants have reduced levels of cell wall chitin and lack septal ring (17). Double mutants lacking Chs1 and Chs2 or Chs1 and Chs3 are viable under certain conditions, but triple mutants lacking all three activities have not been isolated and are presumably non-viable. Chs3 activity requires three genes, *CSD2/CAL1*, *CSD4/CAL2* and *CAL 3* (20). *CSD2* is the structural gene for the catalytic subunit, based on significant homology to *CHS1* and *CHS2*; *CSD4* and *CAL3* encode two ~75 kDa proteins which may link chitin synthesis with morphogenetic events, though their precise functions are unknown (3). All three chitin synthases are proteolytically activated *in vitro*, but their *in-vivo* activation is more complex and may involve both transcriptional and post-translational mechanisms (21).

Chitin synthesis is inhibited competitively by the nucleoside-peptide antibiotics polyoxins and nikkomycins produced by streptomycetes (22, 23) (Figure 2). The compounds act as analogs of the UDP-GlcNAc substrate and inhibit different chitin synthase isozymes to different degrees, generally in the (sub)micromolar range (23-26). This may have chemotherapeutic implications, especially in light of the functional redundancy of the enzymes. Nevertheless, *C. albicans* mutants deficient in Chs3 have been reported to be less virulent than the parental strain in mice (27), suggesting that at least this particular enzyme may be a good chemotherapeutic target. Polyoxins and nikkomycins have modest activity against human pathogens due to transport limitations (they are taken

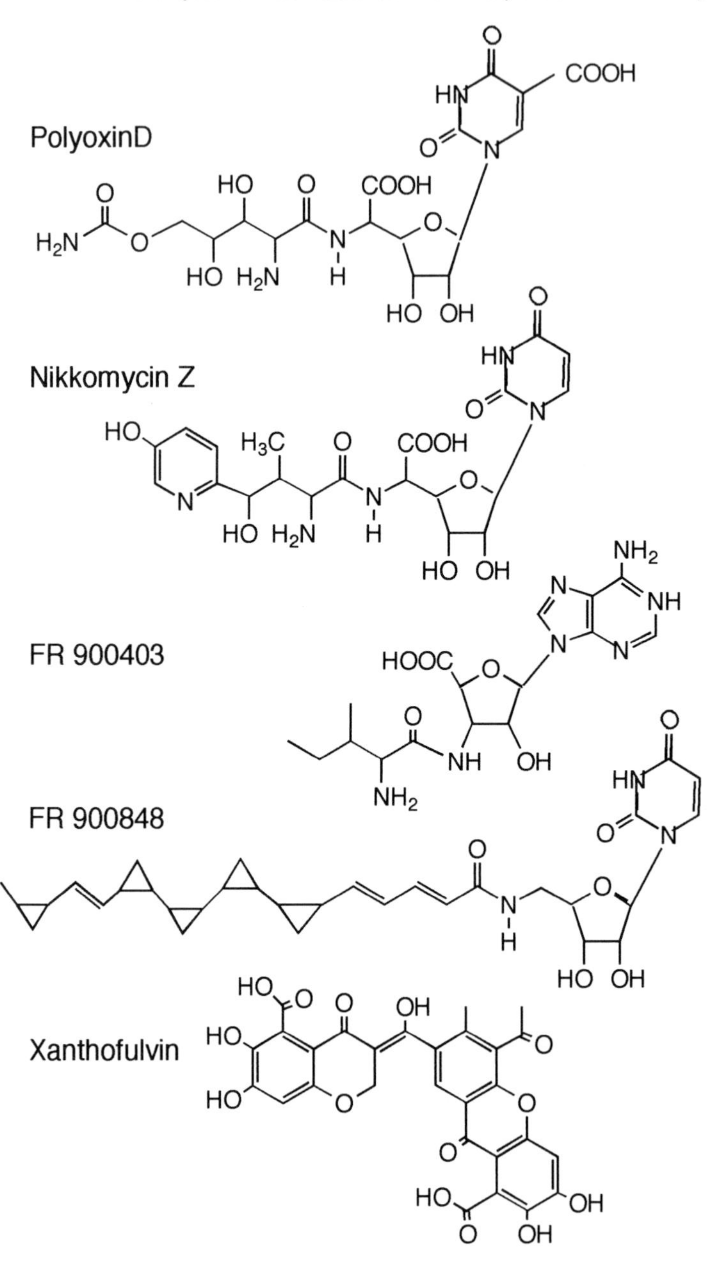

Figure 2 *Chitin synthase inhibitors: polyoxins/nikkomycins and xanthofulvin. FR 900403 and FR 900848 are also shown, though it is not known whether they inhibit chitin synthase*

up by a dipeptide permease (28) and are used exclusively as agricultural fungicides. Bypassing peptide transport is an obvious goal, and appropriate polyoxin derivatives have been synthesized, though none has shown so far promising activity *in vitro* (29). Nikkomycin Z, a natural product, is the only compound of this class which has shown some activity in animal models of fungal infections (30). It may soon enter clinical development as a single agent and in combination with glucan or ergosterol synthesis inhibitors where it has shown synergy (31).

FR 900848 (Figure 2), a nucleoside-fatty acid antibiotic with good activity against filamentous fungi (MICs <0.5 μg/ml) but not yeasts (32) may also be an inhibitor of chitin synthase despite lack of a carboxyl substituent on either fatty acid or sugar moiety. Data on chitin synthase inhibition and transport for this compound, when they become available, may significantly expand structure-activity relationships. Data on chitin synthase inhibition and transport are also needed for FR 900403, a nucleoside-peptide antibiotic reported to have good *in-vitro* activity against *C. albicans* (33).

Xanthofulvin (Figure 2), an aromatic natural product, has been reported to inhibit selectively chitin synthase 2 of *S. cerevisiae* with an IC_{50} of 2 μM (34). Unfortunately, its antifungal activity and inhibition of the corresponding *C. albicans* enzyme were not given. Pentachloronitrobenzene, an aromatic synthetic compound reported to inhibit chitin synthesis in *Agaricus bisporus* at 1 μM (35), did not inhibit chitin synthesis in *C. albicans* up to 1 mM (Georgopapadakou and Bertasso, unpublished results). Several fungicides and insecticides were screened for inhibition of chitin synthesis in *Phycomyces* and cockroach and had higher activity in the latter system (36). However, polyoxin D was only weakly active, casting doubt as to the relevance of the findings to fungal pathogens. Finally, several *cis* unsaturated fatty acids weakly inhibited chitin synthesis in permeabilized *C. albicans* (IC_{50}s, ~ 1 mM) as did citrinin, nalidixic acid and other amphiphilic compounds (Georgopapadakou and Bertasso, unpublished results). To sum up, none of the above compounds appears to be a new lead for non-polyoxin inhibitors, with the possible exception of xanthofulvin.

2.2 Chitin Degradation and its Inhibition

Noncrystalline chitin is hydrolyzed to GlcNAc dimers (diacetylchitobiose) by chitinases, 30- to 50-kDa periplasmic enzymes. The process may contribute to cell wall plasticity during fungal growth and proliferation, particularly cell separation after cell division (37, 38). A single endochitinase has been cloned and sequenced from *S. cerevisiae* (37). Analysis of the derived amino acid sequence suggests that the protein contains a signal sequence, a catalytic domain, a serine/threonine-rich region, and a chitin-binding domain. Disruption of its structural gene results in cell clumping and failure of the cells to separate after cell division (37). A similar effect is produced by the aza-oligosaccharide demethylallosamidine (39), a specific inhibitor of chitinase (40) that functions as an analog of the putative oxazolidinium intermediate (41). Three chitinase isozymes have been reported in *C. albicans*, of which two are preferentially produced in the yeast phase (42). Although allosamidine lacks antifungal activity, further work on chitinases may be necessary to determine whether any of them are potential antifungal targets.

3 GLUCAN

Glucans are polymers (average length, 1,500 units) of β- or α-1,3-linked glucose, with occasional side chains (average length, 150 units) involving β-1,6-linkages (43). They constitute the major structural component of the cell wall in most fungi, determining cell shape and integrity. In addition, some β-1,3-/1,6-glucan heteropolymers may anchor cell wall mannoproteins through covalent linkages (44, 45). So far, most studies have focused on the synthesis of β-1,3-glucan, its physiological regulation and its inhibition by antifungals.

3.1 β-1,3-Glucan Synthesis and its Inhibition

β-1,3-Glucan is distributed uniformly over the cell surface. It is synthesized during cell growth and deposited primarily at sites of new cell-wall: the hyphal apex in filamentous forms and the bud surface in yeast forms. Synthesis is probably turned on and off by regulatory mechanisms linked to the cell cycle. Like chitin synthesis, β-1,3-glucan synthesis is vectorial: UDP-Glc in the cytoplasm is polymerized with concommitant translocation of the linear polymer through the membrane to the outside of the cell. The reaction is catalyzed by at least two isoenzymes whose catalytic subunits are encoded by *FKS1* and *FKS2* genes (46, 47). Fks1p is a 215-kDa protein with 10-16 predicted transmembrane segments (46) that may form a transmembrane pore through which the linear glucan polymer is translocated. Fks2p is a 217-kDa integral membrane protein highly homologous (88% identical) to Fks1p (47). β-1,3-Glucan synthase activity in membrane preparations is stimulated by micromolar concentrations of GTP and can be fractionated, with sodium chloride and detergent, into two components: a still membrane-bound, catalytic component, and a soluble, regulatory component that binds GTP (48). Purification of the soluble component led to the isolation of a 20-kDa, GTP-binding protein (49) subsequently identified as Rho1p (50, 51). Rho1p is essential for proliferation and may interact not only directly with the catalytic subunit of glucan synthase (Fks1p and probably Fks2p), but also with protein kinase C (Pkc1p) (52), linking glucan synthase (Fks2p) transcription to the cell cycle via a phosphorylation-dephosphorylation relay system (53-56). Accordingly, the protein phosphatase calcineurin has been shown to be involved in the regulation of *FKS2*: inhibition of calcineurin by FK506 shuts down glucan synthesis in *fks1* mutants, hence their hypersensitivity to this compound (46).

1,3-β-Glucan synthase is inhibited noncompetitively by the cyclic lipopeptides echinocandins/aculeacins/pneumocandins/mulundocandins and the liposaccharides papulacandins/chaetiacandins (Figure 3) (5, 43, 57). The two classes of compounds may bind to different sites on glucan synthase, since there is incomplete cross resistance between them (58). Interestingly, some echinocandin-resistant mutants show increased susceptibility to chitin synthase inhibitors (59). Papulacandins are no longer pursued as antifungal agents since their *in vitro* activity is limited to *Candida* species and, most importantly, does not translate to *in-vivo* activity (5). Echinocandins, on the other hand, have fungicidal activity *in vitro* and in animal models against *Candida* sp., *Pneumocystis carinii* and, some, *Aspergillus* sp. (60-66). They are inactive against fungi whose glucan is not mainly 1,3-β-linked, such as *Cryptococcus* sp., a major pathogen that contains mostly α-glucan (67). At least two water-soluble semisynthetic derivatives, LY303366 and presumably L-733560, are currently under clinical development as parentereal antifungals (60, 66).

3.2 Other Glucan Synthases

Very little is known about the biochemistry of β-1,6-glucan formation, a minor but

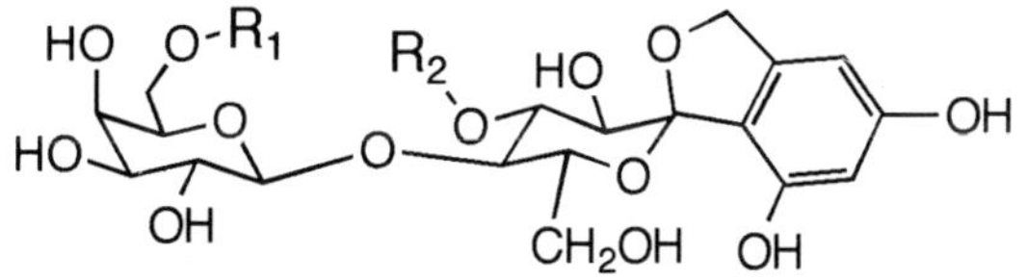

Papulacandin A

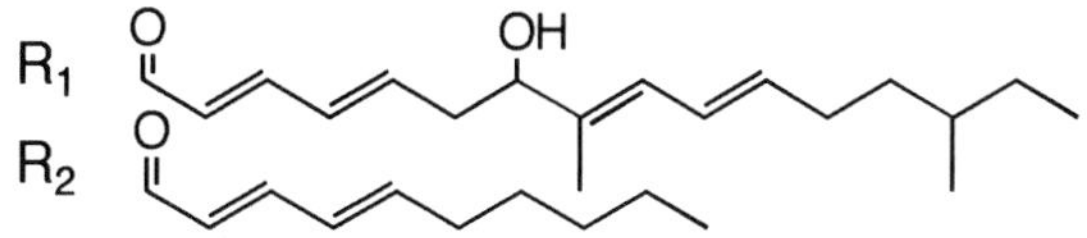

BU-4794F

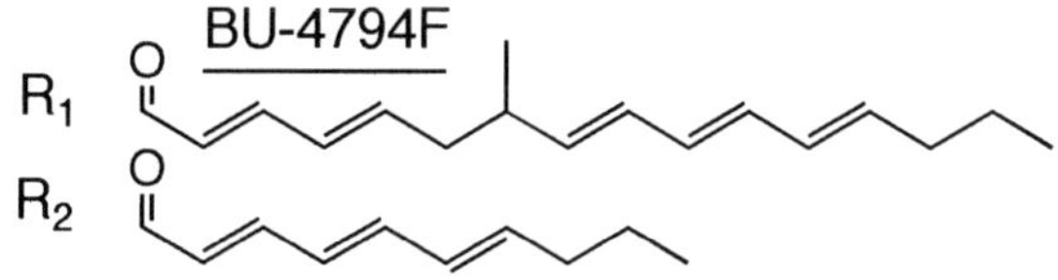

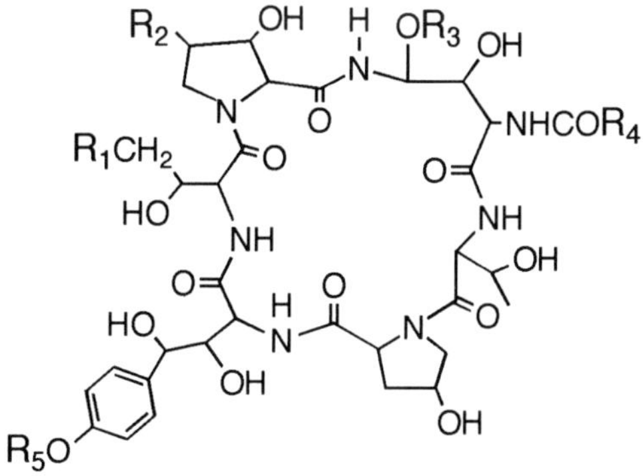

	LY 303366	L-733560
R_1	-H	$-CH_2CH_2NH_2$
R_2	$-CH_3$	-H
R_3	-H	$-CH_2CH_2NH_2$
R_4	$-(Ph)_3-O-(CH_2)_4CH_3$	$-(CH_2)_8CH(CH_3)CH(CH_3)CH_2CH_2$
R_5	-H	-H

	FR 901379	Deoxymulundocandin
R_1	$-CONH_2$	-OH
R_2	$-CH_3$	$-CH_3$
R_3	-H	-H
R_4	$-(CH_2)_{14}CH_3$	$-(CH_2)_{10}CH(CH_2CH_3)CH_3$
R_5	$-SO_3H$	-H

Figure 3 β*(1.3)-Glucan synthase inhibitors: papulacandins and echinocandins*

important component of the glucan heteropolymer responsible for glucan branching and for covalently linking mannoproteins to the cell wall (44, 45, 68, 69). β-1,6-Glucan synthesis has been reported in a crude envelope preparation from *S. cerevisiae* (70) and in a soluble preparation from *C. albicans* (71), but rigorous biochemical studies have yet to appear in the literature. Proteins involved in β-1,6-glucan synthesis have been sought in genetic screens based on the resistance of β-1,6-glucan-deficient mutants to yeast killer toxin (hence the name, *k*iller-*re*sistant (kre) mutants) (4, 72). Several *KRE* genes have been identified and cloned but, with the exception of *KRE2* and *PMT1* whose products are mannoprotein O-transferases (73), the precise function of *KRE* products is unknown. Nonetheless, because of the association of mannoprotein maturation with β-1,6-glucan synthesis, the latter process probably also involves the secretory pathway. This is supported by a recent report on the involvement of an endoplasmic reticulum membrane N-glycoprotein in β-1,6-glucan assembly (74).

Nothing is known about the formation of α-1,3-glucan, a plausible chemotherapeutic target in *Cryptococcus*. It may be analogous to the formation of β-1,3-glucan, a concept supported by the finding of a *FKS1* homolog in this organism (58). However, the lack of inhibitory activity of papulacandins and echinocandins against this organism puts limits to the analogy.

3.3 Glucan degradation

Cell-wall glucans probably undergo some hydrolysis during fungal growth and morphogenesis and both exo- and endo-glucanases have been found (6). The former hydrolyze both 1,3- and 1,6 linkages and release glucose, while the latter are more specific and release a mixture of oligosaccharides with glucose as a minor product. β-Glucanase mutants of *S. cerevisiae* lack a phenotype, suggesting that at least some of these enzymes are functionally redundant and/or dispensable.

4 MANNOPROTEINS

Fungal mannoproteins (average mannan length, 80 units) form the cell wall matrix and their extensive O- and N-glycosylation is generaly similar to that in the mammalian cells. It involves the synthesis of manno-oligosaccharides on a lipid carrier and their transfer to the appropriate aminoacids in proteins. Additional mannan units are added to this core as the protein transverses the secretory pathway, often resulting in branched mannans (75). However, mannoproteins in the outer layer of the fungal envelope form covalent linkages to β-1,3-/β-1,6-glucan through a phosphodiester bridge derived from the glycosyl phosphatidylinositol (GPI)-anchor (69). This last step has no apparent mammalian counterpart and may be a selective target (68). Nonetheless, selective inhibitors of fungal mannoprotein synthesis have not been found.

Mannoproteins undergo degradative turnover (76) and a nonspecific α-mannosidase has been purified and cloned (77, 78). However, deficient mutants lack phenotype suggesting a non-essential function.

4.1 Antifungals Interacting with Mannoproteins

Specific interaction with fungal mannoproteins is the basis for selectivity of the benzonaphthacene quinone antibiotics benanomicins/pradimicins produced by *Actinomadura*

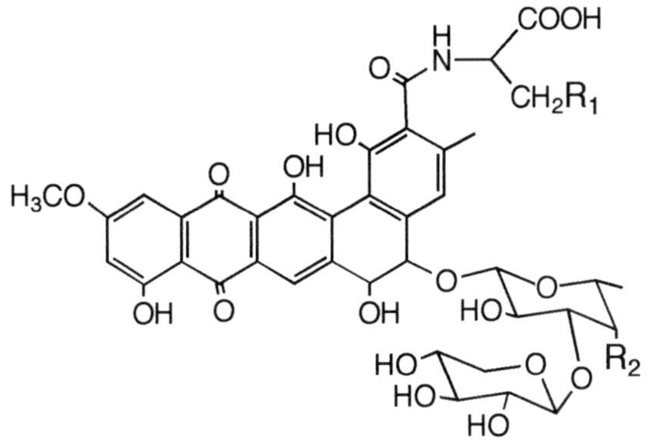

	R_1	R_2
Pradimicin A	H	$NHCH_3$
Benanomicin A	H	OH
BMS-181184	OH	OH

Figure 4 *Antifungals acting through mannoproteins*

sp. (Figure 4) (79, 80). The interaction involves the saccharide portions of mannoproteins and requires calcium (81, 82). The compounds have broad spectrum, fungicidal activity *in vitro* that includes *Candida, Aspergillus* and *Cryptococcus* (83, 84). Their *in-vivo* potencies (murine model) are intermediate of those of ketoconazole and amphotericin B. A water-soluble pradimicin derivative, BMS-181184, is currently under development as a parenteral antifungal (84).

5 EMERGING TARGETS

Relatively unexplored potential cell wall targets are outlined below. Challenges in each case include limited biochemical information on the target, absence of lead compounds and target-specific assays, and uncertainty as to the essential nature of the target, exacerbated by the possibility of functional redundancy. The multiplicity of chitin and glucan synthases, noted earlier in this review, makes a powerful precedent.

The most obvious target is the cross-linking reaction between β-1,6-/β-1,3-glucan heteropolymer and mannoproteins. The reaction is probably catalyzed by periplasmic enzymes and involves a GPI-derived sugar on the protein and β-1,6-glucan on the heteropolymer. Inhibition, even if nonlethal, may result in increased permeability to other antifungals and decreased adherence to the mammalian host since mannoproteins in the outer layer of the cell wall are involved in both processes.

A second target is the polymerization reaction producing β-1,6-glucan. Specific biochemical assays must be developed, in addition to cell-based assays involving killer toxin resistance, both to study the biochemistry of the reaction and for high throughput screening of compounds.

A third target is the polymerization reaction producing α-1,3-glucan. The reaction may be catalyzed by an enzyme similar to β-1,3-glucan synthase, and assays modeled on the latter could be developed using microsomal preparations from *Cryptococcus* sp.

A fourth target is mannan. Mannan-complexing compounds could be used either to decrease virulence by interfering with adherence or, as conjugates, to convey fungal selectivity to an antibiotic (prototype: pradimicins).

The signal pathway that regulates glucan synthesis may also contain chemotherapeutic targets (Pkc1, for example), although homology to the mammalian systems would be a concern for selectivity. Nevertheless, signal transduction is a very active research area in immunology and oncology and compounds generated in those therapeutic areas could also easily be tested for their effects on the relevant fungal enzymes and the cell wall (using in the latter case osmotic instability as an end-point).

Putative transcription factors that regulate the expression of cell wall synthetic and hydrolytic enzymes would be more speculative targets. Again, research advances in both fungi and mammalian cells may reveal specific targets amenable to small molecule intervention.

6 CONCLUSIONS

The increasing frequency of systemic fungal infections in immunocompromised patients underscore the need for broad-spectrum, fungicidal drugs suitable for empiric treatment. Cell-wall targeted antifungals are intrinsically selective, potentially broad-spectrum and fungicidal. Three classes of compounds, targeted respectively to chitin synthase, 1,3-β-glucan synthase and mannoproteins, are currently in clinical development. In addition, significant progress has been made in elucidating cell wall biosynthesis and several new targets have emerged.

References

1. E. Cabib, B. Bowers, A. Sburlati and S.J. Silverman, *Microbiol. Sci.*, 1988, **5**, 370.
2. D. Gozalbo, M.V. Elorza, R. Sanjuan, A. Marcilla, E. Valentin and R. Sentandreu, *Pharmacol. Ther.*, 1993, **60**, 337.
3. C.E. Bulawa, *Annu. Rev. Microbiol.*, 1993, **47**, 505.
4. F.M. Klis, *Yeast*, 1994, **10**, 851.
5. N.H. Georgopapadakou and J.S. Tkacz, *Trends Microbiol.*, 1995, **3**, 98-104.
6. V.J. Cid, A. Duran, F. Del Rey, M.P. Snyder, C. Nombela and M. Sanchez, *Microbiol. Rev.*, 1995, **59**, 345.
7. R.A. Calderone and P.C. Brown, *Microbiol. Rev.*, 1991, **55**, 1.
8. N.H. Georgopapadakou and T.J. Walsh, *Science*, 1994, **264**, 371-373.
9. N.H. Georgopapadakou and T.J. Walsh, *Antimicrob. Agents Chemother.*, 1996, **40**, 279.
10. E. Cabib, *Adv. Enzymol.*, 1987, **28**, 572.
11. N.H. Georgopapadakou, in J. Sutcliffe and N.H. Georgopapadakou (eds.), "Emerging Targets in Antibacterial and Antifungal Chemotherapy" , Chapman and Hall, New York, 1992, p. 476.
12. A. Duran, E. Cabib and B. Bowers, *Science*, 1979, **203**, 363.
13. P.J. Wood, *Carbohydr. Res.*, 1980, **85**, 271.
14. R.L. Roberts, B. Bowers, M.L. Slater and E. Cabib, *Mol. Cell. Biol.*, 1983, **3**, 922.
15. N.H. Georgopapadakou and A.M. Bertasso, in A. Adam, H. Lode, and E. Robinstein (eds.), "Recent Advances in Chemotherapy: Antimicrobial Section II. Proceedings

of the 17th International Congress of Chemotherapy", Futuramed Verlag, Munich, 1992, p. 2208.
16. J. Au Young and P.W. Robbins, *Mol. Microbiol.*, 1990, **4**, 197.
17. J.A. Shaw, P.C. Mol, B. Bowers, S.J. Silverman, M.H. Valdivieso, A. Duran and E. Cabib, *J. Cell Biol.*, 1991, **114**, 111.
18. M. Sudoh, S. Nagahashi, M. Doi, A. Ohta, M. Takagi and M. Arisawa, *Mol. Gen. Genet.*, 1993, **241**, 351.
19. N.A.R. Gow, P.W. Robbins, J.W. Lester, A.J.P. Brown, W.A. Fonzi, T. Chapman and O.S. Kinsman, *Proc. Natl. Acad. Sci. USA*, 1994, **91**, 6216.
20. W.-J. Choi, A. Sburlati and E. Cabib, *Proc. Natl. Acad. Sci. USA*, 1994, **91**, 4727.
21. W.-J. Choi, B. Santos, A. Duran and E. Cabib, *Mol. Cell. Biol.*, 1994, **14**, 7685.
22. K. Isono and S. Suzuki, *Heterocycles*, 1979, **13**, 333.
23. H. Decker, H. Zahner, H. Heitsch, W.A. Konig and H.P. Fiedler, *J. Gen. Microbiol.*, 1991, **137**, 1805.
24. R. Furter and D.M. Rast, *FEMS Microbiol. Lett.*, 1985, **28**, 205.
25. E. Cabib, *Antimicrob. Agents Chemother.*, 1991, **35**, 170.
26. J.P. Gaughran, M.H. Lai, D.R. Kirsch and S.J. Silverman, *J. Bacteriol.*, 1994, **176**, 5857.
27. C.E. Bulawa, *Proc. Natl. Acad. Sci.* USA, 1995, **92**, 10570.
28. J.-C. Yadan, M. Gonneau, P. Sarthou and F. Le Goffic, *J. Bacteriol.*, 1984, **160**, 884.
29. V. Girijavallabhan, A.K. Ganguly A.K. Saksena, A.B. Cooper, A. Lovey, D. Loebenberg, D. Rane, J. Desai, R. Pike, and E. Jao, in P.H. Bentley and R. Ponsford (eds.), "Recent Advances in the Chemistry of Antiinfective Agents", Royal Society of Chemistry, London, 1993, p. 192.
30. R.F. Hector, B.L. Zimmer and D. Pappagianis, *Antimicrob. Agents Chemother.* 1990, **34**, 587.
31. R.F. Hector and K. Schaller, *Antimicrob. Agents Chemother.*, 1992, **36**, 1284.
32. M. Yoshida, M. Ezaki, M. Hashimoto, M. Yamashita, N. Shingematsu, M. Okuhara, M. Kohsaka and K. Horikoshi, *J. Antibiot.*, 1990, **43**, 748.
33. T. Iwamoto, A. Fujie, Y. Tsurumi, K. Nitta, S. Hashimoto and M. Okuhara, *J. Antibiot.*, 1990, **43**, 1183.
34. M. Masubuchi, T. Okuda and H. Shimada, *European Patent Application*, 1993, EP 0537622A.
35. R. Rodewald, L. Hoesch and D.M. Rast, Abstr. 87th Annu. Meet. Am. Soc. Microbiol., 1987, p. 228.
36. T. Leighton, E. Marks and F. Leighton, *Science*, 1981, **213**, 905.
37. M.J. Kuranda and P.W. Robbins, *J. Biol. Chem.*, 1991, **266**, 1758.
38. E. Cabib, S.J. Silverman and J.A. Shaw, *J. Gen. Microbiol.*, 1992, **138**, 97.
39. S. Sakuda, Y. Nishimoto, M. Ohi, M. Watanabe, S. Takayama, A. Isogai and Y. Yamada, *Agric. Biol. Chem.* 1990, **54**, 1333.
40. K. Dickinson, V. Keer, C.A. Hitchcock and D.J. Adams, *J. Gen. Microbiol.*, 1989, **135**, 1417.
41. A.C. Terwisscha van Scheltinga, S. Armand, K.H. Kalk, A. Isogai, B. Henrissat and B.W. Dijkstra, *Biochemistry*, 1995, **34**, 15619.
42. K.J. McCreath, C.A. Specht and P.W. Robbins, *Proc. Natl. Acad. Sci. USA*, 1995, **92**, 2544.
43. J.S. Tkacz, in J. Sutcliffe and N.H. Georgopapadakou (eds.), "Emerging Targets in

Antibacterial and Antifungal Chemotherapy", Chapman and Hall, New York, 1992, p. 495.

44. J.C. Kapteyn, R.C. Montijn, G.J.P. Dijkgraaf, H. Van Den Ende and F.M. Klis, *J. Bacteriol.*, 1995, **177**, 3788.
45. J.C. Kapteyn, R.C. Montijin, E. Vink, J. de la Cruz, A. Llobell, J.E. Douwes, H. Shimoi, P. Lipke and F.M. Klis, *Glycobiology,* 1996, **6**, 337.
46. C.M. Douglas, F. Foor, J.A. Marrinan, N. Morin, J.B. Nielsen, A.M. Dahl, P. Mazur, W. Baginsky, W.L. Li, M. El Sherbeini, J.A. Clemas, S.M. Mandala, B.R. Frommer and B.M. Kurtz, *Proc. Natl. A*cad. Sci. USA, **91**, 5686.
47. P. Mazur, N. Morin, W. Baginsky, M. El-Sherbeini, J.A. Clemas, J.B. Nielsen and F. Foor, *Mol. Cell. Biol.*, 1975, **15**, 5671.
48. M.S. Kang and E. Cabib, *Proc. Natl. Acad. Sci. USA*, 1986, **83**, 5808.
49. P.C. Mol, H.-M. Park, J.T. Mullins and E. Cabib, *J. Biol. Chem.*, 1994, **269**, 31267.
50. J. Drgonova, T. Drgon, K. Tanaka, R. Kollar, G.-C. Chen, R. Ford, C.S.M. Chan, Y. Takai and E. Cabib, *Science*, 1996, **272**, 277.
51. H. Qadota, C.P. Python, S.B. Inue, M. Arisawa, Y. Anraku, Y. Zheng, T. Watanabe, D.E. Levin and Y. Ohya, *Science*, 1996, **272**, 279.
52. H. Nonaka, K. Tanaka, H. Hirano, T. Fujiwara, H. Kohno, M. Umikawa, A. Mino and Y. Takai, *EMBO J.*, 1995, **14**, 5931.
53. Y. Kamada, H. Qadota, C.P. Python, Y. Anraku, Y. Ohya and D.E. Levin, *J. Biol. Chem.*, 1996, **271**, 9193.
54. B. Errede and D.E. Levin, *Curr. Opin. Cell Biol.*, 1993, **5**, 254.
55. K. Irie, M. M. Takase, K.S. Lee, D.E. Levin, H. Araki, K. Matsumoto and Y. Oshima, *Mol. Cell. Biol.*, 1993, **13**, 3076.
56. G. Sherlock and J. Rosamond, *J. Gen. Microbiol.*, 1993, **139**, 2531.
57. M. Debono and R.S. Gordee, *Annu. Rev. Microbiol.,* 1994, **48**, 471.
58. C.M. Douglas, J.A. Marrinan, W. Li and M.B. Kurtz, *J. Bacteriol.*, 1994, **176**, 5686.
59. M. El-Sherbeini and J.A. Clemas, *Antimicrob. Agents Chemother.*, 1994, **39**, 200.
60. G.K. Abruzzo, A.M. Flattery, C.J. Gill, L. Kong, J.G. Smith, D. Krupa, V.B. Pikunis, H. Kropp and K. Bartizal, *Antimicrob. Agents Chemother.*, 1995, **39**, 1077.
61. F.A. Bouffard, R.A. Zambias, J.F. Dropinski, J.M. Balcovec, M.L. Hammond, G.K. Abruzzo, K.F. Bartizal, J.A. Marrinan, M.B. Kurtz, D.C. McFadden, K.H. Nollstadt, M.A. Powles and D.M. Schmatz, *J. Med. Chem.*, 1994, **37**, 222.
62. J.M. Balkovec, R.M. Black, G.K. Abruzzo, K. Bartizal, S. Dreikorn and K. Nollstadt, Bioorg. *Med. Chem. Lett.*, 1993, **3**, 2039.
63. R.A. Zambias, C. James, M.L. Hammond, G.K. Abruzzo, K.F. Bartizal, K.H. Nollstadt, C. Douglas, J. Marrinan and J.M. Balkovec, *Bioorg. Med. Chem. Lett.*, 1995, **5**, 2357.
64. R.A. Zambias, M.L. Hammond, J.V. Heck, K. Bartizal, C. Trainor, G. Abruzzo, D.M. Schmatz and K.M. Nollstadt, *J. Med. Chem.*, 1992, **35**, 2843.
65. K. Bartizal, T. Scott, G.K. Abruzzo, C.J. Gill, C. Pacholok, L. Lynch and H. Kropp, *Antimicrob. Agents Chemother.*, 1995, **39**, 1070.
66. W.L. Current, J. Tang, C. Boylan, P. Watson, D. Zeckner, W. Turner, M. Rodriguez, C. Dixon, D. Ma and J.A. Radding, in G.K. Dixon, L.G. Copping and D.W. Hllomon (eds.),"Antifungal Agents: Discovery and Mode of Action", Bios Scientific Publishers, Oxford, 1995, p. 143.
67. P.G. James, R. Cherniak, R.G. Jones, C.A. Stortz and E. Reiss, *Carbohydr. Res.*, 1990, **198**, 23.

68. H. deNobel and P.N. Lipke, *Trends Cell Biol.*, 1994, **4**, 42.
69. C.-F. Lu, R.C. Montijn, J.L. Brown, F.M. Klis, J. Kurjan, H. Bussey and P.N. Lipke, *J. Cell. Biol.*, 1995, **128**, 333.
70. E. Lopez-Romero and J. Ruiz Herrera, *Biochim. Bioohys. Acta,* 1977, **500**, 372.
71. R.P. Hartland, G.W. Emerson and P.A. Sullivan, *Proc. R. Soc. London Ser. B Biol. Sci.*, 1991, **246**, 155.
72. C. Boone, S.S. Sommer, A. Hensel and H. Bussey, *J. Cell Biol.*, 1990, **110**, 1833.
73. M. Lussier, M. Gentzsch, A.M. Sdicu, H. Bussey and W. Tanner, *J. Biol. Chem.*, 1995, **270**, 2770.
74. B. Jiang, J. Sheraton, A.F.J. Ram, G.J.P. Dijkgraaf, F.M. Klis and H. Bussey, *J. Bacteriol.*, 1996, **178**, 1162.
75. N. Shibata, K. Ikuta, T. Imai, Y. Satoh, R. Satoh, A. Suzuki, C. Kojima, H. Kobayashi, K. Hisamichi and S. Suzuki, *J. Biol. Chem.*, 1995, **270**, 1113.
76. F. Pastor, E. Herrero and R. Sentadreu, *Arch. Microbiol.*, 1982, **132**, 144.
77. M.J. Kuranda and P.W. Robbins, *Proc. Natl. Acad. Sci. USA*, 1987, **84**, 2585.
78. T. Yoshihisa and Y. Anraku, *Biochem. Biophys. Res. Commun.*, 1989, **163**, 908.
79. T. Oki, M. Konishi, K. Tomatsu, K. Tomita, K. Saitoh, M. Tsunakawa, M. Nishio, T. Miyaki and H. Kawaguchi, *J. Antibiot.*, 1988, **41**, 1701.
80. T. Takeuchi, T. Hara, H. Naganawa, M. Okada, M. Hamada, H. Umezawa, S. Gomi, M. Sezaki and S. Kondo, *J. Antibiot.*, 1988, **41**, 807.
81. Y. Sawada, K.-I. Numata, T. Murakami, H. Tanimichi, S. Yamamoto and T. Oki, *J. Antibiot.*, 1990, **43**, 715.
82. T. Ueki, K.I. Numata, Y. Sawada, M. Nishio, H. Ohkuma, S. Toda, H. Kamachi, Y. Fukagawa and T. Oki, *J. Antibiot.*, 1993, **46**, 455.
83. T. Oki, M. Kakushima, M. Hirano, A. Takahashi, A. Ohta, S. Masuyoshi, M. Hator and H. Kamei, *J. Antibiot.*, 1992, **45**, 1512.
84. J.C. Fung-Tomc, B. Minassian, E. Huczko, B. Kolek, D.P. Bonner and R.E. Kessler, *Antimicrob. Agents Chemother.*, 1995, **39**, 295.

13

Strobilurin Analogues as Inhibitors of Mitochondrial Respiration in Fungi

J. M. Clough,* P. J. Fraine, C. R. A. Godfrey and S. B. Rees
ZENECA AGROCHEMICALS, JEALOTT'S HILL RESEARCH STATION, BRACKNELL, BERKSHIRE RG42 6ET, UK

1 INTRODUCTION

Azoxystrobin is Zeneca's new broad spectrum systemic fungicide for use in agriculture, sold for the first time in 1996. Its discovery was inspired by the naturally-occurring strobilurins and oudemansins, which are inhibitors of mitochondrial respiration in fungi. This article gives a brief account of the evolution of ideas which led us from the natural products to azoxystrobin. Full details of this work can be found in references 1-5. A final section discusses the suitability of strobilurin analogues as potential antifungals for use in medicine.

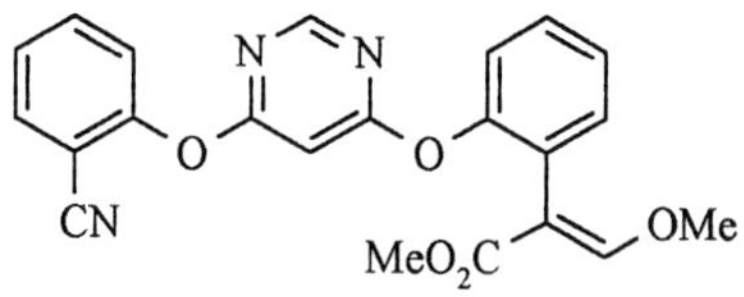

N N O O CN MeO_2C OMe

Azoxystrobin

2 AZOXYSTROBIN, A NOVEL BROAD SPECTRUM SYSTEMIC AGRICULTURAL FUNGICIDE

The strobilurins and oudemansins, represented here by the simplest examples, strobilurin A and oudemansin A, are fungicidal natural products isolated from Basidiomycete fungi such as *Oudemansiella mucida* (for a review of the natural products, see ref. 6). We first became interested in these compounds in 1981 following a publication by Becker, von Jagow, Anke and Steglich[7] which pointed out that they are not only structurally related, but are all fungicidal through the same biochemical mode of action, inhibiting mitochondrial respiration in fungi. Of course, many inhibitors of respiration were already known, but importantly the strobilurins and oudemansins had been shown to bind at a previously unidentified site on cytochrome b. This information strongly implied that if we

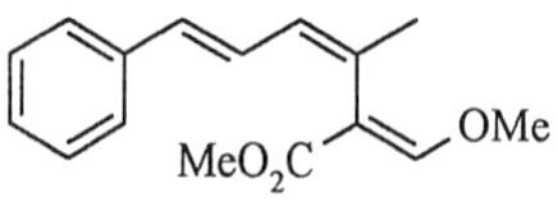

MeO_2C OMe

Strobilurin A

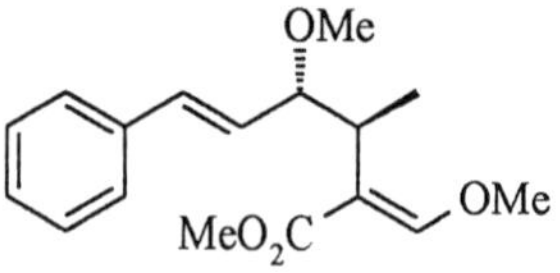

OMe MeO_2C OMe

Oudemansin A

could design compounds of this class with fungicidal activity against important plant pathogens, they would not have cross-resistance to existing agricultural fungicides. In addition, we recognised that a knowledge of the mode of action would allow us to establish an *in vitro* assay, useful for the direction of a programme of synthesis. Finally, although respiration inhibitors have the potential to be toxic in mammals, we were reassured by the low acute toxicity of strobilurin A and oudemansin A, which indicated that toxicities towards fungi and mammals are not inextricably linked in this series of compounds.

Naturally we were keen to test oudemansin A and strobilurin A in our own glasshouse, using commercial fungicides as standards. Professor Anke of the University of Kaiserslautern, Germany, kindly provided us with a sample of oudemansin A, and we obtained a sample of strobilurin A by total synthesis.[8] Oudemansin A showed activity against several fungi growing on plants in the glasshouse when applied as foliar sprays at concentrations of 33 mg/l. In parallel tests, strobilurin A gave a sharply contrasting result, showing no useful activity. It was, however, highly active in a mitochondrial assay, and we were able to demonstrate that its intrinsic activity was not expressed in the glasshouse because of its photochemical instability and relatively high volatility, through which it is rapidly lost from a leaf surface. It was clear that the natural products themselves were not suitable for use as agrochemicals because of unsuitable physical properties, insufficient levels of activity and/or problems associated with their preparation on a large scale and at a suitable price, either through fermentation or synthesis. Nevertheless, we believed that a knowledge of their structures and properties could provide a useful starting point for the synthesis of a new class of synthetic fungicides.

A programme of synthesis was initiated. Its aim was to prepare analogues of the natural products with high levels of activity and suitable physical properties for an agricultural fungicide. The stilbene (1), in which the (*Z*)-olefinic bond of strobilurin A is incorporated within a benzene ring, was an important early compound. Although (1) gave excellent activity in the glasshouse because of its improved photostability and reduced volatility in comparison with strobilurin A, it still showed only moderate activity in the field. Quite independently of our work, Anke and Steglich and their co-workers have also prepared the stilbene (1), as well as other synthetic analogues of the strobilurins.[9]

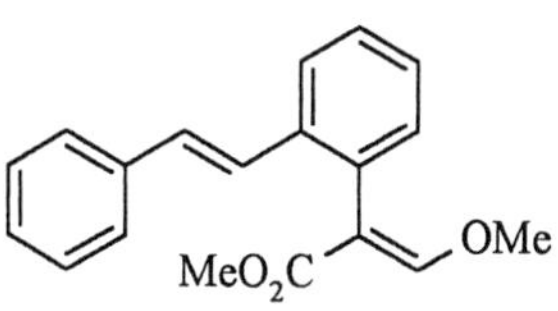

MeO_2C OMe

(1)

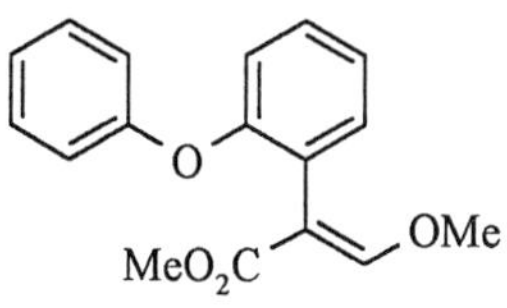

O MeO_2C OMe

(2)

O O MeO_2C OMe

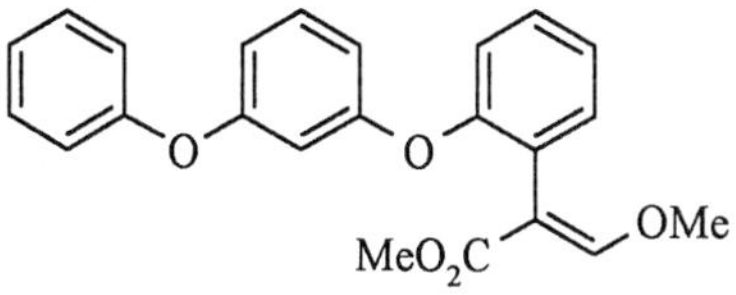

(3)

Photostability was improved still further by replacing the styryl side-chain of the stilbene (1) with a phenoxy group. The resulting diphenyl ether (2) was found to control the growth of a variety of commercially-important fungi in the field. In addition, it has the important property of systemic movement within plants, a key attribute of modern fungicides which improves performance through redistribution of the compound in plant tissues after application.

While examining the scope for the introduction of substituents in the diphenyl ether (2) we discovered that the tricyclic compound (3) gave improved levels of activity. However, (3) is not systemic because it is too lipophilic. Extensive further work showed that certain heterocyclic analogues of (3) retain high activity and, as reflected by their lower partition coefficients, have systemic movement in plants. Azoxystrobin was the culmination of this work. It was clearly the best of about 1400 analogues of strobilurin which we prepared and tested, and met the three key bioefficacy objectives of our programme of synthesis, namely breadth of spectrum, high intrinsic activity and systemicity.

Azoxystrobin is an outstanding systemic fungicide with a novel mode of action and excellent bioefficacy, safety and environmental properties.[10] Extensive field trials against economically important fungal pathogens of more than 60 crops have established that it gives robust disease control in a wide variety of climates and environmental conditions. Because it has a novel biochemical mode of action, azoxystrobin is effective against fungi which have developed reduced sensitivity to other classes of fungicide, such as phenylamides, dicarboximides, benzimidazoles and inhibitors of ergosterol biosynthesis. Azoxystrobin was registered in Germany in early 1996 for use on cereals, and further registrations for use on a broad range of crops around the world are expected during the next three years. We confidently predict that azoxystrobin will have a major impact on the global market for agricultural fungicides.

3 STROBILURIN A AND ANALOGUES AS HUMAN ANTIFUNGALS

Strobilurin A itself (originally known as "mucidin"), obtained by fermentation, has been used as a topical antifungal in human and veterinary medicine under the trade-name "Mucidermin Spofa". Further details and primary references are given in reference 6. Of course, the important modern human antifungals are mainly systemically active compounds, dosed either orally or intravenously. With this in mind, representative examples of our large collection of strobilurin analogues were tested for systemic activity in mouse models of fungal infection, but little or no activity was observed. There is also other information which suggests that strobilurin analogues are not attractive starting points for medicinal chemical research. For example, there is evidence that the spectrum of activity of these compounds *in vitro* has a poor fit with key human fungal pathogens. Furthermore, there is a dearth of patent applications which claim strobilurin analogues *primarily* for use in medicine, in sharp contrast to the large number (more than 320 by June 1996) which claim compounds of this type as agricultural fungicides.

Acknowledgements

We thank our many colleagues at Zeneca Agrochemicals who have participated in the azoxystrobin project. We also thank David Roberts and Graham Dixon at Zeneca Pharmaceuticals for the results and information about our strobilurin analogues as potential human antifungals.

References

1. K. Beautement, J. M. Clough, P. J. de Fraine and C. R. A. Godfrey, *Pestic. Sci.*, 1991, **31**, 499.
2. J. M. Clough, P. J. de Fraine, T. E. M. Fraser and C. R. A. Godfrey, in 'Synthesis and Chemistry of Agrochemicals III', Eds. D. R. Baker, J. G. Fenyes and J. J. Steffens, ACS Symposium Series No. 504, American Chemical Society, Washington, DC, 1992, p. 372.
3. J. M. Clough, D. A. Evans, P. J. de Fraine, T. E. M. Fraser, C. R. A. Godfrey and D. Youle, in 'Natural and Engineered Pest Management Agents', Eds. P. A. Hedin, J. J. Menn and R. M. Hollingworth, ACS Symposium Series No. 551, American Chemical Society, Washington, DC, 1994, p. 37.
4. J. M. Clough, V. M. Anthony, P. J. de Fraine, T. E. M. Fraser, C. R. A. Godfrey, J. R. Godwin and D. Youle, in 'Eighth International Congress of Pesticide Chemistry: Options 2000', Eds. N. N. Ragsdale, P. C. Kearney and J. R. Plimmer, ACS Conference Proceedings Series, American Chemical Society, Washington, DC, 1995, p. 59.
5. J. M. Clough and C. R. A. Godfrey, *Chem. in Brit.*, 1995, **31**, 466.
6. J. M. Clough, *Nat. Prod. Reports*, 1993, **10**, 565.
7. W. F. Becker, G. von Jagow, T. Anke and W. Steglich, *FEBS Letts.*, 1981, **132**, 329.
8. K. Beautement and J. M. Clough, *Tet. Letts.*, 1987, **28**, 475.
9. T. Anke and W. Steglich, in 'Biologically Active Molecules, Identification, Characterization and Synthesis', Ed. U. P. Schlunegger, Springer-Verlag, Berlin and Heidelberg, 1989, p.9.
10. J. R. Godwin, V. M. Anthony, J. M. Clough and C. R. A. Godfrey, *Brighton Crop Prot. Conf.: Pests and Diseases - 1992*, Vol. 1, British Crop Protection Council, Farnham, U.K., 1992, p. 435.

14
Advances in the Chemistry of Novel Broad-Spectrum Orally Active Azole Antifungals: Recent Studies Leading to the Discovery of SCH 56592

A. K. Saksena,* V. M. Girijavallabhan, R. G. Lovey, R. E. Pike, H. Wang, Y. T. Liu, P. Pinto, F. Bennett, E. Jao, N. Patel, J. A. Desai, D. F. Rane, A. B. Cooper and A. K. Ganguly
SCHERING-PLOUGH RESEARCH INSTITUTE, KENILWORTH, NJ 07033, USA

INTRODUCTION

Therapeutic breakthroughs in certain areas of medicine (e.g. diabetes mellitus, cancer chemotherapy, extensive surgery and organ transplantation) have created a sharp increase in the number of immunocompromised hosts highly susceptible to systemic fungal infections. Widespread incidence of life threatening *Candidiasis, Cryptococcosis* and *Aspergillosis* in patients with AIDS has underscored the need for safe and efficacious antifungal agents.

In marked contrast to other antiinfectives, the pace of discovery of newer antifungals for systemic infections has been rather slow. Thus for over forty years, Amphotericin B (AMB) remains the only broad-spectrum drug for the management of most systemic infections. However, severe toxicity of amphotericin B continues to be a major problem not easily dealt with even today.[1]

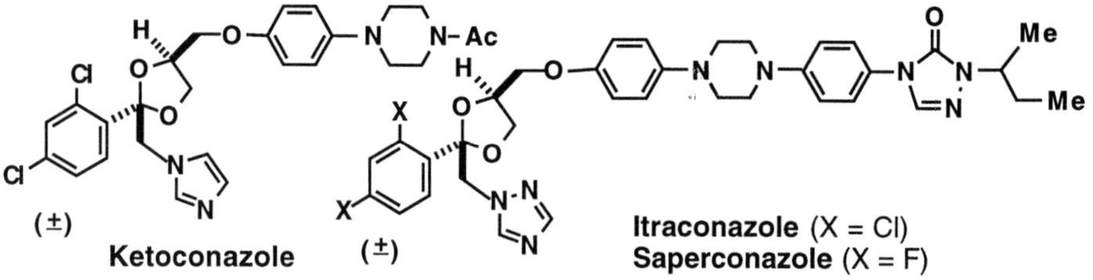

Ketoconazole (±)

Itraconazole (X = Cl)
Saperconazole (X = F)
(±)

The Azole Antifungals: Two decades of early research in the azole area led to only topically active agents until ketoconazole (KTZ, Janssen) the first orally effective agent was introduced into clinical practice. However, some of the side effects associated with prolonged therapy with KTZ (e.g., hepatotoxicity and testosterone biosynthesis inhibition) result from marked inhibition of mammalian cytochrome P-450-dependent enzymes. Due to the very nature of fungal growth inhibition by azoles, specific binding to fungal rather than mammalian cytochrome P-450 is imperative. Itraconazole (ITZ, Janssen), a relatively selective inhibitor of fungal cytochrome P-450, now in clinical use, offers certain advantages over ketoconazole in terms of oral efficacy, broader spectrum and fewer side effects.

ITZ is the most effective azole for the treatment of *Aspergillosis*.[2] Fluconazole (FLZ, Pfizer), the most recently introduced azole is a relatively narrow spectrum agent used extensively against *Candida* and *Cryptococcus* infections.[3] However it lacks efficacy against *Aspergillus* infections. There have been increasing reports of FLZ resistance in AIDS patients with oropharyngeal and/or esophageal *Candidiasis;* resistance to FLZ in *Cryptococcus neoformans* has also been reported.[4a,b]

OBJECTIVES:

At Schering-Plough we sought to develop a novel antifungal agent with broad-spectrum activity against the key pathogens *Candida, Histoplasma, Cryptococcus, Blastomyces* and *Aspergillus* with reduced host toxicity. The new agent should not be a P-450 enzyme inducer, should have superior activity against *Aspergillus* versus ITZ and comparable activity against *Candida* versus FLZ and ITZ. In view of reported emergence of intrinsically FLZ resistant strains of *C. glabrata* and *C. krusei*,[4a,b] activity against these pathogenic strains would also be a very desirable attribute in a new azole antifungal. Further, where oral administration is impractical such as in critically ill AIDS and cancer patients, a water soluble version of the new clinical entity would be of definite advantage.

THE NATURE OF THE PROBLEM

The major features common to the three Janssen azoles viz KTZ, ITZ and SPZ include a trisubstituted 1,3-dioxolane having a defined *cis*-relationship between a azolyl methyl group and a lipophilic phenylpiperazinyl moiety. Early work on new analogs based on Janssen leads in several laboratories centered on a combination of azolylmethyl and phenylpiperizinyl group either in a dithiolane ring [I] or in conformationally mobile acyclic systems [II]. These analogs as well as several hundred isoxazolidines [III] incorporating elements of Janssen azoles did not provide a single analog worthy of preclinical evaluation.[3]

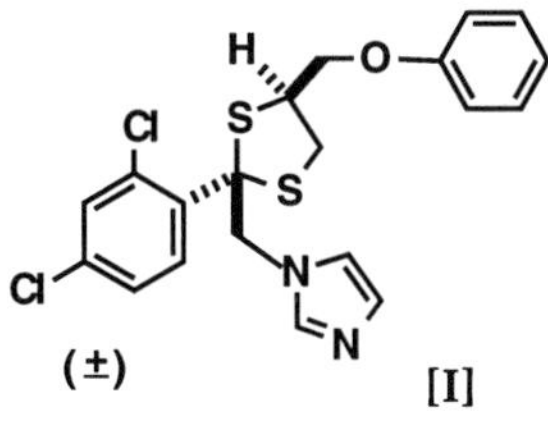

(±) [I] 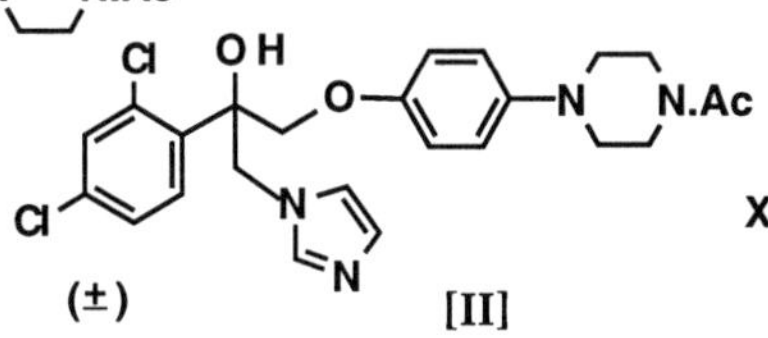 (±) [II] 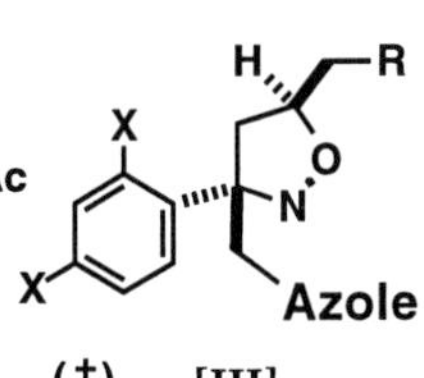 (±) [III]

WORKING HYPOTHESIS:

The ketal functionality present in Janssen azoles has the potential to get at least partially inactivated in the stomach by acid hydrolysis. Any modification that would inhibit such inactivation but retain other physicochemical properties, may provide agents with improved absorption and efficacy.

Early Explorations: [5] In meeting the above criteria, it was of interest to explore whether novel analogs of the type **A**, but especially types **B** to **F** would have certain desirable attributes in terms of oral potency, antifungal spectrum, pharmacokinetics and lack of toxicity. Oxygen containing heterocycles being relatively flexible, it was not apparent at the outset in what manner the subtle conformational differences between these analogs would be reflected in their *in-vivo* (PO) antifungal profile.

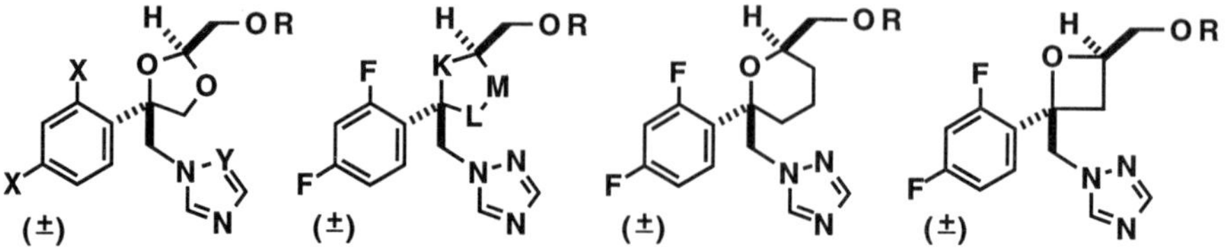

(±) **A:** X = F; Cl; Y = N; CH

(±) **B:** K=O; L=M=CH_2
C: L=O; K=M=CH_2
D: M=O; K=L=CH_2

(±) **E**

(±) **F**

Chemistry involved in the stereoselective construction of these ring systems having a defined *cis* relationship between the triazolylmethyl and the nonligating lipophilic groups (OR") was not well documented. A logical understanding of structure-activity relationship also placed an important constraint that we bring about only one change at a time; in the present context the nature of the central ring.

In early studies we had shown that the acetal analogs of the type **A** were equiactive to KTZ *in-vitro* and *in-vivo* (topical; *Candida*). Although at least two analogs from this series had oral activity comparable to KTZ and ITZ, their level of activity at 100 mpk did not offer ample prospects for any further improvement.

Based on our original premise, the tetrahydrofurans of the types **B-D** were likely to be even more stable in a hydrolytic/metabolic sense. We were encouraged to find that these compounds were clearly more active than the type **A** analogs in mouse *Candida* as well as *Aspergillus* models. Of further interest is the fact that analogs of the types **B** and **C** were more active than KTZ and ITZ at a dose of 10 mpk; the type **C** being the most active. *This interesting result could not have been predicted at the outset.*

We may draw attention to the fact that Janssen azoles did not provide adequate protection at doses below 100 mpk in the same mouse *Candida* infection model. In contrast even our least active type **D** analogs had excellent protection at 50 mpk.[5c]

Importance of the Central Ring Size: While the above modifications were in progress, we showed that increasing or decreasing the central ring size by one carbon (viz 6 and 4-membered analogs) had a profoundly adverse effect on oral activity.[5, 6] Thus, the importance of a central 5-membered ring was amply demonstrated by the lack of *in-vivo* oral activity of analogs of the types **E** and **F.**

Role of Side Chains: For rapid identification of novel side chains, we had in hand a highly stereoselective route to 2,2,5-trisubstituted tetrahydrofurans via aqueous Diels-Alder reactions of electron deficient 2-aryl furans **1**. Use of an aqueous version of the Diels-Alder reaction made possible isolation of previously inaccessible adducts **2** which were successfully elaborated (via **3**) in a general stereocontrolled route to type **B** analogs.[7] As we did not have an efficient stereoselective route to the more active type **C** compounds at the time, we used type **B** template to introduce a host of side chains. This interim strategy proved invaluable later.

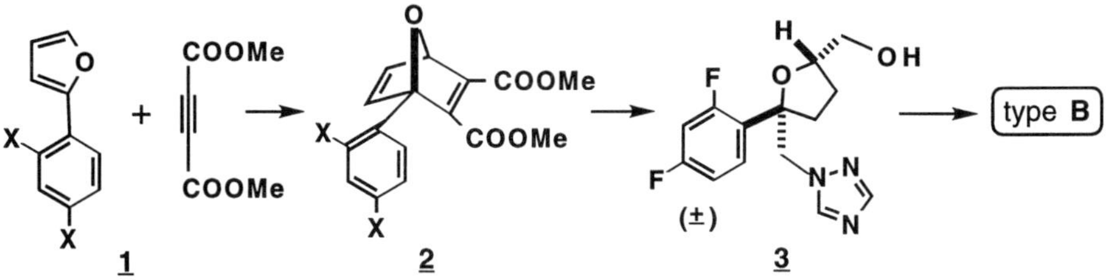

Thus, using type **B** analogs as examples, considerable oral activity was maintained with analogs having diverse side chains. The importance of an aromatic spacer group for oral activity also became clear from these studies. It was also shown that analogs having triazolyl functionality were more active (PO) than those containing imidazolyl group.[5, 7]

INITIAL LEAD: SCH 45012

Following modifications of types **B** and **C** analogs, Sch 45009 and Sch 45012 were synthesized and their activity compared with ITZ and saperconazole (SPZ) in systemic *Candida* and pulmonary *Aspergillus* infection models.[5a, c] Both Sch 45009 and Sch 45012 had comparable broad-spectrum activity *in-vitro*. However, in animal models, Sch 45012 was shown to be more active than Sch 45009, ITZ and SPZ.[5, 8b]

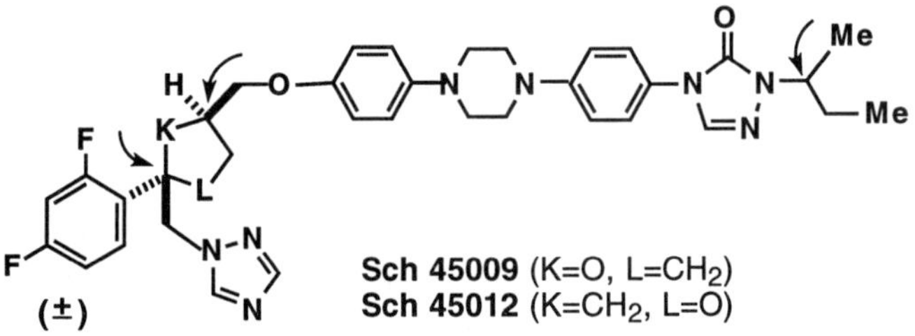

Sch 45009 (K=O, L=CH_2)
Sch 45012 (K=CH_2, L=O)

With its three chiral centers, two in a defined *cis*-stereochemical relationship (arrows), Sch 45012 is a mixture of four optical isomers. Having met our primary objectives, we sought all four stereoisomers of Sch 45012 for detailed biological evaluation and enzyme induction studies.

Chiral Epoxide Route to Sch 45012 Isomers:

Based on our synthesis of the racemic type **C** analogs, the Sharpless-Katsuki asymmetric epoxidation offered a promising approach. Indeed the readily available allyl alcohol **4** could be converted to both **5R** and **5S** epoxy alcohols which acted as key intermediates to provide respectively the **6S** and **6R** *cis*-tosylates from which all four stereoisomers of (±) Sch 45012 were derived.[8a]

Predictable course of the Sharpless-Katsuki asymmetric epoxidation and origin of the chiral side chains from the known (R) and (S)-2-butanols, enabled us to assign relative and absolute stereochemistry to each stereoisomer. Later, the structure of Sch 50002 was also verified by X-ray crystallography.[8d] (Scheme I)

Scheme I

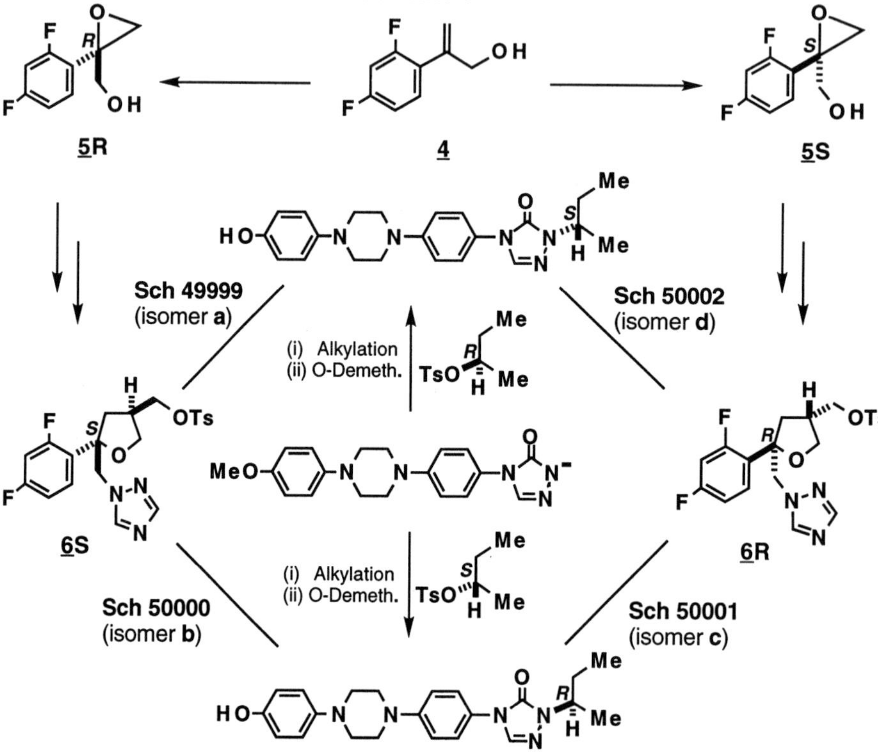

Initial supplies of all four isomers of (±) Sch 45012 provided sufficient material for preliminary *in vitro* and *in vivo* biological evaluation. Results indicated the two (5S)-(+) isomers, Sch 49999 and Sch 50000 to be virtually inactive "dystomers" and the two (5R)-(-) isomers, Sch 50001 and Sch 50002 identified as the two "eutomers" of (±) Sch 45012.[8a, c]

Although we had made significant progress, we were far from a situation to provide sufficiently large quantities of Sch 50001 and Sch 50002 for detailed biological evaluation and enzyme induction studies. *Nonethless, at this point we had discovered an important absolute stereochemistry requirement for optimal oral activity in a given series .*

A Rapid Synthesis of Sch 45012 Isomers via (±) *cis*-Tosylate: The racemic *cis*-tosylate **6**, available in large amounts at the time, when coupled with the phenylpiperazinyl side chains derived from (R) and (S)-2-butanols produced a pair of mixtures. These two-component sets were composed of in each case, the so-called isomers **a**/**d** (Sch 49999/Sch 50002) and **b**/**c** (Sch 50000/Sch 50001) respectively. Unlike the original four-component mixture of Sch 45012 isomers, preparative separation of these two pairs on a Chiralcel® OD column proved relatively trivial. In this manner adequate quantities of all four isomers were quickly made available for detailed biology.

SCH 50001 vs SCH 50002

In terms of *in-vitro* activity both Sch 50001 and Sch 50002 were equiactive. Both compounds were devoid of liver P-450 enzyme inducing capability. *In-vivo* however, Sch 50002 was more efficacious (PO) than Sch 50001 in *Candida* as well as *Aspergillus* infection models, both compounds being clearly more efficacious than ITZ and SPZ.[8]

ACHIRAL SIDE CHAIN SUBSTITUTION:

With the identification of Sch 50002 as an excellent lead structure, concerted efforts were devoted to introduction of a variety of side chains. A number of saturated, unsaturated and polar side chains were shown to have excellent *in vitro* activity. In general, branched alkyl chains appeared to have a desirable effect on *in vivo* activity. The presence of either relatively basic or acidic side chains also resulted in profound loss in activity.[9]

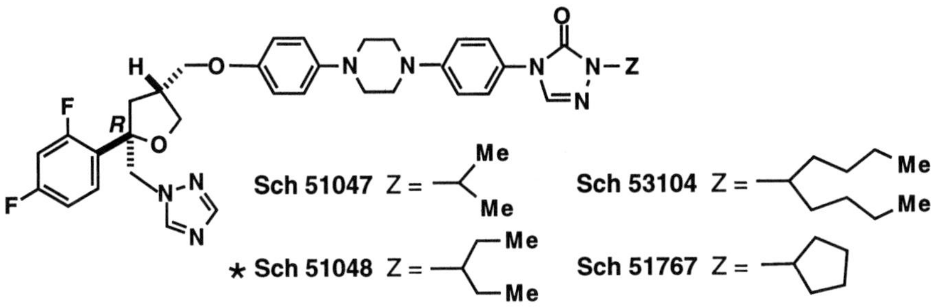

Sch 51047 Z =
Sch 53104 Z =
★ Sch 51048 Z =
Sch 51767 Z =

SCH 51048, A BROAD-SPECTRUM ANTIFUNGAL:

In a most productive study aimed at elimination of chirality of the side chain of Sch 50002, two analogs, Sch 51047 and Sch 51048 were prepared. Interestingly the 3-pentyl analog, Sch 51048 was found to have greatly improved therapeutic potential over Sch 50002 and other existing drugs against a variety of systemic fungal infections in normal and immunocompromised infection models; the isopropyl analog, Sch 51047 was the least active compound in this comparison. Sch 51767, a conformationally rigid cyclopentyl side-chain analog, was also shown to have reduced albeit significant percent survival compared to ITZ. The lack of activity of Sch 53104 a 9-carbon branched chain analog highlighted the importance of optimal carbon chain length.

Biology of Sch 51048 and other analogs has been discussed in some detail elsewhere.[9] In view of this new development, further efforts on Sch 50002 were curtailed in support of this promising new azole. With one chiral center eliminated from its side chain, Sch 51048 would also be a relatively less expensive compound to synthesize.

NEW SYNTHETIC ROUTES TO SCH 51048 AND ANALOGS:

Synthetic problems associated with our original route to the *cis*-tosylate **6R** were mainly associated with our inability to take advantage of the chirality induced at the benzylic carbon via the Sharpless-Katsuki epoxidation, an excellent step by itself. In the meantime a few kg of Sch 51048 was made available for preclinical toxicity by this first route in its original tedious version.

Attempts to overcome these problems required extensive exploratory chemistry culminating in at least three practical enantioselective routes to Sch 51048 and its analogs.

ACYCLIC STEREOSELECTION *VIA* HALOCYCLIZATION :

A conceptually different approach not involving chiral epoxide intermediate was next considered. Halocyclization is an effective way to construct the tetrahydrofuran ring system although stereoselectivity is often difficult to predict or rationalize. The major activity and success in this regard has been in the formation of 2,5-substituted tetrahydrofurans in which the chiral centers in the product bears a 1,3-relationship.[10a] These cyclizations not unexpectedly lead to the thermodyanamically more stable *trans*-products. In contrast Rychnovsky and Bartlett have shown that cyclization of appropriate benzyl ether derivatives can give thermodyanamically less stable *cis*-2,5-substituted tetrahydrofurans with high diastereoselectivity.[10b]

A careful search in the literature led us to conclude that no examples existed of stereoselective halocyclizations leading to the uncommon *2,2,4-cis-trisubstituted tetrahydrofurans* needed in our present studies. In any case we speculated that the bulky 2,4-difluorophenyl substituent on the electrophilic olefinic end may enforce a directive influence in favor of the desired 2,2,4-trisubstituted tetrahydrofurans. Two approaches were investigated which are now described.

(i) 5-Endo-trig-cyclization Approach: Our original idea was to take advantage of halocyclization of a chiral olefin such as 7. Specific conditions to obtain either 2,5-*cis* or *trans* substituted tetrahydrofurans via 5-endo halo- (and seleno) cyclizations of β-hydroxyalkenes have been described. Interestingly depending on the nature of electrophile, either E or Z olefins could be used in these cases as long as their geometrical homogeniety was ensured.[10c] It appeared *2,2,4-cis-trisubstituted* tetrahydrofurans could also be obtained by application of this methodology.

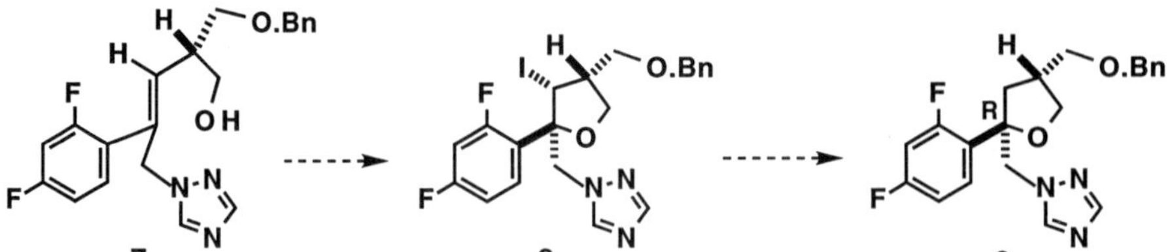

7 ------> 8 ------> 9

Due to lack of energetic barriers, geometrically homogeneous trisubstituted olefins are all the more difficult to prepare. We succeeded in devising a specific methodology to prepare exclusively the Z olefin **12** by Horner-Wittig reaction of the 2,4-difluorophenyl pyruvate ester **10** with the phosphine oxide **11**.[11] Conversion of **12** to the desired olefin **7** could be foreseen as a straightforward

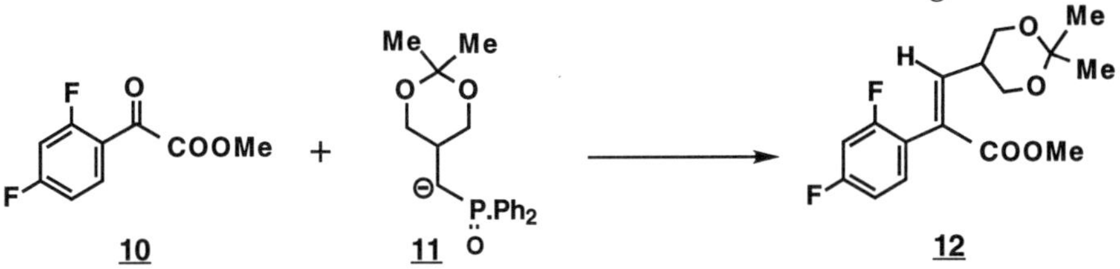

10 + 11 → 12

transformation. But we did not pursue this route, as a more promising approach involving a novel 5-exo type halocyclizations emerged.

(ii) 5-Exo-cyclization Approach: While the above studies were in progress, we discovered a unprecedented yet remarkably simple 2,4-diastereoselective 5-exo halocyclization of 2,2-disubstituted olefins.[12] Synthesis of the key olefinic homochiral diol **15** was accomplished quite simply and in high yields from the allyl alcohol **4**. (Scheme II)

In the very first iodocyclization attempt at room temperature, the *cis:trans* ratio of the iodoalcohols **16** and **17** was an impressive 88:12. This stereoselectivity improved at lower temperatures (0°—>5°C) to better than 90:10 in favor of the desired *cis*-diastereomer **16**. An attractive feature of this 5-exo-cyclization approach is that the resulting iodomethyl functionality presents itself as an electrophile for the direct introduction of the azolyl moiety. The scope and limitations of this 2,4-diastereoselective 5-exo-halocyclization process have been discussed elsewhere. The presence of a 2-aryl substituent on the olefinic end appears to dictate the observed 2,4-diastereoselectivity.[12]

Scheme II

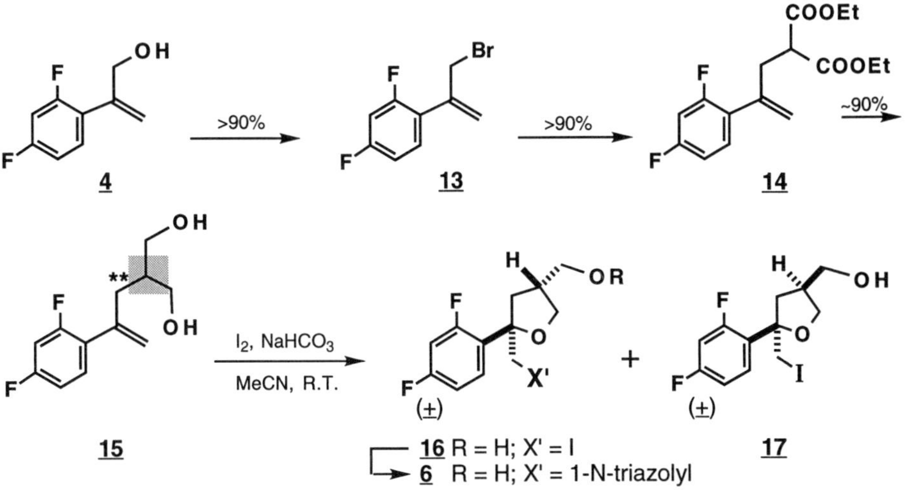

4 → (>90%) 13 → (>90%) 14 → (~90%) 15

15 → (I_2, $NaHCO_3$, MeCN, R.T.) 16 (±) + 17 (±)

16 R = H; X' = I

6 R = H; X' = 1-N-triazolyl

Displacement of iodine in a neopentyl-like system by the highly nucleophilic triazolyl anion posed no problems. In preliminary experiments, protection of the alcohol functionality during triazole anion displacement was considered necessary only as a precaution to prevent intramolecular oxy-anion displacement of iodine. As we will show later this was found quite unnecessary thus saving two steps. In this manner we accomplished the shortest and most efficient synthesis of the racemic *cis*-tosylate **6**.

From this success it appeared that we were close to a breakthrough if only we could devise an efficient synthesis of a desymmetrized derivative of the diol **15**. An elegant enzymatic desymmetrization of **15** will be described later. In this conceptually distinct approach, the chiral induction on the benzylic carbon would follow by the very nature of the iodocyclization process.

ASYMMETRIC SYNTHESIS *VIA* CHIRAL IMIDE-ENOLATES:

Enantioselective chemical routes to **6R** (and **6S**) were viewed as being directly accessible via the chiral imide technology developed independently by the Evans[13a] and Oppolzer[13b] groups. A wide choice of electrophiles, excellent diastereoface selective alkylations, and aldol condensations led us to choose Evans' methodology. However it was not clear at the outset in what manner the 2,2-disubstituted olefin functionality placed β to the site of an enolate would influence overall efficiency of electrophilic substitutions; to the best of our knowledge these type of substrates have not been studied earlier.[14]

(a) From Valinol Derived Chiral Auxilliary: In a one-pot sequence, the readily available allyl alcohol **4** was subjected to Claisen-Johnson orthoester rearrangement (via **18**) providing the olefinic acid **19** in virtually quantitative yield. Our first experiments were conducted with lithium enolates of **20** using benzyloxymethyl bromide as the electrophilic reagent. Although high diastereoselectivity of the desired benzyl ether **21** was observed, the chemical yields were unacceptably low (~30%).

Scheme III

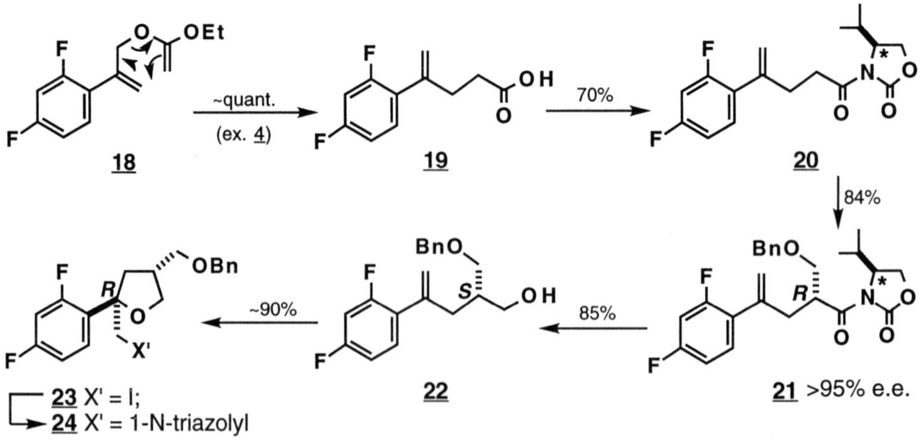

In a marked improvement, alkylation of **20** with benzyloxymethyl chloride via Evans' titanium enolate protocol[13c] gave **21** in excellent yields. Most importantly no isomerization of the 2,2-disubstituted olefin was detected during either lithium or titanium enolate formation, a requirement crucial to success in this chiral imide route to **6R**. Reduction of **21** with LAH in tetrahydrofuran gave the desired (S)-diol monobenzyl ether **22** (85% yield) with the oxazolidinone auxilliary recovered in 70% yield by chromatography. The remaining steps from the iodocyclization onwards proceeded as expected according to Scheme II. Thus the desired (-)-(2R)-*cis*-tosylate **6R** was obtained in excellent yields and over 99% optical purity. (Scheme III)

(b) From Phenylalinol Derived Chiral Auxiliary: The following alternative approach was undertaken as we were keen to avoid yield limiting protection-deprotection operations. The allyl alcohol **4** we had been using so far was prepared in 4-steps from 2,4-difluorobenzene. Thus synthesis of the olefinic acid **19** used above required a total of 6-steps. We were able to obtain **19** in two simple steps by Friedel-Craft reaction of 2,4-difluorobenzene with succinic anhydride to provide the keto acid **25**. Wittig reaction of **25** with 2 equivalents of methylene triphenyl phosphorane then gave **19** in ~60% yield over 2 steps.

Scheme IV

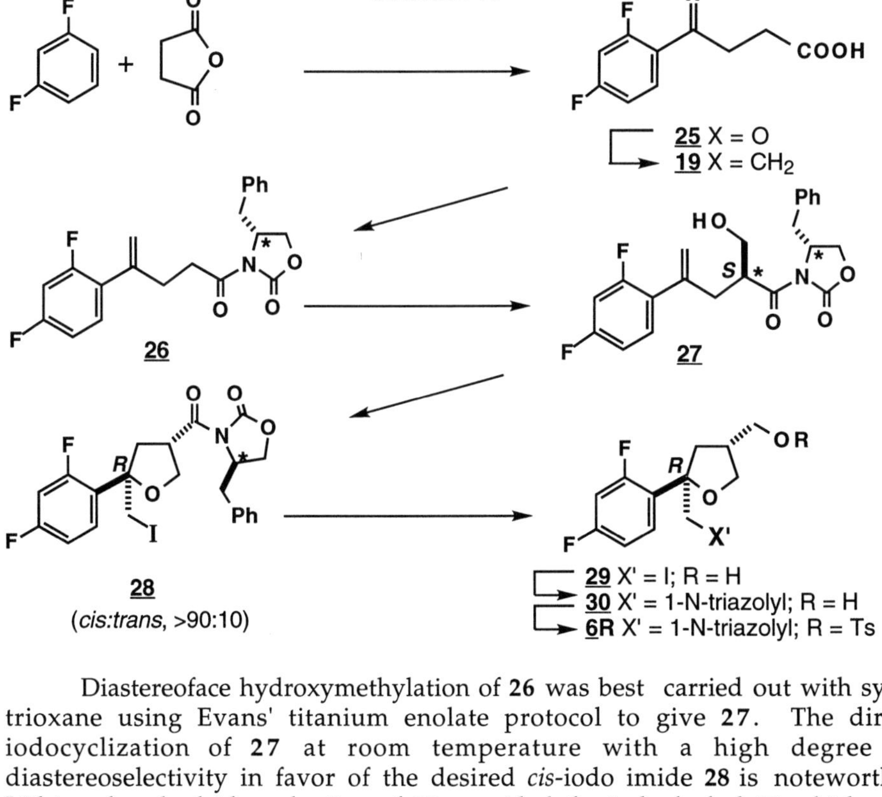

25 X = O
19 X = CH_2

26

27

28
(*cis:trans*, >90:10)

29 X' = I; R = H
30 X' = 1-N-triazolyl; R = H
6R X' = 1-N-triazolyl; R = Ts

Diastereoface hydroxymethylation of **26** was best carried out with sym-trioxane using Evans' titanium enolate protocol to give **27**. The direct iodocyclization of **27** at room temperature with a high degree of diastereoselectivity in favor of the desired *cis*-iodo imide **28** is noteworthy. Lithium borohydride reduction of **28** provided the iodoalcohol **29** which was displaced directly with sodium-triazole followed by tosylation; the **6R** *cis*-tosylate so obtained was identical in all respects to the authentic material. (Scheme IV)

Notably, as we had hoped, no protecting groups were required throughout this entire sequence. This "protecting group free" new route to **6R** offers excellent opportunity for large scale operations.[14]

Dichloro Analogs: No problems arose in the successful application of the above 2,4-diastereoselective 5-exo iodocyclizations combined with Evans' chiral imide enolate chemistry to prepare the dichloro analogs of Sch 50001, Sch 50002 and Sch 51048. As reported earlier,[12] presence of the bulkier ortho-chlorine substituent considerably slowed down the rate of iodocyclization but the *cis:trans* selectivity was still a respectable 78:28. These dichloro analogs had excellent broad-spectrum antifungal activity *in vitro* but they were found to be less active in animal infection models than the corresponding difluoro analogs.

31

Sch 55198, Z = (R)-[CH(Me)Et]
Sch 55200, Z = (S)-[CH(Me)Et]
Sch 55202, Z = $CH(Et)_2$

HYDROXYLATED ANALOGS OF SCH 51048:

In an effort to further improve oral absorption properties of Sch 51048, sythesis of analogs having polar side chains was undertaken. An added incentive behind this investigation was to see if we could prepare water soluble salts and/or derivatives from these polar side chains for possible parenteral use.

Key Triazolone: A synthesis of the triazolone **35** via a linear sequence was reported earlier.[9] In a more direct approach to this key intermediate, the side chain triazolone **32** was first demethylated, then partially N-SEM protected to provide a suitable fragment **33** which was used to alkylate the *cis*-tosylate **6R**. The resulting N-SEM protected intermediate **34** from this convenient source was used to introduce a variety of side chains. (Scheme V)

Sacrificing methodology development and efficiency for speed, the two isomeric olefins **36** and **37** and the allylic alcohol **38** were converted to the mesylates **39**, **40** and **41** which could be coupled to the triazolone core **35** in the presence of sodium hydride. This effort took barely three weeks to give the resulting three pairs, Sch 55801, Sch 55803 and Sch 55805, in amounts sufficient for MICs. Biological evaluation of these diastereomeric pairs provided unexpectedly promising results. *Thus while introduction of relatively basic or acidic side chains* *resulted in reduced activity, incorporation of hydroxyl groups led to greatly improved* *in-vivo* activity. (Scheme VI)

Scheme V

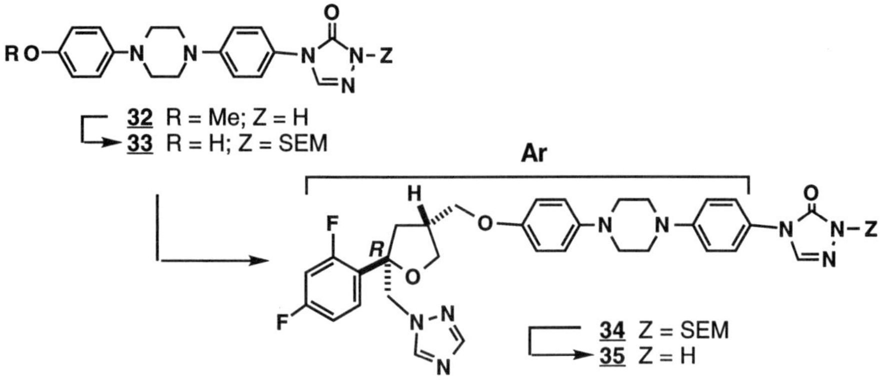

32 R = Me; Z = H

33 R = H; Z = SEM

Ar

34 Z = SEM

35 Z = H

Scheme VI

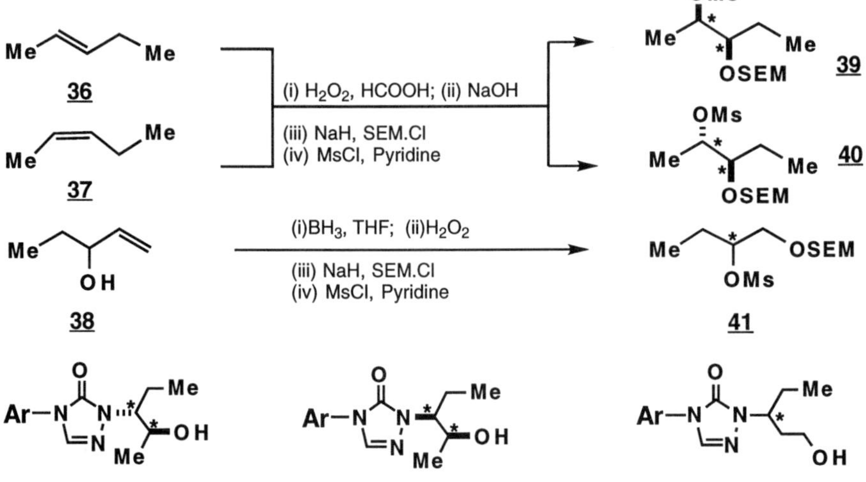

36

37

38

(i) H_2O_2, HCOOH; (ii) NaOH

(iii) NaH, SEM.Cl

(iv) MsCl, Pyridine

(i)BH_3, THF; (ii)H_2O_2

(iii) NaH, SEM.Cl

(iv) MsCl, Pyridine

39

40

41

SCH 55801 (*R,R/S,S) **SCH 55803** (*R,S/S,R) **SCH 55805** (*R/S)

Coupling Step: Compared to the methanesulfonate used above, the bromophenylsulfonate group proved to be a better choice leading to nearly complete reactions and higher isolated yields of N-alkylated products. Analysis of comparative kinetic data for substitution and elimination reactions of various leaving groups led us to surmise that a spatially more bulky group would actually be more desirable here and one with higher acidity of its conjugate acid. At the same time, *use of cesium carbonate*[15] as base in place of sodium hydride or hydroxide significantly improved N-alkylation yields in reactions virtually free of by-products.

With the new procedure to synthesize the triazolone segment **35** and efficient methodology to attach hydroxylated side chains in place, we now describe synthesis of all six hydroxylated analogs of Sch 51048.

(i) Secondary Hydroxy Analogs:

Among the various approaches that we considered, use of (R) and (S)-lactic acids appeared most appropriate as chiral sources to prepare the four secondary hydroxy analogs of Sch 51048. This may be exemplified in the synthesis of Sch 56586 and Sch 56590, the (R),(R) and (R),(S) isomers respectively.

Protection of (R)-methyl lactate as a benzyloxymethyl (BOM) ether followed by DIBAL reduction provided the aldehyde **42**. Grignard reaction of **42** with EtMgBr in tetrahydrofuran gave the (S),(R) and (R),(R) monoprotected diols **43** and **44** (~50:50) which were separated by chromatography. Paradoxically, this lack of stereoselectivity in the presence of tetrahydrofuran by design, proved extremely useful especially since we needed initially both **43** and **44**. The brosylates **45** and **46** derived from these alcohols were then used to N-alkylate the triazolone **35** according to conditions described above. Acid catalysed deprotection of the BOM ether finally gave the targeted Sch 56586 (R),(R) and Sch 56590 (S),(R) analogs.

Scheme VII

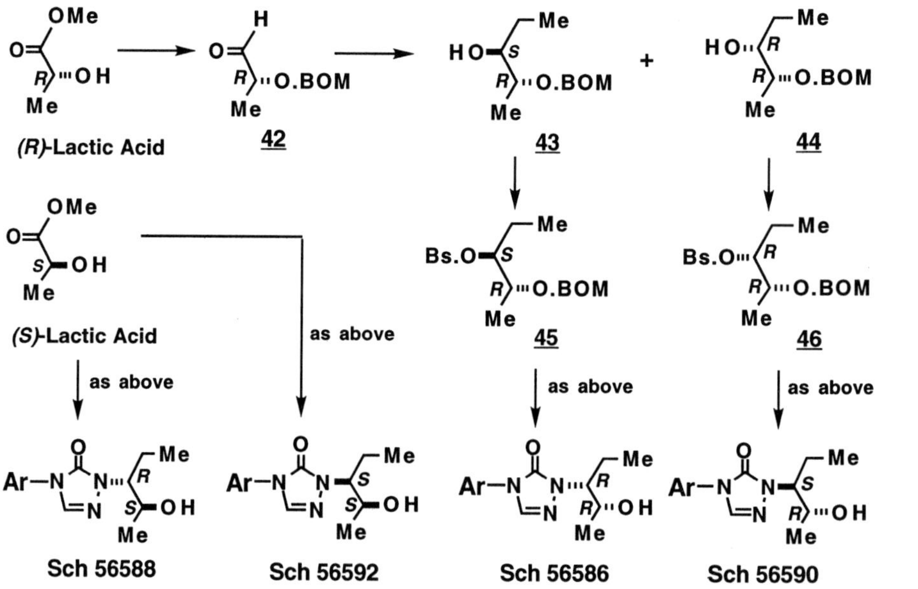

(R)-Lactic Acid
42
43
44
(S)-Lactic Acid
as above
45
46
Sch 56588
Sch 56592
Sch 56586
Sch 56590

Following the above set of reactions using (S)-methyl lactate, the remaining secondary hydroxy analogs, Sch 56588 (R),(S) and Sch 56592 (S),(S) were synthesised.

(ii) Primary Hydroxy Analogs:

The commercially available methyl (R)-3-hydroxyvalerate acted as a common source for the two remaining optical isomers of hydroxylated Sch 51048. LAH reduction of (R)-3-hydroxyvalerate gave the diol, which was selectively protected as the mono SEM ether **47**. A portion of this monoprotected (R)-diol was used to provide the monoprotected (S)-alcohol **49**. Now both (S) and (R)-

MICs of Polar Side Chain Analogs:

(mcg/ml)

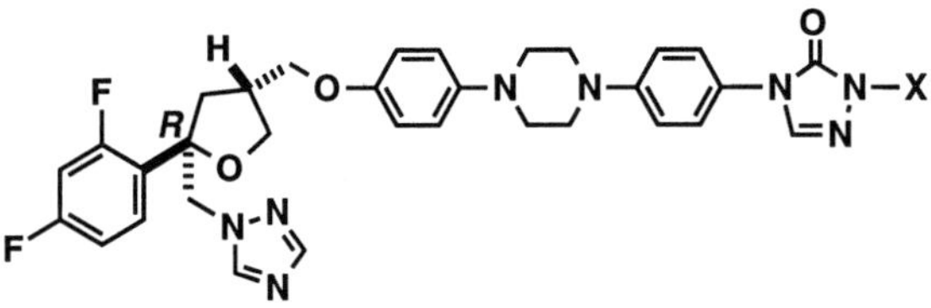

Sch No.	X	*Ca. alb.* N=31	*Asp.* N=37	Sch No.	X	*Ca. alb.* N=31	*Asp.* N=37
54160	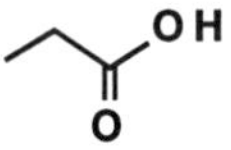	0.790	40.30	57138	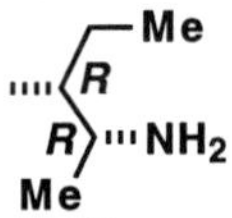	0.010	2.370
56594	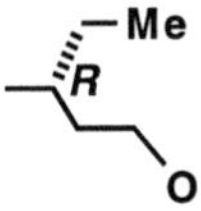	0.021	0.088	56998	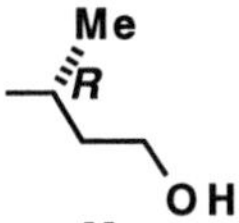	0.012	0.060
56596	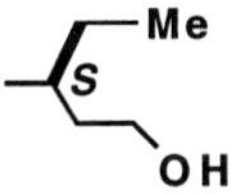	0.025	0.077	56996	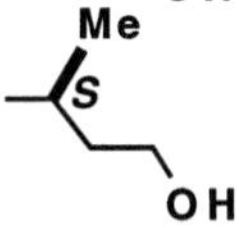	0.009	0.060
57033	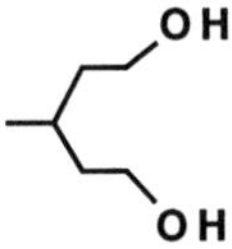	0.198	0.810	56810 (Syn-mix.)	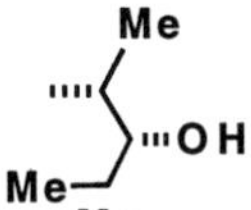	0.006	0.090
56586	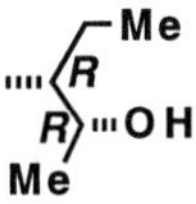	0.014	0.077	56986	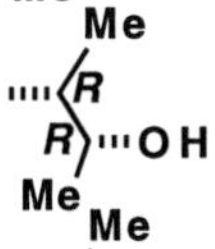	0.006	0.090
56588	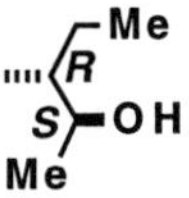	0.005	0.130	57276	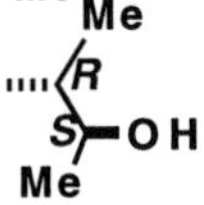	0.030	0.080
56590	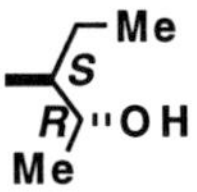	0.019	0.086	56984	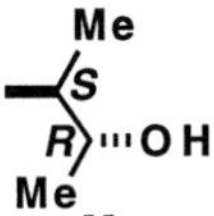	0.007	0.050
56592	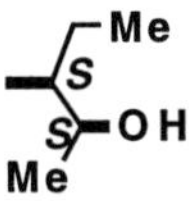	0.018	0.048	57270	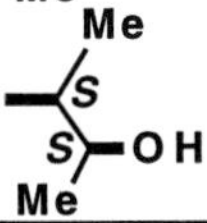	0.020	0.060

alcohols were converted to the brosylates **48** and **50** respectively. Finally, N-alkylation of the triazolone **35** with these brosylates under the above established conditions followed by deprotection, completed synthesis of Sch 56594 and Sch 56596, the two primary hydroxy isomers.

Scheme VIII

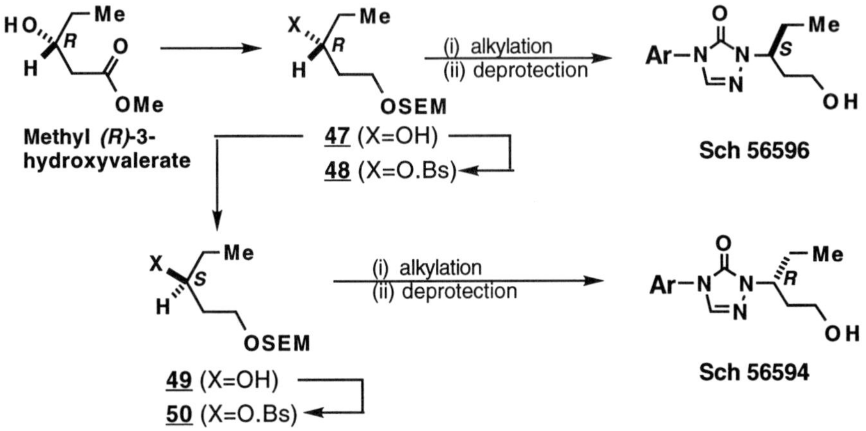

Methyl *(R)*-3-hydroxyvalerate

(i) alkylation
(ii) deprotection

47 (X=OH)
48 (X=O.Bs)

Sch 56596

(i) alkylation
(ii) deprotection

49 (X=OH)
50 (X=O.Bs)

Sch 56594

This successfully completed synthesis of *all six* optical isomers of hydroxylated Sch 51048 for *in vivo* evaluation. Although presented in only its essential details, the entire effort was a major undertaking on our part involving considerable methodology development.

(iii) Synthesis of Additional Hydroxy Analogs:

Synthetic methodology devised in the course of preparing the above six hydroxy analogs was successfully applied to synthesis of a variety of rational hydroxylated analogs of Sch 51048 within a short period of time. For example all eight hydroxylated analogs of Sch 50001/Sch 50002 were synthesized. In all, over 50 hydroxylated analogs were synthesized for biological evaluation. Without going into a detailed discussion of synthetic operations, we are giving MICs of some of the more significant analogs. One purpose of this investigation was to look for more accessible hydroxy side chains which may show at least comparable antifungal activity *in vivo* to the original six hydroxylated analogs of Sch 51048.[18]

SELECTED COMPOUND: SCH 56592

Preliminary biological evaluation (*in vitro*) showed most hydroxylated analogs of Sch 51048 as having excellent broad-spectrum antifungal activity against *Candida albicans* (31 strains) and *Aspergillus* (38 strains). The *in vitro* activity of a selected number of analogs was compared with ITZ, FLZ and AMB against 285 strains of yeasts and filamentous fungi, comprising 19 different genera and 37 different species, including FLZ resistant strains.[19] In order to select one compound with the best overall biological profile for further preclinical evaluation Sch 56588, Sch 56592 and Sch 56984 were subjected to detailed *in vivo* evaluation in several

Biology

C. albicans C72 Systemic infection in Mice

RX: PO- 1xDx4D- 2.5 MG/KG

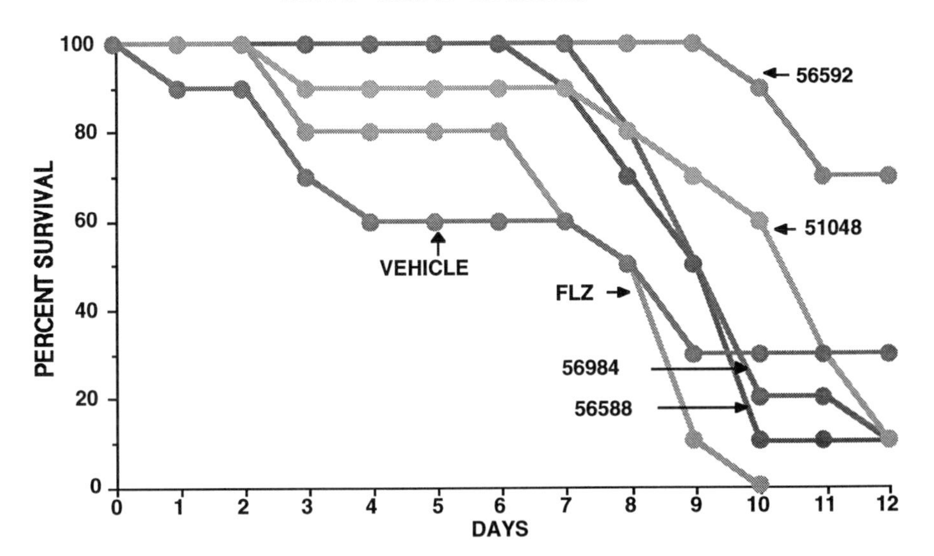

A. fumigatus, ND159 Pulmonary infection in Mice

RX: PO-1xDx4D- 10 MG/KG

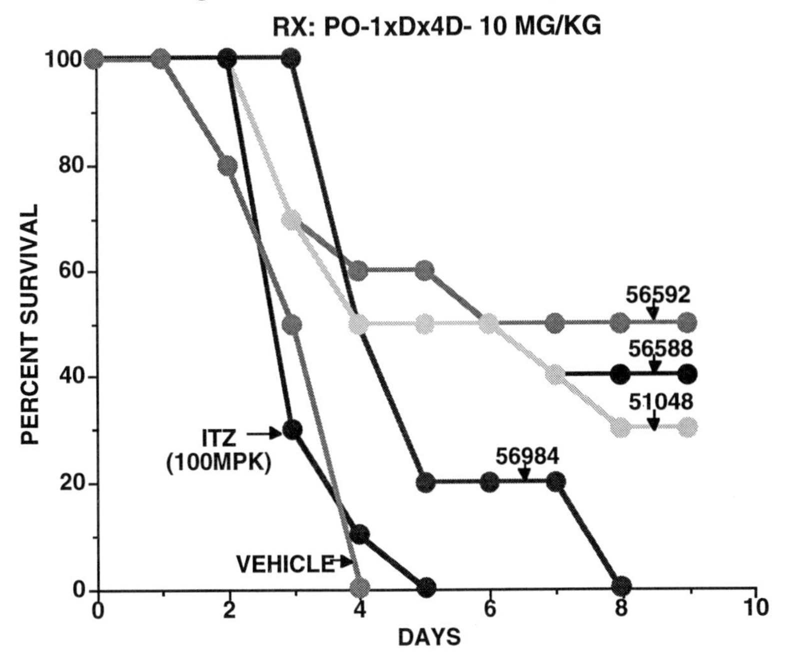

Mouse Dose Response (PO)
HPLC

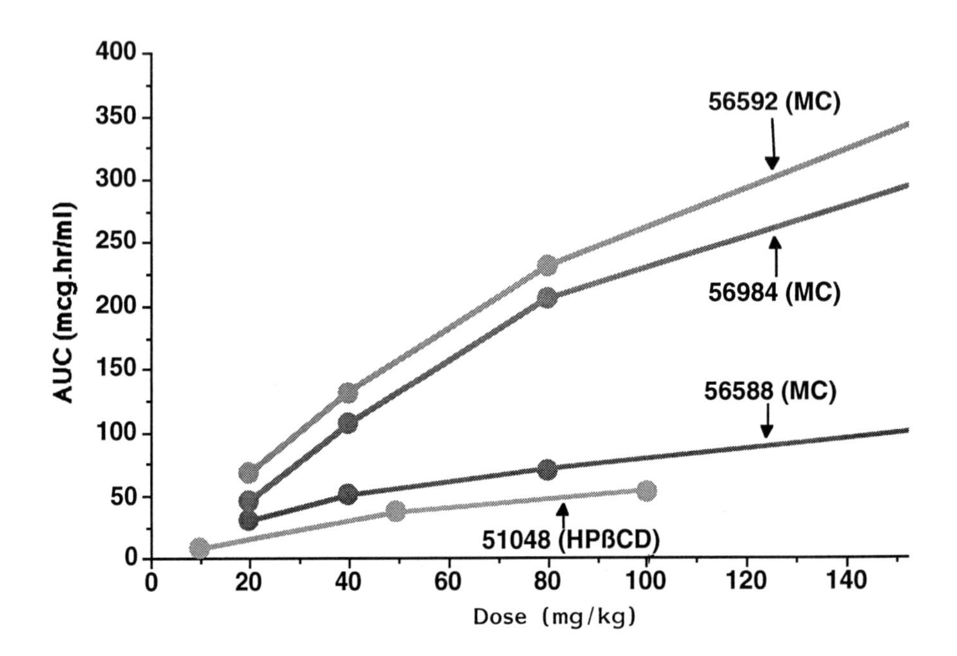

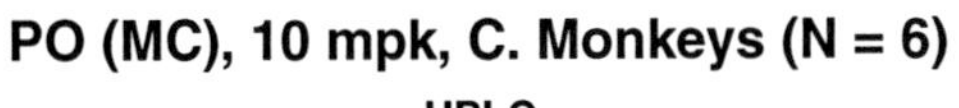

HPLC

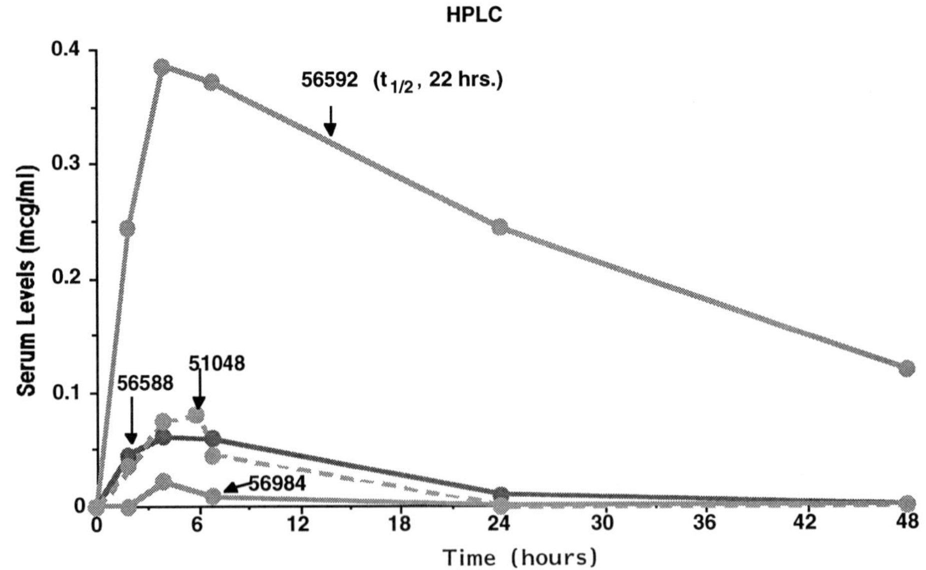

normal and immunocompromised infection models (*Candida* and *Aspergillus*).[5c] A comparision of *in vivo* efficacy and pharmacokinetic profile of these three compounds is shown. Based on the overall efficacy and bioavailability in 5 animal species, Sch 56592 was recommended for further development.

PRACTICAL SYNTHESIS OF SCH 56592:

Enzymatic desymmetrization[17] of the homochiral diol **15** was considered an attractive possibility from the outset. Successful model reactions in the presence of PPL convinced us that the optical purity of a desymmetrized (R)-monoester of **15** carried all the way to the **6S** *cis*-tosylate. But it took a systematic screening of 169 hydrolases to make it a practical proposition. Out of these hydrolases only 6 displayed desirable reactivity and selectivity and only Novo SP 435 showed enhanced pro-(S) selectivity under the conditions of the screen. Complete consumption of the diol **15** was imperative to ensure optical integrity of the final product. Thus, under the optimised conditions <10% diester of **15** was also formed; this did not interfere with iodocyclization and subsequent triazolyl anion displacement reaction. The most practical route to Sch 56592 now conducted on over 30 kg batches was accomplished as follows.[12, 18]

Scheme IX

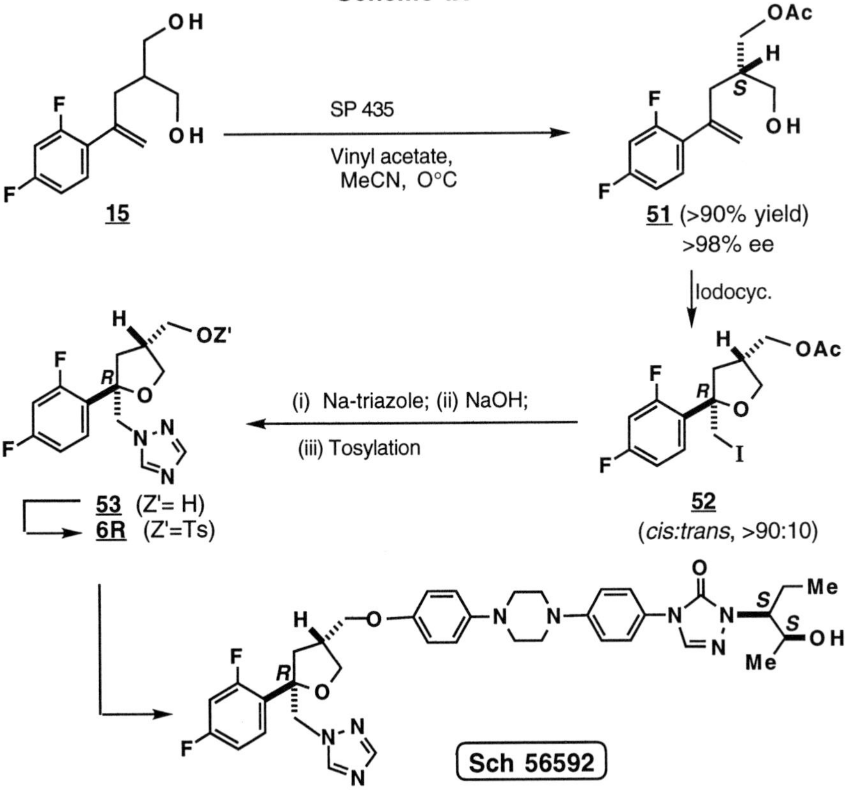

15 → 51: SP 435; Vinyl acetate, MeCN, O°C

51 (>90% yield) >98% ee

51 → 52: Iodocyc.

52 (*cis:trans*, >90:10)

52 → 53: (i) Na-triazole; (ii) NaOH; (iii) Tosylation

53 (Z'= H)

6R (Z'=Ts)

Sch 56592

CONCLUSION:

We have described here a significant effort leading to the discovery of Sch 56592 a potent antifungal agent now undergoing phase I clinical trials. Its excellent broad-spectrum activity (P.O.) and superiority over most existing antifungal agents has been confirmed by outside investigators. Detailed biology of Sch 56592 has been discussed elsewhere.[19,20]

Systematic studies directed at understanding of stereo and regiochemical requirements for optimal oral activity led to the discovery of tetrahydrofurans with specific placement of the ring oxygen (cf. type **C**). *This significant result could not have been predicted at the outset.*

Efficient stereo and enantioselective synthesis of *2,4-cis-trisubstituted* tetrahydrofurans was critical to our success in the rapid synthesis of analogs. In the course of these studies we discovered a unprecedented *2,4-cis-selective* halocyclization of 2,2-disubstituted olefins, which provided a powerful methodology utilized in three most practical routes to Sch 56592 and its analogs.

ACKNOWLEDGEMENTS:

We would like to cordially thank Dr. George Miller's group for carrying out all the biology referred to in this presentation; and our colleagues in the Chemical Development and Preparatory Groups for providing many bulk intermediates critical to our need. Indeed our profound appreciation to Drs. Birendra Pramanik and Mohindar Puar for their invaluable analytical support.

REFERENCES AND NOTES:

1. Reviews: J. A. Cuomo, W. E. Dismukes, *The New England Journal of Medicine*, **1994**, 330, p. 263; R. J. Hay, "Recent Advances in Chemistry of Antiinfective Agents", *Royal Society of Chemistry, Special Publication No. 119*, **1993,** p. 163.
2. K. De Beule, *International J. of Antimic. Agents*, **1996**, 6, p.175.
3. K. Richardson, "Recent Advances in Chemistry of Antiinfective Agents", *Royal Society of Chemistry, Special Publication No. 119*, **1993,** p. 182.
4. Minireviews: (a) J. H. Rex, M. G. Rinaldi and M. A. Pfaller, *Antimicrob. Agents Chemother.* , **1995**, 39, p. 1; (b) N. H. Georgopapadakou and T. J. Walsh, *Antimicrob. Agents Chemother.* , **1996**, 40, p. 279. DeVries, J. G. and Kellog, R. M. *J. Am. Chem. Soc.*, **1979**, 101, 2759.
5. (a) A. K. Saksena, , V. M. Girijavallabhan, D. F. Rane, R. E. Pike, J. A. Desai, A. B. Cooper, E. Jao, A. K. Ganguly, D. Loebenberg, R. S. Hare and R. Parmegiani, *9th International Symposium on Future Trends in Chemotherapy, Geneva, Switzerland*, 26-28, March **1990**, Abstract No. 128; (b) V. M. Girijavallabhan, A. K. Ganguly, A. K. Saksena, A. B. Cooper, R. Lovey, D. F. Rane, R. E. Pike, J. A. Desai and E. Jao, "Recent Advances in Chemistry of Antiinfective Agents", *Royal Society of Chemistry, Special Publication No. 119*, **1993,** p. 191; (c) For a detailed description of these infection models, see: D. Loebenberg, A. Cacciapuoti, R. Parmegiani, E. L. Moss, F. Menzel, B. Antonacci, C. Norris, T. Yarosh-Tomaine, R. S. Hare, and G. H. Miller, *Antimicrob.Agents Chemother.*, **1992**, 36, p. 64.
6. D. F. Rane, V. M. Girijavallabhan, M. S. Puar, A. K. Saksena, D. Loebenberg and R. M. Parmegiani, *Bioorg. Med. Chem. Lett.*, **1994**, 4, p. 1313.

7. A. K. Saksena, V. M. Girijavallabhan, D. F. Rane, R. E. Pike, J. A. Desai, Y-T. Chen, E. Jao, A. K. Ganguly, *Heterocycles*, **1993**, 35, p. 129; Aqueous Diels-Alder: R. Breslow and D. C. Rideout, *J. Am. Chem. Soc.*, **1980**, 102, p. 7816
8. (a) A. K. Saksena, V. M. Girijavallabhan, R. G. Lovey, R. E. Pike, J. A. Desai, A. K. Ganguly, R. S. Hare, D. Loebenberg, A. Cacciapuoti and R. M. Parmegiani, *Bioorg. Med. Chem. Lett.*, **1994**, 4, p. 2023; (b) SPZ is Janssen's investigational azole reported to have better activity compared to ITZ against a wide range of *Aspergillus* strains *in vitro* and *in vivo*; (c) The terms "eutomer" and "dystomer" were coined by Prof. E. J. Ariens (University of Nijmegen, The Netherlands), to denote the active and the "other" enantiomer respectively; (d) X-ray crystallography of Sch 50002 was performed by Prof. Andrew T. McPhail (Duke University).
9. A. K. Saksena, V. M. Girijavallabhan, R. G. Lovey, R. E. Pike, J. A. Desai, A. K. Ganguly, R. S. Hare, D. Loebenberg, A. Cacciapuoti and R. M. Parmegiani, *Bioorg. Med. Chem. Lett.*, **1995**, 5, p. 127; R. A. Fromtling and J. Castaner, *Drugs of the Future*, 1995, 20 (3), p. 241.
10. (a) Reviews: P. A. Bartlett in " Asymmetric Synthesis", Vol. 3, Academic Press, Inc. **1984**, 411; G. Cardillo and M. Orena, *Tetrahedran*, **1990**, 46, p. 3321; (b) S. D. Rychnovsky and P. A. Bartlett, *J. Am. Chem. Soc.*, **1981**, 103, p. 3963; (c) B. H. Lipshutz and J. C. Barton, *J. Am. Chem. Soc.*, **1992**, 114, p. 1084.
11. A. K. Saksena and R. E. Pike, unpublished.
12. A. K. Saksena, V. M. Girijavallabhan, R. G. Lovey, R. E. Pike, H. Wang, and A. K. Ganguly, *Tetrahedron Lett.*, **1995**, 36, p. 1787.
13. (a) Review: D. A. Evans, *Aldrichimica Acta*, 15, 23 (1982); (b) W. Oppolzer, J. Blagg, I. Rodrigues, and E. Walther, *J. Am. Chem. Soc.*, **1990**, 112, p. 2767; (c) D. A. Evans, F. Urpi, T. C. Somers, J. S. Clark, and M. T. Bilodeau, *J. Am. Chem.* Soc., **1990**, 112, p. 8215.
14. A. K. Saksena, V. M. Girijavallabhan, H. Wang, Y-T. Liu, R. E. Pike and A. K. Ganguly, *Tetrahedron Lett.*, **1996**; accepted for publication.
15. J. G. DeVries and R. M. Kellog, *J. Am. Chem. Soc.*, **1979**, 101, p. 2759.
16. O. Mitsunobu, *Synthesis*, **1981**, p. 1; S. F. Martin and J. A. Dodge, *TetrahedronLett.*, **1991**, 32, p. 3017.
17. P. G. Hultin, F.-J. Muesseler, J. B. Jones, *J. Org. Chem.*, **1991**, 56, p. 5375; and cited references.
18. A. K. Saksena, V. M. Girijavallabhan, R. G. Lovey, F. Bennett, R. E. Pike, H. Wang, P. Pinto, Y.T. Liu, N. Patel and A. K. Ganguly, *35th Interscience Conference on Antimicrobial Agents and Chemotherapy (ICAAC), San Francisco*, 17-20 September 1995 (abstract nos. F 61 and F 83); we shall describe a very practical methodology to introduce Sch 56592 side chain at a future date.
19. *35th Interscience Conference on Antimicrobial Agents and Chemotherapy (ICAAC), San Francisco*, 17-20 September 1995: (a) D. Loebenberg, A. Cacciapuoti, R. Parmegiani, E. L. Moss, F. Menzel, B. Antonacci, C. Norris, T. Yarosh-Tomaine, R. S. Hare, G. H. Miller, (abst. nos. F 62, F 66 and F 67); (b) A. Nomeir, P. Kumari, M. J. Hilbert, D. Loebenberg, A. Cacciapuoti, F. Menzel, E. L. Moss, F. Menzel, R. S. Hare, G. H. Miller, M. N. Cayen and C. C. Lin (abs. no. F 68); (c) J. N. Galgiani, M. L. Lewis and T. Peng (abs. no. F 63); (d) J. R. Perfect and W. A. Schell (abs. no. F 64); (e) A. M. Sugar and X-P Liu (abs. no. F 65).
20. A. M. Sugar and X-P Liu, *Antimicrob.Agents Chemother.*, **1996**, 40, p. 1314.

Advances in Antivirals

15

4′-Thio-2′-Deoxyribonucleosides, Their Chemistry and Biological Properties – A Review

R. T. Walker

SCHOOL OF CHEMISTRY, UNIVERSITY OF BIRMINGHAM, EDGBASTON, BIRMINGHAM B15 2TT, UK

1 INTRODUCTION

To the best of my knowledge, nature has chosen in general not to use sugar moieties in which the heteroatom in a furanose or pyranose ring is sulfur instead of oxygen. In the pages which follow, we may begin to understand some of the reasons for this choice. In this review, I have chosen, when naming compounds, not to split the word 2-deoxyribose, so that it is obvious when 2'-deoxynucleosides are the subject of discussion.

1.1 Early Syntheses of Ring-Sulfur Containing Sugars

The first sugar to be synthesized containing sulfur in the ring was 5-thio-**D**-xylopyranose (1), the synthesis of which in 1961-62 was almost simultaneously reported by three groups.[1-3] One group came from the laboratory of R.L. Whistler who has surveyed this early work.[4] The first nucleoside analogue of such a sugar was reported in 1962 when the adenine nucleosides of 4-thio-**D**-xylose (2) and 4-thio-**D**-arabinose (3) were synthesized.[5] Subsequently, 4-thio-**D**-ribofuranose (4)[6,7] and 4-thio-**L**-ribofuranose[8] (5) were synthesized as were many other configurational isomers.

The first 4'-thioribonucleoside to be synthesized was 4'-thioadenosine (6) in 1964.[7] Further examples were reported in 1970[9] but the problem of stereocontrol in the condensation reaction, even with a participating group at C2' was not mentioned until the synthesis of 5-fluoro-4'-thiouridine (7).[10] Subsequently, several additional syntheses appeared,[11] which include a synthesis of 2,3,5-tri-*O*-benzyl-4-thio-1-*O*-acetyl-**D**-ribofuranose (8)[11] from **L**-lyxose in an overall yield of 21%. Following the publication of an efficient synthesis of a suitable derivative of 4-thio-2-deoxyribose (9) in 1991,[12] a similar method was used for the preparation of the corresponding ribo-analogue with equal success in 1994.[13] Other 4'-thionucleoside analogue preparations of interest include the synthesis of 4'-thio-2',3'-dideoxynucleosides (10),[14,15] 4'-thio-2',3'-didehydro-2',3'-dideoxynucleosides (11)[15], 4'-thio-3'-C (hydroxymethyl)-2',3'-dideoxynucleosides (12)[16] and the use of 2-(tert-butyldimethylsiloxy)thiophenes (13) as precursors to both enantiomers of 4'-thio-2',3'-dideoxycytidine.[17]

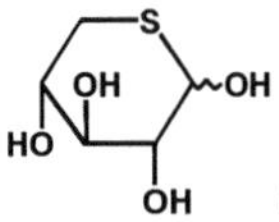

(1)

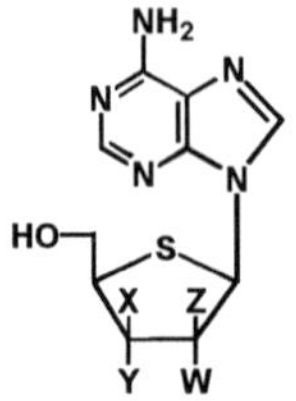

(2) X = W = OH
Y = Z = H

(3) Y = Z = OH
X = W = H

(6) W = Y = OH
X = Z = H

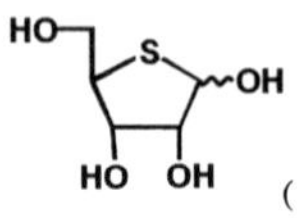

(4)

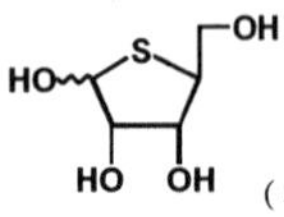

(5)

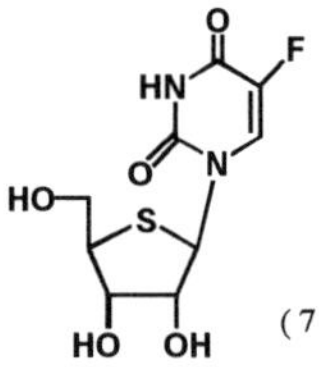

(7)

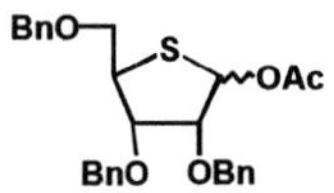

(8)

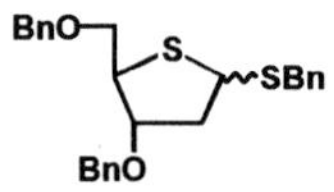

(9)

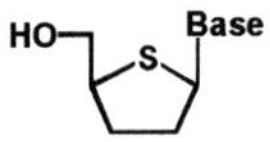

(10)

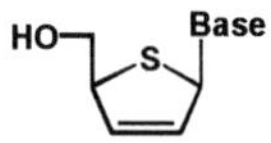

(11)

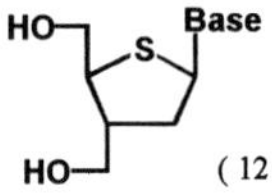

(12)

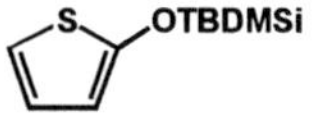

(13)

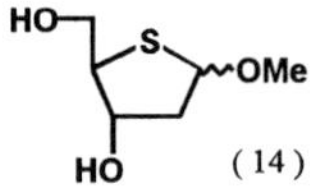

(14)

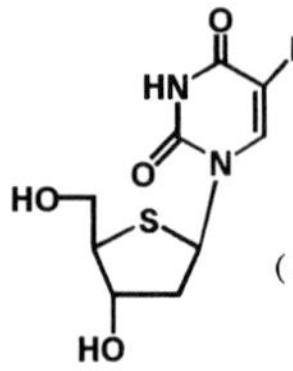

(15)

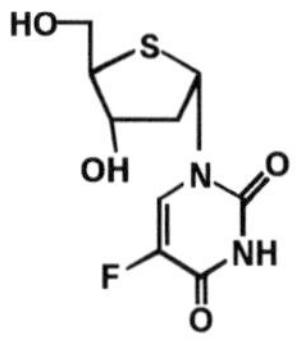

(16)

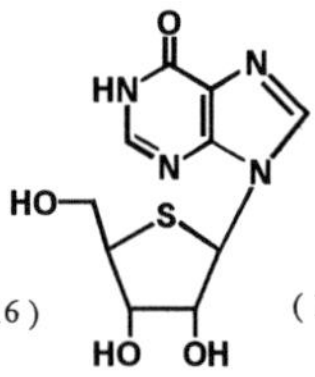

(17)

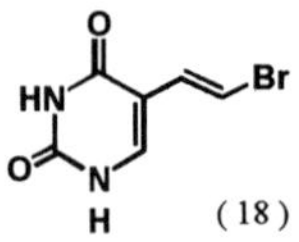

(18)

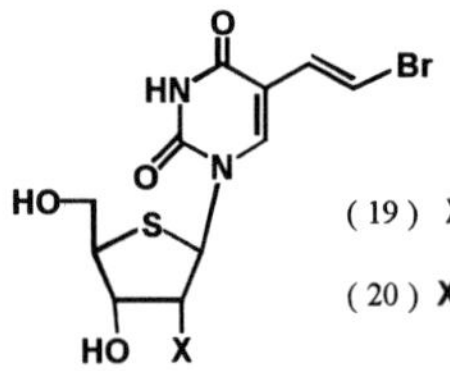

(19) X = OH

(20) X = H

2 SYNTHESIS OF 4'-THIO-2'-DEOXYRIBONUCLEOSIDES

2.1 The Original Synthesis of a 4'-Thio-2'-deoxyribofuranoside

It was not until 1970 that methyl 2-deoxy-4-thio-**D**-*erythro*-pentofuranoside (14) was first synthesized by Nayak and Whistler.[18,19] Starting from **D**-glucose, the total synthesis was achieved following a total of 15 steps with an overall yield of about 10%. However, several of the steps are inconvenient to say the least and it was possible for Fu and Bobek to claim in 1976[20,21] that a 14 step synthesis from **L**-arabinose, with an overall yield of around 4%, had a distinct advantage over the previous method. Thus in 1976[22] and finally published with a full experimental in 1978,[23] Fu and Bobek were able to report the first synthesis of 4'-thio-2'-deoxyribonucleosides which were the α- and β-anomers of 5-fluoro-4'-thio-2'-deoxyuridine (15,16).[16] Unfortunately these heroic efforts apparently left the scientists completely exhausted and the difficulty in scaling up the synthesis was such an energy barrier, that it was nearly 15 years before the next move in this area occurred. Then as often happens, two groups independently and simultaneously published syntheses of further analogues in this series.[24,25]

With hindsight, it was unfortunate that Fu and Bobek chose to make the 5-fluorouracil derivative of their hard-won sugar although at the time as they were working in a cancer institute and antiviral chemotherapy had not yet come of age, the choice was perfectly logical. The biological results which were obtained from the 5-fluorouracil analogue (15) were typical of many *in vitro* results being generated and which were soon to be severely criticized by Schabel in 1979.[26] Also it was not known at that time that these nucleoside analogues were stable to nucleoside phosphorylase and therefore comparison of 5-fluoro-4'-thio-2'-deoxyuridine with 5-fluoro-2'-deoxyuridine itself, when the latter is almost certainly initially a pro-drug of 5-fluorouracil, is not strictly valid. The result was that the biological effects seen were not exciting enough for chemists to spend the necessary effort to synthesize further analogues.

2.2 Rationale for Synthesis of 4'-Thio-2'-deoxynucleosides

The rationale for the synthesis of this series of nucleoside analogues, apart from the challenge, was that evidence was accumulating that the presence of sulfur in the sugar ring, stabilized the N-glycosidic bond to phosphorolysis.[27] Thus 4'-thioinosine (17) was known to be resistant to phosphorolytic cleavage, which is a normal metabolic pathway of nucleoside catabolism. We had for many years been aware of the problem of degradation of potentially clinically useful anti-herpesvirus nucleoside analogues by phosphorolysis in cell lines and particularly in animals. Thus the realization that *in vivo*, both *(E)*-5-(2-bromovinyl)uracil (18) and *(E)*-5-(2-bromovinyl)uridine (19) were pro-drugs of *(E)*-5-(2-bromovinyl)-2'-deoxyuridine (20, BVDU) itself because of their interconversion through phosphorolysis, led to the demonstration that BVDU was an excellent substrate for pyrimidine phosphorylase.[28] For this drug, and many similar analogues, to exert their true effect *in vivo*, this line of catabolism had to be prevented.

2.3 Modern Synthetic Methods

The next significant step in attempts to improve the synthesis of 4'-thio-2'-deoxyribonucleosides came in a series of publications in 1991[12,29-31] which reported work published in the thesis of M.R. Dyson in 1989.[24]

An improved synthesis of benzyl 3,5-di-*O*-benzyl-2-deoxy-1,4-dithio-**D**-*erythro*-pentofuranoside (9), an intermediate which could be used in the synthesis of 4'-thio-2'-deoxyribonucleosides was announced. Starting from 2-deoxy-**D**-ribose, the desired product could be obtained in 7 steps in 11% overall yield (Scheme 1). At this stage the yields had not been optimized and following work in Process Development at the Wellcome Foundation, this pathway was used to produce greater than 100 Kg quantities of product, involving no chromatography and with a yield of around 50%. It is against this standard that subsequent syntheses, many of which have been claimed to be superior, have to be judged.

There are two key steps involved in the synthesis: (1) the inversion of configuration at C4 using the Mitsunobu reaction[32] and (2) the final cyclization of the dithioacetal accompanied by further inversion at C4, to give a product of the desired configuration. This cyclization was first described by Harness and Hughes.[33] Optimization of the conditions which includes using DIAD for the Mitsunobu reaction and changing the solvent to methyl ethyl ketone in the final ring closure reaction, means that both of these key steps are now almost quantitative. This method will hereafter be referred to as the preferred synthesis. The group of Secrist has proposed an alternative high yielding method by going directly to the final product from the inverted alcohol from the Mitsunobu reaction using triphenylphosphine, iodine and imidazole in THF.[34]

Simultaneously and independently of the work in Birminingham, U.K., the group of Secrist in Birmingham Alabama was using the method of Fu and Bobek to synthesize a 4-thio sugar suitable for converting into 4'-thio-2'-deoxyribonucleosides.[35,36] It was then immediately obvious[12,36] that interesting biological activity could be expected in this series of nucleoside analogues and further syntheses of suitable sugar moieties quickly became available.

The first of several synthetic routes from acyclic precursors using a stereocontrolled synthesis was announced by Uenishi *et al.*[37-40](Scheme 2). A novel ring opening reaction of an optically active epoxy alcohol with a xanthate anion gives an α-hydroxy cyclic xanthate which is a precursor to a 4-thio-2-deoxyribose derivative[37] from which the corresponding nucleosides can be made.[39] This elegant method has a further advantage in that the stereocontrol is determined by Sharpless epoxidation conditions[41] and hence it is possible to produce stereoselectively either enantiomer, one of which leads to the **D**-sugar and the other to an **L**-4-thio-2-deoxyribose derivative.[42,43] This synthesis was opportune because it had just been realized that herpesvirus thymidine kinase (and to a lesser extent, the DNA polymerase) is incapable of distinguishing between many **D**- and **L**-deoxynucleosides.[44] Thus, **L**-thymidine inhibited the proliferation of HSV-1 in HeLa cells. Although 4'-thiothymidine is toxic both to the virus and the host cells (see below) unfortunately **L**-4'-thiothymidine has no obvious antiviral activity and only shows a low toxicity to L1210 cells.[44]

"A facile synthesis of 4'-thio-2'-deoxypyrimidine nucleosides" proved to be anything but.[45] A key intermediate, 3,5-di-*O*-benzyl-2-deoxy-**D**-*erythro*-pentose dithiobenzylacetal

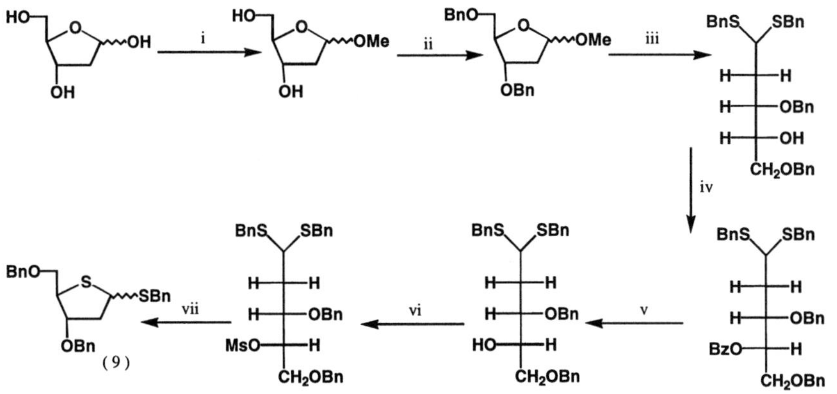

Scheme 1 *Reagents and conditions* : i, MeOH/HCl ; ii, NaH, TBAI, BnBr, THF ; iii HCl (conc), BnSH ; iv, triphenylphosphine, benzoic acid, DEAD ; v, NaOMe, MeOH ; vi, $MeSO_2Cl$, pyridine ; vii, NaI, $BaCO_3$, acetone.

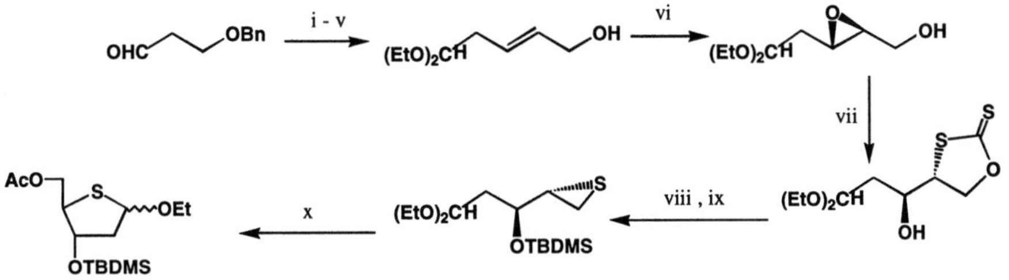

Scheme 2 *Reagents and conditions* : i, $(EtO)_3CH$, $ZnCl_2$; ii, Li, liq. NH_3, -78 °C; iii, Swern oxidation; iv, Ph_3PCHCO_2Me, benzene, sep^n of *E* and *Z* - isomers; v, diisopropylaluminium hydride, CH_2Cl_2, -78 °C; vi, Bu^tO_2H, $Ti(OPr^i)_4$,(+)-diethyl tartrate, CH_2Cl_2, -20 0 °C; vii, KH, CS_2, THF, -78 - 40 °C; viii, $TfOSiMe_2Bu^t$, 2,6-lutidine, CH_2Cl_2; ix, K_2CO_3, MeOH; x, AcONa, AcOH, 100 °C.

(21) was cyclised with triphenylphosphine, iodine and imidazole and the product incorrectly identified as the D-*erythro*-pentofuranoside (9) whereas two subsequent publications from the groups of Secrist[46] and Imbach[47] showed conclusively that the product had the L-threo configuration (22) and was in fact identical to an intermediate previously reported by Walker and co-workers.[29]

Not surprisingly, 3'-azido-4'-thio-2',3'-dideoxythymidine (23, 4'-thio-AzT) had been a primary target of early syntheses[12,14] but it proved to have no significant biological activity.[12] A convergent synthesis of 3'-azido-4'-thio-2',3'-dideoxynucleosides was described by Mackenzie and co-workers starting from D-xylose (Scheme 3).[48,49] The method has the advantage of making the suitably derivatized sugar moiety so that a range of 3'-azido nucleosides can be made. However, with an overall yield of 1%, it can hardly compare in efficiency with the optimized conditions already described if a known target is desired. Mackenzie and co-workers have subsequently devised a further convergent synthesis of C3-substituted sugars with the 2-deoxy-4-thio-D-*erythro*-pentofuranose configuration from L-arabinose (Scheme 4) in 10 steps.[50] Whether this has any advantage over methods already described, has yet to be seen but with a claimed overall yield of 17%, it does give some flexibility in the choices of protecting groups on the final product and has the advantage of using cheap starting material. A similar method has been published by scientists from the Wellcome Foundation (see below) and the main disadvantage of this strategy would appear to be the number of steps involved.

A very interesting recent development has been described by Matsuda and co-workers[51-53] which involves a Pummerer rearrangement (Scheme 5). The presence of sulfur in the ring gives an obvious chance to derivatize the position α- to it following oxidation to the sulfoxide. Indeed, such a transformation was discussed by McCormick and McElhinney in 1976[54] and subsequently applied to the synthesis of acyclic nucleosides,[55] but no rearrangement of a 'normal' 4'-thionucleoside had appeared. Similar work by O'Neil and co-workers[56,57] clearly had the same goal but as yet, the final stage has not been completed. The Japanese synthesis starts from diisopropylideneglucose to finally yield the 2'-methylene- (24) or 2',2'-difluoro-4'-thio-2'-deoxycytidine derivatives (25) 19 steps later.[53] The sugar moiety is produced in about 18% overall yield and the β-nucleosides in 2% overall yield. No doubt the individual steps can be optimized but the length and complexity of the synthesis will mean that few will follow the trail so imaginatively blazed by Matsuda and his colleagues.

A recent publication details a specific preparation of the potent broad-spectrum antiviral agent 5-ethyl-4'-thio-2'-deoxyuridine (26) in an attempt to avoid the use of large quantities of thiol reagents and consecutive double inversion of the stereochemistry at C4 of the sugar moiety.[58] The key step (Scheme 6) uses the observation that 4-thiopyranoses can be produced from thiocarbonates by radical-induced rearrangement, thus achieving the required double inversion in a single step.[59] A pyranose-furanose rearrangement would then be required.[60] It is difficult from the data given to calculate an overall yield but optimistically, 5-ethyl-4'-thio-2'-deoxyuridine could be isolated in about 10% yield starting from 500g of 2-deoxy-D-ribose. The chemistry is elegant and individual steps have not been optimized but there are many side reactions and generation of isomers, only one of which can be used in subsequent steps.

A preliminary communication outlines another strategy starting from 2-deoxy-D-ribose to give a key D-*erythro* chiron, which is a precursor to 4-thio sugars of the L-*erythro* configuration.[61] The equivalent L-*erythro* chiron could not be conveniently made

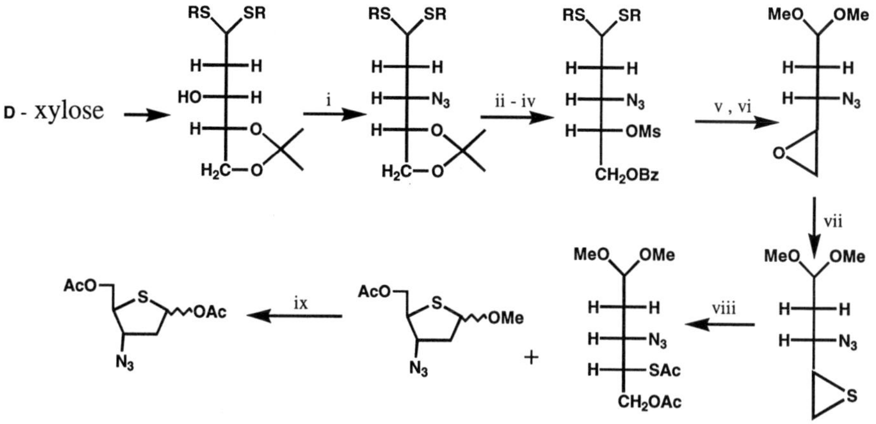

Scheme 3 *Reagents and conditions* : i, $[Zn(N_3)_2.2Py]$, Ph_3P, DIAD, toluene; ii, H^+; iii, PhCOCl, pyridine; iv, MsCl, Et_3N, CH_2Cl_2; v, HgO, $HgCl_2$, MeOH; vi, MeONa, MeOH; vii, NH_2CSNH_2; viii,AcONa, Ac_2O, AcOH; ix, Ac_2O, AcOH, H_2SO_4.

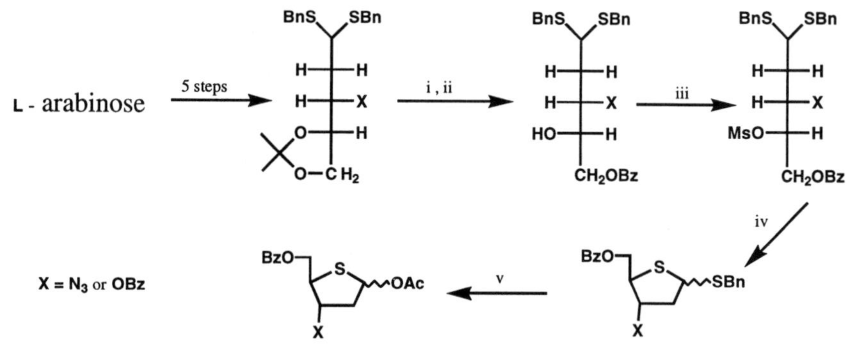

Scheme 4 *Reagents and conditions* : i, H^+, EtOH; ii, BzCl, pyridine; iii, MsCl, Et_3N, CH_2Cl_2; iv, n-Bu_4NF, $BaCO_3$, DMF; v, $Hg(OAc)_2$, AcOH.

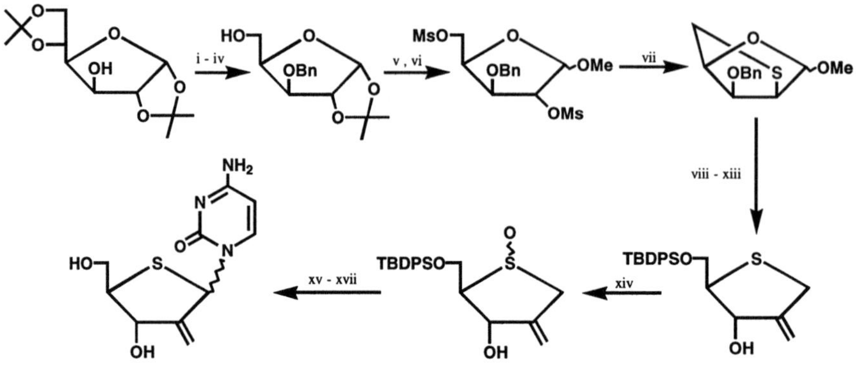

Scheme 5 *Reagents and conditions* : i, BnBr, NaH, DMF, THF; ii, 2M HCl, THF; iii, $NaIO_4$, H_2O, MeOH; iv, $NaBH_4$, MeOH; v, 5% HCl/MeOH; vi, MsCl, pyridine; vii, Na_2S, DMF, 100 °C; viii, 4M HCl, THF; ix, $NaBH_4$, MeOH; x, TBDPSCl, imidazole, DMF; xi, DMSO, Ac_2O; xii, $Ph_3P^+CH_3Br^-$, NaH, *t*-amyl alcohol, THF; xiii, BCl_3 CH_2Cl_2, -78 °C, then MeOH, pyridine; xiv, *m*-CPBA, CH_2Cl_2, -78 °C; xv, silylated *N*-acetylcytosine, TMSOTf, $ClCH_2CH_2Cl$, 0 °C; xvi, TBAF, THF; xvii, aqueous NH_3, MeOH.

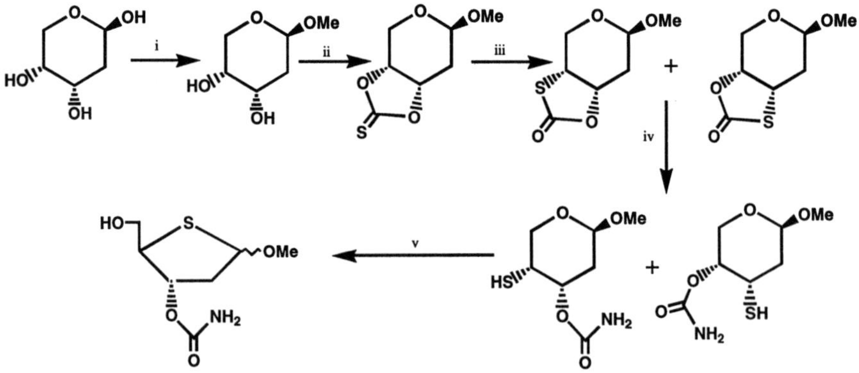

Scheme 6 *Reagents and conditions* : i, HCl/MeOH; ii, $CSCl_2$; iii, $tBu_4N^+Br^-$/diglyme, 150 °C; iv, NH_3/MeOH; v, Dowex H^+, MeOH.

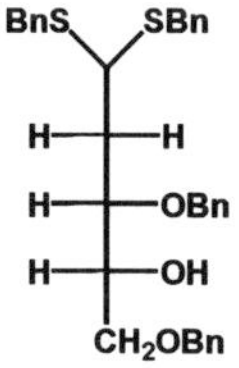

BnS SBn H H H OBn H OH CH_2OBn (21)

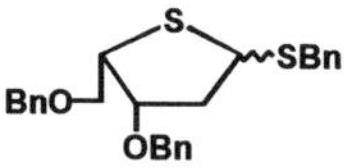

S SBn BnO OBn (22)

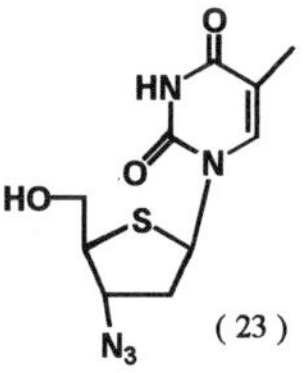

O HN O N HO S N_3 (23)

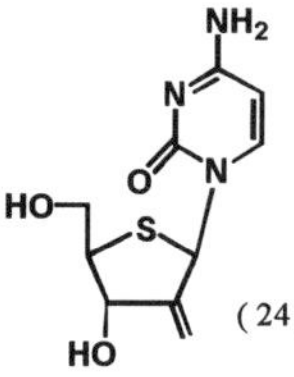

NH_2 N O N HO S HO (24)

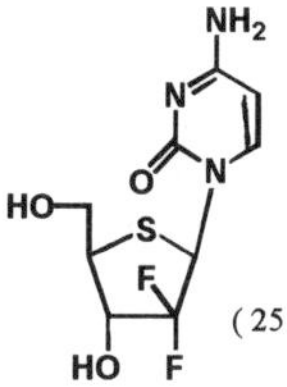

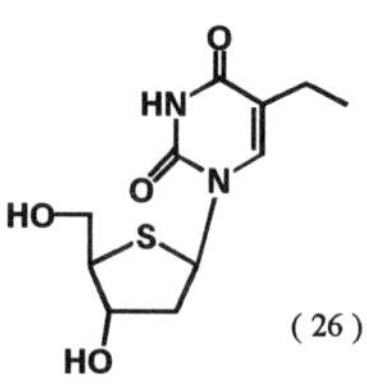

NH_2 N O N HO S F HO F (25)

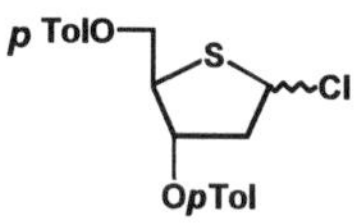

O HN O N HO S HO (26)

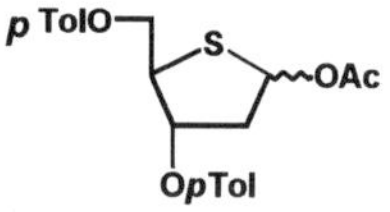

p TolO S Cl OpTol (27)

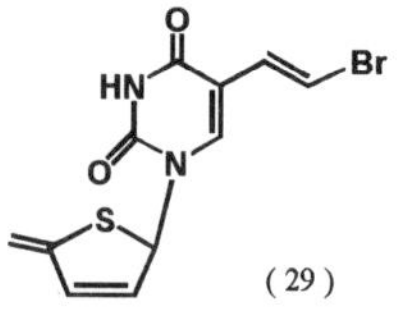

p TolO S OAc OpTol (28)

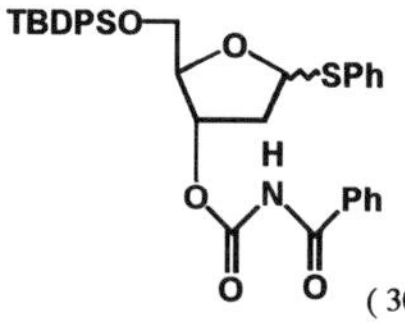

O HN Br O N S (29)

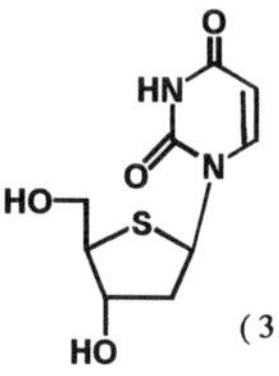

TBDPSO O SPh H O N Ph O O (30)

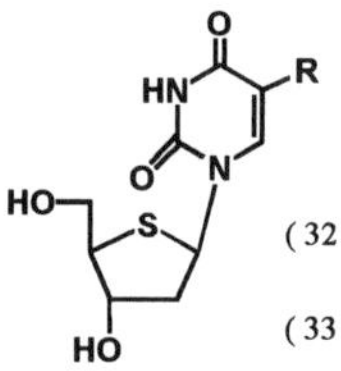

O HN O N HO S HO (31)

O HN R O N HO S HO

(32) R = CH_3

(33) R = H

NH_2 N O N HO S HO (34)

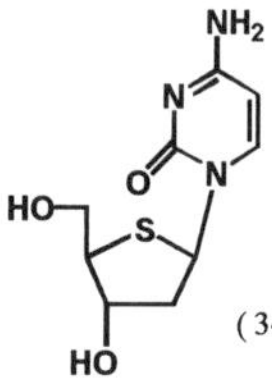

O HN O N HO S HO (35)

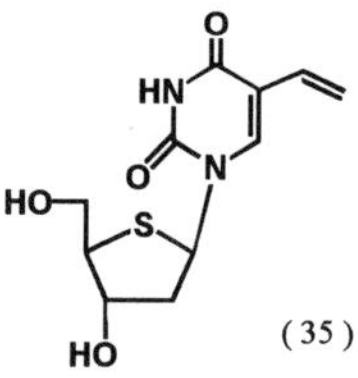

O HN I O N BnO S OBn (36)

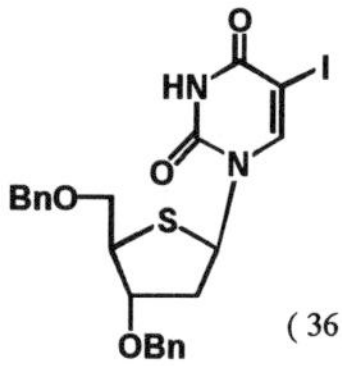

S O HN N O N N HO S HO (37)

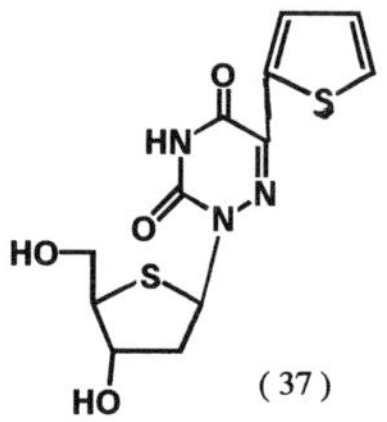

O HN O N O O HO S HO (38)

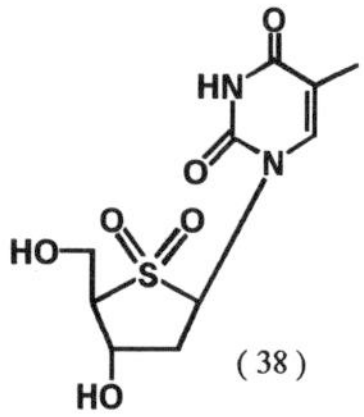

from the prohibitively costly 2-deoxy-**L**-ribose and so was formed from **L**-arabinose. The purpose of this route was to address two problems encountered in the preferred route[12] in which harsh deprotection conditions (benzyl group removal) are required and the α:β ratio of nucleosides formed in the condensation reaction (at best 1:1 and often 2:1) is unsatisfactory. Starting from 2-deoxy-**D**-ribose, the synthesis of a suitably protected sugar appears to be 9 steps (and from **L**-arabinose is even longer) with an overall yield below 20%. However the precursor to nucleoside synthesis so formed (Scheme 7) gives much more acceptable α:β ratios (1:3) in the condensation step.[62] Thus when (if) this route is ever optimized, it could be a competitor for the currently preferred method but the number of steps and the cost of silyl protection may prove to be disadvantageous.

3 FACTORS AFFECTING NUCLEOSIDE YIELD IN CONDENSATION REACTIONS

3.1 Condensation Conditions

Two main factors need to be considered here: (1) the overall yield of nucleoside and (2) the α:β ratio produced. At present, it appears unlikely that a practical chemical stereospecific synthesis of a β-4'-thio-2'-deoxynucleoside will be devised and therefore anomer separation will play a large part in determining the best synthetic strategy to use. Thus it may possibly be advantageous to use protecting groups which yield a poorer α/β ratio but which enable the anomers readily to be separated (by crystallization or chromatography) rather than necessarily to take the best α/β ratio available. Also if a preparation is to be scaled up much beyond the gram scale, it is advantageous to be able to separate the anomers (either protected or deprotected) by crystallization rather than by column chromatography.

Unfortunately, all the available evidence so far is that when condensing a heterocyclic base with a 4-thiosugar, control of stereochemistry is lost. This is even seen to a limited extent when 4'-thioribonucleosides are synthesized when a potential neighbouring group should participate from the 2-position.[10] This is in contrast to the situation found in normal nucleoside synthesis where even in the synthesis of 2'-deoxynucleosides which have no such participating group, control can be maintained, so that normally, exclusively β-anomer formation can be obtained.[63]

The sugar moiety (9) obtained from the preferred synthesis is of course an anomeric mixture (α:β, 1:4)[64] which can easily be separated. However, each individual anomer when used in a condensation reaction gives an identical α:β ratio of nucleoside.[64]

3.1.1 Pyrimidine Nucleosides. The first 4'-thio-2'-deoxynucleoside synthesis used the 3',5'-di-*O*-p-toluoyl chloro sugar (27) which had been made *in situ*. Whenever these halo-sugars have been prepared, the word unstable[12,23,36] is used, although it is not clear whether this instability is an inherent property or whether the compound has never been produced on a sufficient scale such that one anomer could be obtained crystalline, as is the case with the equivalent oxygen-containing sugar.

The original synthesis[23] was Lewis acid-catalyzed as the base used, 5-fluorouracil, is particularly unreactive and it was claimed that the α/β ratio depended upon the catalyst (mercuric oxide - mercuric bromide favouring the β-anomer and stannic chloride favouring the α-anomer). It should be remembered that one equivalent of "catalyst" or more is used in these condensation reactions and examination of the experimental

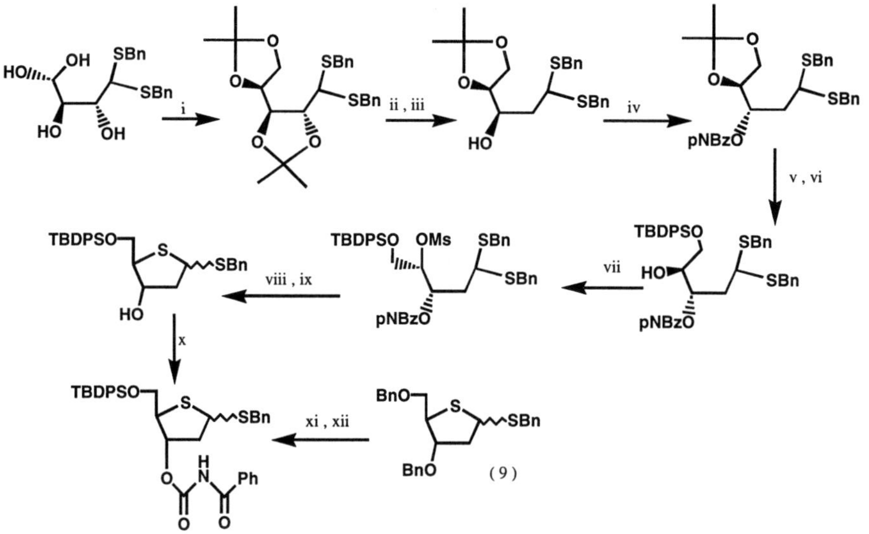

Scheme 7 *Reagents and conditions* : i, $H_2SO_4/CuSO_4$; ii, KOtBu, THF, DMSO ; iii, $LiAlH_4$, THF ; iv, pNBzOH, DIAD, PPh_3, THF ; v, Dowex H^+, MeOH ; vi, TBDPSO, DMAP, imidazole, CH_2Cl_2 ; vii, MsCl, DMAP, CH_2Cl_2 ; viii, NaI, Et_3N, 2-butanone ; ix, NaOMe, MeOH ; x, PhCONCO, toluene ; xi, $TiCl_4$, CH_2Cl_2 ; xii, TBDPSCl, DMAP, imidazole, CH_2Cl_2.

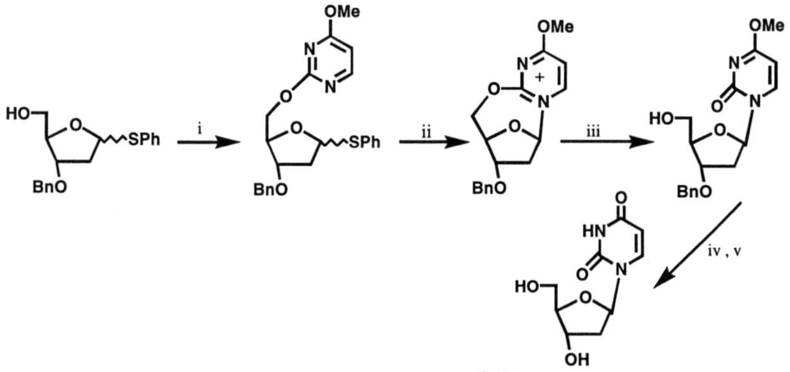

Scheme 8 *Reagents and conditions* : i, NaH, [2-chloro-4-methoxypyrimidine structure], DMF; ii, $Me_2S(SMe)BF_4$, CH_3CN, MS 4A, -20 °C; iii, OH^-; iv, Dowex 50, EtOH, H_2O; v, BCl_3, -78 °C.

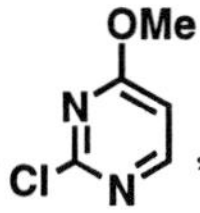

conditions suggests that in the former case, the α:β ratio was at best 1:1. Using the bromo sugar, and *(E)*-5-(2-bromovinyl)uracil as the base, catalysis was unnecessary but the α:β ratio was only 1.8:1.[30] These appear to be the only two occasions on which a halosugar has been used.

The 4-thio-2-deoxyribose moiety as first synthesized by Bobek[21] and subsequently used by Secrist[36] is initially prepared as the methyl glycoside (14). With the sugar hydroxyl groups protected, this analogue fails to react in base condensation reactions[35,36] but can be easily acetolyzed to the acetate (28) which couples in the presence of a Lewis acid (trimethylsilyl triflate[36,65] is often used) in anhydrous acetonitrile. Condensation yields are high, deprotection under basic conditions is straightforward but at best an α:β ratio of 1:1 is obtained.

The carbohydrate moiety initially produced by the preferred synthesis of a 4-thio-2-deoxy sugar is the S-benzyl glycoside (9). The initial nucleoside condensation method attempted was the method of Horton and Markovs[66] which involves heating a mixture of the sugar, silylated base, an equivalent of mercuric bromide and 3 equivalents of cadmium carbonate in toluene for 24h. Not surprisingly, the yield was low (50%) and the α:β ratio poor (2.8:1; [NMR]).[12] This method cannot be recommended and gives even lower yields with less reactive bases (such as *(E)*-5-(2-bromovinyl)uracil) where the majority product has a sugar moiety containing a diene structure (29).[30] However in 1991, Sugimura *et al.*,[67,68] described a novel glycosylation reaction using N-bromosuccinimide to effect a condensation between a silylated base and thioglycoside to give high β-selectivity. In our hands, this method has always given excellent overall yields but with 4-thio sugars, the α:β ratio is not favourable. However, on balance, this is still our method of choice providing that the nucleoside so produced is compatible with the deprotection conditions needed to remove the benzyl groups (see below).

Attempts to improve the overall yield and β-selectivity have not yielded spectacular results, which only show that N-iodosuccinimide is better than N-bromosuccinimide (the use of NCS results in reaction with the pyrimidine ring[69]) and acetonitrile is the preferred solvent. Under these conditions, one can expect to obtain a nucleoside yield of >80% with an α:β ratio of no better than 1:1.[64]

As can be imagined, considerable effort has been expended to try to exert stereocontrol over the condensation process. It does appear that variables like the leaving group, the catalyst, temperature and solvent have essentially no effect and it is the kinetic product from reaction at a centre with no sterechemistry which is produced. In the normal sugar field, intramolecular glycosylations have been achieved with complete stereoselectivity.[70,71] An example is given in Scheme 8. Adapting these conditions to 4-thiosugars resulted in no desired product being formed and some starting material remained but considerable decomposition had also occurred.[72]

In 1994, scientists at the Wellcome Foundation showed that by manipulating the protecting groups on the sugar hydroxyl groups in a normal 2-deoxy sugar, anomer-selective synthesis was possible using S-phenyl as the leaving group with NBS catalysis.[73] The combination of protecting groups found to give the most favourable α:β ratio is as shown (30). The carbamoyl group is designed to hinder the lower face of the sugar ring and hence encourage attack from the β-face. As previously mentioned, a synthesis of the necessary thio precursor became available from L-arabinose[61] and when applied to 4'-thio-2'-deoxynucleoside synthesis, good yields with a much better α:β ratio (1:~4) were obtained. It is not clear that the strategy works entirely for the reasons suggested but as a minimum, the substituent on the 3'-OH group has to exert a stereoelectronic rather than a purely steric effect. Thus it would seem logical to reduce the size of the protecting group on the 5'-OH group but this is counterproductive. However these results do suggest that eventually this problem of steric control could be solved.

Some years ago we published a method for the anomerization of 2'-deoxynucleosides.[74,75] Thus addition of a catalytic amount (0.3 mol %) of sulfuric acid and acetic anhydride to a solution of 3',5'-di-*O*-acetylthymidine resulted in the immediate production of an equilibrium mixture of α- and β-anomers in the ratio of 2:1. With time, substantial quantities of a diastereoisomeric mixture of fully acetylated open-chain nucleosides were formed (Scheme 9). When applied to 4'-thiothymidine, more drastic conditions (at least one equivalent of 'catalyst') and a longer time were needed before an equilibrium was established (α:β, 2:1). Also it proved much quicker to establish an equilibrium starting from the β-nucleoside than it did starting from the α-nucleoside. No open-chain product was ever seen, which is not surprising if the postulated mechanism of formation is correct.[64] Attempts to establish an equilibrium at a concentration using 5-ethyl-4'-thio-2'-deoxyuridine such that the β-anomer would crystallize from the solution, thus displacing the equilibrium failed, but further work in this area might yield useful results.

3.1.2 Purine Nucleosides. If the situation with 4'-thio-2'-deoxypyrimidine nucleoside synthesis is bad, the synthesis of the equivalent purine nucleosides is far worse. As is usual in the purine field, further derivatization of the heterocycle is required following condensation and purine nucleosides are acid-labile. Therefore benzyl protection of sugar hydroxyl groups (which usually cannot be removed by catalytic hydrogenation, see below) is not appropriate. However the overwhelming problem is the unfavourable α:β ratio which can be as bad as 9:1.[34] There are rumours that certain purines give rather better α:β ratios than others (even as good as 2.5:1!)[76] but of course there is also the possibility of reaction with N7 and N9 to be faced. Added to this, is the expectation that purine nucleoside analogues are likely to show biological activity in the anti-cancer field where substantial quantities of compounds are required even for preliminary tests and thus it is not too surprising that there are few heroes (or masochists) active in this field. Normal condensation conditions for purine nucleosides require the presence of the carbohydrate precursor and an appropriate purine base in acetonitrile at 0° in the presence of stannic chloride.[34] Overall yields are 70-80% with the unrepeatable α:β ratio cited above. Thus the few scientists working in this area have tended to change their strategy.

Stereospecific synthesis of 4'-thio-2'-deoxypurine nucleosides is catalyzed by the enzyme trans-N-deoxyribosylase (E.C. 2.4.2.6).[77] Enzymatic transfer reactions have previously been used successfully in the preparation of a wide variety of sugar-modified purine nucleosides.[78-80] The method has particular appeal in the current case because it is not necessary even to separate the α- and β-anomers of the donor pyrimidine nucleoside and only β-4'-thio-2'-deoxypurine nucleoside will be prepared, so that the final isolation and purification is relatively trivial. The donor molecule is an α,β mixture (3:2) of 4'-thio-2'-deoxyuridine prepared by the preferred method.[30] Attempts to use a mixture of *Escherichia coli* thymidine phosphorylase and purine nucleoside phosphorylase failed, which is none too surprising given the resistance of 4'-thionucleosides to phosphorolytic enzymes.[27] However, trans-N-deoxyribosylase proved to be very satisfactory (Scheme

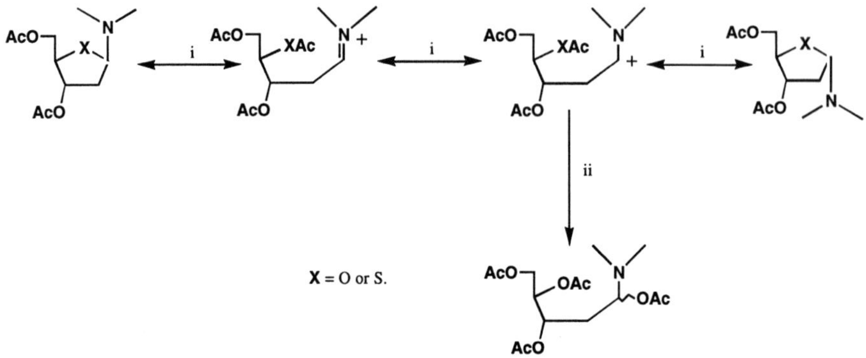

X = O or S.

Scheme 9 *Reagents and conditions* : i, Ac_2O, AcOH, H_2SO_4 ($\equiv$ 'CH_3C^+O'); ii, AcOH.

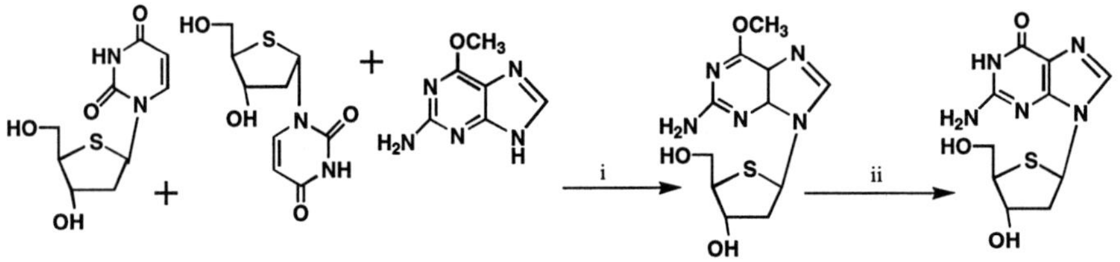

Scheme 10 *Reagents and conditions* : i, *trans-N*-deoxyribosylase; ii, adenosine deaminase.

10) and effected the complete transfer of the 4'-thiosugar from the β-anomer of the pyrimidine nucleoside to a variety of purine bases, to give exclusively the β-anomer. Yields are variable (5-48%) but simple purines (2-amino-6-chloropurine, 48%); 2-amino-6-methoxypurine, 34%; based on the purine used) gave satisfactory yields. The latter product is a good substrate for adenosine deaminase, treatment with which gives 4'-thio-2'-deoxyguanosine in 87% yield. Compounds can be made in gram quantities.[77]

Another potential solution to the disadvantageous α:β ratio in purine nucleoside synthesis has been suggested by Secrist[34,81] who has explored the ribo → 2'-deoxyribo conversion by deoxygenation at C2' (Scheme 11). The synthesis of the required sugar with the ribo-configuration is essentially identical to the preferred synthesis of compound (9) and it has the advantage that the starting material (D-ribose) is very cheap. Benzyl protecting groups are incompatible with the stability of the glycosyl bond in the purine nucleosides and so the synthesis starts from the penta acetate of 4-thioribofuranose. The overall isolated yield of 5% is still acceptable when one considers that the theoretical maximum in the condensation step alone in the normal synthetic route is only 10% but the method still leaves some room for improvement.

3.2 Protecting Group Strategy

In the original synthesis reported by Bobek, the sugar synthesized was an anomeric mixture of methyl 4-thio-2-deoxy-D-*erythro*-pentofuranosides (14)[21] and thus the hydroxyl groups could be protected in a way which was compatible with the further reactions involved. Thus p-toluoyl (which has the advantage that the compound can be detected on TLC under a UV light) or acetyl (which often gives a cleaner NMR spectrum) are usually the derivatives of choice and both can be removed using sodium methoxide.

Our preferred synthesis provides a sugar with benzyl protection (9). Early attempts to deprotect the deoxynucleosides produced using catalytic hydrogenation were not satisfactory. The sulfur-containing compounds appeared to poison the catalyst such that increasingly drastic conditions were required, which partially hydrogenated the 5,6-double bond in the pyrimidine moiety (31).[24,30] On a small scale, by far the most efficient method is the use of BCl_3[65] or BBr_3.[81] Our experience is that it is essential to keep the reaction mixture as cold as possible (-80 to -90°C) and to quench the reaction while maintaining that temperature. Primary benzyl groups are deprotected faster than secondary and the secondary benzyl group on a β-nucleoside is deprotected far faster than the corresponding benzyl group on an α-nucleoside. Thus if deprotection of an α,β-mixture has to be performed, BCl_3 is wasted in deprotecting α-anomer which is probably not required and while the α-anomer is being deprotected, the deprotected β-anomer will be decomposing. Although the huge excesses of boron trichloride cited in the literature[12,30,65] are not normally required, the purity of the reagent varies and it is also very expensive. Thus on a large scale, titanium tetrachloride[62] deprotection at 0°C is preferable as the regioselectivity is not so pronounced and the products are more stable under the reaction conditions.

However we have two other considerations to bear in mind before a decision on hydroxyl group protection is made. Firstly, it is almost inevitable that at some stage in the synthesis, α- and β-anomers will have to be separated. How this can best be achieved is largely a matter of experience. Thus 5-ethyl-4'-thio-2'-deoxyuridine (26) preferentially crystallizes out in close to quantitative yield from a solution which contains an α:β ratio

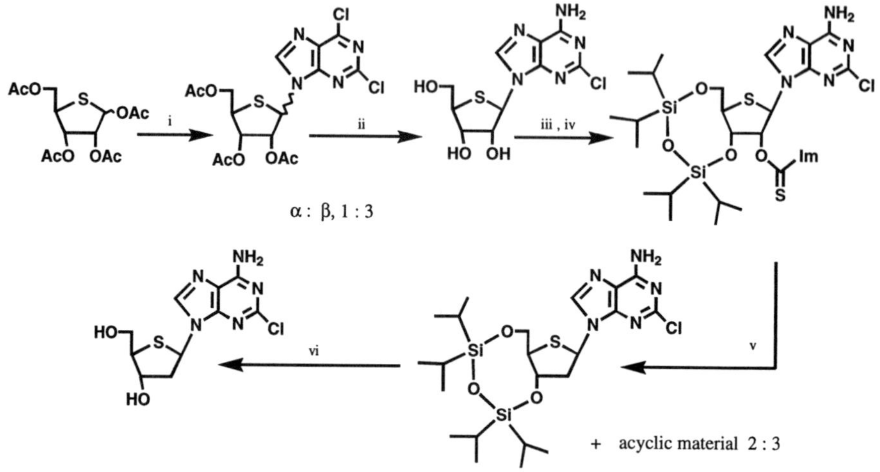

α : β, 1 : 3

+ acyclic material 2 : 3

Scheme 11 *Reagents and conditions* : i, 2,6-dichloropurine, $SnCl_4$, CH_3CN; ii, $EtOH/NH_3$; iii, $[ClSi(Pr^i)_2]_2O$, pyridine; iv, Im–C(=S)–Im ; v, nBu_3SnH, AlBN, toluene; vi, TBAF.

Scheme 12 *Reagents and conditions* : i, BBr_3 or BCl_3, CH_2Cl_2 or $TiCl_4$; ii, *p*-TolCl, pyridine; iii, $Hg(OAc)_2$, HOAc.

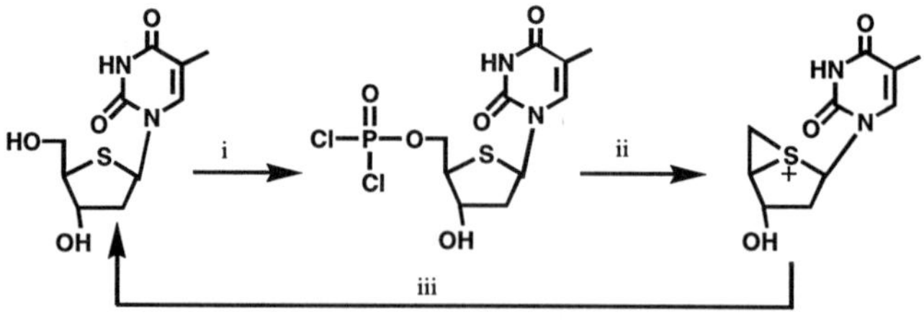

Scheme 13 *Reagents and conditions* : i,ii $POCl_3$, $(MeO)_3P{=}O$; iii, H_2O

of 2:1.[65] This lucky break depends upon the fact that the α-anomer has so far never crystallized. On the other hand, the anomers of 5-iodo-4'-thio-2'-deoxyuridine separate very easily by chromatography or by crystallization when protected as the dibenzyl ether.[65] The anomers of 4'-thio-2'-deoxycytidine have so far proved to be impossible to separate no matter how or whether the sugar hydroxyl groups are protected.[36]

Thus, currently our preferred methods used for anomer separation of the 4'-thio-2'-deoxynucleosides of thymine (T,32), uracil (U,33), and cytosine (C,34) are: T, separate by crystallization or chromatography at the protected (benzyl) stage; U, N3-benzoylate the mixed protected (benzyl) anomers and separate by chromatography, debenzoylate using $MeOH/NH_3$ and remove the benzyl protecting groups in the usual way; C, N^4-benzoylate and separate the anomers by crystallization and deprotect the separated anomers in the usual way. Overall yields in the condensation step of isolated β-anomer are 20-30% which corresponds to 50-80% of the β-anomer produced.[82]

The second consideration concerning the selection of sugar hydroxyl protecting groups is that the stability of many pyrimidine analogues (and all the purines) are not compatible with deprotection conditions used for removing benzyl groups. Thus the preparation of 4'-thio-5-vinyl-2'-deoxyuridine (35)[65] involved anomer separation of 3',5'-di-*O*-benzyl-5-iodo-4'-thio-2'-deoxyuridine (36) followed by deprotection, reprotection with acyl groups and subsequent conversion of 5-iodo- to 5-vinyl.[83]

Other syntheses require the protecting groups to be changed before condensation with a base. Hence we[65] and Secrist have used a similar scheme[34] (Scheme 12).

Upon reprotection of the 3- and 5-hydroxyl groups as the p-toluoyl esters, the thiobenzyl glycoside is almost totally inert when condensed with a silylated base in the presence of NIS. We have no explanation for this. Thus the leaving group has to be changed to acetoxy and condensation achieved under Lewis acid 'catalysed' conditions (using over one equivalent of catalyst) with subsequent deprotection using methoxide. The result of this series of reactions is the addition of 3 further steps with an overall yield of no more than 50% and the condensation yields and α:β ratio are not improved. Thus there is some incentive, particularly when using acid-sensitive base analogues, to preplan a sugar synthesis which has compatible protecting groups[62] as our preferred synthesis of the sugar inevitably leads to the presence of benzyl protecting groups.

4 CHEMICAL REACTIONS OF 4'-THIO-2'-DEOXYNUCLEOSIDES

Trivial protection and deprotection of sugar hydroxyl groups will not be covered, nor will simple exemplification reactions of purine and pyrimidine moieties be discussed as they are essentially identical to the similar reaction conditions used for normal nucleosides.

4.1 Early Observations

The first sign, that the chemistry of 4'-thio-2'-deoxynucleosides would be affected due to the presence of the sulfur in the sugar ring, appeared as soon as attempts were made to remove the benzyl protecting groups by catalytic hydrogenation as previously mentioned.[24,30] However, reduction of a 5-propynyl derivative to the corresponding 5-propyl analogue[65] can be achieved in good yield using an excess of catalyst (5% Pd/C). Early attempts to shortcut the synthesis of O^2,3'-anhydronucleosides by treating 4'-thiothymidine with the Yarovenko reagent[84] also revealed that the regiospecificity seen when used with a normal 2'-deoxynucleoside was not maintained with the 4'-thio-2'-deoxynucleosides.. Substantial quantities of O^2,5'- as well as the required O^2,3'-anyhdro-4'-thio-2'-deoxynucleoside were formed.[12,30]

4.2 Synthesis of 4'-Thio-2'-deoxynucleoside 5'-triphosphates

Synthesis of 5'-phosphates and hence 5'-triphosphates of 4'-thio-2'-deoxynucleosides until recently has not been possible in reasonable yield, if one discounts the publication by Chinese scientists[85] whom we now know had wrongly identified their starting material. Attempts by us and by others to phosphorylate 4'-thiothymidine using standard conditions had resulted in very low yields or no yield at all. It became clear that the presence of a good leaving group at C5' (which occurs when the dichloridate is formed, Scheme 13) enables the sulfur to participate in the formation of a bicyclic episulfonium intermediate which would then be converted back into starting material during the aqueous work-up.[86] The problem has finally been solved in a recent publication when the 5'-phosphate was made in 56% isolated yield using β-cyanoethylphosphate and DCC in pyridine.[87] Unlike the phosphorodichloridate, the nucleotide is perfectly stable and could be converted into the nucleoside triphosphate using standard conditions (Scheme 14).

An alternative method of synthesis of the 5'-phosphates of 4'-thio-2'-deoxynucleosides is to use the phosphoryl transferase system described by Shugar.[88,89] 4'-Thio-2'-deoxynucleosides are excellent substrates for the enzyme isolated from wheat shoots and actually give higher yields (> 80%) than do the normal nucleosides.[90] This could well be the result of increased stability of the sulfur-containing nucleosides to phosphorolysis, as the phosphotransferase used is a crude extract and for high yields, long periods of incubation are required. The method is easily adaptable to intermediate (multi-gram) scale quantities.

The indication that it would not be possible to functionalize the 5'-hydroxyl group of a 4'-thio-2'-deoxynucleoside with a good leaving group was the starting point for a series of reactions to make the 5'-tosyl ester.[86] Comparative studies with thymidine showed that indeed it was not possible to isolate the 5'-tosyl ester of 4'-thiothymidine and using a suitably protected derivative of 4'-thiothymidine (Scheme 15), the identity of some minor products was established, although the majority product isolated was starting material.

Oxidation of 5-ethyl-4'-thio-2'-deoxyuridine with sodium metaperiodate gives a separable mixture of the two diastereoisomers of the sulfoxides. Stereoselective oxidation can also be achieved but the relative cost and ease of separation of the diasterioisomers means that the periodate oxidation method has a distinct advantage. One diastereoisomer from 5-ethyl-4'-thio-2'-deoxyuridine has been crystallized and its structure determined by X-ray analysis.[91] Thus in future, it should be possible to identify

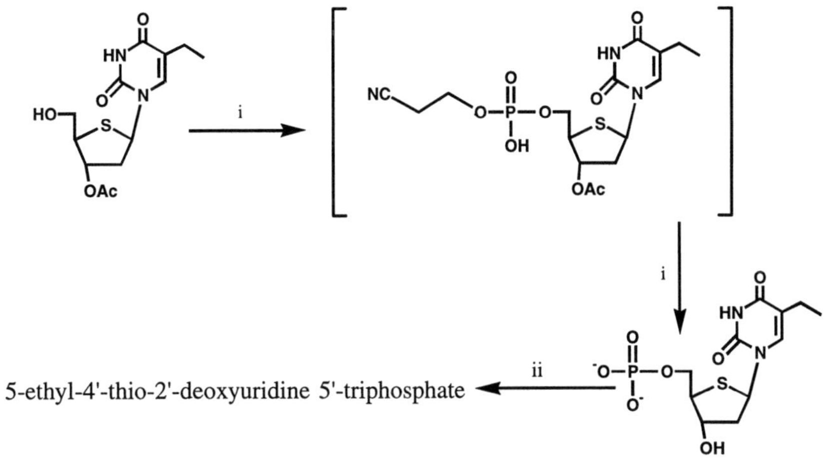

5-ethyl-4'-thio-2'-deoxyuridine 5'-triphosphate

Scheme 14 *Reagents and conditions* : i, β- cyanoethylphosphate, pyridine, DCC; ii, *N,N*-carbonyldiimidazole, bis(tri-*n*-butylammonium)pyrophosphate, DMF.

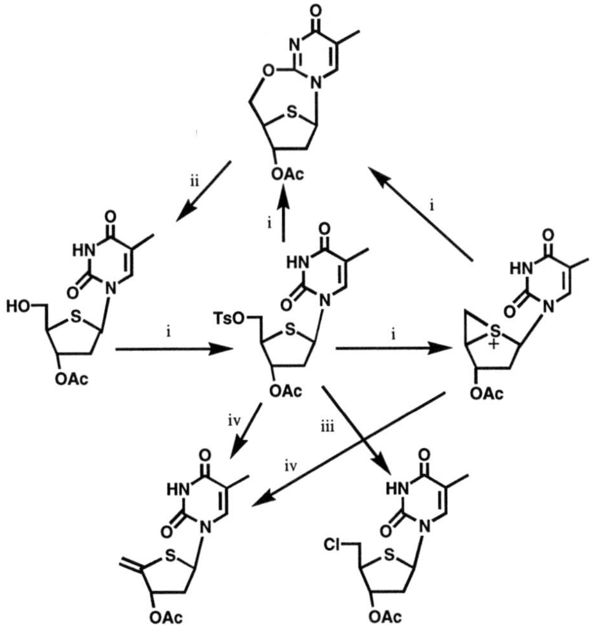

Scheme 15 *Reagents and conditions* : i, TsCl, pyridine; ii OH^-; iii, Cl^-; iv, Base.

the two isomers from NMR data. Oxidation with m-chloroperoxybenzoic acid gives the sulfone in high yield. An X-ray structure of the sulfone(38)of 4-thiothymidine has been obtained[92] and although the conformation of the sugar ring is very similar to that found in thymidine and 4'-thiothymidine, the glycosidic torsion angle is 85.5° as a result of steric interactions between the sulfone oxygen atoms and the pyrimidine ring. The H4'-proton in these compounds is quite acidic and when treated with aqueous base in D_2O gives an epimeric mixture (at C4') of the monodeuterated products.[91]

Tritylation or dimethoxytritylation of 4'-thio-2'-deoxynucleosides always seems to be slower than for the corresponding normal nucleosides. To achieve quantitative and selective 5'-tritylation, it is important to ensure the absolute dryness of the reagents and to be patient.[93,94] Subsequent phosphitylation occurs very easily and the products can be used on a commercial DNA synthesizer without modification of the program for incorporation into oligonucleotides (see below). Anecdotal evidence (T. Brown, Oswel) is that yields of multi-substituted oligonucleotides (e.g. $T^{S}{}_{18}T$) are at least as high if not higher than those normally obtained. There is no evidence for the oxidation of the sulfur during multiple rounds of synthesis and the only immediate effect noticeable on the properties of such oligomers is an increase in lipophilicity.

Thus the chemistry of 4'-thio-2'-deoxynucleosides is still in its infancy but already some interesting reactivities have been identified and no doubt others will follow when the starting materials are more readily available.

5 ENZYMOLOGY

5.1 Phosphorylases

The original rationale for the synthesis of 4'-thio-2'-deoxynucleosides was to produce an analogue with a glycosidic bond which had increased stability to phosphorolysis. For all compounds in this series so far tested (all 5-substituted uracil analogues), no degradation by thymidine phosphorylase has been seen and nucleosides have been stable for over 25h when incubated with intact human blood platelets; conditions which degrade completely *(E)*-5-2(bromovinyl)-2'-deoxyuridine in 1h.[95]

5.2 Kinases

There is so far little kinetic data available to show the ability of 4'-thio-2'-deoxynucleosides to act as substrates for a range of nucleoside kinases. The most comprehensive set of data is available for 5-ethyl-4'-thio-2'-deoxyuridine: K_m (μM) for HSV-1 TK= 1.02, HSV-2 TK = 1.5, VZV TK = 0.22 and for cellular kinases > 50.[96] The selective effect of phosphorylation of 5-ethyl-4'-thio-2'-deoxyuridine in virus-infected cells compared with non-infected cells can be seen from a comparison of the levels of nucleoside triphosphate found (pmols/million cells): uninfected cells 0.032; HSV-1 infected cells, 138.4.[97]

A somewhat larger range of analogues has been tested in a competition assay using the TK from HSV-1 and from HeLa cells.[95] The assay uses [^{3}H]-labelled thymidine and an IC_{50} value for such analogues is given in Table 1. These values clearly show that apart from 4'-thiothymidine, none of the analogues is likely to be phosphorylated to any significant extent by cellular kinases and yet while the 5-substituent remains fairly small, the analogues are capable of inhibiting the phosphorylation of thymidine and are likely to be acting as substrates.[95] Indeed in two cases which have been checked, 5-isopropyl- and 5-cyclopropyl-4'-thio-2'-deoxyuridines are both good substrates for HSV-1 TK and the former is a competitive inhibitor with a Ki of 0.09 μM.[95] No information is available about the subsequent conversion of the mononucleotides to the corresponding nucleoside 5'-diphosphates.

Table 1

IC_{50} (μM) values of 5-substituted 4'-thio-2'-deoxyuridines

	HSV-1 TK	HeLa TK
[^{3}H]-Thymidine in the assay	0.8 μM	2 μM
Substituent at 5-position		
Methyl	0.09	6.26
Ethyl	0.12	> 200
Isopropyl	0.35	> 100
Cyclopropyl	0.53	60
tert-Butyl	0.14	> 200
Iodo	0.14	21
(E)-5-(2-bromovinyl)	0.17	> 200

5.3 Polymerases

Little information is available concerning the substrate specificity of 4'-thio-2'-deoxynucleoside 5'-triphosphates for polymerizing enzymes. The Ki values (μM) for the inhibition of various viral DNA polymerases for 5-ethyl-4'-thio-2'-deoxyuridine are: HSV-1, 0.2; HSV-2, 0.09; VZV, 0.20.[97] The Km for human placental DNA polymerase α = 2.6 μM (dTTP = 4.8 μM)[87] suggests that if phosphorylated, 5-ethyl-4'-thio-2'-deoxyuridine (as the 5'-triphosphate) would be a good substrate for viral and cellular DNA polymerases. Terminal deoxynucleotidyl transferase is capable of extending a tetradecanucleotide by 20-30 nucleotides in the presence of 5-ethyl-4'-thio-2'-deoxyuridine 5'-triphosphate but the modified nucleotide is much less efficient as an acceptor than its normal counterpart and thus oligonucleotide chains so extended are markedly shorter than those containing only natural nucleotide residues. The reverse transcriptases of AMV and HIV do not appear to distinguish between 5-ethyl-4'-thio-2'-deoxyuridine 5'-triphosphate and thymidine 5'-triphosphate.[87]

As was immediately recognized by both groups working in the field,[12,35] 4'-thiothymidine is surprisingly toxic to a range of viruses and also to the host cells. It is now clear that 4'-thiothymidine is a good substrate for herpesvirus-encoded and cellular thymidine kinases (Km for TK from L1210 cells ~ 10 μM, V max ~ 400 pmol/min/mg; thymidine 1.2 μM, Vmax ~ 500 pmol/min/mg) and 4'-thiothymidine has been shown to be incorporated into L1210 cells where it is stable and not immediately removed by DNA repair enzymes.[98] No effect of 4'-thiothymidine on protein or RNA synthesis is seen and the data is consistent with the toxicity being due to the inhibition of DNA synthesis and the toxicity can be directly correlated with the percentage incorporation of analogue. However, cell growth, following incubation with 4'-thiothymidine, is very rapidly inhibited, which suggests that functions that are impaired by the incorporation of 4'-thiothymidine into DNA must occur soon after its incorporation. The precise events which are affected have not yet been identified but could they be so, it might go a long way to explaining why other more selective nucleoside analogues inhibit viral replication.

6 ANTIVIRAL AND ANTITUMOUR ASSAYS

The following tables are an attempt to combine all the testing results so far published together with a limited number of unpublished results. As it is often not meaningful to compare results obtained by different laboratories, particularly for antitumour assays; the original reference should be consulted to find the precise conditions for each assay.

6.1 Antiviral Assays

See Tables 2 and 3

The following analogues have been shown not to possess any significant antiviral activity (or toxicity): (1) All 6-aza-4'-thio-2'-deoxyuridines tested including the 5-methyl, 5-cyclopropyl (α/β) and 5-thienyl derivatives;[100] (2) 4'-Thio-3'-azido-2',3'-dideoxythymidine;[12] (3) The sulfones and sulfoxides of 5-methyl- and 5-ethyl-4'-thio-2'-deoxyuridine;[99] (4) All 4'-thio-β-L-2'-deoxynucleosides tested.[99]

6.2 Antitumour Assays

See Tables 4 and 5

7 CONFORMATION AND STRUCTURE

A very recent review[101] contains a comprehensive coverage of NMR data on 4'-thio-2'-deoxynucleosides as it is relevant to configurational (α/β) assignment. Here it is only intended to cover the one publication[102] which uses NMR data to assign conformational (S/N) ratios in solution and the surprisingly large amount of X-ray crystal data which is available.

7.1 NMR Studies

A detailed study of the conformations adopted by some pyrimidine 4'-thio-2'-deoxynucleosides (5-methyl- and *(E)*-5-(2-bromovinyl)-) uses vicinal proton-proton NMR coupling constants and nuclear Overhauser contacts.[102] Significant adaptations of the conventional routines for J-coupling analysis in nucleoside structures were necessary due to the presence of sulfur instead of 04' in the thiofuranose ring. The sugar ring was shown to have a preference for a South-type (C2'-endo; C3'-exo) puckered conformation with an S/N ratio of about 70:30 at ambient temperature. The pyrimidine base adopted an

Table 2

Antiviral activities and cytotoxicities of 5-substituted pyrimidine 4'-thio-2'-deoxynucleosides

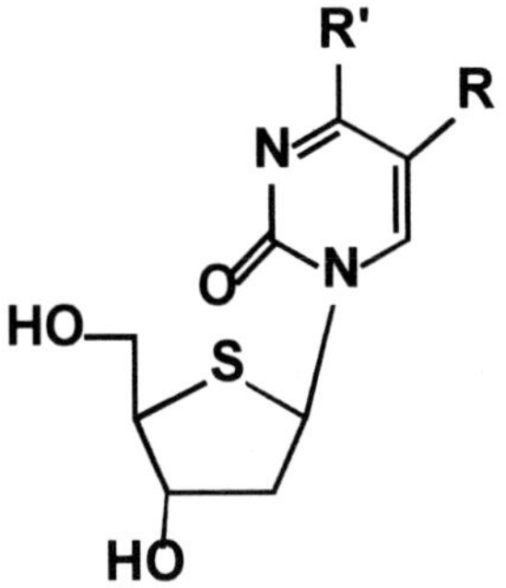

Ref.	R	R'	α/β	HSV-1	IC_{50} (μM) HSV-2	VZV	HCMV	$CCID_{50}$ (μM)
30	Me	OH	α	> 100	> 100	> 100	ND	> 100
12	Me	OH	β	0.37	2.3	10	0.98	7.1
65	Et	OH	α	ND	> 100	> 100	> 100	> 100
65	Et	OH	β	0.17-0.5	2-5	0.99	138	> 500
65	n-Pr	OH	β	6.8	> 10	3	> 100	> 100
65	i-Pr	OH	β	2.2	> 100	4.1	> 100	> 500
65	c-Pr	OH	β	1.6	> 100	0.5	> 100	> 500
65	CH(Me)Et	OH	α/β	ND	> 100	ND	> 100	> 500
65	t-Bu	OH	β	ND	> 100	ND	> 100	475
65	adamantyl	OH	β	> 100	> 100	T40	> 100	> 125
65	$CH{=}CH_2$	OH	β	1.4	> 100	1.1	93	> 500
65	$C(=CH_2)Me$	OH	β	3.4	> 100	< 40	> 100	> 500
65	(E)-CH=CHMe	OH	β	> 2	> 20	0.46	> 100	> 100
65	ethynyl	OH	β	6.6	> 10	5.5	9.3	33
65	propynyl	OH	β	> 100	> 100	1.4	> 100	492
65	CH_2CH_2OMe	OH	β	ND	> 100	> 10	> 100	ND
65	CH(OH)Me	OH	α/β	40	> 100	12	> 100	> 500
99	CH_2OH	OH	β	ND	> 100	ND	> 100	> 500
65	CH_2OMe	OH	α/β	> 10	> 10	> 10	> 100	ND
65	CH(OMe)Me	OH	α/β	> 100	> 100	> 40	> 100	> 500
65	COMe	OH	α/β	> 10	> 10	> 10	> 100	351
65	CF_3	OH	α	ND	> 100	ND	> 100	> 500
65	CF_3	OH	β	< 1.6	15	ND	> 10	139

					IC_{50} (μM)			$CCID_{50}$ (μM)
Ref.	R	R'	α/β	HSV-1	HSV-2	VZV	HCMV	
65	CH_2CH_2F	OH	β	0.7	> 100	0.12	> 100	>500
65	CH_2CH_2Cl	OH	α/β	0.15	> 100	0.3	ND	>200
65	CH_2CF_3	OH	β	ND	> 100	>40	> 100	>500
65	$CH{=}CF_2$	OH	β	ND	> 100	<10	> 100	>500
12	(E)-CH=CHBr	OH	α	> 500	> 500	ND	ND	> 500
12	(E)-CH=CHBr	OH	β	0.6	~ 10	0.08	ND	> 500
65	(E)-CH=CHCl	OH	β	0.5	> 100	> 10	> 100	> 500
65	$CF{=}CCl_2$	OH	β	> 50	ND	> 100	> 100	> 500
65	CH_2SMe	OH	α/β	ND	> 100	ND	> 100	> 500
65	OMe	OH	α/β	> 100	> 100	> 100	> 100	ND
65	F	OH	α/β	> 10	> 10	> 100	> 100	92
65	Cl	OH	β	> 100	> 100	> 40	0.2	59
65	Br	OH	β	5.3	> 100	> 40	0.44	> 500
65	I	OH	β	0.8-1	10	3.5	4.0	> 100
65	CN	OH	β	> 100	> 100	> 100	> 100	> 500
65	NO_2	OH	α	> 100	> 100	> 100	ND	ND
65	NO_2	OH	β	18	27	9-13	ND	476
65	$5,6(CH_2)_3$	OH	α/β	ND	> 100	> 40	> 100	> 500
99	H	NH_2	α/β	1.6	7.2	0.38	0.1	< 30
99	Et	NH_2	β	2.9	37	< 40	> 100	> 500
99	Cl	NH_2	β	> 100	> 100	> 40	2.5	251
99	Br	NH_2	α/β	ND	> 100	T40	ND	> 500
99	I	NH_2	α	> 100	> 100	> 100	> 100	> 500
99	I	NH_2	β	1.06	37.7	< 40	4.8	> 500
99	F	NH_2	β	2.2	0.85	T100	0.1	> 500
99	CH_2CH_2Cl	NH_2	β	> 100	> 100	ND	> 100	> 500
99	CN	NH_2	β	> 100	> 100	< 40	> 100	> 500

ND = Not determined

Table 3

Antiviral activities and cytotoxicities of purinyl-4'-thio-2'-deoxynucleosides

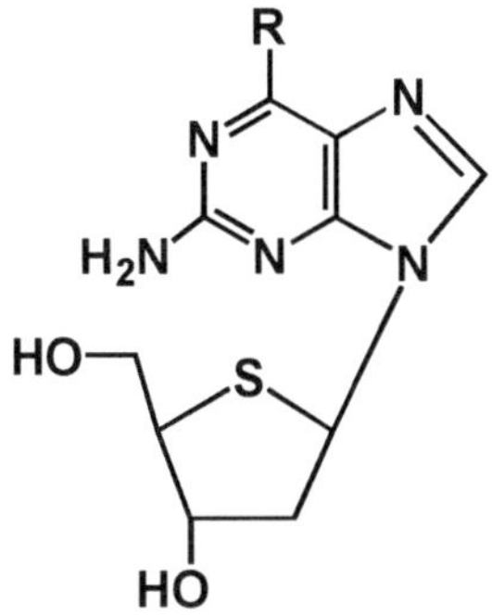

Ref.	R	α/β	HSV-1	HSV-2	IC_{50} (μM) HBV	HMCV	$CCID_{50}$ (μM)
77,99	OH	β	ND	ND	0.002	0.06	ND
99	$N(CH_3)_2$	β	> 10 > 100	> 100	< 4	< 100	> 500
77,99	piperidino	β	> 100	> 100	4	6	> 500
77,99	pyrrolidino	β	> 100	> 100	0.85	10	> 500
77,99	SCH_3	β	> 10 < 100	> 100	0.45	2	> 500
77,99	NHPr	β	> 100	> 100	0.061	2	> 500
77,99	N(Et)Me	β	> 10 < 100	> 100	0.19	2	> 500
77,99	$N(CH_3)cPr$	β	> 10 < 100	> 100	0.3	2	> 500
77,99	OCH_3	β	0.26	< 1.6	0.0025	0.62	2.7
77,99	NHcPr	β	< 100	> 100	0.0072	0.2	> 500
77	Cl	β	ND	ND	0.001	0.1	*
77	OCH_2cPr	β	ND	ND	0.035	0.6	*
77	NHallyl	β	ND	ND	0.058	1.5	*
77	NHiPr	β	ND	ND	0.3	4	*
99	4'-thio-2'-deoxy A	α	> 100	> 100	ND	> 100	420
99	4'-thio-2'-deoxy A	β	37	> 100	ND	> 100	> 500

ND = Not determined

*The toxicities listed are for the antiherpesvirus assays in Vero or MRC-5 cells. Toxicities in the cell lines used in HBV and HCMV assays are usually much higher (i.e. lower μM figure) and the original reference should be consulted.

Table 4

Antitumour assays of some 4'-thio-2'-deoxynucleosides

(All figures taken from reference 43)

		IC_{50} (μM)	
Configuration	Compound (4'-Thio)	L1210	KB
D	Deoxyuridine	> 100	> 100
L	Deoxyuridine	> 100	> 100
D	Thymidine	0.033	10.4
L	Thymidine	10.4	> 100
D	Deoxycytidine	0.62	12.8
L	Deoxycytidine	> 100	> 100
D	5-Trifluoromethyl dC	0.0026	21.1

Table 5

Antitumour assays of some 4'-thio-2'-deoxynucleosides

(All figures taken from reference 34)

		IC_{50} (μM)	
α/β	Compound (4'-Thio)	H.Ep.2 (epidermoid)	CCRF-CEM (leukaemic)
β	Thymidine	0.06	0.66
α	Deoxyuridine	60	> 160
β	Deoxyuridine	1.5	> 4
β	Deoxycytidine	0.2	4
β	5-Aza dC	> 80	80
β	2-Chloro dA	< 0.17	< 1.7

anti-conformation. Using the parameters established, it should now be possible to determine the sugar conformations of a range of 4'-thio-2'-deoxynucleosides in solution.

7.2 X-ray Studies

X-ray information invariably gives an exaggerated idea of the conformations which can be adopted in solution and the conformation seen by X-ray may or may not have much relevance to the situation in solution and possibly even less relevance to the conformation required for the analogue to be processed enzymatically and hence to show biological activity.

The initial X-ray structures for 4'-thiothymidine[39,102] (32) and *(E)*-5-(2-bromovinyl)-4'-thio-2'-deoxyuridine[102] (20) indeed confirmed the exclusive presence of the S-conformer in the crystal. The bases were in the anti-conformation and apart from the C-S bonds being longer and the C4'-S-C1' angle being larger than the corresponding values in a normal nucleoside, the overall shape of 4'-thio-2'-deoxynucleosides and their natural counterparts is remarkably similar. Indeed it would appear (and has been since confirmed; see below), that the disruption caused by the presence of such an analogue in an oligonucleotide would be minimal, as the bond lengths and torsion angles involved in the formation of the phosphodiester backbone are essentially unaffected. The crystal structure of 5-ethyl-4'-thio-2'-deoxyuridine (Figure 1a) fits this pattern[103] and one further structure which is available and will be published shortly is that of 6-aza-5-thienyl-4'-thio-2'-deoxyuridine (37) which also has a C2'-endo (5), anti, conformation.[104]

Four α-nucleoside X-ray structures of α-4'-thio-2'-deoxyuridine,[43] α-4'-thiothymidine,[91] α-4'-thio-2'-deoxycytidine[82] and α-5-adamantyl-4'-thio-2'-deoxyuridine[105] have also been determined and all have been shown to have the pyrimidine base in the anti-conformation and the sugar in a C2'-endo (S) conformation. The X-ray structure of the sulfone of 4'-thiothymidine (38)[92] again surprisingly showed that the C2'-endo (S) conformation was still preferred. However, because of the steric interactions between the sulfone oxygen atoms and the pyrimidine ring, the glycosidic torsion angle S4'-C1'-N1-C6 is 85.5° compared with usual values which are in the range of 33-59°. It is likely that this unusual torsion angle is partly responsible for the lack of any biological activity (toxicity) shown by this compound. The availability of an X-ray structure of this sulfone should in future enable the identity of the two diastereoisomers to be distinguished from NMR data[91] (Figure 1b).

A recent publication has confirmed that the overall effect of incorporating 4'-thio-2'-deoxynucleosides into an oligonucleotide on the structure of the oligonucleotide is minimal.[106] Thus a self-complementary dodecamer d(CGCGAAT^{S}T^{S}CGCG) where T^S = 4'-thiothymidine (32), was crystallized and the structure compared with that of the unmodified duplex.[107] The major differences between the two structures is a change in the conformation of the sugar-phosphate backbone in the regions at and adjacent to the positions of the modified nucleosides. This adjustment is no doubt an attempt to position the pyrimidine bases accurately for base-pairing. The thiosugars adopt a C3'-exo conformation rather than the approximate C1'-exo conformation found for the corresponding sugar moieties in the unmodified oligomer.

One other X-ray structure of a co-crystal of EcoRV endonuclease and an oligonucleotide containing a 4'-thiothymidine residue in a (resistant) recognition site[108] has demonstrated that the conformation adopted by the analogue-containing oligomer was

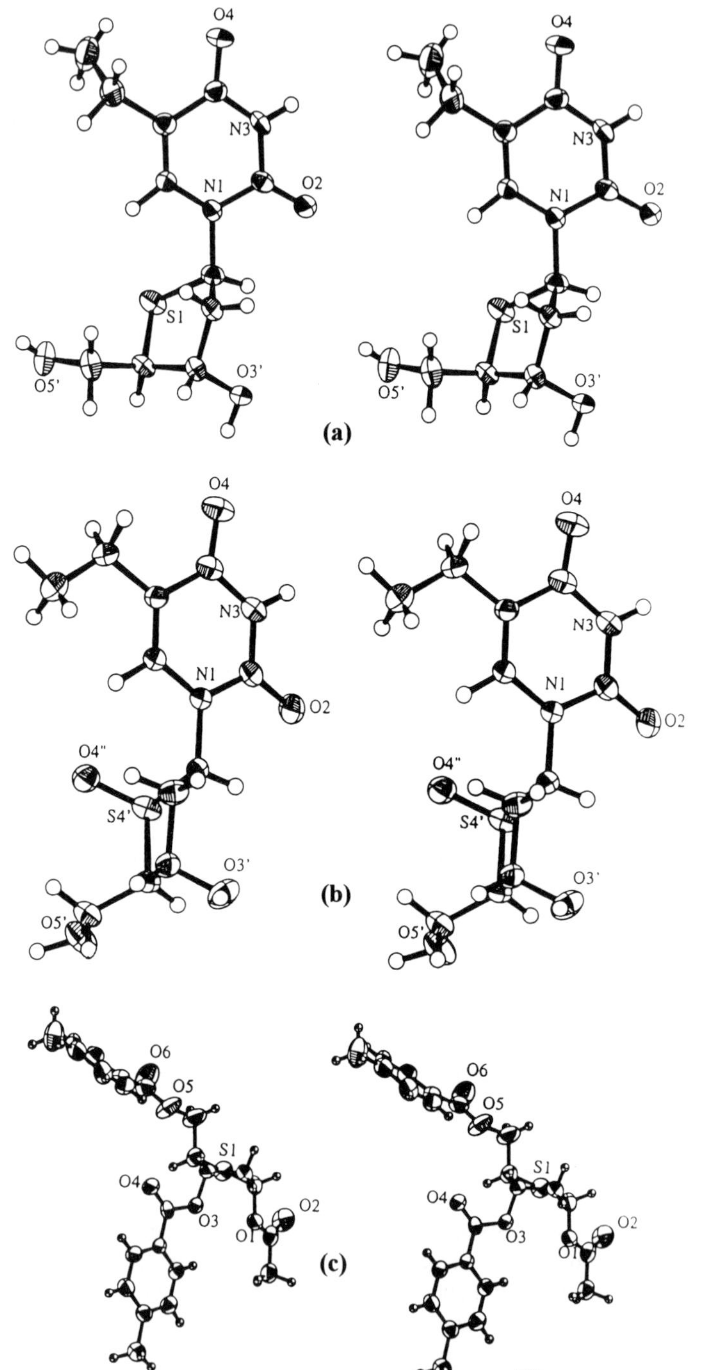

Figure 1. Stereopair of (a) 5-ethyl-4'-thio-2'-deoxyuridine,[103] (b) a sulfoxide of 5-ethyl-4'-thio-2'-deoxyuridine,[91] and (c) α-1-O-acetyl-3',5'-di-O-p-toluoyl-4'-thio-2'-deoxyribofuranose.[110]

unchanged from the distorted structure seen in the co-crystal of the enzyme and a natural oligomer.[109]

Thus it appears that apart from the obvious impact on the immediate bond lengths and angle size, the replacement of O4' by S4' in 2'-deoxynucleosides has a minimal effect on the conformations which can be adopted. The only X-ray structure available for a 4'-thio-2'-deoxysugar is for the α-anomer of compound 28 [110] (Figure1c).

8 OLIGODEOXYNUCLEOTIDE STABILITY AND STRUCTURE

There is very little data published on oligonucleotides containing 4'-thio-2'-deoxynucleosides and this short overview will therefore also outline briefly some finished but unpublished work and work which is in progress, so that the possible uses of these analogues become more widely known.

As has previously been mentioned, the synthesis of the suitably protected phosphoramidites of the 4'-thio-2'-deoxynucleosides of a series of pyrimidine derviatives [uracil (33), cytosine (34), thymine (32) and 5-ethyluracil (26)] have been made and can be incorporated into oligonucleotides using standard automated methodology. The effects of such incorporation on the structure of self-complementary duplexes appears to be minimal as has been shown by X-ray analysis of a crystal of d(CGCGAAT^{S}T^{S}CGCG).[106] T_m measurements against complementary DNA of a wide range of analogue-containing oligomers show a loss of 1°C per analogue incorporated.[82,93] I am unaware of any purine derivatives having been incorporated into an oligonucleotide either enzymatically or chemically and this is certainly due to (a) the difficulty until recently[87] in making 5'-triphosphates and (b) the adverse α:β ratio when performing purine deoxynucleoside condensations. These two factors limit the amount of starting material available but there is surely no fundamental problem with synthesizing such oligomers.

In a detailed study of the recognition of specific DNA sequences by the restriction endonuclease EcoRV and its associated methylase, it was found that replacement of thymidine by 4'-thiothymidine next to the scissile bond resulted in complete inhibition of both activities. When a thymidine residue positioned in the recognition site but not adjacent to the scissile bond was replaced, the K_m for the endonuclease remained unaltered but the K_{cat} was only 0.03% of that obtained for the unmodified sequence. Similar results were obtained for the methylase. Binding of EcoRV to a cognate sequence is accompanied by a distortion in the DNA of ~ 90° [109] and this distortion is also seen in the analogue-containing oligomer.[108] The reason for endonuclease resistance appears to be an absence of the essential magnesium ions in the crystal structure despite their presence in solution. So far, no entirely satisfactory explanation has been advanced for this absence of magnesium ions but this was the first example of a now growing series of 4'-thio-2'-deoxynucleoside-containing oligonucleotides which have anomalous interactions with endo-enzymes. Thus it is likely that the Hha I methylase does not methylate 4'-thio-2'-deoxycytidine in the recognition site and a 4'-thiothymidine-thymidine photodimer is not cleaved by T4 endonuclease V. The ability of a uracil DNA-glycosylase to excise uracil from a 4'-thio-2'-deoxyuridine-containing oligomer is currently under investigation.

Oligonucleotides of sequence d(GCGXAGC) adopt extremely stable ($T_m \geq 70$°C) hair-in structures (X can be any of the four natural deoxynucleosides) in which each residue is a member of one of two hydrogen-bonded stacks consisting of residues 1-4 and

5-7.[111] The base of the fourth nucleotide is a member of the first stack but the torsion angles between residues 4 and 5 are very different from those found in a typical B-form structure. The T_m of such a hairpin with the fourth residue of thymidine is 70°C and when 4'-thiothymidine is present, the T_m is 69°C showing that the sulfur atom does not prevent the necessary distortion required.[112]

In view of the high stability of DNA duplexes containing 4'-thiothymidine, an on-going study is attempting to assess the suitability of such oligomers for use in antisense technology. The primary concern here is to produce a metabolically stable oligonucleotide which binds with high affinity and specificity to its complementary RNA sequence and induces RNase H cleavage of the RNA strand. A brief summary of the current situation (in the course of which an oligonucleotide containing 18 consecutive 4'-thiothymidine residues has been made in excellent yield), shows that such oligonucleotides are stable to endo- but not exo-nucleolytic degradation, that they form specific and stable (no lowering of T_m) duplexes with complementary RNA but that the duplex is considerably more resistant to RNase H than is a normal RNA-DNA duplex.[82] Studies on triplex formation are in progress.

9 CONCLUSIONS

In summary, the last 7 years have seen a rapid increase in work on 4'-thio-2'-deoxynucleosides. Methods are now available to make the appropriate sugar derivatives so that nucleosides (particularly pyrimidine analogues) can be made without too much difficulty. The outstanding problem faced is the adverse anomer ratio produced in purine nucleoside synthesis and this is likely to be circumvented by the use of enzymatic methods.

Many of the deoxynucleosides so formed have interesting biological activities although an analogue has yet to be identified which has a potentially clinically useful activity which is accompanied by an acceptably low toxicity in long-term, high-dose studies.

Structural determination studies have shown that the presence of sulfur in the sugar ring makes little difference to the structure of the monomers or to the polymers in which they are incorporated.

Thus one might expect interaction with a wide range of metabolic enzymes either as substrates or inhibitors and a wide range of biological properties may be expected. They certainly can be kinase substrates, the triphosphates can be polymerase substrates but the nucleosides appear to be stable to phosphorolysis. The availability of 4'-thio-2'-deoxynucleosides as a further weapon in the armoury of analogues which can be used to investigate the mode of action of a range of nucleic acid metabolizing enzymes is to be welcomed and even the limited experiments performed so far have unearthed a wealth of unexpected results.

10 ACKNOWLEDGEMENTS

Much of the funding for the work described here which was performed in the laboratory in Birmingham (U.K.) has come from the Wellcome Foundation. Additional sources have been provided by the Leverhulme Trust, Amgen Inc. and the SERC and BBSRC.

At the present early stage in the development of this area, the work described here has come from just a few laboratories and it has been my pleasure and privilege to know personally and to respect the work of the major participants, starting with the initial work of Dr. Bobek over 20 years ago. In particular it is a pleasure to acknowledge the friendly rivalry and help of Dr. Secrist in Birmingham, Alabama; Dr. Mackenzie in Hull, U.K., Dr. Spadari in Pavia, Italy and Japanese scientists Drs. Sugimura, Uenishi and Matsuda. Scientists at the Wellcome Research Laboratories have also contributed to our scientific knowledge in this area and while there are too many to acknowledge individually and most can be found listed as authors of several of the papers in the references, I would like to acknowledge the special contributions of Dr. R.J. Young and Dr. S.G. Rahim and the support given over many years by Dr. Dorothy Purifoy, Dr. Ken Powell, Dr. Graham Darby and Mr. Laurence Jenkins who have been encouraged by Mr. Paul Sadler who oversees Intellectual Property Rights at Birmingham University.

None of this work would have been possible without the dedication and skill of a number of postgraduates and graduate students in the Birmingham laboratory since 1986. Again I would only like to select out Dr. Michael Dyson who started this project while studying for his Ph.D. His contribution and those of others will be found in the papers already published and those which are being prepared.

Work outside the immediate competence of an organic chemistry laboratory has been accomplished through a series of collaborations, among the most successful of which have been with Professor B.A. Connolly (Newcastle), Professor T. Brown (Edinburgh and Southampton), Drs. W.N. Hunter and G. Leonard (Manchester), Dr. F.K. Winkler (Basel), Dr. G. Beaton (Amgen, Boulder, CO.), Professor M.H. Caruthers (Boulder, CO.), Professor A.A. Krayevsky (Moscow), Dr. R.J. Roberts (New England Bioscience), Dr. K.-H. Altmann (Ciba-Geigy, Basel) and many excellent scientists at ISIS Pharmaceuticals (Carlsbad, CA.).

11 REFERENCES

1. J.P.C. Schwarz and K.C. Yule, *Proc. Chem. Soc.*, 1961, 417.
2. T.J. Adley and L.N. Owen, *Proc. Chem. Soc.*, 1961, 418.
3. D.L. Ingles and R.L. Whistler, *J. Org. Chem.*, 1962, **27**, 3896.
4. R.L. Whistler and A.K.M. Anisuzzaman, *Amer. Chem. Soc. Symp. Ser.*, 1976, **399**, 33.
5. E.J. Reist, A. Benitez, L. Goodman, B.R. Baker and W.W. Lee, *J. Org. Chem.*, 1962, **27**, 3274.
6. R.L. Whistler, W.E. Dick, T.R. Ingle, R.M. Rowell and B. Urbas, *J. Org. Chem.*, 1964, **29**, 3723.
7. E.J. Reist, D.E. Gueffroy and L. Goodman, *J. Amer. Chem. Soc.*, 1964, **86**, 5658.
8. E.J. Reist, D.E. Gueffroy and L. Goodman, *J. Amer. Chem. Soc.*, 1963, **85**, 3715.
9. M. Bobek, R.L. Whistler and A. Bloch, *J. Med. Chem.*, 1970, **13**, 411.
10. M. Bobek, A. Bloch, R. Parthasarathy and R.L. Whistler, *J. Med. Chem.*, 1975, **18**, 784.
11. L. Bellon, J.-L. Barascut and J.-L. Imbach, *Nucleosides and Nucleotides*, 1992, **11**, 1467 and references cited therein.
12. M.R. Dyson, P.L. Coe and R.T. Walker, *J. Chem. Soc. Chem. Commun.*, 1991, 741.
13. C. Leydier, L. Bellon, J.-L. Barascut, J. Deydier, G. Maury, H. Pelicano, M.A. Elalaoui and J.-L. Imbach, *Nucleosides and Nucleotides*, 1994, **13**, 2035.
14. J.A. Secrist, R.M. Riggs, K.M. Tiwari and J.A. Montgomery, *J. Med. Chem.*, 1992, **35**, 533.
15. R.J. Young, S. Shaw-Ponter, J.B. Thomson, J.A. Miller, J.G. Cumming, A.W. Pugh and P. Rider, *Bioorg. Med. Chem. Lett.*, 1995, **5**, 2599.
16. J. Brånalt, I. Kvarnström, G. Niklasson, S.C.T. Svensson, B. Classon and B. Samuelsson, *J. Org. Chem.*, 1994, **59**, 1783.
17. G. Rassu, P. Spanu, L. Pinna, F. Zanardi and G. Casiraghi, *Tetrahedron Lett.*, 1995, **36**, 1941.
18. U.G. Nayak and R.L. Whistler, *J. Org. Chem.*, 1969, **34**, 3819.
19. U.G. Nayak and R.L. Whistler, *Leibigs Ann. Chem.*, 1970, **741**, 131.
20. Y.-L. Fu and M. Bobek, *J. Org. Chem.*, 1976, **41**, 3831.
21. Y.-L. Fu and M. Bobek, "Nucleic Acid Chemistry", L.B. Townsend and R.S. Tipson, ed., John Wiley and Sons, New York, 1978, 183.
22. Y.-L. Fu, M. Bobek and A. Bloch, 172nd ACS National Meeting, San Francisco, 1976, *Med. Chem. Abstract* # 69.
23. Y.-L. Fu and M. Bobek, "Nucleic Acid Chemistry", L.B. Townsend and R.S. Tipson, eds., John Wiley and Sons, New York, 1978, 317.
24. M.R. Dyson, Ph.D. Thesis, University of Birmingham, 1989.
25. J.A. Montgomery, *Nucleic Acids Res. Symp. Series*, 1989, **21**, 109.
26. F.M. Schabel, "Nucleoside Analogues: Chemistry, Biology and Medical Applications", R.T. Walker, E. De Clercq and F. Eckstein, eds., Plenum, New York, 1979, NATO ASI Series, **A26**, 363.
27. R. Parks, J. Stoeckler, C. Cambor, T. Savarese, G. Crabtree and S. Chu, "Molecular Actions and Targets for Cancer Chemotherapeutic Agents", A. Sartorelli, T. Lazo and J. Bertino, eds., Academic Press, New York, 1981, 229.

28. C. Desgranges, G. Razaka, M. Rabaud, H. Bricaud, J. Balzarini and E. De Clercq, *Biochem. Pharmac.*, 1983, **32**, 3583.
29. M.R. Dyson, P.L. Coe and R.T. Walker, *Carbohydrate Res.*, 1991, **216**, 237.
30. M.R. Dyson, P.L. Coe and R.T. Walker, *J. Med. Chem.*, 1991, **34**, 2782.
31. M.R. Dyson, P.L. Coe and R.T. Walker, *Nucleic Acids Res. Symp. Series*, 1991, **24**, 1.
32. O. Mitsunobu, *Synthesis*, 1981, **7**, 1.
33. J. Harness and N.A. Hughes, *J. Chem. Soc. Chem. Commun.*, 1971, 811.
34. J.A. Secrist, W.B. Parker, K.N. Tiwari, L. Messini, S.C. Shaddix, L.M. Rose, L.L. Bennett and J.A. Montgomery, *Nucleosides and Nucleotides*, 1995, **14**, 675.
35. K.N. Tiwari, J.A. Secrist and J.A. Montgomery, 199th ACS National Meeting, Boston, 1990.
36. J.A. Secrist, K.N. Tiwari, J.M. Riordan and J.A. Montgomery, *J. Med. Chem.*, 1991, **34**, 2361.
37. J. Uenishi, M. Motoyama, Y. Nishiyama and S. Wakabayashi, *J. Chem. Soc. Chem. Commun.*, 1991, 1421.
38. J. Uenishi, M. Motoyama and K. Takahashi, *Nucleic Acids Res. Symp. Series,* 1992, **27**, 77.
39. J. Uenishi, K. Takahashi, M. Motoyama and H. Akashi, *Chem. Lett.*, 1993, 255.
40. J. Uenishi, H. Kawanami and Y. Kubo, *Nucleic Acids Res. Symp.* 1993, **29**, 37.
41. C.H. Behrons and K.B. Sharpless, *Aldrichimica Acta*, 1983, **16**, 67 and references cited therein.
42. J. Uenishi, M. Motoyama and K. Takahashi, *Tetrahedron Asymmetry,* 1994, **5**, 101.
43. J. Uenishi, K. Takahashi, M. Motoyama, H. Akashi and T. Sasaki, *Nucleosides and Nucleotides*, 1994, **13**, 1347.
44. S. Spadari, G. Maga, F. Focher, G. Ciarrocchi, R. Manservigi, F. Arcamone, M. Capobianco, A. Carcuro, F. Colonna, S. Iotti and A. Garbesi, *J. Med. Chem.*, 1992, **35**, 4214.
45. B. Huang and Y. Hui, *Nucleosides and Nucleotides*, 1993, **12**, 139.
46. K.N. Tiwari, J.A. Montgomery and J.A. Secrist, *Nucleosides and Nucleotides,* 1993, **12**, 841.
47. L. Bellon, C. Leydier, J.-L. Barascut and J.-L. Imbach, *Nucleosides and Nucleotides*, 1993, **12**, 847.
48. B. Tber, N. Fahmi, G. Ronco, G. Mackenzie, P. Villa and G. Ville, *Coll. Czech. Chem. Commun.,* 1993, **58**, 18.
49. B. Tber, N. Fahmi, G. Ronco, P. Villa, D.F. Ewing and G. Mackenzie, *Carbohydrate Res.*, 1995, **267**, 203.
50. H. Ait-sir, D.F. Ewing, N. Fahmi, G. Goethals, G. Mackenzie, G. Ronco, B. Tber and P. Villa, Nucleosides and Nucleotides, 1995, **14**, 359; H. Ait-sir, N. Fahmi, G. Goethals, G. Ronco, B. Tber, P. Villa, D.F. Ewing and G. Mackenzie, *J. Chem. Soc., Perkin I*, 1996, In the press.
51. Y. Yoshimura, H. Satoh, M. Watanabe, S. Sakata, S. Miura, M. Tanaka, T. Sasaki and A. Matsuda, *Nucleic Acids Res. Symp. Series,* 1995, **34**, 161.
52. Y. Yoshimura, M. Watanabe, S. Sakata, N. Ashida, S. Miyazaki, H. Machida and A. Matsuda, *Nucleic Acids Res. Symp. Series*, 1995, **34**, 21.
53. Y. Yoshimura, K. Kitano, H. Satoh, M. Watanabe, S. Miura, S. Sakata, T. Sasaki and A. Matsuda, *J. Org. Chem.*, 1996, **61**, 822.

54. J.E. McCormick and R.S. McElhinney, *J. Chem. Soc. Perkin I*, 1976, 2533.
55. J.E. McCormick and R.S. McElhinney, *J. Chem. Res.*, 1981, **(5)** 310; **(M)** 3601.
56. I.A. O'Neil and K.M. Hamilton, *Synlett.*, 1992, 791.
57. I.A. O'Neil, K.M. Hamilton and J.A. Miller, *Synlett.*, 1995, 1053.
58. K.S. Jandu and D.L. Selwood, *J. Org. Chem.*, 1995, **60**, 5170.
59. Y. Tsuda, K. Kanemitsu, K. Kakimoto and T. Kikuchi, *Chem. Pharm. Bull. (Jpn).*, 1987, **35**, 2148.
60. G.D. Kini, C.R. Petrie, W.J. Hennen, N.K. Dalley, B.E. Wilson and R.K. Robins, *Carbohydrate Res.*, 1987, **159**, 81.
61. S. Shaw-Ponter, P. Rider and R.J. Young, *Tetrahedron Lett.*, 1996, **37**, 1871.
62. S. Shaw-Ponter, G. Mills, M. Robertson, R.D. Bostwick, G.W. Hardy and R.J. Young, *Tetrahedron Lett.*, 1996, **37**, 1867.
63. A.J. Hubbard, A.S. Jones and R.T. Walker, *Nucleic Acids Res.*, 1984, **12**, 6827.
64. S.R. Pedley, M.Sc. Thesis, University of Birmingham, 1992.
65. S.G. Rahim, N. Trivedi, M.V. Bogunovic-Batchelor, G.W. Hardy, G. Mills, J.W.T. Selway, W. Snowden, E. Littler, P.L. Coe, I. Basnak, R.F. Whale and R.T. Walker, *J. Med. Chem.*, 1996, **39**, 789.
66. D. Horton and R.A. Markovs, *Carbohydrate Res.*, 1980, **80**, 356.
67. H. Sugimura, K. Osumi, T. Yamazaki and T. Yamaya, *Tetrahedron Lett.*, 1991, **32**, 1813.
68. H. Sugimura, K. Sugino and K. Osumi, *Tetrahedron Lett.*, 1992, **33**, 2515.
69. A. Kumar, M. Lewis, S.-I. Shimizu, R.T. Walker, R. Snoeck and E. De Clercq, *Antiviral Chem. Chemotherapy*, 1990, **1**, 35.
70. H. Sugimura, K. Sujino and M. Motegi, *Nucleic Acids Res., Symp. Ser.*, 1993, **29**, 29.
71. M.E. Jung and C. Castro, *J. Org. Chem.*, 1993, **58**, 807.
72. H. Sugimura, personal communication.
73. R.J. Young, S. Shaw-Ponter, G.W. Hardy and G. Mills, *Tetrahedron Lett.*, 1994, **35**, 8687.
74. D.I. Ward, P.L. Coe and R.T. Walker, *Coll. Czech. Chem. Commun.*, 1993, **58**, 1.
75. D.I. Ward, S.M. Jeffs, P.L. Coe and R.T. Walker, *Tetrahedron Lett.*, 1993, **42**, 6779.
76. N. Fahmi, G. Goethals, G. Ronco, P. Villa, D.F. Ewing and F. Mackenzie, Personal Communication.
77. N.A. Van Draanen, G.A. Freeman, S.A. Short, R. Harvey, R. Jansen, G. Szczech and G.W. Koszalka, *J. Med. Chem.*, 1996, **39**, 538.
78. T.A. Krenitsky, G.W. Koszalka and J.V. Tuttle, *Biochemistry*, 1981, **20**, 3615.
79. T.A. Krenitsky, G.W. Koszalka, J.V. Tuttle, J.L. Rideout and G.B. Elion, *Carbohydrate Res.*, 1981, **87**, 139.
80. C.L. Burns, M.H. St. Clair, L.W. Frick, T. Spector, D.R. Averett, M.L. English, T.J. Holmes, T.A. Krenitsky and G.W. Koszalka, *J. Med. Chem.*, 1993, **36**, 378.
81. K.N. Tiwari, J.A. Secrist and J.A. Montgomery, *Nucleosides and Nucleotides*, 1994, **13**, 1819.
82. G.D. Jones, Ph.D. Thesis, University of Birmingham, 1996.
83. P. Herdewijn, L. Kerremans, P. Wigerink, F. Vandendriessche and A. Van Aerschot, *Tetrahedron Lett.*, 1991, **32**, 4397.
84. N.N. Yarovenko and M.A. Raksha, *Chem. Abstr.*, 1960, **54**, 9724h.

85. B.-G. Huang, Y.-Z. Hui, Y.-Y. Lu and G.-F. Hong, *Chinese Sci. Bull.*, 1993, **38**, 1177.
86. E.L. Hancox and R.T. Walker, *Nucleosides and Nucleotides*, 1996, **15**, 135.
87. L.A. Alexandrova, D.G. Semizarov, A.A. Krayevsky and R.T. Walker, *Antiviral Chem. Chemotherapy*, 1996, In the press.
88. J. Giziewicz and D. Shugar, *Acta Biochim. Polon.*, 1975, **22**, 87.
89. J. Giziewicz and D. Shugar, *Nucleic Acid Chemistry*, L.B. Townsend and R.S. Tipson, eds., John Wiley and Sons, New York, 1978, 955.
90. D. Shugar, Personal communication.
91. A. MacCulloch and R.T. Walker, Unpublished results.
92. E.L. Hancox, T.A. Hamor and R.T. Walker, *Tetrahedron Lett.*, 1994, **35**, 1291.
93. E.L. Hancox, B.A. Connolly and R.T. Walker, *Nucleic Acids Res.*, 1993, **21**, 3485.
94. G.D. Jones, M.Sc. Thesis, University of Birmingham, 1995.
95. S. Spadari, Personal communication.
96. R.T. Walker, R.F. Whale, M.R. Dyson, P.L. Coe, W. Alderton, P. Collins, P. Ertl, D. Lowe, G. Rahim, B.W. Snowden and E. Littler, *Nucleic Acids Res., Symp. Series*, 1994, **31**, 9.
97. E. Littler, D. Lowe, B.W. Snowden, P. Ertl, W. Alderton and P. Collins, 34th ICAAC Meeting, Orlando, 1994, Abstract H53.
98. W.B. Parker, S.C. Shaddix, L.M. Rose, K.M. Tiwari, J.A. Montgomery, J.A. Secrist and L.L. Bennett, *Biochem. Pharmacol.*, 1995, **50**, 687.
99. Unpublished results, Wellcome Research Labs.
100. I. Basnak and R.T. Walker, Unpublished results.
101. D.F. Ewing and G. MacKenzie, *Nucleosides and Nucleotides*, 1996, **15**, 809.
102. L.H. Koole, J. Plavec, H. Liu, B.R. Vincent, M.R. Dyson, P.L. Coe, R.T. Walker, G.W. Hardy, S.G. Rahim and J. Chattopadhyaya, *J. Amer. Chem. Soc.*, 1992, **114**, 9936.
103. C. Cooper, T.A. Hamor and R.T. Walker, Unpublished results.
104. I. Basnak, T.A. Hamor and R.T. Walker, Unpublished results.
105. I. Basnak, M. Sun, P.L. Coe and R.T. Walker, *Nucleosides and Nucleotides*, 1996, **15**, 121; M. Sun, I. Basnak, T.A. Hamor and R.T. Walker, *Acta Cryst.*, 1996, **C52**, In the press.
106. T.J. Boggon, E.L. Hancox, K.E. McAuley-Hecht, B.A. Connolly, W.N. Hunter, T. Brown, R.T. Walker and G.A. Leonard, *Nucleic Acids Res.*, 1996, **24**, 951.
107. R. Wing, H.R. Drew, T. Takano, C. Broka, S. Tanaka, K. Itakura and R.E. Dickerson, *Nature*, 1980, **287**, 755.
108. D. Kostrewa, E.L. Hancox, R.T. Walker, B.A. Connolly and F.K. Winkler, Unpublished results quoted in reference 106.
109. F.K. Winkler, D.W. Banner, C. Oefner, D. Tsernoglou, R.S. Brown, S.P. Heathman, R.K. Bryan, P.D. Martin, K. Petratos and K.S. Wilson, *EMBO J.*, 1993, **12**, 1781.
110. G. Otter, T.A. Hamor and R.T. Walker, Unpublished results.
111. I. Hirao, G. Kawai, S. Yoshizawa, Y. Nishimura, Y. Ishido, K. Watanabe and K.-I. Miura, *Nucleic Acids Res.*, 1994, **22**, 576.
112. I. Hirao, S. Yoshizawa and R.T. Walker, Unpublished results.

16
The Invention and Development of Crixivan®: An HIV Protease Inhibitor

Bruce D. Dorsey,* Joseph P. Vacca and Joel R. Huff
DEPARTMENT OF MEDICINAL CHEMISTRY, MERCK RESEARCH LABORATORIES, WEST POINT, PA 19486, USA

1 INTRODUCTION

The alarming spread of human immunodeficiency virus (HIV), the etiologic agent of the acquired immunodeficiency syndrome (AIDS),[1] has initiated an urgent pursuit to comprehend and control this disease. Advances in molecular, viral and cell biology have defined numerous targets for potential drug intervention. The virally encoded homodimeric aspartyl protease,[2] which is responsible for processing the *gag* and *gag/pol* gene products that allow for the organization of core structural proteins and release of viral enzymes, is one such target. Inhibition of this protease enzyme prevents the maturation and replication of the virus in cell culture.[3] Recently, we and others have described antiviral effects of protease inhibitors in human clinical trials.[4] These results confirm the importance of HIV protease (HIV-PR) inhibitors as another weapon in the arsenal needed to confront AIDS.

A number of reports have described very potent HIV-PR inhibitors based on the transition state isostere concept.[5] In general, these peptidomimetic compounds retain a substantial amount of peptide character and as a result they possess poor aqueous solubility and inadequate oral bioavailability in animal models. These difficulties have limited their usefulness as therapeutic agents.[6] To circumvent these difficulties, Lam and coworkers[7] have taken a different tack. Through a series of 3D chemical database searches in combination with molecular modeling, they have developed a series of nonpeptide cyclic ureas which possess reasonable pharmacokinetic properties. On the other hand, we have chosen to optimize a peptidomimetic lead structure by manipulating its physical properties. This has provided a novel series of potent HIV-PR inhibitors which maintain desirable pharmacokinetics. We would like to report the invention and development of a novel class of HIV-1 protease inhibitors which possess a high degree of intrinsic potency and inhibit the spread of the virus in infected cells at concentrations of less than 100 nM. One of these inhibitors, L-735,524 (CRIXIVAN®, indinavir sulfate), has shown excellent effects on surrogate markers, reduction in viral RNA and elevations of CD4 cells, in HIV infected patients. The drug design rational, the development of the medicinal chemistry, and the presentation of human clinical results will be the focus of this lecture.

1.1 Design Rationale

A series of hydroxyethylene dipeptide isostere inhibitors of HIV-PR, represented by L-685,434 (**2**) in Figure 1, have been previously described from these laboratories.[9] Although very potent, the optimized molecules of this series lacked aqueous solubility and an acceptable pharmacokinetic profile.[10] Researchers at Hoffmann-La Roche published a series of potent hydroxyethylamine HIV-PR inhibitors, exemplified by Ro 31-8959 (**1**).[11] We hypothesized that incorporation of a basic amine into the backbone of the L-685,434 series might improve the bioavailability of this series of compounds. Therefore, replacement of the P_2/P_1 ligands, the *tert*-butyl carbamate and Phe moieties, of compound **2** with the P_2'/P_1' ligands, the decahydroisoquinoline *tert*-butyl amide, of **1** would generate a novel class of hydroxylamine pentaneamide (HAPA) isosteres.[12] Incorporation of the decahydroisoquinoline *tert*-butyl amide should provide two advantages. The amine would provide much needed aqueous solubility and enclosure of the amine into a ring should limit the conformational freedom of this P_1/P_2 ligand thereby decreasing the entropy change upon binding to HIV-PR. This unique class of transition state inhibitors is represented by compound **3**.

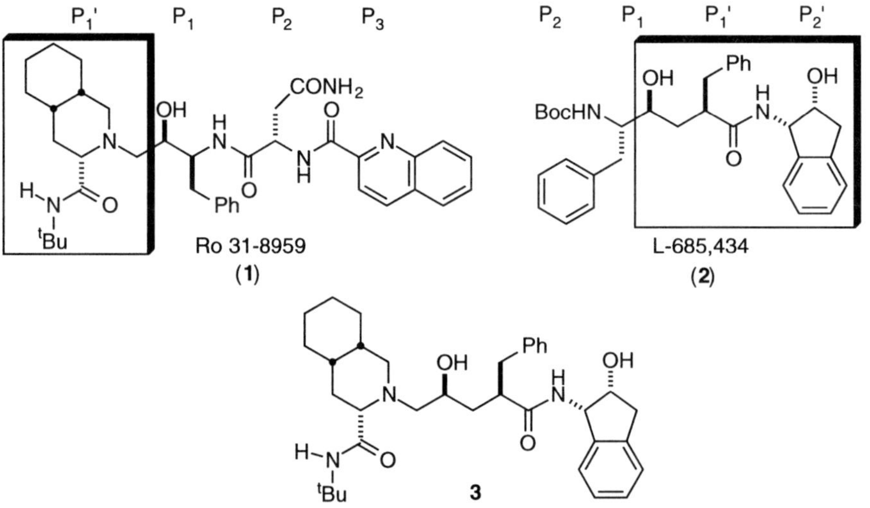

Figure 1 *Design Concept*

To support the concept that compound **3** could serve as a novel HIV-PR inhibitor, computer-assisted modeling studies were initiated.[13] These results show both inhibitors fill the same hydrophobic binding pockets from S_2 through S_2' with compound **1** extending into the S_3 domain of the HIV-PR with the quinaldic amide group. The P_2 and P_1' carbonyl moieties in both inhibitors are in position to maintain a critical hydrogen bond to a water molecule which is found in most reported X-ray crystal structures of inhibitors bound to HIV-1 PR.[14] Replacement of the Boc and Phe ligands of compound **2** with the decahydroisoquinoline amide then generates our hybrid structure **3**. The P_1' and P_2' groups of each of these inhibitors almost superimpose. The P_1 (Phe) moiety and P_2 (Boc) group of compound **2** occupy the same three dimensional space as the P_1 decahydroisoquinoline and P_2 *tert*-butyl amide in **3**. The position of the P_2 and P_1' carbonyl moieties of compound **3** maintain the proper alignment to hydrogen bond to water, which in turn hydrogen bonds to Ile_{50} and Ile_{250} on the flaps of the HIV-1 PR. With these promising molecular modeling results in mind, we turned our attention to the synthesis of compound **3**.

1.2 Synthetic Chemistry

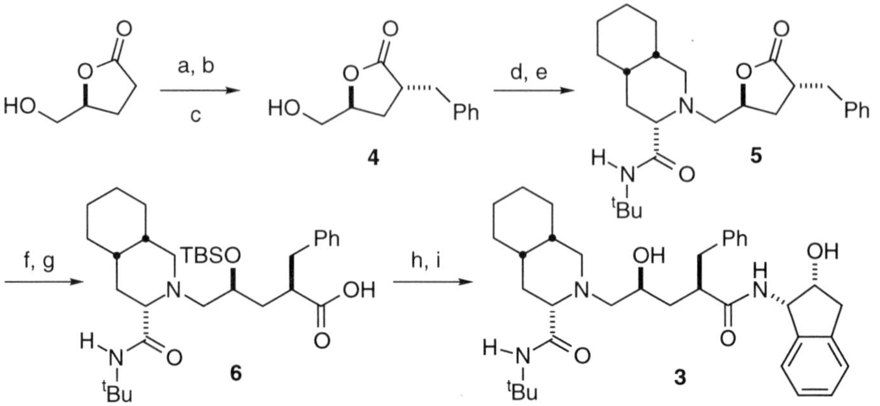

Reagents and conditions: (a) *tert*-butyldimethylsilyl chloride, imidazole, DMF, RT. (b) n-BuLi, diisopropylamine, THF; lactone (-78°); benzyl bromide. (c) HF, CH_3CN, RT, 1h. (d) MsCl, CH_2Cl_2, Et_3N; 66% yield for four steps. (e) xylenes, K_2CO_3, decahydroisoquinoline *tert*-butyl amide, 140°C; 35% yield . (f) 1 M LiOH, DME, RT, 3 h. (g)TBSCl, imidazole, DMF, RT. (h) EDC, HOBt, aminohydroxyindan, pH = 8-9, DMF; 61% for three steps (i) tetrabutylammonium fluoride, THF, RT, 20h; 85% yield.

Scheme 1 *Representive Synthesis of a HIV-PR Inhibitor*

As illustrated in Scheme 1, commercially available (S)-(+)-dihydro-5-(hydroxymethyl)-2(3H)-furanone was converted into the *tert*-butyldimethylsilyl ether under standard conditions.[15] The protected lactone was deprotonated with LDA and alkylated with benzyl bromide to afford a 6:1 mixture of diastereomers that were separated via flash column chromatography. The major diastereomer was then treated with HF and resulting alcohol **4** was treated with methanesulfonyl chloride and triethylamine to provide the activated lactone. Displacement of the mesylate moiety with *tert*-butyl (3(S), 4a(R), 8a(S))-decahydroisoquinoline-3-carboxamide in hot xylenes afforded lactone **5** in 35% yield. Hydrolysis of the lactone with LiOH was followed by treatment with excess TBSCl. Upon workup with water, the silyl ester hydrolyzed to provide protected carboxylic acid **6**. Standard coupling with EDC, HOBt and 1(S)-amino-2(R)-hydroxyindan[9b] yielded the amide which was followed by silyl ether deprotection with tetrabutylammonium fluoride to provide target molecule **3**. This sequence provided the desired product in nine steps in 12% overall yield.

Besides the efficiency, another major advantage of this synthetic route was the flexibility to introduce various P_1, P_2, P_1' and P_2' ligands onto target molecules. For example, modification of the P_1 and P_2 ligands could be effected by substituting a variety of cyclic secondary amino *tert*-butylcarboxylic amides for the decahydroisoquinoline carboxamide. Similarly, P_1' ligands could be modified by replacement of electrophiles other than benzyl bromide and the P_2' moieties could be varied by the coupling of amines other than 1(S)-amino-2(R)-hydroxyindan. Initially, L-proline, L-pipecoline, *cis*-4-(2-naphyloxy) L-proline, and N-4-protected L-piperazine *tert*-butyl amides were explored as replacements for the decahydroisoquinoline *tert*-butylamide, as illustrated in Table 1. These targets were obtained following the procedure outlined in Scheme 1. However, because of the low yield in the mesylate displacement/coupling step it was difficult to obtain gram quantities of these materials needed for biological testing. Therefore, an optimized procedure was developed, as illustrated in Scheme 2 with the synthesis of L-735,524.

The known (S)-2-piperazine carboxylic acid bis-(S)-(+)-camphorsulfonic acid salt **7**[16] was converted into the differentially protected piperazine following the procedure of Bigge and Hays.[17] The resulting acid was then coupled with *tert*-butylamine following standard peptide coupling procedures. Hydrogenolysis then removed the N1 benzyloxycarbonyl protecting group to provide amine **8** in 78% yield for the three steps. At this point, all attempts made to optimize the mesylate displacement reaction by varying solvents, temperature and bases were unsuccessful. However, modification the leaving group provided a significant improvement in yield. Reaction of lactone **4** (scheme 1) with triflic anhydride and 2,6-lutidine provided the very stable, crystalline lactone **9** in 96% yield. Displacement of triflate with amine **8** at room temperature in isopropanol with N, N-diisopropylethylamine now provided the desired coupled lactone in 83% yield. Hydrolysis, protection, amine coupling, and final deprotection then provided penultimate amine **10**. This was converted into L-735,524 through the reaction with 3-picolyl chloride in DMF with triethylamine. This sequence could provided the desired target molecules in ten steps, from (S)-(+)-dihydro-5-(hydroxymethyl)-2(3H)-furanone, generally in 35% overall yield. Most importantly, ten gram batches could be conveniently processed in this manner.

Table 1 *Exploration of the S_1 and S_3 Domains*

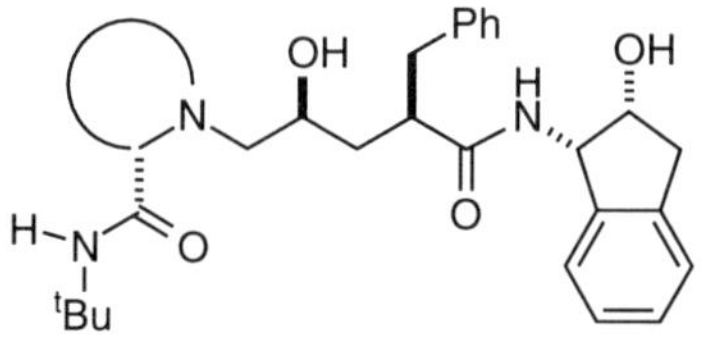

Compd	R	IC_{50} (nM)[a]	CIC_{95} (nM)[a]
3	DIQ[b]	7.8	400
11	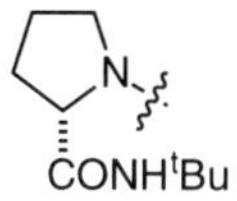 CONH^{t}Bu	347	1500
12	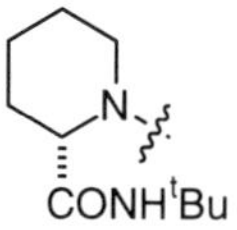 CONH^{t}Bu	80	nd[c]
13	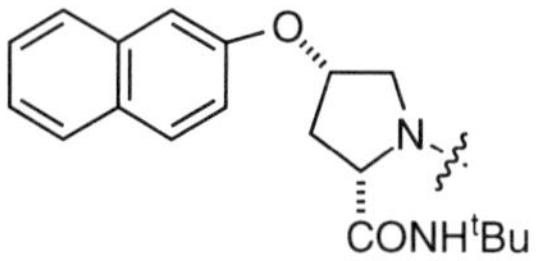 O; CONH^{t}Bu	15	>400
14 **L-732,747**	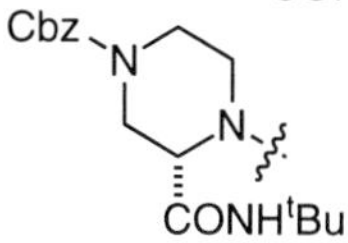 Cbz; N; CONH^{t}Bu	0.35	100

[a] For each determination n = 1. [b] DIQ-N-*tert*-butyl-(4aS,8aS)-(decahydroisoquinoline)-3(S)-carboxamide; [c] not determined

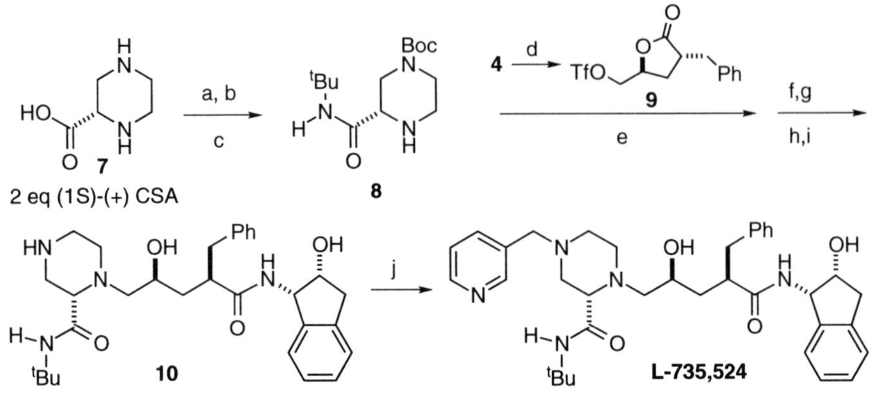

Reagents and conditions: (a) BocON, H_2O pH = 11.0; CbzCl, pH = 9.5 (96%yield). (b) EDC, HOBt, *tert*-butylamine, Et_3N, DMF (85% yield). (c) H_2, Pd/C 10%, MeOH (96% yield). (d) triflic anhydride, 2,6-lutidine, CH_2Cl_2 (96% yield). (e) isopropanol, diisopropylethylamine (83% yield). (f) 1 M LiOH, DME, RT, 3 h. (g) TBSCl, imidazole, DMF, RT(96% for two steps). (h) EDC, HOBt, hydroxyamino indane, pH = 8-9, DMF; (i) 8N HCl, isopropanol (78% yield). (j) 3-picolyl chloride hydrochloride, Et_3N, DMF (76% yield).

Scheme 2 *Synthesis of L-735,524*

1.3 Results and Discussion

Despite the relatively high intrinsic potency of compound **3** (IC_{50} = 7.6 nM), the inhibition of the spread of viral infection in MT4 human T-lymphoid cells infected with the IIIb isolate was relatively weak (CIC_{95} = 400 nM). However, because **3** did possess a favorable pharmacokinetic profile,[18] when compared to L-685,434, other analogs were pursued. Initially, L-proline, *cis*-4-substituted L-proline and L-pipecolinic *tert*-butylamides were explored as replacements for the decahydroisoquinoline *tert*-butylamide, as shown in Table 1. These N-terminal analogs provided no improvement in potency. Next, 2-*tert*-butylcarboxamide-4-substituted-piperazines were examined. The piperazine analogs would provide two potential advantages over the decahydroisoquinoline *tert*-butylamide or the substituted L-proline *tert*-butylamides. First, the nitrogen in the four position of the piperazine ring could be easily functionalized. This would allow for the introduction and optimization of a P_3 ligand which could balance both the hydrophobic and hydrophilic requirements of our target molecules. Second, the additional amine in the piperazine ring should provide improved aqueous solubility which might improve oral bioavailability.

One of the first compounds prepared in the piperazine series (**14**), possessed a benzyloxycarbonyl moiety attached to the N4 position of the piperazine ring. This compound showed an improvement in both intrinsic potency and in the ability to inhibit viral spread in infected cells. To better understand the significant increase in potency of compound **14** (L-732,747), a co-crystallization with HIV1-PR was undertaken.[8e] Consistent with the modeling observations, the ligands from P_2 to P_2' tightly bind into the S_2 to S_2' region of the HIV1-PR. Also observed is the critical water molecule bridging the two carbonyl moieties of the P_2 and P_1' ligands. The most important observation was that the benzyloxycarbonyl moiety fills the lipophilic S_3 domain of

HIV1-PR. These interactions combined to generate the first subnanomolar compound in the HAPA isostere series.

Replacement of the benzyloxycarbonyl moiety with an acyl or sulfonyl moiety resulted in, for most examples, compounds with subnanomolar potency (Table 2). However, this did not always translate into increased potency in the cell based assay, as exemplified by **17** and **18**. One critical factor in cell based potency is the ability of the inhibitors to cross a cell membrane. In these examples, this ability to penetrate the cell membrane appears to have been severely restricted.

Concurrently with the above series, alkylated piperazine analogs were also pursued. Structure-activity data from this set of examples revealed that a variety of arylmethyl substitutions increased potency in the cell based assay over the previously presented acylated or sulfonylated piperazines by 2-3 fold. The ability to modify the P_3 ligands without adversely affecting potency proved crucial in our search for an orally bioavailable inhibitor. Large and highly lipophilic P_3 ligands, i.e. phenylmethyl and 2-(benzyloxy)ethyl, were effective at increasing potency but also significantly decreased aqueous solubility. Smaller P_3 ligands improved solubility but lost potency (**10** and **21**). The 3-pyridylmethyl group provided both lipophilicity for binding to the protease and a weakly basic nitrogen which increased aqueous solubility. This combination of the 3-pyridylmethyl and the piperazine basic amines proved successful in improving oral bioavailability.

1.4 Pharmacokinetics

A representative set of compounds which possessed a range of aqueous solubilities and LogP values was examined for oral bioavailability in dogs and the results are shown in Table 3. Compound **14,** which was very insoluble at pH 7.4, showed no appreciable plasma levels when administered to dogs as a citric acid solution (10 mg/Kg). The same result was obtained for sulfonamide **16**. Acylation or sulfonylation completely removes the basic character of the N4 nitrogen and also decreases the basicity of the N1 nitrogen of the piperazine ring. This lack of basicity translated into a lack of aqueous solubility which was found to be detrimental to bioavailability. Adequate plasma levels were obtained with the slightly soluble difluorophenylmethyl **24**. Still further improvement occurred when the lipophilic aromatic moiety was replaced by a more soluble pyridine derivative. Of the variety of ligands explored, the most exciting results were obtained with L-735,524. The maximum plasma concentration levels achieved with an oral dose of 10 mg/Kg in 0.05 M citric acid solution for dogs (n = 4) was 11.4 ± 2.3 mM. The C_{max} levels for rat (n = 4, 20 mg/Kg) and monkey (n = 4, 10 mg/Kg) after oral dosing as a citric acid solution was found to be 2.80 ± 1.05 mM and 0.71 ± 0.24 mM, respectively (not shown). The oral bioavailability for this compound in the three animal species was 70%, 22% and 13%, respectively, when compared to iv studies. It should be noted that in all animal models examined the plasma concentrations after six hours were twice the levels needed to completely stop viral growth in the cell based assay. Solid dosage formulations with the crystalline free base, although more variable, gave comparable levels to the solution formulations. An improvement in formulation was found with the sulfate salt of L-735,524. A crystalline sulfate salt was prepared with both improved aqueous solubility (>450 mg/mL) and consistency of bioavailability in the solid dosage formulation studies.

Table 2 *Acylated, Sulfonylated and Alkylated Piperazines*

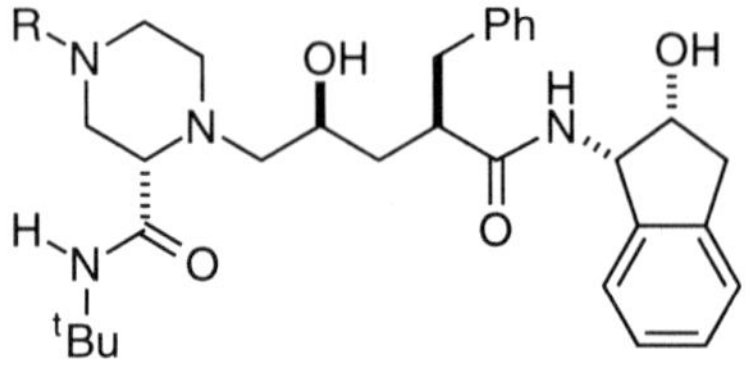

Compd	R	IC_{50} (nM)[a]	CIC_{95} (nM)[a]
10	hydrogen	38	3000
14	benzyloxycarbonyl	0.36	100
15	coumarin-3-carbonyl	0.13	50
16	8-quinoline sulfonyl	0.013 ± 0.03 n = 2	12.5 - 50 n = 2
17	2,5-dimethyl thiazole -4-sulfonyl	0.16	200
18	3-pyridyl sulfonyl	0.30	200
19	2-(benzyloxy)ethyl	1.03	100
20	(*t*-butyloxycarboxyl)methyl	0.81	100
21	ethyl	6.1	> 400
L-735,524	3-pyridylmethyl	0.56 ± 0.2 n = 89	25 - 100[b] n = 59
22	4-pyridylmethyl	0.39	50 - 100 n = 2
23	phenylmethyl	0.30 ± 0.05 n = 2	50 n = 2
24	2,4-difluoro-3-phenylmethyl	0.31 ± 0.05 n = 2	25 -50 n =2
25	4-bromo-2-thiophenemethyl	0.06 ± 0.02 n = 2	12 - 25 n =2
26	2-thiophenemethyl	0.54	25

[a] All entries are for n = 1 except where noted, (mean $\pm$ range). [b] Some variability was observed in the antiviral potency for L-735,524 due to inherent difficulties in the cell based assay. However, for n = 59 the average determination was 50.4 nM.

Table 3 *Bioavailability in dogs*

Compd	R	C_{max}(μM)[a]	Sol pH=7.4 (5.2) mg/mL	LogP (± 0.05)
14	benzyloxycarbonyl	< 0.10	< .001	4.67
16	8-quinoline sulfonyl	< 0.10	< .001	3.70
24	2,4-difluoro-3-phenylmethyl	0.73 ± 0.15	.0012 (0.03)	3.69
L-735,524	3-pyridylmethyl	11.4 ± 2.3	0.07 (0.69)	2.92

[a] Each compound was delivered orally in 0.05 M citric acid. For all cases n = 2, except L-735,524 (n = 4), (mean ± range). C_{max}, maximum plasma concentration.

2 POTENTIAL BACKUP CANDIDATES

As defined in our earlier work,[8e] modifications of the N4 position of the piperazine ring were generally well tolerated. The most preferred ligands for the S3 pocket tended to be large lipophilic heterocycles. Although potent, these ligands reduced aqueous solubility and this resulted in poor oral bioavailability. To improve the solubility of the target molecules containing potent lipophilic P3 ligands, a synthesis was designed to replace the P_1' Phe moiety with a pyridyl methyl group. This moiety would provide an additional weakly basic amine and therefore should improve aqueous solubility.[19]

2.1 Chemistry

To test this hypothesis, a synthesis introducing a 3-pyridyl methyl was developed. As illustrated in Scheme 3, commercially available trans-3(3-pyridyl)acrylic acid was first activated as the mixed anhydride and coupled to 1(S)-amino-2(R)-hydroxyindan,[9b] followed by protection and reduction of the olefin to provide the amide **26** in 67% yield for the three steps. The preparation of the key epoxide **27** was based on applying the elegant chemistry first developed by Askin and co-workers[8f] in their synthesis of L-735,524. Generation of the amide enolate of **26** with lithium hexamethyldisilazide (LHMDS) and condensation with (2S)-glycidyl tosylate afforded the epoxide **27** in 68% isolated yield after flash column chromatography to remove minor isomers. Thus, the protected aminohydroxyindan was able to direct the facial selective epoxide opening in a highly diastereoselective manner (>95:5 de). The epoxide was then reacted with Boc-protected piperazine **8**[8e] in hot isopropyl alcohol to afford the coupled product **28**. Deprotection of both the acetonide and t-butyl carbamate was effected with trifluoroacetic acid in methylene chloride and final reductive amination with thieno[2,3-b]thiophen-2-carboxaldehyde[20] provided amine **29** (L-748,496) in 22% overall yield in only seven steps. Besides efficiency, another advantage of this synthetic route was the flexibility to introduce various P1' and P3 ligands. In a manor similar to that described above, 4-pyridyl methyl P1' analogs could be synthesized by starting with trans-3(4-pyridyl)acrylic acid.[21] Also, a variety of P3 ligands were prepared by varying the final step in the sequence. The three modifications were simple N-alkylation with an appropriate arylmethyl chloride, sulfonylation, or reductive amination with various aldehyde partners.

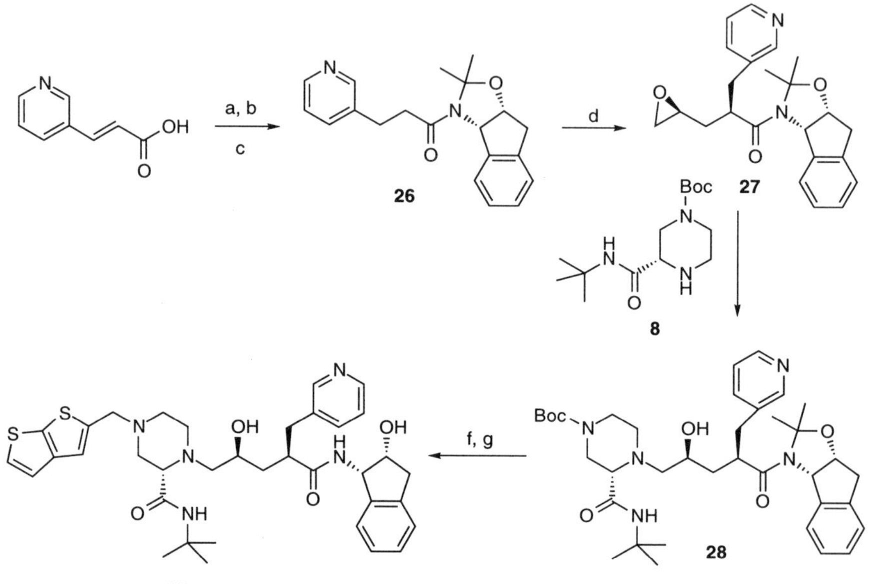

Reagents and conditions: a) pivaloyl chloride, Et_3N, THF,1(S)-amino-2(R)-hydroxyindan b) dimethoxypropane, CSA, CH_2Cl_2 c) $Pd(OH)_2$ / C, H_2, EtOH, THF 67% three steps d) (2S)-(+)-glycidyl tosylate, LHMDS, THF 68% yield e) ^{i}PrOH, 80°C, 16h 75% yield f) TFA, CH_2Cl_2 g) $NaB(OAc)_3H$, AcOH, $ClCH_2CH_2Cl$,thieno[2,3-b]thiophen-2-carboxaldehyde 65% yield for two steps

Scheme 3

2.2 Results and Discussion

A summary of the structure-activity relationships, solubility, and concentration maxima after oral dosing in dogs for the 3-pyridyl and 4-pyridyl series of compounds is shown in Table **4**. In general, the two series are equipotent with respect to HIV PR inhibition (IC_{50}) and cellular inhibition (IC_{95}). Interestingly, the bis-pyridyl analogs **30** and **31** and the sulfonamide targets **32** and **33** show good potency in the peptide cleavage assay but very weak potency in the cell based assay. One explanation for this is that analogs which are highly water soluble, or completely insoluble, can not penetrate the cell membrane. However, molecules that maintain the proper balance of lipophilicity and aqueous solubility (**34-41**) show a reasonable correlation of potency in the peptide cleavage assay and potency in the cell based assay.

As shown in the table, modification of the P3 ligand resulted in gradual improvements in cellular potency. Similar to our work in the L-735,524 series, the difluorobenzyl analogs (**34** and **35**) were potent. Compound **35** was found to reach high plasma levels in our dog model. However, this compound was only equipotent to L-735,524 and it was dropped from further considerations. An improvement in potency was found with the bromothiophene **37**, but again this compound did lack an acceptable pharmacokinetic profile. Replacement of the bromothiophene with the [2,3-b] or [3,2-b] thienothiophene,[20] as demonstrated by Kim and Guare in another HIV PR inhibitor series,[22] resulted in the first analogs with reasonable aqueous solubility combined with picomolar potency in our peptide cleavage assay. Also, these bicyclic heterocycles showed tremendous potency in the cell based assay (6-12 nM). In the 4-pyridyl series, compounds **39** and **41** lacked a reasonable oral bioavailability, obtaining plasma concentration levels of under 2 mM. Rewardingly, analog **40** (L-748,496) was found to reach peak plasma levels of over 7 mM in dogs (n = 4) and was determined to be 65% orally available when compared to IV dosage. The estimated oral bioavailability of this compound in rats was determined to be 70% (not shown).

Table 4 3-Pyridyl and 4-Pyridyl P1′ Analogs

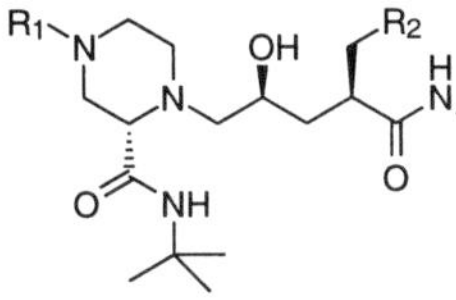

Compd	R_1	R_2	IC_{50} (nM)	IC_{95} (nM)	Sol. (mg/mL) pH 7.4/5.2	CLogP[11,c]	C_{max} (μM)[a]	Dose (mg/kg) In Citric Acid
30	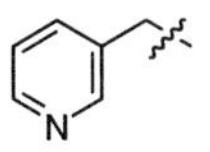	3-pyridyl	1.7	>200	na	0.969	na	na
31		4-pyridyl	1.7	>200	1.04 1.22	0.969	414	10
32	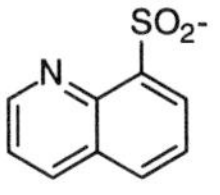	3-pyridyl	0.03	>200	poor	1.755	na	na
33		4-pyridyl	0.15	>200	na	1.755	na	na
34	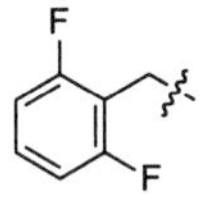	3-pyridyl	0.14	12-25	na	2.732	1010	7
35		4-pyridyl	0.48	25-50	0.04 0.37	2.732	8063	10
36	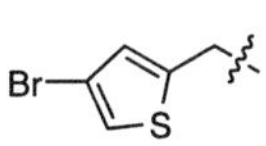	3-pyridyl	0.03	25	0.015 0.06	3.039	633	10
37		4-pyridyl	0.11	12-25	0.005 0.14	3.039	2200	7.5
38	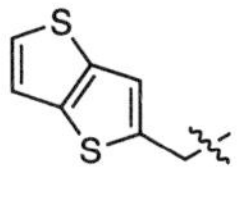	3-pyridyl	0.09	6-12	0.0003 0.024	3.332	3290	7
39		4-pyridyl	0.08	6-12	na	3.332	1586	10
40	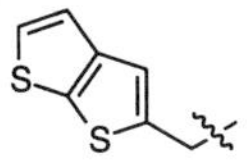	3-pyridyl	0.12	6-12	0.0003 0.025	3.122	7294[b]	10
41		4-pyridyl	0.08	6-12	0.0002 0.012	3.122	1726	10

a) average of n = 2 dogs b) average of n = 4 dogs c) CLogP for L-735,524 was 2.40 and the measured LogP was 2.92

2.2 Conclusion

In summary, a novel series of HIV-PR transition state isosteres has been developed. Starting from an initial peptide renin screening lead, Vacca[9a] and Lyle[9b] developed the highly potent and selective hydroxyethylene isostere L-685,434 (**2**). Based on this achievement, we were able to incorporate a basic amine into the backbone of this series to provide a novel HAPA isostere series of HIV-PR inhibitors. By modifying the physical properties of this series of inhibitors (i.e. solubility and lipophilicity) and concurrently maintaining potency we were able to design L-735,524.

L-735,524 is potent and competitively inhibits HIV-1 PR and HIV-2 PR with K_i values of 0.52 and 3.3 nM, respectively. This compound effectively halts the spread of HIV-1_{IIIb} infected MT4 lymphoid cells at concentrations of 50 nM. Also, this compound prevented the spread of viral infection in the genetically diverse SIV_{mac251} infected cells at concentrations of less than 100 nM. L-735,524 is a selective inhibitor of HIV-1 PR, showing no inhibition against a variety of proteases including human renin, human cathepsin D, porcine pepsin, and bovine chymosin at concentrations exceeding 10 mM. No serious safety liability was observed in several animal experiments.

Initial safety, tolerability and pharmacokinetic studies of L-735,524 free base and sulfate salt in 48 healthy HIV-1 seronegative and 24 HIV-1 infected human subjects has been undertaken.[4b] The multi-dose studies of 400, 800 and 1000 mg of sulfate salt q6h and 200 and 400 mg free base q6h were well tolerated, with the sulfate salt formulation providing the most consistent pharmacokinetic profile. L-735,524 (CRIXIVAN®) was recently cleared for marketing by the FDA for the treatment of HIV infection in adults when antiretroviral therapy is warranted.

References

1. (a) F. Barre-Sinoussi, J.-C. Chermann, F. Rey, M. T. Nugeyre, S. Chemaret, J. Gruest, C. Dauguet, C. Axler-Blin, F. BrunVezinet, C. Rouzioux, W. Rozenbaum and L. Montagnier, *Science* 1983, **220**, 868. (b) M. Popovic, M. G. Sarngadharan, E. Read and R. C. Gallo, *Science* 1984, **224**, 497. (c) R. C. Gallo, S. Z. Salahuddin, M. Popovic, G. M. Shearer, M. Kaplan, B. F. Haynes, T. J. Plaker, R. Redfield, J. Oleske, B. Safai, G. White, P. Foster and P. D. Markham, *Science* 1984, **224**, 500. (d) R. C. Gallo and L. Montagnier, *Sci. Am.* 1988, **259**, 40.
2. (a) L. Ratner, W. Haseltine, R. Patarca, K. J. Livak, B. Starcich, S. F. Josephs, E. R. Doran, J. A. Rafalsi, E. A. Whitehorn, K. Baumeister, L. Ivanoff, S. R, Petteway, Jr., M. L. Pearson, J. A. Lautenberger, T. S. Papas, J. Ghrayeb, N. T. Chang, R. C. Gallo and F. Wong-Staal, *Nature* 1985, **313**, 277. (b) L. H. Pearl, W. R. Taylor, *Nature* 1987, **329**, 351. (c) M. A. Navia, P. M. D. Fitzgerald, B. M. McKeever, C.-T. Leu, J. C. Heimbach, W. K. Herber, I. S. Sigal, P. L. Darke and J. P. Springer, *Nature* 1989, **337**, 615. (d) A. Wlodawer, M. Miller, M. Jaskolski, B. K. Sathyanarayana, E. Baldwin, I. T. Weber, L. M. Selk, L. Clawson, J. Schneider and S. B. H. Kent, *Science* 1987, **245**, 616.
3. (a) N. E. Kohl, E. A. Emini, W. A. Schleif, L. J. Davis, J. C. Heimbach, R. A. F. Dixon, E. M. Scolnick and I. S. Sigal, *Proc. Natl. Acad. Sci, U.S.A* . 1988, **85**, 4686. (b) H. G. Gottlinger, J. G. Sodroski and W. A. Haseltine, *Proc. Natl. Acad. Sci. U.S.A.* 1989, **86**, 5781. (c) C. Peng, B. K. Ho, T. W. Chang and N. T. Chang, *J. Virol.* 1989, **63**, 2550.
4. (a) J. F. Delfraissy, D. Sereni, F. Ren-Vezinet, E. Dussaix, A. Krivine and J. A. Dormont IX International Conference on AIDS; Berlin, Germany, June 6-11,

1993. (b) H. Teppler, R. Pomerantz, T. Bjornsson, J. Pientka, B. Osborne, E. Woolfe, K. Yeh, P. Deutsch, E. Emini, K. Squires, M. Saag and S. Waldman, Session 77-L8 Presented at The First National Conference on Human Retroviruses and Related Infections, December 12-16, 1993. Washington D. C.

5. (a) J. R. Huff, *J. Med. Chem.* 1991, **34**, 2305 and references sited therein. For other reviews see: (b) D. W. Norbeck and D. J. Kempf, *Annual Reports in Medicinal Chemistry*; Bristol, J. A., Ed.; Academic Press: New York, 1991; Vol. 26, pp 141. (c) A. G. Tomasselli, W. J. Howe, T. K. Sawyer, A. Wlodawer, and R. L. Heinrikson, *Chim. Oggi.* 1991 (May), 6. (d) C. Debouck, *AIDS Res. Human Retroviruses* 1992, **8** (2), 153. (e) J. A. Martin, *Antiviral Res.* 1992, **17**, 265. (f) T. D. Meek, *J. Enzyme Inhib.* 1992, **6**, 65. (g) M. Clare, *Perspect. Drug Discovery Des.* 1993, **1**, 49. (h) M. L. Moore, G. B. Dreyer, *ibid*, 1993, **1**, 85.
6. W. Greenlee, *Med. Res. Rev.* 1990, **10**, 173.
7. P. Y. S. Lam, J. K. Prabhakar, C. J. Eyermann, C. N. Hodge, Y. Ru, L. T. Bacheler, J. L. Meek, M. J. Otto, M. M. Rayner, Y. N. Wong, C.-H. Chang, P. C. Weber, D. A. Jackson, T. R. Sharpe and S. Erickson-Viitanen, *Science*, 1994, **263**, 380.
8. (a) J. P. Vacca, B. D. Dorsey, J. P. Guare, M. K. Holloway, R. W. Hungate, World Patent WO 9309096, 1993. (b) J. P. Vacca, B. D. Dorsey, W. A. Schleif, R. B. Levin, S. L. McDaniel, P. L. Darke, J. Zugay, J. C. Quintero, O. M. Blahy, E. Roth, V. V. Sardana, A. J. Schlabach, P. I. Graham, J. H. Condra, L. Gotlib, M. K. Holloway, J. Lin, I.-W. Chen, K. Vastag, D. Ostovic, P. S. Anderson, E. A. Emini and J. R. Huff. IX International Confernce on AIDS; Berlin, Germany, June 6-11, 1993. (c) B. D. Dorsey, R. B. Levin, S. L. McDaniel, J. P. Vacca, P. L. Darke, J. A. Zugay, E. A. Emini, W. A. Schleif, J. C. Quintero, J. L. Lin, I. W. Chen, D. Ostovic, P. M. D. Fitzgerald, M. K. Holloway, P. S. Anderson and J. R. Huff, MEDI-6, 206th ACS National Meeting, Chicago, IL, August 22-27, 1993. (d) J. P Vacca, B. D. Dorsey, W. A. Schleif, R. B. Levin, S. L. McDaniel, P. L. Darke, J. Zugay, J. C. Quintero, O. M. Blahy, E. Roth, V. V. Sardana, A. J. Schlabach, P. I. Graham, J. H. Condra, L. Gotlib, M. K. Holloway, J. Lin, I.-W. Chen, K. Vastag, D. Ostovic, P. S. Anderson, E. A. Emini and J. R. Huff, *Proc. Natl. Acad. Sci, USA* 1994, **91**, 4096. (e) B. D. Dorsey, R. B. Levin, S. L. McDaniel, J. P. Vacca, J. P. Guare, P. L. Darke, J. A. Zugay, E. A. Emini, W. A. Schleif, J. C. Quintero, J. H. Lin, I.-W. Chen, M. K. Holloway, P. M. D. Fitzgerald, M. G. Axel, D. Ostovic, P. S. Anderson and J. R. Huff, *J. Med. Chem.* 1994, **37**, 3443. (f) D. Askin, K. K. Eng, K. Rossen, R. M. Purick, K. M. Wells, R. P. Volante and P. J. Reider, *Tetrahedron Lett.* 1994, **35**, 673.
9. (a) J. P. Vacca, J. P. Guare, S. J. deSolms, W. M. Sanders, E. A. Giuliani, S. D. Young, P. L. Darke, J. Zugay, I. S. Sigal, W. A. Schleif, J. C. Quintero, E. A. Emini, P. S. Anderson and J. R. Huff, *J. Med. Chem.* 1991, **34**, 1225. (b) T. A. Lyle, C. M. Wiscount, J. P. Guare, W. J. Thompson, P. S. Anderson, P. L. Darke, J. A. Zugay, E. A. Emini, W. A. Schleif, J. C. Quintero, R. A. F. Dixon, I. S. Sigal and J. R. Huff, *J. Med. Chem.* 1991, **34**, 1228.
10. W. J. Thompson, P. M. D. Fitzgerald, M. K. Holloway, E. A. Emini, P. L. Darke, B. M. McKeever, W. A. Schleif, J. C. Quintero, J. A. Zugay, T. J. Tucker, J. E. Schwering, C. F. Homnick, J. Nunberg, J. P. Springer and J. R. Huff, *J. Med. Chem.* 1992, **35**, 1685.
11. (a) N. A. Roberts, J. A. Martin, D. Kinchington, A. V. Broadhurtst, J. C. Craig, I. B. Duncan, S. A. Galpin, B. D. Handa, J. Kay, A. Krohn, R. W. Lambert, J. H. Merrett, J. S. Mills, K. E. B. Parkes, S. Redshaw, A. J. Ritchie, D. L. Taylor, G. J. Thomas and P. J. Machin, *Science* 1990, **248**, 358. (b) A. Krohn, S. Redshaw, J. C. Ritchie, B. J. Graves, M. H. Hatada, *J. Med. Chem.* 1991, **34**, 3340.

12. A concurrent report of HAPA isosteres have been reported. See M. P. Trova, R. E. Babine, R. A. Byrn, W. T. Casscles, R. C. Hastings, G. C. Hsu, M. R. Jiroiusek, B. D. Johnson, S. S. Kerwar, S. R. Schow, A. Wissner, N. Zhang and M. M. Wick, *Bioorg. Med. Chem. Lett.* 1993, **3**, 1595.
13. Methods used to model all structures are described in reference 11.
14. M. Miller, J. Schneider, B. K. Sathyanarayana, M. V. Toth, G. R. Marshall, L. Clawson, L. Selk, S. B. H. Dent and A. Wiodawer, *Science* 1989, **246**, 1149. (b) P. M. D. Fitzgerald, B. M. McKeever, J. F. VanMiddlesworth, J. P. Springer, J. C. Heimbach, C. T. Leu, W. K. Herber, R. A. F. Dixon and P. L. Darke, *J. Biol. Chem.* 1990, **265**, 14209. (c) J. Erickson, D. J. Neidhart, J. VanDrie, D. J. Kempf, X. C. Wang, D. W. Norbeck, J. J. Plattner, J. W. Rittenhouse, M. Turon, N. Wideburg, W. E. Kohlbrenner, R. Simmer, R. Helfrich, D. A. Paul and M. Knigge, *Science* 1990, **249**, 527. (d) A. L. Swain, M. M. Miller, J. Green, D. H. Rich, J. Schneider, S. B. H. Kent and A. Wlodawer, *Proc. Natl. Acad. Sci. U.S.A.* 1990, **87**, 8805.
15. E. J. Corey and A. Venkateswarlu, *J. Am. Chem. Soc.* 1972, **94**, 6190.
16. E. Felder, S. Maffei, S. Pietra, D. Pitre, *Helv. Chim. Acta* 1960, 888.
17. C. F. Bigge, S. J. Hays, P. M. Novak, J. T. Drummond, G. Johnson, T. P. Bobovski, *Tetrahedron Lett.* 1989, **30**, 5193.
18. Unpublished results of J. Lin and I-W. Chen.
19. B. D. Dorsey, S. L. McDaniel, R. B. Levin, J. P. Vacca, P. L. Darke, J. A. Zugay, E. A. Emini, W. A. Schleif, J. H. Lin, I-W. Chen, M. K. Holloway, P. S. Anderson and J. R. Huff, *Bioorg. Med. Chem. Lett.* 1994, **4**, 2769.
20. J. D. Prugh, G. D. Hartman, P. J. Mallorga, B. M. McKeever, S. R. Michelson, M. A. Murcko, H. Schwam, R. L. Smith, J. M. Sondey, J. P. Springer and M. F. Sugrue, *J. Med. Chem.* 1991, **34**, 1805.
21. C. S. Marvel, L.E. Coleman and G. P. Scott, *J. Org. Chem.* 1955, **20**, 1785.
22. B. M. Kim, J. P. Guare, J. P. Vacca, S. R. Michelson, P. L. Darke, J. A. Zugay, E. A. Emini, W. Schleif, J. H. Lin, I. W. Chen, K. Vastag, P. S. Anderson and J. R. Huff, *Bioorg. Med. Chem. Lett.* 1995, **5**, 185.

17

Design, Synthesis and Evaluation of Some Novel Nucleotides as Inhibitors of HIV

C. McGuigan, D. Cahard, A. Salgado, L. Bidois, S. Velazquez, C. J. Yarnold, K. Turner, P. Sutton. O. Wedgewood, H.-W. Tsang, S. J. Turner, Y. Wang, G. O'Leary, N. Mahmood,[1] A. Hay,[1] J. Balzarini,[2] and E. De Clercq[2]

WELSH SCHOOL OF PHARMACY, UNIVERSITY OF WALES CARDIFF, KING EDWARD VII AVENUE, CARDIFF, CF1 3XF, UK

[1]MRC COLLABORATIVE CENTRE, MILL HILL, LONDON NW7 1AD, UK

[2]REGA INSTITUTE FOR MEDICAL RESEARCH, KATHOLIEKE UNIVERSITEIT LEUVEN, B3000 LEUVEN, BELGIUM

1. INTRODUCTION

Nucleoside analogues have found widespread use in the chemotherapy of viral infections, most notably herpes simplex and HIV.[1] All nucleoside analogues share a common mechanistic feature which, in almost every case, may limit the therapeutic efficacy of the compounds as clinical agents. Thus, nucleoside analogues are almost always effective only after kinase-mediated phosphorylation to their 5'-phosphate forms. Most usually the triphosphate is the active material, though occasionally it may be the monophosphate [as in the case of the anti-neoplastic fluorinated pyrimidines FU and FUDR[2]] or the diphosphate [as in the case of ara nucleosides such as araC[3]] which is responsible, at least in part, for the therapeutic effect.

In most cases little or no benefit accrues from the dependence of the nucleoside analogues on kinase-mediated activation. The exceptions are the few cases where judicious design of the nucleoside permits selective activation by viral-coded nucleoside kinase enzymes [notably herpes simplex thymidine kinase]. In these cases, which include acycloguanosine and its analogues,[4] and E-5-[2-bromovinyl]-2'-deoxyuridine[5] and related compounds, the nucleoside analogue is selectively activated only in the presence of the viral kinase; in un-infected cells little or no phosphorylation takes place. It appears that these represent the minority of cases, where kinase dependence leads to a real advantage for nucleoside analogues. However, most cases do not fit this model. There are two major reasons for this. Firstly, most viruses do not encode for [useful] nucleoside kinases; in these cases activation can only be mediated by host enzymes, and nucleosides requiring viral kinase phosphorylation will be inactive. Secondly, many nucleoside analogues may be poor substrates for both viral and host kinases. Such cases - which may be the majority - will lead to poor antiviral potency.

1.1 Nucleotides and blocked nucleotides.

Much evidence has now arisen to support the suggestion that free nucleotides do not easily permeate living cells, and any antiviral action observed from such compounds results primarily from extracellular cleavage to the nucleoside analogue, membrane transport and re-phosphorylation.[6] Efforts to circumvent this membrane impermeability of free [charged] nucleotides have spanned almost 4 decades and involved many strategies.In outline, the strategy is straightforward. An un-charged, blocked nucleotide ["triester"] (1) is used as an inactive, membrane-soluble pro-drug for the free monophosphate (2):

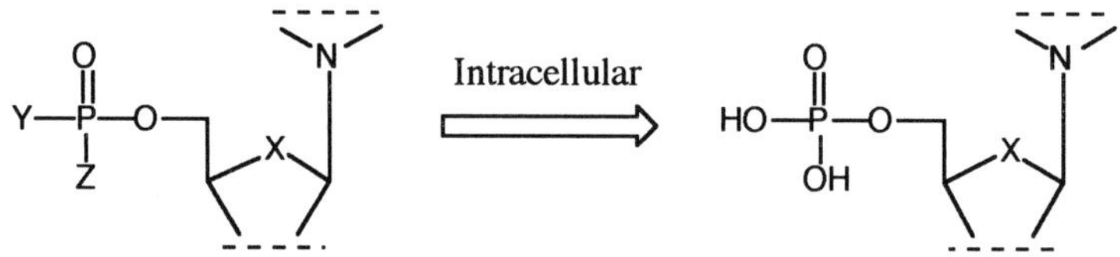

Figure 1. *Phosphate pro-drugs.*

The nature of the phosphate blocking groups [Y and Z in (1)] has been varied greatly in attempts to optimise the intracellular delivery of nucleoside analogues as their pre-activated 5'-monophosphate forms. The nature of the nucleoside [X, 2',3' groups, base etc] used in phosphate delivery has also been varied considerably.

Walker and co-workers have used both phosphate heterocycles and nucleotide dimers as potential delivery vehicles[7] whilst Meier has principally used the latter methodology[8] and Imbach[9] has used beta-substituted alkyloxy phosphates such as the bis-SATE group. In most cases, the blocked phosphates retained antiviral action in vitro but they were rarely more active than the parent nucleoside analogues from which they were derived. Moreover, in most cases, biological activity could be explained simply by release [extracellular or intracellular] of the free nucleoside analogue. Only recently has evidence arisen of the successful intracellular delivery of free nucleotides by the pro-drug approach.

2. RESULTS AND DISCUSSION

2.1 Diaryl phosphates of AZT.

We have previously noted that dialkyl phosphates (3) of AZT are inactive against HIV replication in vitro where the phosphate blocking group is a simple alkyl chain, whilst substituted dialkyl phosphates (4) are active.[10] The explanation for this was suggested to be the enhanced hydrolytic lability of the haloalkyl - phosphate chains.

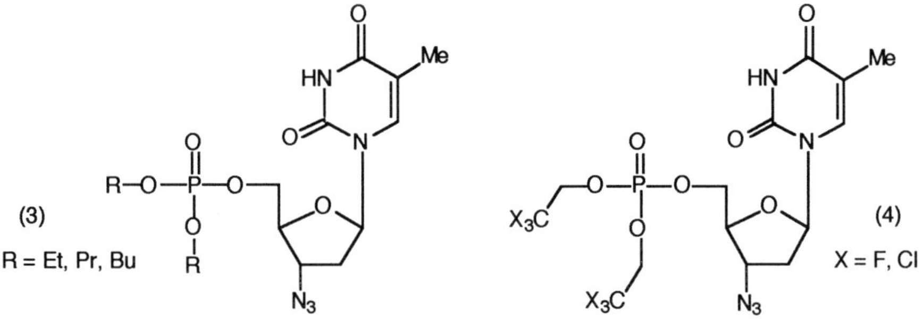

(3) R = Et, Pr, Bu

(4) X = F, Cl

Figure 2. *Dialkyl phosphates derived from AZT.*

More recently, we noted that diaryl phosphate esters (5) of AZT were especially potent inhibitors of HIV, particularly if the aryl moieties were substituted with electron withdrawing groups [Graph 1]:[11]

Graph 1. Anti-HIV potency of diaryl phosphates of AZT.

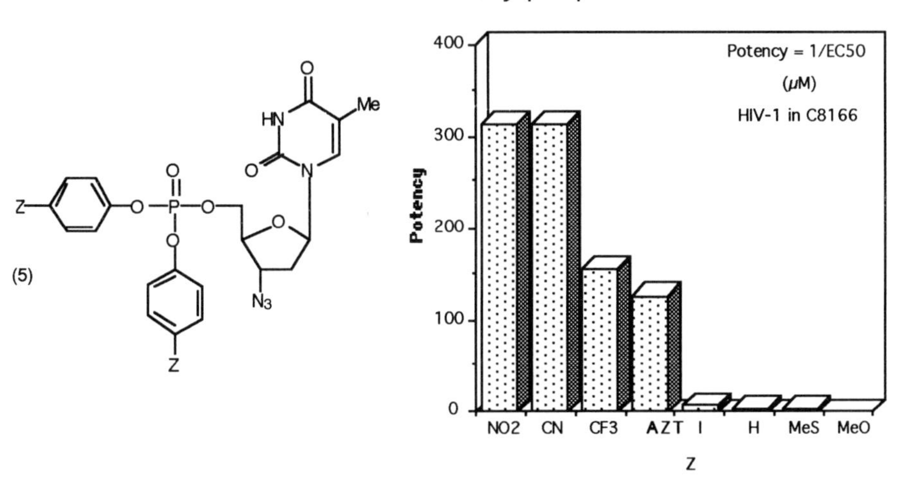

The most obvious interpretation of these data was that electron withdrawing groups in the diaryl phosphate unit lead to enhanced intracellular hydrolysis to liberate the free monophosphate. However, studies in cells which were resistant to the effects of AZT, but which were anticipated to be susceptible to intracellular AZT-monophosphate [AZTMP] indicated that the diaryl phosphates entirely failed to deliver AZTMP intracellularly.[12] Thus, although the potency of the parent nucleoside had been increased some 3-fold by the addition of an appropriately [p-NO_2 or -CN] substituted diaryl phosphate unit, this had been achieved purely through the intracellular delivery of the free nucleoside, rather than the delivery of AZTMP, and the intended by-pass of the nucleoside kinase.

2.2. Phosphoramidates.

Given the failure of diaryl phosphates to achieve intracellular nucleotide delivery, we turned our attention to more highly modified phosphate structures. In particular, we suggested phosphoramidates as potential phosphate delivery motifs. More specifically, we wondered if phosphoramidates derived from natural amino acids might be effective, by virtue of the potential for enzyme-induced intracellular hydrolysis. Indeed, we noted that amino acid related materials such as (6) and (7) were potent anti-HIV agents.[13-14]

(6) (7)

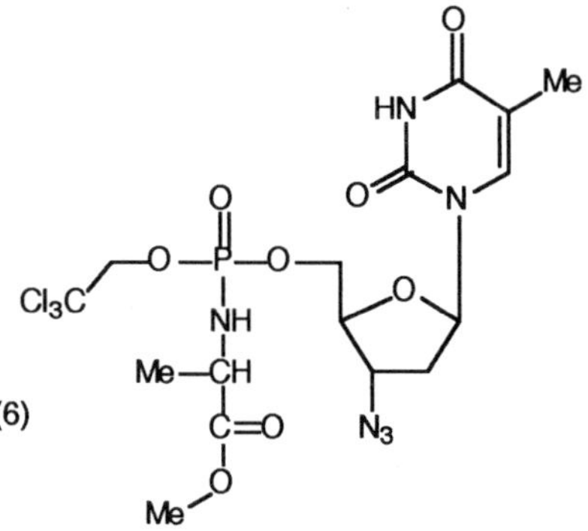

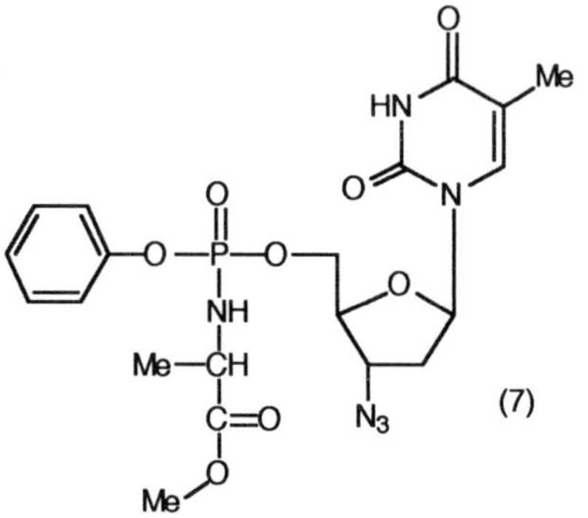

Figure 3. *Phosphoramidates derived from AZT.*

Moreover, we found that aryloxy derivatives such as (7) were effective in thymidine-kinase deficient cells, wherein the parent nucleoside AZT was virtually inactive.[15] This was taken as good evidence that aryloxy phosphoramidates such as (7) had indeed achieved the sought-after intracellular delivery of the free phosphate AZTMP.

2.3. d4T Phosphoramidates

Our recent application of the phosphoramidate delivery technology to the anti-HIV nucleoside analogue d4T (8) has lead to the most striking demonstration of the effectiveness of nucleotide delivery. Thus, data for the phenyl phosphoramidate (9) versus parent d4T (8) demonstrate the clear advantage of (9) [Table 1]:[16]

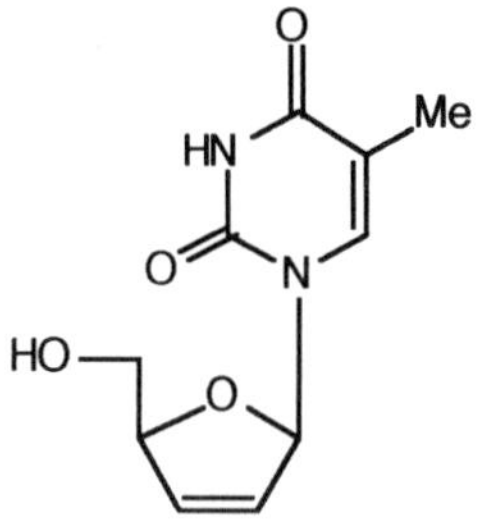

(8) 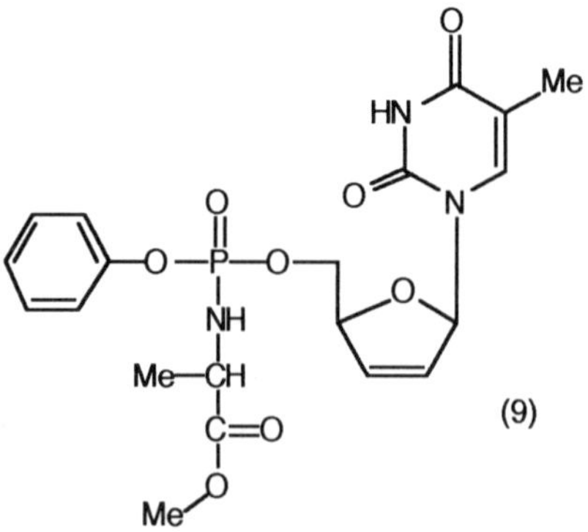 (9)

Figure 4. *Phosphoramidates derived from d4T.*

	EC_{50} /μM	CC_{50} /μM	SI
Cell:	MT-4	MT-4	MT-4
Virus:	HIV-1	HIV-1	HIV-1
Strain:	[III_B]	[III_B]	[III_B]
d4T [8]	0.65	4	6.1
[9]	0.066	>100	1515
Improvement	10-fold	>25-fold	252-fold

Table 1. *Antiviral activity, cytotoxicity and potency of d4T [8] and phosphoramidate [9] against HIV-1 in MT-4 cells.*

Particularly striking is the retention of activity by (9) in thymidine kinase-deficient [TK^-] cells, by comparison to the loss of activity for (8) [Table 2]:

	EC_{50} /μM	EC_{50} /μM	% Retention activity in TK^-
Cell:	CEM-0	CEM-TK^-	
Virus:	HIV-2	HIV-2	
Strain:	[ROD]	[ROD]	
d4T [8]	0.78	25	3%
[9]	0.20	0.075	267%
Improvement	4-fold	333-fold	

Table 2. *Antiviral activity of d4T [8] and phosphoramidate [9] against HIV-2 in CEM and CEM-TK^- cells.*

The most obvious interpretation of this data is the efficient intracellular delivery of free nucleotide by (9), making the activity of the compound entirely independent of thymidine kinase. Indeed, by the use of radio-labelled (8) or (9) we have been able to directly observe the intracellular levels of mono- [d4TMP], di- [d4TDP] and triphosphate [d4TTP] formed from equimolar extracellular doses of either compound [Table 3]:[17]

Levels of:	d4T [8] @ 5μM	[9] @ 5μM	Improvement
[8] or [9]	15	10	
d4TMP	1.6	2.4	
d4TDP	0.31	0.95	
d4TTP	3.1	7.8	2.5-fold

[24h incubation of CEM cells with methyl-^{3}H drug at 5μM:
figures are for drug / metabolite levels in nmole / 10^9 cells]

Table 3. *Intracellular levels of parent drug, and mono-, di- and triphosphate resulting from either d4T [8] or phosphoramidate [9] treatment of CEM cells.*

The ca. 3-fold increase in d4T triphosphate levels generated from (9) versus (8) under these conditions correlates very well with the approximately 4-fold reduction in EC_{50} for the phosphoramidate noted under these conditions [Table 2].

2.4 Structure Activity Relationships.

Because of the notable efficacy of the phenyl phosphoramidate (9) derived from d4T (8) we have pursued the Structure Activity Relationships in operation for compounds of this type. Firstly, it is notable that an amino acid, or at least a substituted amine, is vital for activity; analogues, such as (10) and (11) which are derived from simple alkyl amines, are virtually inactive as antivirals: [18]

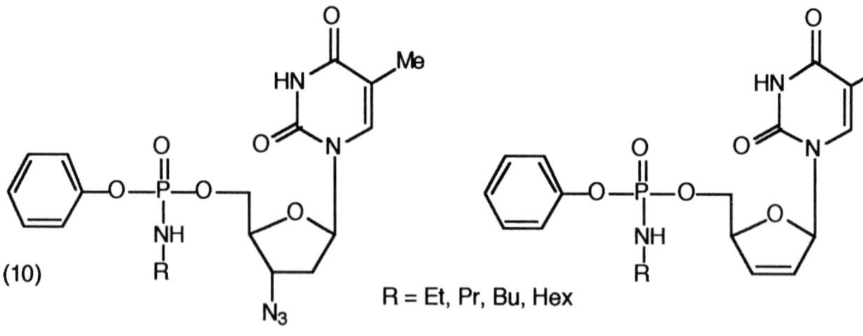

(10) (11) R = Et, Pr, Bu, Hex

Figure 5. *Simple phosphoramidates derived from AZT and d4T.*

Furthermore, small changes in the structure of the amino acid side chain lead to marked changes in activity; L-alanine is the most effective amino acid studied, whilst the ostensibly rather similar amino acids glycine and L-valine are extremely poor [Graph 2]:

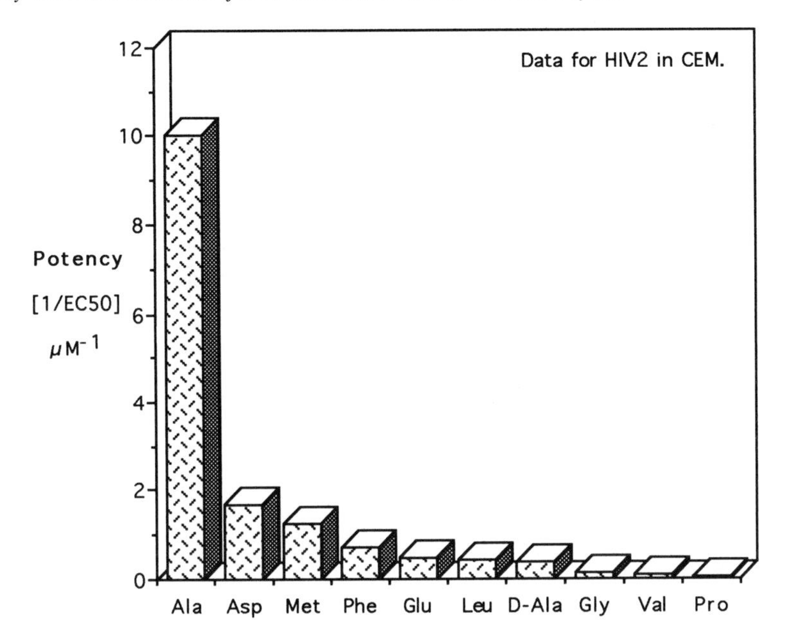

Graph 2. *The anti-HIV potency of compounds related to phosphoramidate (9) with varying amino acid side-chains.*

Particularly notable amongst these data is the observation that the D-alanine compound is ca. 30-fold less potent than that derived from the (natural) L-isomer.[19] Clearly, both the amino acid side-chain and its stereochemistry have a significant impact on the potency of these materials. The origins of these Structure Activity Relationships remain unclear, but may largely represent the relative efficiencies of intracellular nucleotide delivery. The fact that the amino acid stereochemistry is relatively important, and that small changes in amino acid side-chain can have significant effects on potency further support the notion that the release of nucleotides from phosphoramidates such as (9) is enzyme controlled. However, we have not yet identified the enzyme activities responsible for this conversion intracellularly.

Acknowledgements

We gratefully acknowledge the financial support of the MRC, the European Community, Glaxo-Wellcome Research, the Leverhulme Trust, and the Nuffield Foundation.

References

1. 'British National Formulary', Number 31, March 1996, section 5.3. Pub. British Medical Association and Royal Pharmaceutical Society of Great Britain, ISSN-0260-535-X; "Nucleoside Analogues: Chemistry, biology and medical applications", R.T. Walker, E De Clercq and F. Eckstein [Eds], NATO Advanced Study Report A26, Plenum Press, New York & London, 1979.

2. C.E. Myers, *Pharm. Rev.*, 1981, **33**, 1.

3. K. Uchida, and W. Kreis, *Biochem. Pharmacol.*, 1969, **18**, 1115; K. Uchida, W. Kreis, and D.J. Hutchinson, *Proc. Amer. Assoc. Cancer Res.*, 1968, **9**, 72.

4. P.A. Furman, M.H. St. Clair, J.A. Fyfe, J.L. Rideout, P.M. Keller, and G.B. Elion, *J. Virol.*, 1979, **32**, 72.

5. E. De Clercq, *Nucleosides Nucleotides*, 1994, **13**, 1271.

6.Th. Posternak, *Ann. Rev. Pharmacol.*, 1974, **14**, 23; J. Lichtenstein, H.D. Barner, and S.S. Cohen, *J. Biol. Chem.*, 1960, **235**, 457; K.C. Liebman, and C. Heidelberger, *J. Biol. Chem.*, 1955, **216**, 823.

7.S.N. Farrow, A.S. Jones, A. Kumar, R.T. Walker, J. Balzarini, and E. De Clercq, *J. Med. Chem.*, 1990, **33**, 1400; A. Kumar, P.L. Coe, A.S. Jones, R.T. Walker, J. Balzarini, and E. De Clercq, *J. Med. Chem.*, 1990, **33**, 2368.

8. C. Meier, L.W. Habel, J. Balzarini and E. De Clercq, *Liebigs Ann.*, 1995, 2203.

9. C. Perigaud, G. Gosselin, I. Lefebvre, L.-L. Girardet, S. Benzaria, I. Barber, and J.-L. Imbach, *BioMed. Chem. Lett.*, 1993, **3**, 2521.

10. C. McGuigan, S.R. Nicholls, T.J. O'Connor, and D. Kinchington, *Antiviral Chem. Chemother.*, 1990, **1**, 25; C. McGuigan, T.J. O'Connor, S.R. Nicholls, C. Nickson, and D. Kinchington, *Antiviral Chem. Chemother.*, 1990, **1**, 355.

11. C. McGuigan, M. Davies, R. Pathirana, N. Mahmood, and A.J. Hay, *Antiviral. Res.*, 1994, **24**, 69.

12. C. McGuigan, R.N. Pathirana, M.P.H. Davies, J. Balzarini and E. De Clercq, *BioMed. Chem. Lett.*, 1994, **4**, 427.

13. C. McGuigan, K.G. Devine, T.J. O'Connor, and D. Kinchington, *Antiviral Res.*, 1991, **15**, 255.

14. C. McGuigan, R.N. Pathirana, N. Mahmood, K.G. Devine, and A.J. Hay, *Antiviral . Res.*, 1992, **17**, 311.

15. C. McGuigan, R.N. Pathirana, J. Balzarini, and E. De Clercq, *J. Med. Chem.*,1993, **36**, 1048.

16. C. McGuigan, D. Cahard, H. M. Sheeka, E. De Clercq and J. Balzarini, *BioMed. Chem. Lett*, 1996, **6**, 1183.

17. C. McGuigan, D. Cahard, H.M. Sheeka, E. De Clercq, and J. Balzarini, *J. Med. Chem.*, 1996, **39**, 1748; J. Balzarini, A. Karlsson, S. Aquaro, C.-F. Perno, D. Cahard, L. Naesens, E. De Clercq and C. McGuigan, *Proc. Natl. Acad. Sci. USA*, 1996, **93**, 7295-7299.

18. C. McGuigan, D. Cahard, A. Salgado, E. De Clercq, and J. Balzarini, *Antiviral Chem. Chemother.*, 1996, **7**, 31.

19. C. McGuigan, A. Salgado, C. Yarnold, T.Y. Harries, E. De Clercq, and J Balzarini, *Antiviral Chem. Chemother.*, 1996, **7**, 184-188.

18

An Overview of Tsao-T Derivatives, A "Peculiar" Class of HIV-1 Specific RT Inhibitors

María-José Camarasa,[*,1], María-Jesús Pérez,[1] Sonsoles Veláquez,[1] Ana San-Félix,[1] Rosa Alvarez,[1] Simon Ingate,[1] María-Luisa Jimeno,[1] Erik De Clercq[2] and Jan Balzarini[2]

[1]INSTITUTO DE QUÍMICA MÉDICA (CSIC), JUAN DE LA CIERVA, 3.28006 MADRID, SPAIN
[2]REGA INSTITUTE FOR MEDICAL RESEARCH, KATHOLIEKE UNIVERSITEIT LEUVEN, B-3000, LEUVEN, BELGIUM

1 INTRODUCTION

AIDS is without any doubt one of the most important challenges for the Medical World in the 20th Century. There is currently no cure nor an efficient treatment of this disease. To date, a variety of drugs are available, five of which are officially approved for treatment [i.e. AZT (zidovudine), ddC (zalcitabine), ddI (didanosine), d4T (stavudine) and 3TC (lamivudine)][1,2]. However, their efficacy is limited, and most, if not all of them are endowed with dose-limiting serious side effects.[2,3,4] All these drugs represent dideoxy nucleoside (ddN) analogues and require metabolic activation to interfere with their target enzyme, the HIV-reverse transcriptase. The latter enzyme represents a virus-encoded protein catalysing a crucial enzymatic step in the replication cycle of HIV. Efficiently blocking this enzyme avoids incorporation of retroviral DNA into the genome of the target cells, and thus blocks the infection (replication) of HIV at a crucial stage.

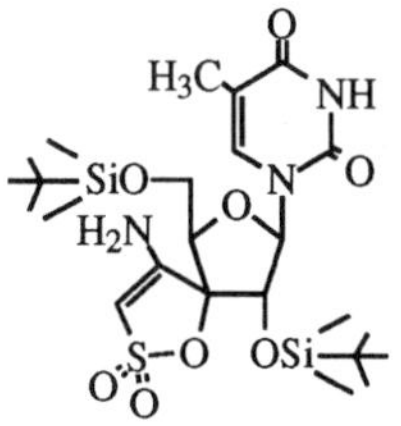

TSAO-T

In a collaborative effort between our group and the group of virologists at the Rega Institute, in 1992 we reported on an entirely novel class of highly functionalized nucleoside derivatives that were highly specific and potent inhibitors of human immunodeficiency virus type 1 (HIV-1) replication.[5-8] The prototype compound is [1-[2',5'-bis-*O*-(*tert*-butyldimethylsilyl)-β-D-ribofuranosyl]thymine]-3'-spiro-5"-(4"-amino-1",2"-oxathiole-2",2"-dioxide) designated TSAO-T. They do not require intracellular metabolism. They are specifically targeted at the HIV-1-encoded reverse transcriptase (RT) with which they interact at an allosteric non-substrate binding site,[9,10] a hydrophobic pocket located in the vicinity of the polymerase active site. This contrasts with the behaviour of the ddNTPs, which competitively inhibit the incorporation of natural substrates into the DNA. In this respect they behave like the non-nucleoside HIV-1-specific RT inhibitors (NNRTIs) (ie. HEPT, TIBO, Nevirapine, BHAP, Quinoxaline and α-APA ...).[11-13]

This overview is focussed on the synthesis, conformational studies, antiviral properties of TSAO derivatives, as well as their mechanism of antiviral action and the molecular basis of the rapid selection of resistant HIV-1 strains that emerge in cell culture in the presence of TSAO derivatives.

2 SYNTHESIS OF TSAO DERIVATIVES

The synthesis of TSAO-T was initially carried out by reaction of *O*-mesylcyanohydrins of furanosid-3'-ulosylthymine under basic conditions (Scheme 1). The method involves abstraction of one proton from the mesylate methyl group, followed by nucleophilic attack of the carbanion thus formed at the nitrile carbon atom. Thus, treatment of the 3'-ketonucleoside **1**[7] with sodium cyanide followed by mesylation (mesyl chloride/pyridine) of the corresponding 3'-cyanohydrin epimers obtained, gave the respective 3'-*C*-cyano-3'-*O*-mesyl-β-D-xylo- and -ribofuranosyl thymine nucleosides **2** and **3**. The major diastereomer was the *xylo* nucleoside **2**, resulting by the attack of the CN^- ion from the sterically less hindered α-face of the furanose ring.[14] Reaction of cyanomesylates **2** and **3** with Cs_2CO_3 gave the *xylo*- and *ribo*-spiro nucleosides **4** and **5**, respectively.[7]

Scheme 1

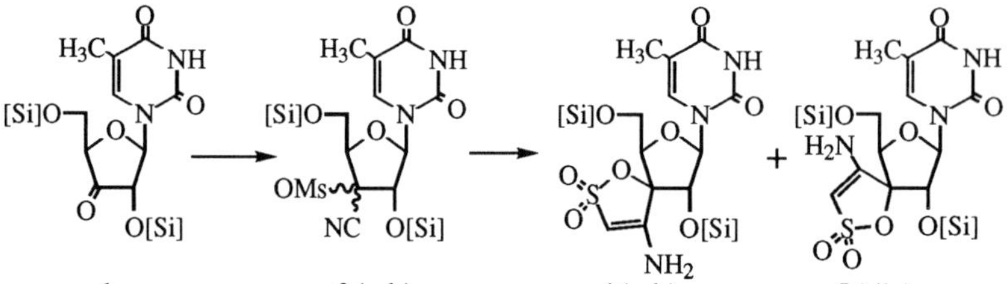

1   **2** (*xylo*) **3** (*ribo*)   **4** (*xylo*)   **5** (*ribo*) **(TSAO-T)**

[Si] = TBDMS

Since, the *ribo* -spiro nucleoside **5**, obtained as minor product, exhibited a potent and selective inhibition of HIV-1 replication *in vitro*.[5,7] whereas its *xylo isomer* was devoid of any anti-retroviral activity, we designed a new synthetic route which exclusively provided the active *ribo* derivatives.[8] Thus, *ribo* TSAO analogues were stereoselectively prepared (Scheme 2) by glycosylation[15] of a tertiary cyanomesylate of ribose with persilylated heterocyclic bases followed by basic treatment (Cs_2CO_3) of the cyanomesyl nucleosides obtained to give exclusively the β-D-*ribo*-spiro nucleosides. The *ribo* configuration of the TSAO nucleosides (**7** and **8**) was determined by the configuration of the starting cyanohydrin used in the preparation of the cyano mesylate of ribose **6.**[16,17] Subsequent deprotection and silylation of these derivatives (**8**) gave the *ribo* TSAO-analogues (**9**). Following this method a variety of TSAO analogues of pyrimidines and purines were prepared.[8,16,18] Selective N-3 alkylation of TSAO-pyrimidine derivatives,[8] or N-1-alkylation of the TSAO-purine derivatives gave the corresponding N-3 alkyl pyrimidine or N-1-alkyl purine nucleosides, respectively.[16]

Scheme 2

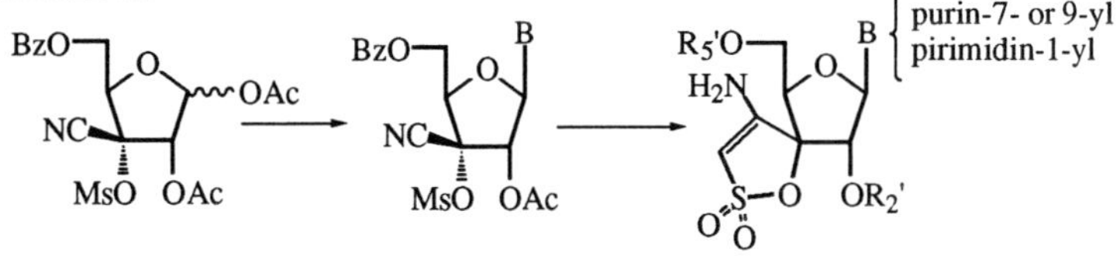

Also, a series of 1,2,3-triazole-TSAO derivatives (Scheme 3) were stereoselectively prepared by 1,3-dipolar cycloaddition of a suitably functionalized and protected ribofuranosyl azide intermediate **10** to differently substituted acetylenes, to give β-D-ribo-spironucleosides.[19] Cycloaddition of azide **10** to unsymmetrical acetylenes gave a mixture of the corresponding 4- and 5-substituted isomers **11** and **12**, whereas cycloaddition of azide **10** to symmetric acetylenes gave difunctional 1,2,3-triazole nucleosides **13**.

Scheme 3

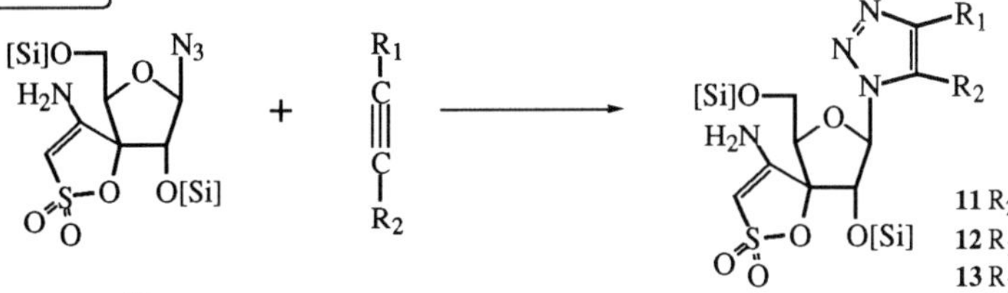

Several other TSAO derivatives modified at the sugar moiety have been prepared. Thus, we prepared 2'-spiro-5"-(4"-amino-1",2"-oxathiole-2",2"-dioxide)-TSAO-T analogues,[20] allofuranosyl-TSAO-T derivatives,[21] TSAO-T nucleosides in which the 3'-spiromoiety was replaced by other spiro rings such as 4-amino-2-oxazolone or 4-amino-1,2,3-oxatiazole-2,2-dioxide,[22] TSAO-T derivatives bearing L-sugars ,[23,24]. Also modifications on the protecting groups at positions 2', and 5' of TSAO-T were carried out.[25] Thus we removed one or both *tert*-butyldimethylsilyl protecting groups, we prepared TSAO-T analogues in which one (2' or 5') or both (2' and 5') silyl ether protecting groups were replaced by other groups that may mimic either their lipophilic or steric properties. Finally, 2' or 5' deoxy TSAO-T derivatives were also prepared.[7,24]

3 STRUCTURE-FUNCTION RELATIONSHIP STUDIES

TSAO-T is endowed with potent anti-HIV-1 activity (EC_{50} = 0.058 μM),[5] the antiviral activity of TSAO-T does not vary markedly from one cell line to another (EC_{50} range: 0.017-0.058μM).[5] TSAO derivatives are not inhibitory to HIV-2 strains, simian immunodeficiency virus (SIV), Moloney murine sarcoma virus and a broad range of DNA and RNA viruses at subtoxic concentrations.[5] TSAO-T does not act as a DNA chain terminator. Interaction of TSAO-T with the enzyme is non-competitive with respect to both the natural substrate (dGTP) and the template/primer [poly(rC).oligo(dG)]. TSAO-T, like the other HIV-1-specific inhibitors, interferes with a non-substrate binding site at HIV-1 RT that seems not to be present at other DNA polymerases including HIV-2 RT.[9]

TABLE 1. *Anti-HIV-1 activity of some TSAO-T analogues*[a]

Compound[b]	EC_{50}[c] (μM)	CC_{50}[d] (μM)	SI[e]
TSAO-T	0.06	14	217
TSAO-m^3T	0.059	240	4088
TSAO-e^3T	0.123	123	1000
TSAO-a^3T	0.233	≥ 330	≥ 1418
TSAO-U	0.019	14	74
TSAO-C	0.76	≥ 360	460
TSAO-m^5C	0.127	30	250
TSAO-dm^4m^5C	0.16	62	388
TSAO-A	0.278	13	47
TSAO-m^6A	0.146	14	96
TSAO-Hx	0.158	14	89
TSAO-7Hx	0.173	15	86
TSAO-m^1Hx	0.514	> 150	> 292
TSAO-$7m^1$Hx	0.201	156	776
TSAO-$7e^1$Hx	0.604	> 150	> 242
TSAO-mes^4Tr	0.602	≥ 150	≥ 249
TSAO-mes^5Tr	0.90	128	142
TSAO-dmesTr	0.48	111	231
TSAO-mam^5Tr	0.16	28	175
TSAO-$dmam^5$Tr	0.056	20	357

[a] Data taken from refs. 7, 8, 16, 18 and 19. [b] T=thymine; m^3T, e^3T, a^3T=3-methyl-, 3-ethyl-, 3-allyl-thymine; U=uracil; C=cytosine; m^5C=5-methylcytosine; dm^4m^5C=N^4-dimethyl-5-methylcytosine; A=adenine; m^6A=6-methyladenine; Hx=hypoxanthine; 7Hx=hypoxanthin-7-yl; m^1Hx=1-methylpoxanthine; ^{1}Hx, $7e^1$Hx=1-methyl-, 1-ethyl-hypoxanthin-7-yl; mes^4Tr, mes^5, dmesTr=4-methylester-, 5-methylester-, 4,5-dimethylester-1,2,3-triazole; mam^5Tr, $dmam^5$Tr=methylamido-, 5-dimethylamido-1,2,3-triazole. [c] 50% effective concentration. [d] 50% cytotoxic concentration. [e] selectivity index, or ratio of CC_{50} to EC_{50}.

Extensive structure-activity relationship (SAR) studies have been conducted with a large variety of TSAO derivatives, and these studies have revealed that within this class of compounds stringent requirements exist with regard to the structural determinants for optimal anti-HIV activity in cell culture. The presence of *tert*-butyldimethylsilyl (TBDMS) groups at both C-2' and C-5' positions of the ribose moiety is a prerequisite for anti-HIV-1 activity.[6,8] The fully or partially deprotected compounds or the 2'or 5'-deoxy derivatives were completely inactive at subtoxic concentrations (EC_{50}: 30μM-1000 μM)[7]. The 5'-silyl protecting group seems to be more critical for activity than the 2'-silyl protecting group.[25] Thus, replacement of the silyl moiety at the C-5' position by other groups that mimic either the lipophylic or the steric properties of TBDMS, results in antivirally inactive TSAO-T derivatives. However, a similar replacement of the silyl moiety at the C-2' position leads to compounds with only 2- to 10- fold reduced anti-HIV-1 activity.[25] The presence of the unique 3'-spiro group [3'-spiro-5"-(4'-amino-1",2"-oxathiole-2",2"-dioxide)], in nucleosides having a D-*ribo* configuration, is also a prerequisite for antiviral activity.[6-8]

The D-*xylo* or L-*ribo* isomers render the TSAO-T molecule completely inactive (EC_{50} >10 μM).[7] Replacement of the 3'-spiro group by other 3'-spiro moieties or change of this spiro group from position 3' to 2' of the sugar moiety in TSAO-T, results in annihilation of the antiviral activity of TSAO-T.[20,22]

In contrast to the stringent structural requirements of the sugar part of the TSAO derivatives, the nature of the heterocyclic base is less critical for anti-HIV-1 activity. The thymine moiety of TSAO-T can be replaced by a number of other pyrimidines, purines and 1,2,3-triazoles without marked decrease of antiviral efficacy[7,8,16,18,19] (TABLE 1). The TSAO-purine derivatives are in general 3- to 5-fold less effective than the most active TSAO-pyrimidine derivatives. Among the TSAO-1,2,3-triazole compounds, several derivatives with various substitutions at C-4 or C-5 of the triazole ring, show potent anti-HIV-1 activities. In particular, the 5-substituted amido-, methylamido- and dimethylamido-1,2,3-triazole derivatives were the most potent inhibitors of HIV-1 replication in these series with activities (EC_{50} = 0.056-0.52 μM), comparable to that of TSAO-T. Interestingly, the cytotoxicity of the TSAO-pyrimidine or TSAO-purine analogues becomes significantly attenuated (10- to 20- fold) , without affecting the anti-HIV-1 activity, by introduction of an alkyl or alkenyl group at N-3 (pyrimidines) or N-1 (purines) of the base moiety.[5,6,8,16]

4 CONFORMATION OF TSAO-T IN SOLUTION

The conformation in solution of TSAO-T was studied by ^{1}H- and ^{13}C-NMR spectroscopy.[26] The analysis of our data was based on the following features (a) the conformation around the glycosidic bond, (b) the conformation of the furanose ring, and (c) the relative orientation of C4'-C5' side chain.

The conformation around the glycosidic bond was calculated from vicinal $^{3}J_{C2,H1'}$ and $^{3}J_{C6,H1'}$ coupling constants, using the modified Karplus relations. Then, the glycosyl torsion angle (χ = C2-N1-C1'-O4') was derived from these vicinal coupling constants determined from the proton-coupled and low-power selective proton decoupled ^{13}C spectra. By using the concept of pseudorotation and by a modification of PSEUROT method (to calculate the pseudorotational parameters), we obtained information about the geometry of the furanose ring from the five vicinal coupling constants $J_{H1',H2'}$, $J_{C4',H2'}$, $J_{C2',H4'}$, $J_{C4',H1'}$ and $J_{C1',H4'}$. Finally, in order to complete the solution conformational picture of the TSAO-T under study, the conformation of the exocyclic C4'-C5' bond was analyzed. Information on the conformation of the side chain was obtained on the basis of the experimental vicinal proton-proton coupling constants $^{3}J_{H4',H5'a}$ and $^{3}J_{H4',H5'b}$. From these coupling constants the rotamer populations (g^+, g^-, t) were calculated. The overall conformation resulting from these studies shows, for the *ribo*-TSAO-T a preferred *anti* orientation of the thymine base (X_{anti} ≈ 0.7), a geometry of the furanoid ring corresponding to an O4'-*endo* envelope puckering (P ≈ 90°, τ = 30-50°), in which the C1'-C2'-C3'-C4' atoms are in the same plane and the O4' atom is located above the plane, while the side chain (C4'-C5') preferentially adopts a g^+ conformation.

The O4'-*endo* envelope conformation is extremely rare among HIV-inhibitory nucleosides, its occurrence on the TSAO molecules may be due to the high degree of functionalization and specially to the presence, at the 3'-position of the ribose moiety, of the spiroaminooxathiole dioxide ring.

5 TSAO RESISTANT MUTATIONS AT HIV-1 RT

Rapid emergence of TSAO-resistant HIV-1 strains occurred when HIV-1-infected cells were exposed to TSAO derivatives.[27,28] These TSAO-resistant HIV-1 strains proved cross-resistant to the other TSAO-pyrimidines, -purines and -triazole derivatives, but not cross-resistant to most if not all other classes of HIV-1 RT specific inhibitors, as well as, to the 2',3'-dideoxynucleosides (AZT, DDI, DDC, D4T, 3TC etc.) and to the acyclic nucleoside phosphonates [PMEA, (*S*)FPMPA, (*R*)PMPDAP].[27-30]

Molecular characterization from at least 10 different TSAO-resistant HIV-1 strains revealed a single mutation affecting the amino acid at position 138 of RT. Invariably, glutamic acid was replaced by lysine at position 138 of all HIV-1 RT mutants. This amino acid change that renders complete inefficacy of TSAO derivatives [28] must play a crucial role in the recognition of TSAO derivatives by the enzyme. Based on these observations together with our SAR studies on TSAO derivatives and site-directed mutagenesis studies on HIV-1 RT containing amino acids other than glutamic acid at position 138 (i.e. Arg, Lys),[28,31] we postulate that the 4"-amino group of the 3'-spiro moiety is most likely responsible for the specific interaction of the TSAO molecule with the carboxylic acid group of 138-Glu of the p51 subunit.[32] The fact that the glutamic acid-138 is not present in HIV-2 and SIV, further corroborates this hyphothesis and may explain why HIV-2 or SIV are not sensitive to the antiviral action of the TSAO derivatives. There is crystallographic evidence that Glu-138 is located at the top of the finger domain of the p51 subunit of HIV-1 RT, which closely approaches the binding pocket of the HIV-1-specific RT inhibitors at the p66 subunit, and/or may even be part of this pocket.[33,34]

The fact that other mutations in the RT at amino acid positions 181 (Tyr $\rightarrow$ Cys), 188 (Tyr $\rightarrow$ Cys) and 106 (Val $\rightarrow$ Ala) also lead to markedly reduced sensitivity of HIV-1 RT to TSAO derivatives strongly suggest the presence of additional interaction points, at the p66 subunit, that may serve either as direct attachment sites for the TSAO compounds or at least help in maintaining the desirable conformation of the TSAO binding sites in the RT heterodimer.[10]

6 COMBINATION THERAPY

A potential drawback of the HIV-1-specific RT inhibitors is the rapid emergence of drug-resistant HIV-1 strains. HIV-1 strains that are resistant to the nonnucleoside RT inhibitors are still sensitive to AZT or ddI, and, vice versa, HIV-1 strains selected for resistance against AZT or ddI are still sensitive to the HIV-1-specific nonnucleoside RT inhibitors.[35] In fact, HIV-1 strains resistant to AZT or ddI show amino acid substitutions in the HIV-1 RT[36] that are clearly distinct from those reported in nonnucleoside-resistant HIV-1 strains.[35,37] It appears that single-drug therapy will never be able to efficiently tackle the virus. Combination of anti-HIV agents is now being explored as therapeutic modalities to prevent emergence of virus-drug resistance.[38] Three advantages are generally associated with the combined use of different anti-HIV drugs: (1) diminished toxicity, due to a reduction in the dosage of the individual compound; (2) reduced risk of virus-drug resistance development; and (3) synergistic antiviral activity.

A rational approach toward drug combination may be based on the choice of drugs that lead to mutually noncomplementary or antagonistic drug-resistance mutations in the HIV-1 RT.[39]

6.1 Combinations TSAO + BHAP

Among the different classes of HIV-1-specific inhibitors TSAO-T and BHAP derivatives consistently select for HIV-1 mutant strains that contain amino acid changes in their RT which are located in a region other than the 179-to-190 region.[40] When used individually, the compounds led to the emergence of drug-resistant virus strains, containing the Glu-138 →Lys mutation for TSAO-resistant strains and Leu-100→Ile for BHAP-resistant strains. The BHAP-resistant virus was fully sensitive to TSAO whereas the TSAO-resistant virus remained fully sensitive to the inhibitory effects of BHAP.[38] Therefore it might be expected that when the two were combined, BHAP would suppress the emergence of the TSAO-specific mutation at position 138(Glu→Lys) and vice versa. However, combination of both compounds lead to the rapid emergence of resistant virus. The resistant virus showed the Tyr-181→Cys mutation, although neither compound did so when used individually. This RT mutation is a highly resistant viral strain not only to TSAO and BHAP compounds, but also to other HIV-1 RT specific inhibitors.[38]

6.2 Combinations TSAO + UC42

Thus, considering that most of HIV-1-specific RT inhibitors loose much of their antiviral potential against Tyr-181→Cys mutant virus, we considered this mutation as crucial mutation that should, if at all possible, be avoided or suppressed during anti-HIV chemotherapy. Therefore we searched for HIV-1-specific RT inhibitors that could efficiently suppress this mutant virus. The tiocarboxanilide derivatives clearly fulfilled this premise. One of the lead compounds in this serie, UC42 inhibits not only the 181-Cys virus mutant but also other drug-resistant virus mutants (i.e. RT/138 Lys, RT/106 Ala, RT/100 Ile and RT/103 Asn mutant strains). TSAO-T looses antiviral activity against mutants containing 181Cys, 138 Lys and 106 Ala in their RT, while retaining substantial inhibitory efficacy against RT/100 Ile ant RT/103 Asn virus mutants (that are resistant to UC42). Thus the antiviral/resistance spectra of TSAO-T and UC42 that are complementary to each other led us to the design of combination experiments using both compounds. Combination of both compounds dramatically delayed the emergence of resistant virus or even cleared the infected culture of virus particles, at much lower concentrations, than if the compounds were used individually. Moreover, the addition of BHAP to the combination allowed the use of lower concentrations and equal afficacy.[41]

6.3 Combinations TSAO + 3Tc

3TC, the most recently discovered nucleoside analogue, is an unusual and highly potent inhibitor of HIV replication, targeted to RT after its conversion to the 5'-triphosphate. In contrast to the other nucleoside analogues, 3TC rapidly selects for resistant virus strains that consistently contain 184-Ile or Val mutation of the RT,[42] these strains are sensitive to the inhibitory effects of NNRTIs, as well as to AZT and d4T. Therefore, 3TC should also be considered a candidate compound for inclusion in a combination cocktail.

When 3TC was combined with TSAO-T, UC42 , BHAP or MKC-442 marked potentiation of the anti-HIV-1 activity was observed and breakthrough of the virus in cell culture was delayed or even suppressed at concentrations that were ≥1-2 orders of magnitude lower than when the drugs were used individually. For these drug combinations, the concentrations of the individual drugs could be lowered by ≥25-50-fold to suppress virus breakthrough compared with the individual use of the compounds. The concomitant presence of the Lys-138 and Ile/Val-184 mutations was found in RT of the mutant viruses that emerged with combination therapy of the lowest concentrations of 3TC with the lowest concentrations of TSAO-T. These virus strains retained high sensitivity to the other NNRTI. Triple drug combination of TSAO-T+3TC+ UC42 or MKC-442 resulted in complete suppression of virus breakthrough in HIV-1-infected CEM cell cultures at drug concentrations that readily allowed virus replication when single drugs were administered.[43]

7 HETERODIMERS [ddN]-$(CH_2)_n$-[NNRTI] AS POTENTIAL MULTIFUNCTIONAL INHIBITORS OF HIV-1 RT

The binding of the NNRTIs to their hydrophobic pocket of the HIV-1 RT does not interfere with the binding of the dNTPs but slows down the rate of incorporation of the dNTPs (as dNMPs) in the DNA product.[44] Because the cooperative interaction between the substrate-binding site and nonsubstrate (NNRTI)-binding-site, combination of the functionalities of a non-nucleoside and a nucleoside type RT inhibitor has been postulated to result in a very tight binding to the HIV-1 RT.[44]

One approach to combination therapy, would be the use of heterodimers resulting from the linking of a NNRTI and a 2',3'dideoxynucleoside (ddN) through an appropriate spacer, in an attempt to combine the inhibitory capacity of these two different classes of molecules.[34a] Due to the NNRTI the heterodimers might be highly specific to HIV-1 RT and might lower the speed of the emergence of virus-drug resistance.

In an attempt to combine the HIV-inhibitory capacity of ddN analogues and NNRTI we have prepared and evaluated for their anti-HIV-activity several heterodimers, of general formula [ddN]-$(CH_2)_n$-[NNRTI], which combine in their structure a ddN such as AZT and a NNRTI such as TSAO-T or HEPT.[45] These two classes of inhibitors (ddN and NNRTI) were linked at the N-3 of the thymidine base of each compound by an appropriate spacer. As spacer, we used a carbon aliphatic linkage of different lenghts in order to obtain a dimer possessing an optimum distance between both active principals (ddN and NNRTI).

The heterodimers were prepared by basic treatment of the first key nucleoside (AZT or dThd) with the appropriate dibromoalkyl reagent followed by reaction of the N-3-bromoalkyl-AZT or N-3-bromoalkyl-dThd nucleoside intermediate with the second key nucleoside (TSAO-T or HEPT), to give the N-3,N-3-alkyl heterodimers.

The [TSAO-T]-$(CH_2)_n$-[AZT] heterodimers proved markedly inhibitory to HIV-1. When AZT was replaced by thymidine in the heterodimer molecules, potent anti-HIV-1 activity was observed. However, although the compounds proved inhibitory to HIV-1, they were less potent inhibitors than the parent compounds from which they were derived. None of the dimers were endowed with anti-HIV-2 activity. There was a clear trend toward decreased antiviral potency with lengthening the methylene spacer in the [TSAO-T]-$(CH_2)_n$-[AZT] dimers.[45]

8 CONCLUSIONS

TSAO derivatives represent a unique class of nucleosides that are specifically targeted at HIV-1 RT. The TSAO compounds were the first HIV-1-specific RT inhibitors for which a well defined part of the molecule (i.e. the 4"-amino group of the 3'-spiro moiety) was identified as an essential pharmacophore interacting with a well-defined moiety (the -COOH group of Glu-138) of the RT target enzyme. The p51 subunit of RT plays an important role in the sensitivity/resistance of HIV-1 strains to TSAO derivatives. Combinations of TSAO derivatives with other HIV RT inhibitors prevent emergence of viral resistance. Considering that the behaviour of TSAO derivatives is constant and complementary to other HIV RT inhibitors this makes of TSAO derivatives very attractive candidates for the design of combined therapies of two or more compounds.

Acknowledgements. We thank the Spanish CICYT, the Plan Regional de Investigación de la Comunidad de Madrid, the NATO (Collaborative Research Programme) and the Biomedical Research Programme and the Human Capital and Mobility Program of the European Community for financial support.

References

1. (a) M.A. Fischl, D.D. Richman, M.H. Grieco, and the AZT Collaborative Working Group. *New Engl. J. Med.*, 1987, **317,** 185. (b) E. De Clercq, *J. Acquir. Immun. Defic. Syndrom.*, 1991, **4,** 207.
2. (a) R. Yarchoan and S. Broder, *Pharm. Ther.*, 1989, **40,** 329. (b) R. Yarchoan, H. Mitsuya, R.V. Thomas, J.M. Pluda, N.R. Hartman, C.F. Perno, K.S. Marczyk, J.P. Allain, D.G. Johns and S. Broder, *Science* , 1989, **245,** 412. (c) J.A.V. Coates, N. Cammack, H.J. Jenkinson, I.M. Mutton, B.A. Pearson, R. Stover, J.M. Cameron and C.R. Penn, *Antimicrob. Agents Chemother.*, 1992, **36**, 202.
3. (a) D.D. Richman, M.A. Fischl, M.H. Grieco and the Collaborative Working Group *New Engl. J. Med.*, 1987, **317**, 192. (b) B.A.Q. Larder, S.D. Kemp, *Science*, 1989, **246**, 1155.
4. (a) R. Yarchoan, J.M. Pluda, R.V. Thomas, H. Mitsuya, P. Browers, K.M. Wyvill, N.R. Hartman, D.G. Johns and S. Broder, *Lancet* , 1990, **336,** 526. (b) H. Mitsuya, S. Broder, *Proc. Natl. Acad. Sci. USA* , 1986, **83**, 1911.
5. J. Balzarini, M.J. Pérez-Pérez, A. San-Félix, D. Schols, C.F. Perno, A.M. Vandamme, M.J. Camarasa and E. De Clercq, *Proc. Natl. Acad. Sci. USA* , 1992, **89**, 4392.
6. J. Balzarini, M.J. Pérez-Pérez, A. San-Félix, S. Velázquez, M.J. Camarasa and E. De Clercq, *Antimicrob. Agents Chemother.*, 1992, **36**, 1073.
7. M.J. Camarasa, M.J. Pérez-Pérez, A. San-Félix, J. Balzarini and E. De Clercq, *J. Med. Chem.*, 1992, **35**, 2721-2727.
8. M.J. Pérez-Pérez, A. San-Félix, J. Balzarini, E. De Clercq, M.J. Camarasa, *J. Med. Chem.* 1992, **35**, 2988.
9. J. Balzarini, M.J. Pérez-Pérez, A. San-Félix, M.J. Camarasa, I.C. Bathurst, P.J. Barr, and E. De Clercq, *J. Biol. Chem.*, 1992, **267**, 11831.
10. J. Balzarini, M.J. Camarasa and A. Karlsson, *Drugs of the Future*, 1993, **18**, 11043.
11. E. De Clercq, *Clin. Microbiol. Rev.*, 1995, **8**, 200.
12. E. De Clercq, *J. Med. Chem.*, 1995, **38**, 2491.
13. E. De Clercq, *Med. Res. Rev.*, 1996, **16**, 125.
14. H. Hayakawa, H. Tanaka, N. Itoh, M. Nakajina, T. Miyasaka, K. Yamaguchi and Y. Iitaka, *Chem. Pharm. Bull* ., 1987, **35**, 2605.
15. H. Vorbrüggen, K. Kolikiewicz and B. Bennua, *Chem. Ber.*, 1981, **114**, 1234.
16. S. Velázquez, A. San-Félix, M.J. Pérez-Pérez, J. Balzarini, E. De Clercq and M.J. Camarasa, *J. Med. Chem.*, 1993, **36**, 3230.
17. A. Calvo-Mateo, M.J. Camarasa, A. Díaz-Ortíz and F.G. de las Heras, *J. Chem. Soc., Chem. Commun.*, 1988, **11**, 14.
18. A. San-Félix, S. Velázquez, M.J. Pérez-Pérez, J. Balzarini, E. De Clercq and M.J. Camarasa, *J. Med. Chem.*, 1994, **37**, 453.
19. R. Alvarez, S. Velázquez, A. San-Félix, S. Aquaro, E. De Clercq, C.F. Perno, A. Karlsson, J. Balzarini and M.J. Camarasa, *J. Med. Chem.*, 1994, **37**, 4185.
20. S. Velázquez, M.L. Jimeno, M.J. Camarasa and J. Balzarini, *Tetrahedron*, 1994, **50**, 11013.

21. R. Alvarez, A. San-Félix, E. De Clercq, J. Balzarini and M.J. Camarasa, *Nucleosides and Nucleotides,* 1996, **15**, 349.
22. R. Alvarez and M.J. Camarasa, *Unpublished results.*
23. S. Ingate, M.J. Camarasa, E. De Clercq and J. Balzarini, *Antiviral Chem. and Chemother.*, 1995, **6**, 365.
24. S. Ingate, E. De Clercq, J. Balzarini and M.J. Camarasa, *Antiviral Res.*, 1996, In press.
25. S. Ingate, M.J. Pérez-Pérez, E. De Clercq, J. Balzarini and M.J. Camarasa, *Antiviral Res.*, 1995, **27**, 281.
26. M.L. Jimeno and M.J. Camarasa, *Unpublished results.*
27. J. Balzarini, S. Velázquez, A. San-Félix, A. Karlsson, M.J. Pérez-Pérez, M.J. Camarasa and E. De Clercq, *Mol Pharmacol*., 1993, **43**, 109.
28. J. Balzarini, A. Karlsson, A.M. Vandamme, M.J. Pérez-Pérez, L. Vrang, B. Öberg, A. San-Félix, S. Velázquez, M.J. Camarasa and E. De Clercq, *Proc. Natl. Acad. Sci. USA,* 1993, **90**, 6952.
29. J. Balzarini, A. Karlsson, M.J. Pérez-Pérez, L. Vrang, B. Oberg, A.M. Vandamme, M.J. Camarasa and E. De Clercq, *Virology* , 1993, **192**, 246.
30. J. Balzarini, A. Karlsson, M.J. Pérez-Pérez, M.J. Camarasa, W.G. Tarpley, and E. De Clercq, *J Virol*., 1993, **67**, 5353.
31. J. Balzarini, J.P. Kleim, G. Riess, M.J. Camarasa, E. De Clercq and A. Karlsson, *Biochem. Biophys. Res. Commun.,* 1994, **201**, 1305.
32. H. Jonckheere, J.M. Taymans, J. Balzarini, S. Velázquez, M.J. Camarasa, J. Desmyter, E. De Clercq and J. Anné, *J. Biol. Chem.,* 1994, **269**, 25255.
33. A. Jacobo-Molina, J. Ding, R.G. Nanni, A.D.Jr Clark, X. Lu, C. Tantillo, R.L. Williams, G. Kamer, A.L. Ferris, P. Clark, A. Hizi, S.H. Huges and E. Arnold, *Proc Natl Acad Sci USA* , 1993, **90,** 6320.
34. (a) R.G. Nanni, J. Ding, A. Jacobo-Molina, S.H. Hughes and E. Arnold, *Perspect. Drug. Dis. Desi,* 1993, **1**, 129. (b) S.J. Smerdon, J. Jäger, J. Wang, T.L.A. Kohlstaedt, A.J. Chirino, J.M. Friedman, P.A. Rice and T.A. Steitz, *Proc. Natl. Acad. Sci. USA* , 1994, **91**, 3911.
35. (a) D. Richman, A.S. Rosenthal, M. Skoog, R.I. Eckner, T.C. Chou, J.P. Sabo and V.J. Merluzzi, *Antimicrob. Agents. Chemother.,* 1991, **35**, 305. (b) D. Richman, C.K. Shih, I. Lowy, J. Rose, P. Prodanovich, S. Goff and J. Griffin, *Proc. Natl. Acad. Sci.* USA, 1991, **88**, 11241.
36. (a) B.A. Larder and S.D. Kemp, *Science*, 1989, **246**, 1155. (b) P. Kellam, C.A.B. Boucher and B.A. Larder, *Proc. Natl. Acad. Sci.* USA, 1992, **89**, 1934.
37. (a) V.V. Sardana, E.A. Emini, L. Gotlib, D.I. Graham, D.W. Lineberger, W.J. Long, A.J. Schalabach, J.A. Wolfgang and J.H. Condra, *J. Biol. Chem.,* 1992, **267**, 17526. (b) J. Balzarini, A. Karlsson, M.J. Pérez-Pérez, L. Vrang, J. Walbers, H. Zhang, B. Öberg, A.M. Vandamme, M.J. Camarasa and E. De Clercq, *Virology* , 1993, **192**, 246.
38. J. Balzarini, A. Karlsson, M.J. Pérez-Pérez, M.J. Camarasa, W.G. Tarpley and E. De Clercq, *J. Virol.,* 1993, **67**, 5353.
39. E. de Clercq, *Biochem. Pharmacol.*, 1994, **47**, 155.
40. J.W. Mellors, B.A. Larder and R.F. Schinazi, *Internat. Antivir. News.*, Special supplement, 1995.
41. J. Balzarini, M.J. Pérez-Pérez, S. Velázquez, A. San Félix, M.J. Camarasa, E. De Clercq and A. Karlsson, *Proc. Natl. Acad. Sci.* USA, 1995, **92**, 5470.
42. R,F, Schinazi, R.M.Jr Lloyd, M.H. Nguyen, D.L. Cannon, A. McMillan, N. Ilksoy, C.K. Chu, D.C. Liotta, H.Z. Bazmi and J.W. Mellors, *Antimicrob. Agents Chemother.*, 1993, **37**, 875.
43. J. Balzarini, H. Pelemans, M.J. Pérez-Pérez, A. San Félix, M.J. Camarasa and E. De Clercq, *Mol. Pharmacol.*, 1996, **49**, 882.
44. R.A. Spence, N.M. Kati, K.S. Anderson and K.A. Johnson, *Science*, 1995, **267**, 988.
45. S. Velázquez, R. Alvarez, A. San Félix, M.L. Jimeno, E. De Clercq, J. Balzarini and M.J. Camarasa, *J. Med. Chem.,* 1995, **38**, 1641.

19
New Sialidase Inhibitors for the Treatment of Influenza

Paul W. Smith,[*,1] Peter C. Cherry,[1] Peter D. Howes,[1] Steven L. Sollis,[1] and Neil R. Taylor
DEPARTMENT OF [1]ENZYME MEDICINAL CHEMISTRY II, [2]BIOMOLECULAR STRUCTURE, GLAXOWELLCOME MEDICINES RESEARCH CENTRE, GUNNELS WOOD ROAD, STEVENAGE, HERTFORDSHIRE SG1 2NY, UK

Introduction

Influenza is a disease which has brought misery to mankind throughout history, with documented cases dating as far back as the plague of Athens in 430 BC. It is a much more serious illness than is widely appreciated producing considerable morbidity and mortality. Excess mortality due to the virus has been documented from 1889, and since that time vast numbers of flu related deaths have been recorded. The most dramatic illustration of this occurred during the 1918-19 Spanish flu outbreak where a staggering 20 million people died as a result of infection. More recently it has been estimated that an average of 10,000 excess deaths have occurred in the US in each of the 19 epidemics between 1959 and 1986. Influenza is particularly serious in the elderly with approximately 80-90% of excess deaths occurring in people over 65. As well as high mortality, substantial morbidity is also seen, resulting in absenteeism from work and school, increased visits to physicians (approximately 10-15% of all visits to the doctor are for `flu-like` illnesses), and increased hospital admissions. Thus the economic costs of influenza cannot be over-emphasised.

The influenza viruses are negative stranded RNA viruses which belong to the orthomyxovirus family and are classified into three subtypes A, B and C. The 3 types are distinguished by differences in their internal proteins. Type A is the most prevalent and produces the most severe widespread epidemics and pandemics, type B is less common but can still produce a severe illness. Type C produces a somewhat milder and less serious form of the disease. The typical symptoms of influenza include cough, catarrh, headache, fever, chills and myalgia (muscle aches), and are characterised by their abrupt onset. The systemic nature of many of the symptoms distinguishes influenza from the `common cold` where symptoms are restricted to the upper airways. The disease is highly contagious and infection occurs when the airborne virus invades and replicates in the epithelial lining of the respiratory tract.

Currently vaccination is widely used as the first line of defence to protect `at-risk` groups within the community. Existing vaccines are made from inactivated viruses produced from virus grown in hens eggs. The vaccine relies on the generation of an immunological response to the antigenic haemagglutinin (HA) and sialidase (neuraminidase NA) proteins on the inactivated virus surface. Different strains and subtypes of the virus are characterised by differences in the sequences of these surface proteins. Each year the composition of the vaccine is altered to correspond with the existing live virus strains known to be circulating. Unfortunately, however, mutations in the antigenic NA and HA proteins can occur at any time producing new virus strains which can evade the immune response, thus rendering vaccination ineffective. For this reason vaccinations have to be updated annually and protection from infection cannot be guaranteed.

Few effective anti-influenza virus drugs currently exist, and those available have had little success largely due to side-effects and lack of perceived benefit from therapy. Amantadine and rimantadine are available for both prophylaxis and therapy. However, both compounds are selective for flu A virus and following the initiation of treatment the virus rapidly generates resistance to these drugs [1,2]. For the vast majority of patients with flu, 2-3 days bedrest, an analgesic and an anti-pyretic drug such as paracetamol (to reduce fever) are all that the physician currently has to offer. Thus there is still a real need to discover new and effective treatments.

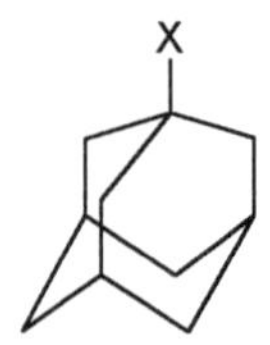

$X = NH_2$ Amantadine

$X = CH(Me)NH_2$ Rimantadine

Influenza Replication Cycle and Sialidase

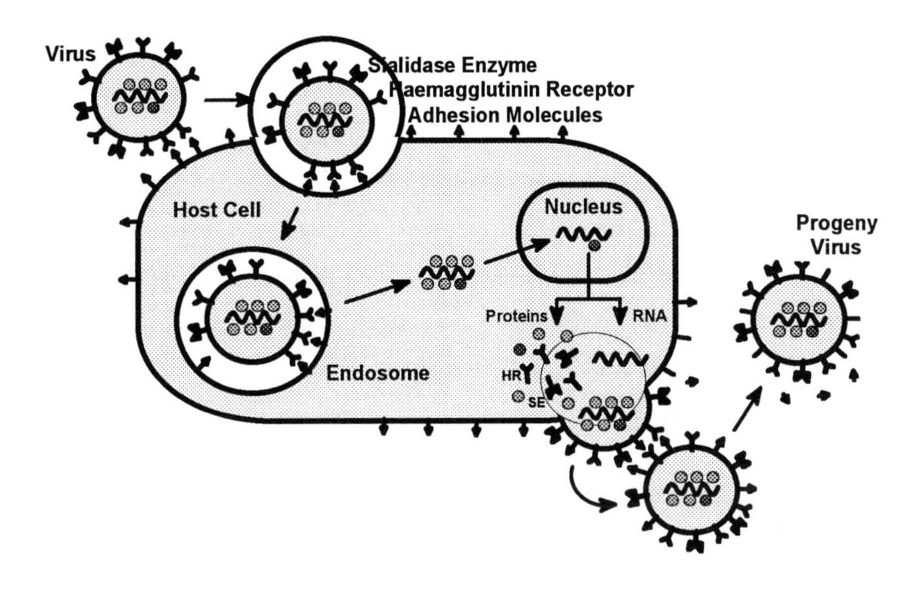

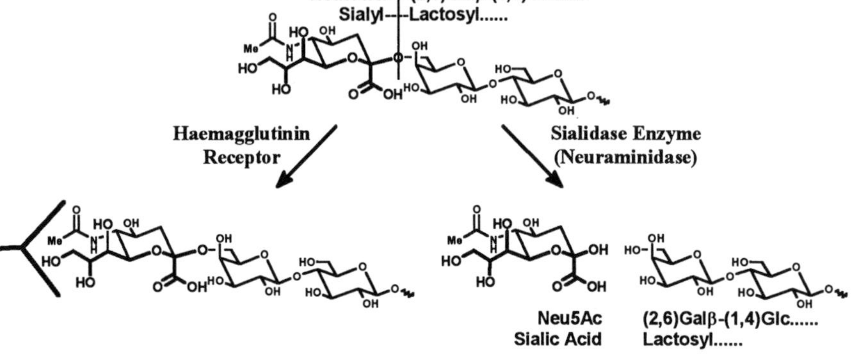

Figure 1: Schematic Representation of influenza virus replication indicating the roles of haemagglutinin and sialidase

Infection is initiated when the virus becomes attached to an epithelial cell through the non-covalent interaction of the haemagglutinin receptor with terminal sialic acid residues on cell surface glycoproteins (Figure 1). This attachment is followed by internalisation of the virus encapsulated within endosomes. Following a virus induced pH change within the endosome the haemagglutinin undergoes a conformational change which allows fusion of the viral membrane with the endosome releasing the viral RNA into the host cell [3]. Following replication of the viral genome and protein synthesis, the mature virions re-assemble and begin to bud from the host cell membrane. At this stage new virions are trapped on the surface of the infected cell by the non-covalent haemagglutinin-sialic acid interactions. Sialidase (also known as neuraminidase) now plays a pivotal role. It cleaves the terminal sialic acid residues from the host cell surface and thus facilitates release of the new infectious virions thus allowing further spread of the infection. In addition to controlling the release of mature virions it has also been proposed that sialidase plays an additional key role in infection by facilitating the movement of the virus through mucus [4].

Early Inhibitors of Sialidase: DANA and FANA

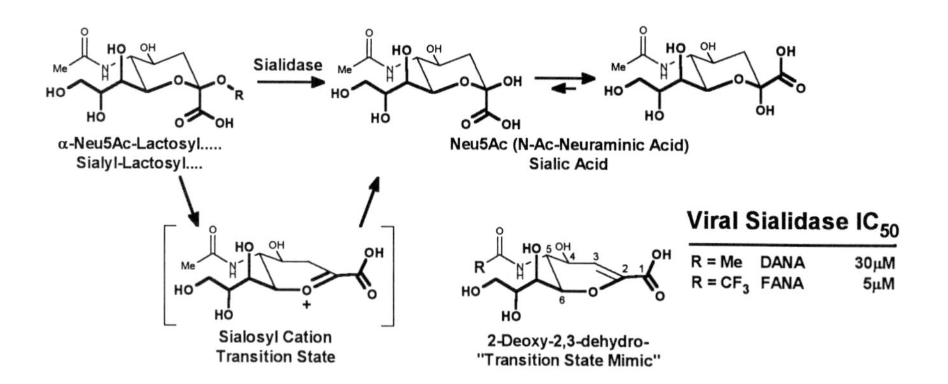

Figure 2: Hydrolysis mechanism and the dihydropyrans DANA and FANA

Having established the key role of sialidase in the virus lifecycle it was appreciated early on that inhibition of this enzyme represented a possible strategy for anti-viral therapy. The sialoside hydrolysis by the enzyme is postulated to proceed via a sialosyl cation like transition state, and in the 1970's Meindl and Tuppy described the analogues 2,3-didehydro-N-acetyl neuraminic acid (DANA) and 2,3-didehydro-N-trifluoroacetyl neuraminic acid (FANA) in which the dihydropyran ring geometrically mimics the transition state of the hydrolysis reaction [5] (Figure 2). These compounds were found to be micromolar inhibitors of the viral enzyme and even showed weak anti-viral activity *in vitro*. However, they lacked any selectivity over other sialidases and showed poor *in vivo* activity in animal models of influenza infection following systemic administration. It was several years before further progress was made in inhibitor design.

Discovery of GG167

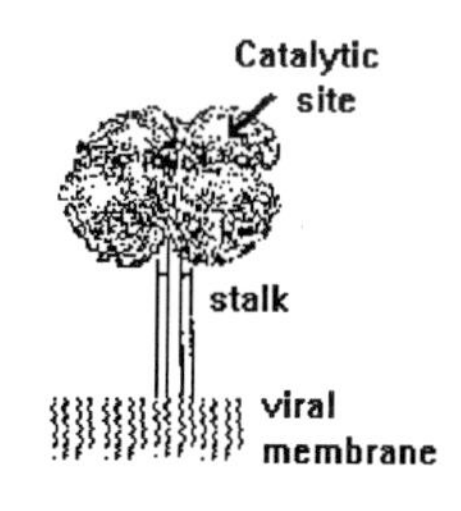

Influenza Sialidase

Influenza sialidase is a tetramer in which the four units associate to form a mushroom shaped `spike` which projects out from the viral membrane (see below). The four catalytically active sites contained within each tetramer reside in the `head` portion of the mushroom which is then embedded into the viral membrane via a long hydrophobic `stalk` region. During the 1980's the group of Laver in Australia successfully used trypsin to cleave the stalk and isolated pure catalytically active `head` protein [6]. Subsequent crystallisation of this protein with sialic acid bound into the active site was achieved by Colman, Varghese and Laver [7]. Analysis of the crystal structure of this complex revealed a highly hydrophilic active site which was not completely filled by the sialic acid. Calculations with the programme GRID [8] and modelling suggested that replacement of the 4-hydroxyl group in DANA with basic substituents would improve affinity of the dihydropyran analogues for the enzyme. This analysis prompted von Itzstein to prepare the 4-guanidino analogue of DANA which turned out to be 1000 fold more active than DANA *in vitro*. This compound, now known as GG167, also turned out to be highly selective for the viral enzymes over other sialidases and displayed outstanding anti-viral activity in animal models of influenza infection [9]. GG167 is now in phase II clinical trials in man for the treatment of both influenza A and B infections [10].

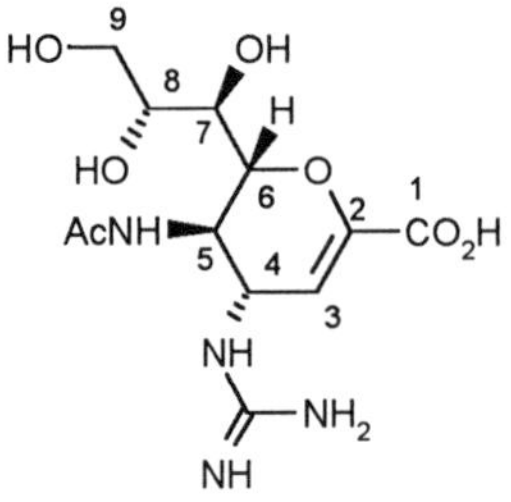

GG167

<u>*Enzyme Inhibition* (IC_{50})</u>
- Flu A sialidase 5 nM
- Flu B sialidase 4nM

<u>*In vitro Anti-viral activity*</u>
(Plaque reduction IC_{50})
- Flu A 14nm, Flu B 5nM

<u>*In vivo Anti-viral activity (intranasal)*</u>
(ED AUC_{10} flu A)
- Mouse 0.03 mg/kg; ferret 0.02 mg/kg

Analogues of GG167: Early work

A schematic representation of the interactions between GG167 and influenza sialidase determined from X-ray crystallography [11] is shown below (figure 3).

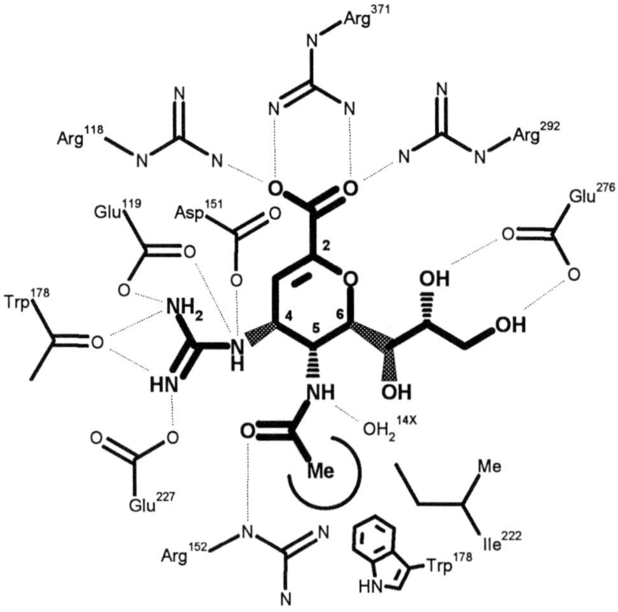

Figure 3: GG167 Interactions with Sialidase

An extensive amount of work has been carried out to investigate the relative importance of each of the dihydropyran substituents in the binding of GG167 [12-17]. The results of these investigations are briefly summarised below:

C-2: The carboxylic acid forms three salt bridges and is essential for biological activity. Esters and amides are much weaker inhibitors, presumably because they cannot form electrostatic interactions with the three adjacent arginine residues, and there is no space for additional substituents.

C-4: The unsubstituted basic guanidino group is much the preferred substituent at this position. It forms salt bridging interactions with Asp 151 and Glu 227 as well as a strong pair of hydrogen bonds with the backbone carbonyl of Trp 178. The only other compounds with activity which exceed that of DANA are those which contain a small, hydrophilic basic group.

C-5: The amide appears to be important for activity forming a hydrogen bond with the guanidino sidechain of Arg 152. Complete removal of this substituent reduces activity by 10^5 fold. Other small amides and sulfonamides are moderately active, but the acetamido group of GG167 appears to represent the optimum fit in the small lipophilic pocket formed by the Ile and Trp sidechains.

C-6: The C-8 and C-9 hydroxyl substituents are hydrogen bonded to the sidechain carboxylate of Glu 276. Sequential deletion of the hydroxymethyl groups from the glycerol sidechain results in a stepwise and dramatic reduction in inhibitory activity.

Dihydropyran Ring: A limited number of cyclohexene analogues have been prepared which have equivalent biological activity to the corresponding dihydropyrans. Saturation of the dihydropyran ring abolishes activity. (It is noteworthy that recently workers at Gilead and Biocryst have shown that the dihydropyran ring can be replaced with a benzene ring without completely abolishing inhibitory activity [18,19]. Interestingly, these compounds bind to the active site of the viral sialidase in a different orientation).

In conclusion, all the substituents make important contributions to the binding of GG167, and it is not possible to remove any of them without significantly reducing inhibitory activity.

Analogues of GG167: Replacement of the Glycerol Sidechain

GG167 shows outstanding *in vitro* inhibitory activity against influenza sialidase and in animal models of influenza infection it is a highly effective anti-viral agent when administered by the intranasal route. However, it is a highly polar compound and, unsurprisingly, is rapidly eliminated from the systemic circulation by renal excretion. Thus it has much reduced systemic efficacy, and the preparation of new agents with modified physicochemical properties would be desirable in order to evaluate their potential as systemically active anti-viral agents.

Analysis of the GG167-sialidase complex reveals that, whereas there is little space in the enzyme to allow modifications at the C-2, 4 and 5 positions, the C-6 glycerol sidechain does not fit as tightly into the protein. Thus the possibility exists to identify alternative substituents to occupy this region of the sialidase active site.

Strategically two synthetic approaches to this type of analogue can be envisaged:- de novo synthesis of dihydropyran rings containing modified C-6 substituents and degradation of the glycerol sidechain followed by reconstruction of alternative sidechains from GG167-derived intermediates.

1. De novo Synthesis via Inverse Demand Hetero Diels Alder Cyclisation

R O CO_2H AcNH NH NH_2 NH → R AcNH **2** + O CO_2R' P NH **1**

The inverse demand hetero-Diels Alder approach involves the coupling of an appropriately substituted and protected heterodiene **1** with a substituted dienophile **2**. The former intermediate, protected with acid labile groups, was readily prepared in four stages on multigram scale from oxalyl chloride (scheme 1). This cis-substituted heterodiene cyclised with simple vinyl ethers, under both thermal and Lewis acid conditions, but did not react with other olefins. Thus synthesis by this approach appears limited to R substituents linked to the dihydropyran through oxygen.

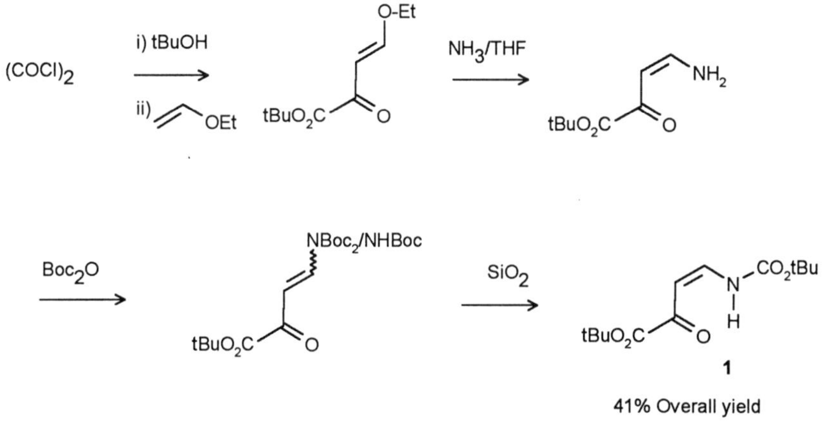

Scheme 1: Synthesis of the Heterodiene **1**

Synthesis of cis-2-acetamido-methyl vinyl ether (Z)-**2** (R = OMe) was achieved in two stages from commercially available amino acetaldehyde dimethyl acetal (Scheme 2). Despite extensive investigation, the elimination of methanol from the acetal proceeded in very low yield (~5%) and produced only the cis-isomer. Reaction of (Z)-**2** (R = OMe) with the heterodiene **1** under Lewis acid conditions with tin (IV) chloride proceeded in moderate yield (35%) with poor selectivity and produced a mixture of three of the four possible isomeric products (ratio 1.8:1.1:1). These were elaborated to GG167 analogues (Scheme 2), but none of the target compounds displayed any useful sialidase inhibitory activity.

Further optimisation of this approach could yet lead to analogues of GG167 with useful biological properties. However, given the apparently limited scope of the 6-substituent, moderate stereoselectivity of the cyclisation and disappointing biological activity of the initial targets, the emphasis of our research concentrated on the alternative strategy.

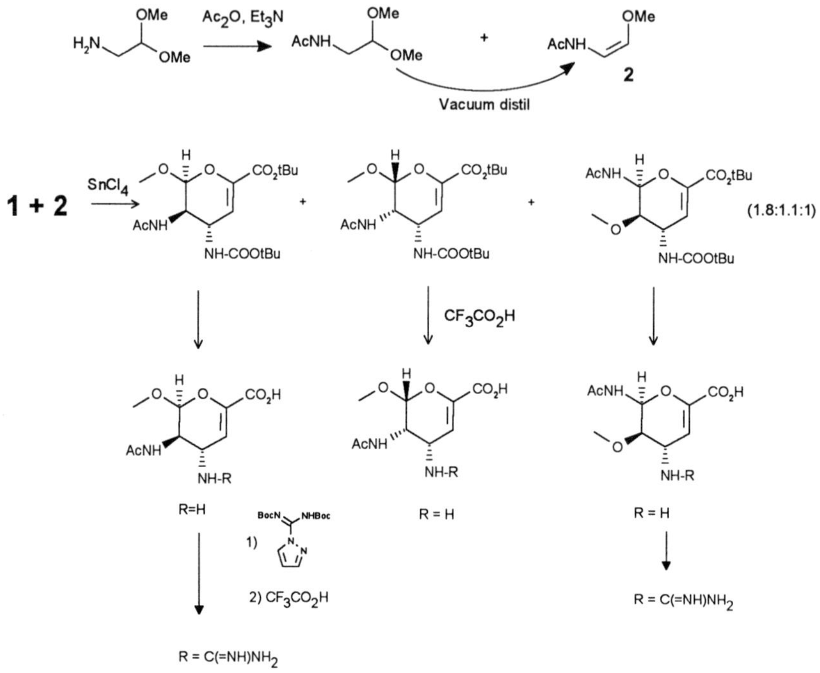

Scheme 2: Synthesis of the dienophile **2** and coupling with **1**

2. Glycerol Cleavage - Sidechain Reconstruction

Sodium periodate cleavage of the glycerol sidechain of otherwise protected GG167 affords an aldehyde which can either be reduced to the alcohol **3**, or oxidised to the 6-carboxylic acid intermediate **4**. Subsequent reactions at these functional centres offer potential starting points for simple chemistries to explore incorporation of other substituents into the glycerol binding pocket.

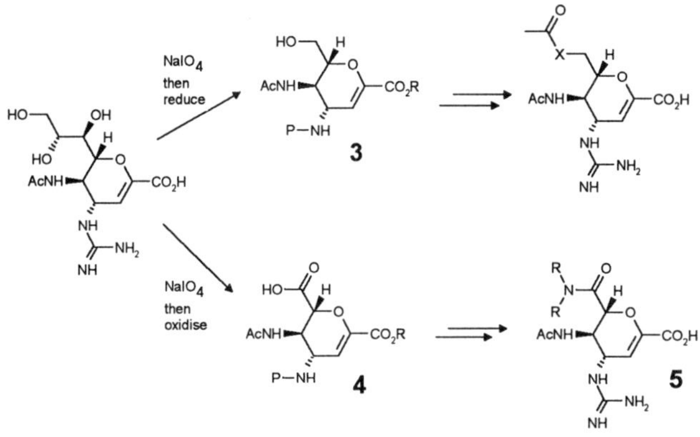

Discovery of 6-Carboxamides

In the case of the carboxylic acid **4**, the preparation of a series of 6-carboxamides **5** was initially envisaged in which hydrophilic amines would be coupled with the acid in order to introduce R groups which might interact with Glu-276 (figure 4). In order to ensure that a fully representative array of amides with diverse functionality and properties was prepared, this initial set was expanded to include several simple hydrophobic secondary and tertiary alkylamide targets.

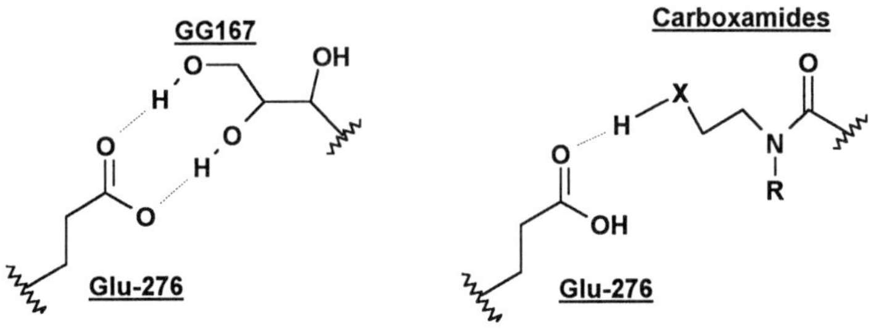

Figure 4: Initial Hypothesis for preparing carboxamides

The general synthesis of the carboxamides is shown in scheme 3. Utilisation of the 4-amino intermediate **6** as starting material allowed the synthesis of both 4-guanidino (**5**) and 4-amino (**7**) 6-carboxamide analogues of GG167.

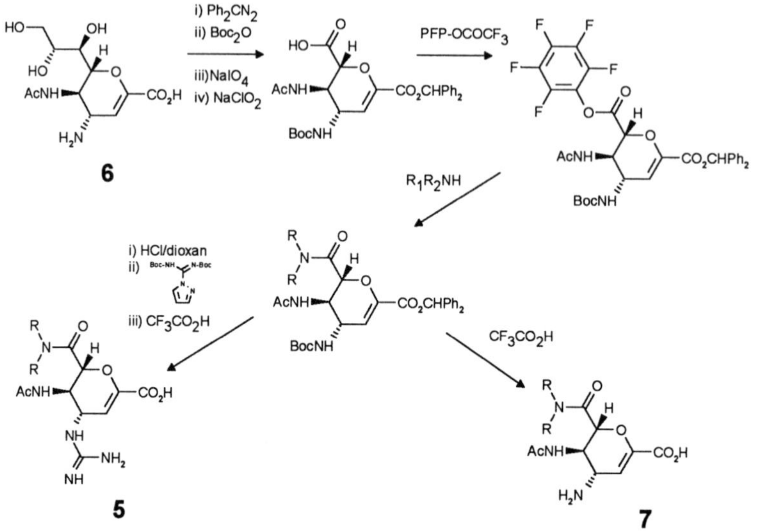

Scheme 3: Synthesis of Carboxamides

Remarkably, and contrary to initial expectations, the hydrophilic secondary amides displayed very modest sialidase inhibitory activity whilst simple alkyl tertiary amides were considerably more active, especially against influenza A sialidase (Table 1). The methylpropylamide (**5**: R_1 = propyl, R_2 = methyl) emerged from this initial work as the most promising lead compound. Its activity against the influenza A sialidase was equivalent to that of GG167. However, it was 1000 times less active against the flu B enzyme.

Table 1: Inhibitory activity of early carboxamides $IC_{50}(\mu M)$

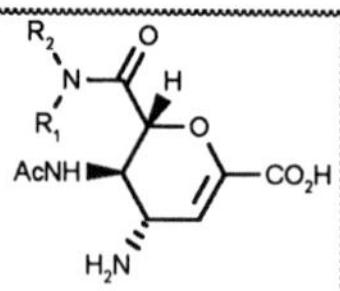

		7		5	
R_1	R_2	Flu A	Flu B	Flu A	Flu B
$-CH_2CH_2NH_2$	-H	>390	-	20	-
$-CH_2CH_2OH$	-H	420	-	-	-
$-CH_2CH_2CH_3$	-H	19	50	0.5	44
$-CH_3$	$-CH_3$	2.4	61	0.025	11
$-CH_2CH_2CH_3$	**$-CH_3$**	**0.18**	**23**	**0.004**	**4.5**

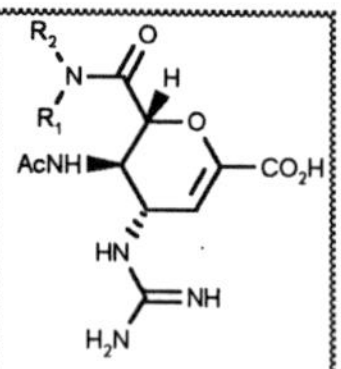

Further Optimisation and SAR of 6-Carboxamides

The chemistry outlined in above was now utilised to prepare a library of 80 target 4-amino-6-carboxamides (7) using a diverse set of different primary and secondary amines. This exercise rapidly established a pattern of SAR within the new series. Further synthesis of analogues merely re-confirmed these general conclusions which are summarised below: (Representative data to illustrate these trends is shown in table 2).

1. *Only tertiary amides are highly potent inhibitors of influenza A sialidase.* In order to explain the biological results, the two R groups on the amide are considered separately as either Rcis- or Rtrans- depending on their position about the amide bond. The Rcis-amide substituent is critical for enhancing the activity of the tertiary amides against influenza A sialidase since all secondary amides (Rcis = H) showed only modest or poor inhibitory activity against both influenza A and B sialidases. However, tertiary amides where Rcis = Me, Et, or propyl were much more active against the influenza A enzyme. If Rcis becomes larger than propyl, then inhibitory activity is diminished.

2. *When Rcis is an ethyl or propyl group, even the 4-amino carboxamides (7) are highly potent inhibitors of influenza A sialidase.* These compounds are the first nanomolar influenza sialidase inhibitors which do not contain a 4-guanidino substituent.

3. *The Rtrans-substituent can be modified considerably without significant loss in inhibitory activity.* It appears that the Rtrans substituent does not form appreciable interactions with the enzyme.

4. *Activity trends are similar against both influenza A and B sialidases, but all the potent carboxamides are selective for the influenza A enzyme*

Table 2: Further Carboxamide SAR

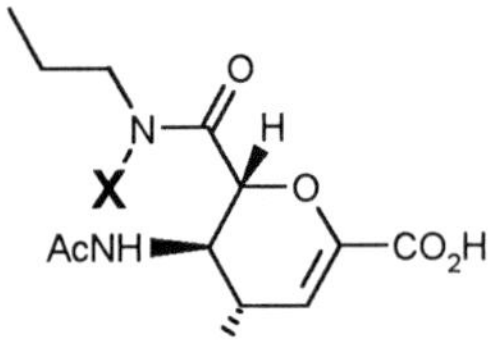

	Sialidase Inhibition (μM)			
	Amine		**Guanidine**	
	Flu A	**Flu B**	**Flu A**	**Flu B**
X				
H	**19**	**50**	**0.5**	**4.4**
Methyl	**0.18**	**23**	**0.004**	**4.5**
Ethyl	**0.003**	**0.41**	**-**	**-**
n-Propyl	**0.003**	**2**	**0.002**	**0.54**
Y				
Ph	**0.005**	**2**	**0.005**	**0.84**
$Me(CH_2)_6$	**0.023**	**37**	**-**	**-**
Glycerol sidechain	**0.32**	**0.41**	**0.005**	**0.004**

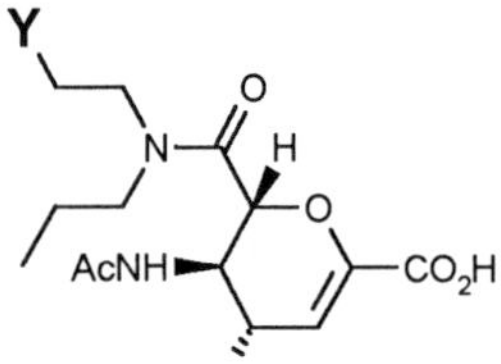

Amine/Guanidine

Amine/Guanidine

Structural Analysis of Carboxamide-Sialidase Complexes

X-ray crystal structures of the sialidase complexes of the 4-amino-phenethylpropyl carboxamide (7: Rtrans = $PhCH_2CH_2$, Rcis = $CH_2CH_2CH_3$) with both influenza A and B sialidase (influenza A N9 and B Beijing) have been determined. The results from these studies, together with molecular modelling calculations have provided an explanation for the activity and selectivity of the carboxamides.

The dihydropyran portion of the carboxamide binds to both the A and B enzymes in essentially the same manner as that observed for GG167 (Figure 5). However, a significant difference from GG167 and other previously determined structures is observed in the region into which the carboxamide sidechain binds. Whilst in GG167 the glycerol moiety forms intermolecular hydrogen bonds with Glu 276, these interactions are no longer available for the carboxamide. Instead, the Glu 276 sidechain adopts an entirely different conformation, in which the carboxylate group forms an intramolecular salt bridge with the guanidino sidechain of Arg 224 (the position of the latter is not significantly altered from previous structures). The result of this salt bridge formation is that a previously unrecognised lipophilic pocket becomes available for the Rcis substituent to occupy. The size of this new pocket appears optimal for an ethyl or propyl group, in accord with the observed structure-activity relationships. The Rtrans phenethyl substituent lies in an extended lipophilic cleft on the enzyme surface formed between the sidechains of Ile 222 and Ala 246 (Figure 5). Inspection of this region shows that the cleft can accommodate a variety of Rtrans substituents. This is also consistent with the observed structure-activity relationships.

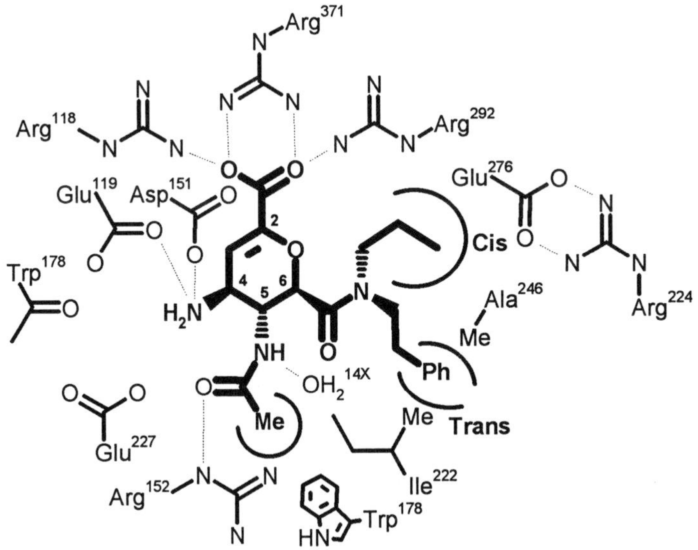

Figure 5. N-phenethyl-N-propylamide interactions with sialidase.

Selectivity for Influenza A Sialidase

Comparison of the X-ray crystal structures of native influenza A and B sialidases shows a very high correspondence between backbone position and side chain torsion angles of the conserved active site residues. There are, however, some significant differences in the region binding the glycerol sidechain of GG167 and particularly in the position and sidechain conformation of Glu 276. Despite these differences, GG167 is able to bind to the active site of both the A and B enzymes with little or no distortion of the native structures, forming similar hydrogen bonds between the 8- and 9- hydroxyl groups and the Glu 276 carboxylates (Figure 6a and b). Figure 6a shows that the formation of the salt bridge between Glu 276 and Arg 224 which occurs upon binding of the carboxamide to the A enzyme is accomplished by changes in the torsion angles of the Glu 276 sidechain, and that this requires little or no distortion of the protein backbone. However, the small difference in the position of Glu 276 in influenza B sialidase means that in order to form a similar salt bridge upon binding of the carboxamide, a significant distortion of the protein backbone must occur (Figure 6b). Distortions in the protein structure of influenza B sialidase also arise around the Rtrans phenethyl substituent - high temperature factors are observed in the loop that forms one side of the lipophilic cleft (the loop containing Ala 246). Average temperature factors are observed in the complex with influenza A sialidase. It thus appears that the structural changes which accompany the binding of carboxamides are energetically less favourable in the B enzyme and this accounts for the observed selectivity of the carboxamides for influenza A sialidase.

Summary

New inhibitors of influenza virus sialidases have been identified through the replacement of the glycerol sidechain of GG167. An inverse demand hetero Diels Alder approach has been utilised to prepare dihydropyran analogues of GG167 containing ether linked 6-substituents. Through degradation of the glycerol and reconstruction a new series of 6-carboxamide inhibitors have been discovered. Some tertiary amides are exceptionally potent inhibitors of influenza A sialidase, but all show much weaker inhibitory activity against flu B sialidase. The activity and selectivity has been rationalised through examination of the X-ray crystal structures of carboxamide-sialidase complexes. A conformational change within the sialidases involving formation of a previously unobserved salt bridge between the sidechains of Arg224 and Glu 276 explains both the activity and observed selectivity of these new inhibitors.

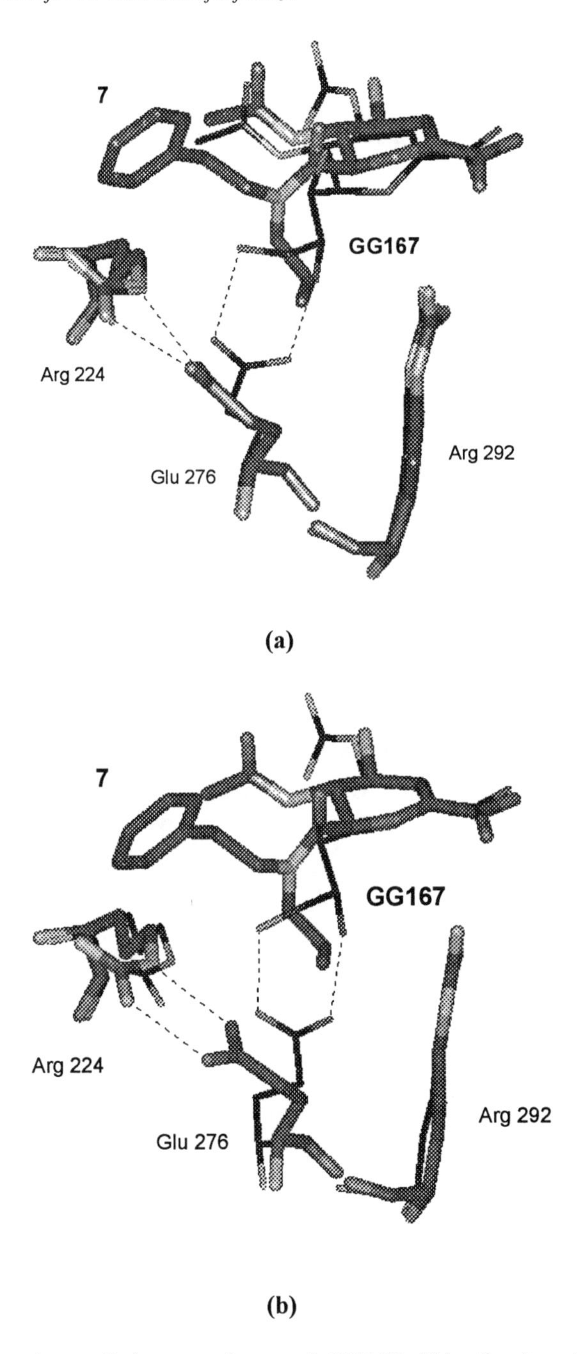

Figure 6. Comparison of the complexes of GG167 (thin lines) and **7** (Rtrans = $PhCH_2CH_2$, Rcis = propyl; thick lines) in (a) influenza A (N9) sialidase and (b) influenza B (Beijing) sialidase.

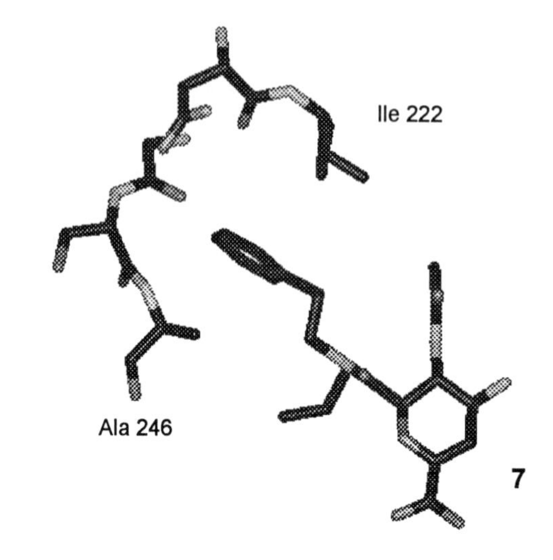

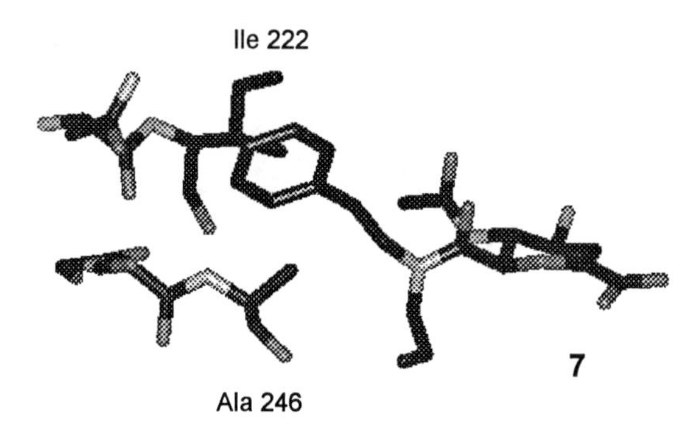

Figure 7. The phenyl ring occupies a lipophilic cleft defined by Ala 246 and Ile 222 sidechains (shown here in influenza A sialidase).

References

1. Hay, A.J.; Wolstenholme, A.J.; Skehel, J.J.; Smith, M.H. (1985) EMBO J. 4, 3021.
2. Pinto, L.H.; Holstinger, L.J.; Lamb, R.A. (1992) Cell 517.

3. Wharton, S.A.; Weis, W.; Skehel, J.J.; Wiley, D.C. (1989) In The Influenza Virus. (Krug, R.M (ed)) pp 153 - 173. Plenum Press, New York

4. Klenk, H.D.; Rott, R. (1988) Adv.Virus.Res 34, 247.

5. Meindl, P.; Bodo, G.; Palese, P.; Tuppy, H. (1974) Virology, 58, 457.

6. Varghese, J.N.; Laver W.G.; Colman, P.M. (1983) Nature, 303, 35.

7. Varghese, J.N.; Colman, P.M J.Mol.Biol (1991), 221, 473.

8. Goodford, P.J. (1985) J.Med.Chem 28, 849.

9. vonItzstein, M.; Wu, W-Y.; Kok, G.B.; Pegg, M.S.; Dyason, J.C.; Jin, B.; Phan, T.V.; Smythe, M.G.; White, H.F.; Oliver, S.W.; Colman, P.M.; Varghese, J.N.; Ryan, D.M.; Woods, J.M.; Bethell, R.C.; Hotham, V.J.; Cameron, J.M.; Penn, C.R. (1993) Nature 363, 418.

10. Hayden, F.G; Treanor, J.J; Betts, R.F; Lobo, M.; Esinhart, J.; Hussey, E. *J.Amer.Med.Assoc* **1996**, *275*, 295.

11. . Varghese, J.N.; Epa, V.C.; Colman, P.M. *Protein Sci.* **1995**, *4*, 1081

12. Bamford, M.J.; Chandler, M.; Conroy, R.; Lamont, B.; Patel, B.; Patel, V.K.; Steeples, I.P.; Storer, R.; Weir, N.G.; Wright, M.; Williamson, C. *J.Chem.Soc.Perkin Trans I.* **1995**, 1173.

13. von Itzstein, M.; Wu, W-Y.; Jin, B. *Carbohydrate Res.* **1994**, *259*, 301

14. Starkey, I.D.; Mahmoudian, M.; Noble, D.; Smith, P.W.; Cherry, P.C.; Howes, P.D.; Sollis, S.L. *Tetrahedron Lett.* **1995**, *36*, 299

15. Smith, P.W.; Starkey, I.D.; Howes, P.D.; Sollis, S.L.; Keeling, S.; Cherry, P.C.; vonItzstein, M.; Wu, W-Y.; Jin, B. *Eur.J.Med.Chem.* **1996**, *31*, 143.

16. Bamford, M.J.; Castro-Pichel, J.; Patel, B.; Storer, R.; Weir, N.G. *J.Chem.Soc.Perkin Trans I.* **1995**, 1181

17. Chandler, M.; Conroy, R.; Cooper, A.W.J.; Lamont, B.; Scicinski, J.J.; Smart, J.E.; Storer, R.; Weir, N.G.; Wilson, R.D.; Wyatt, P.G. J.Chem.Soc.Perkin Trans I (1995), 1189.

18. Singh, S.; Jedrzejas, M.J.; Air, G.M.; Luo, M.; Laver, W.G.; Brouillette, W.J. *J.Med.Chem.* **1995**, *38*, 3217.

19. Williams, M.; Bischofberger, N.; Swaminathan, S.; Kim, C.U. *Bioorg.Med.Chem.Lett.* **1995**, 5, 2251.

20
Viral Proteinases as Targets for Inhibitors in Viruses Other Than HIV-1

J. S. Mills

DEPARTMENT OF MOLECULAR VIROLOGY, ROCHE RESEARCH CENTRE, WELWYN GARDEN CITY, HERTFORDSHIRE AL7 3AY, UK

1 INTRODUCTION

Proteinases are recognised as being valid targets for the development of inhibitors for use as therapeutics in a number of serious human diseases. Interleukin-1-β converting enzyme and tumour necrosis factor convertase are such targets in inflammatory disease. For hypertension, renin, angiotensin converting enzyme and endothelin converting enzyme represent proteinase targets. Additional examples include neutrophil elastase in emphysema and bronchitis and the matrix metalloproteinases in the arthropathies and metastases. Virus encoded proteinases have also been the focus of much attention. The recent approval by the US regulatory authorities of the HIV proteinase inhibitors Invirase™ (Hoffmann-La Roche), Crixivan™ (Merck and Co) and Norvir™ (Abbott) for the therapy of AIDS has added impetus to the characterisation of other viral proteinases with the aim of identifying inhibitors for therapeutic use in other human viral diseases.

2 VIRAL PROTEINASES AS TARGETS FOR DRUG DISCOVERY

Viral proteinases are recognised as particularly good targets for drug discovery for several reasons:

2.1 **"Achilles' Heel"**

Viral proteinases usually have an essential role in viral replication and for this reason alone they represent an attractive target. Furthermore, the viral enzymes are usually sufficiently different from cellular proteinases to allow them to avoid the host cells' proteinase inhibition control mechanisms. The viral proteinase can, therefore, be seen as the "Achilles' heel" of the virus and, as such, a weakness that can be exploited in the search for antiviral drugs.

2.2 **Knowledge Database**

Historically, cellular proteinases have been classified into four groups based upon their catalytic mechanism and the geometry of their active site. The "classical" cellular proteinases (aspartic, metallo, serine and cysteine) have been intensively studied and there is now an excellent database of information available which includes amino acid sequence, substrate specificity data, details of catalytic mechanism and inhibitor profiles. In addition, three dimensional atomic and X-ray crystallographic structural information is available for a steadily increasing number of these enzymes. This database of information is very useful in the early stages of research on poorly characterised viral proteinases. By comparing the amino acid sequence, diagnostic inhibitor profile and biochemical properties of the viral enzyme being investigated with the database, it is often possible to classify the viral enzyme into one of the four classes. This can be an important first step in the inhibitor design process.

2.3 Drug Discovery Strategy

A further advantage of viral proteinases as therapeutic targets concerns the overall strategy for drug discovery. In most cases, a clear and well defined screening cascade and strategy to identify a candidate for clinical evaluation can be planned. Cloning, expression and purification of the target viral proteinase enables biochemical, enzymological and structural studies to be initiated and *in vitro* assays for inhibitor screening to be established. Inhibitors may be designed "rationally" or identified from high throughput screening of compound libraries and natural products. Lead compounds can then be analysed in cell based antiviral assays and suitable animal models if these are available for the target virus. An inhibitor with the desired properties for clinical evaluation will, hopefully, be identified after the completion of pharmacodynamic and toxicological studies.

2.4 HIV Paradigm

The publication of the HIV-1 proviral genomic nucleotide sequence[1] prompted a focused research effort on the virally encoded proteinase. Saquinavir (Hoffmann-La Roche) was the first HIV-1 proteinase inhibitor to enter clinical trials in 1991 and the details of how this compound was identified and developed[2,3] can serve as a paradigm. This model demonstrates that viral proteinases are excellent molecular targets for the development of antiviral therapeutics and indicates that the investigation of proteinases in other viruses of clinical significance is a valid approach in the search for antiviral drugs.

Of the many viral proteinases which have been described and reviewed in the literature[4], the most significant opportunities for developing proteinase inhibitor drugs in the future are to be found in the respiratory viruses, hepatitis C virus and the herpes viruses.

3 RESPIRATORY VIRUSES

Two of the major respiratory diseases of viral aetiology are influenza and the common cold syndrome. The viruses that cause influenza do not encode a proteinase; a cellular serine proteinase is required to process the viral haemagglutinin precursor, a step essential for infectivity. Although a cellular proteinase can be considered as an antiviral target, it is less desirable than a virally encoded enzyme as one must consider the possible detrimental effects of inhibiting a cellular enzyme. The common cold syndrome is caused by infection with a range of respiratory pathogens. Fifty per cent of common colds are caused by rhinovirus infection, approximately twenty per cent by coronavirus infection and the remainder are due to infection with one of the following: coxsackievirus A and B, adenovirus, parainfluenza virus, reovirus respiratory syncytial virus. Proteinase targets in viruses that cause the common cold are summarised in Table 1 and discussed further below. Virally encoded proteinases have not been identified in parainfluenza virus or in respiratory syncytial virus.

3.1 Rhinovirus, Coxsackievirus A and B

Rhinovirus and coxsackievirus A and B are members of the Picornaviridae and the polyprotein encoded by their genomic RNA is proteolytically processed by two proteinases designated 2A and 3C. The 2A proteinase is required to separate the capsid proteins from the non-structural proteins and the 3C enzyme is required for further processing of the viral polypeptide into its mature products. Both enzymes have been defined as cysteine proteinases. Analysis of the X-ray crystallographic structure of rhinovirus 3C proteinase, however, reveals an active site geometry similar to that of a trypsin-like serine proteinase but with a cysteine residue as the active site nucleophile.[5] Assays suitable for inhibitor screening have been described, including a fluorescence-quench peptide substrate assay.[6] First generation rhinovirus proteinase inhibitors based upon chloromethyl ketone peptides[7] and glutamine derived aldehydes[8] have been generated. A high throughput screening strategy has also led to the identification of several inhibitors of this viral proteinase.[9]

3.2 Coronavirus

Coronavirus makes extensive use of virus encoded proteolytic activities during genome replication. The 27kb RNA genome encodes eight functional open reading frames, one of which is translated into a 750KDa polypeptide containing two papain-like cysteine proteinases and a picornavirus 3C-like enzyme. Progress in cloning, expression and biochemical characterisation of the murine[10] and human[11] coronavirus 3C-like proteinase has been described and it is apparent that this enzyme represents an attractive antiviral target for coronavirus disease.

Table 1 *Proteinase Targets in Viruses that cause the Common Cold*

Virus	*Proteinase*	*Proteinase Class*
Rhinovirus	2A	Cysteine
	3C	Cysteine*
Coxsackie A and B	2A	Cysteine
	3C	Cysteine*
Coronavirus	3 enzymes	1 Serine, 2 Cysteine
Adenovirus	23Kda	Cysteine**
Reovirus	?	Serine?
Parainfluenza	No	-
Respiratory Syncytial Virus	No	-

* "trypsin-like" serine proteinase
** new subclass of cysteine proteinase family

3.3 **Adenovirus**

The adenovirus proteinase has been shown to be involved in several processes during viral replication including virus entry into cells,[12] viral DNA replication[13] and virus maturation[14]. Analysis of sequence alignments, mutagenesis and inhibitor studies indicate that the adenovirus proteinase represents a new subclass of the cysteine proteinase family.[15,16] Efforts to generate inhibitors based on substrate-like tetrapeptide nitriles have been reported[17] and progress towards generating an X-ray crystallographic structure of the proteinase[18] augers well for the future identification of more potent inhibitors.

3.4 **Reovirus**

In reovirus, the outer protein shell of the virion is composed of two structural proteins, μl and σ3. The cleavage of μl, the capsid protein precursor, involves formation of a complex with σ3. The presence of amino acid sequence motifs in σ3 that are found in the picornaviral proteinases[19] led to the hypothesis that σ3 may possess a serine-like proteinase activity responsible for the cleavage of μl. Mutagenesis studies, however, indicated that these sequences were not required for the proteolytic processing.[20] The precise details of the proteolytic processing pathway have yet to be elucidated for this virus.

A “cure for the common cold” is often cited as the “Holy Grail” in viral disease therapy and it is apparent that the proteinases encoded by rhinovirus, coronavirus adenovirus and coxsackie A and B, represent valid targets. The full potential of these targets, however, remains to be exploited as there are several factors which have influenced the research in this area. A specific viral proteinase inhibitor is likely to be ineffective in many and possibly the majority of patients due to the multiple aetiological agents associated with the common cold syndrome. The development of effective therapeutics would require the concomitant development of rapid, inexpensive, home-based diagnostics to ensure the use of the appropriate virus specific drug. An additional consideration is that the disease is usually mild and self-limiting; most symptoms probably result from an inflammatory response induced by viral infection and replication. This may limit any proteinase inhibitor to prophylactic use.

4 VIRAL HEPATITIS

The vast majority of cases of acute and chronic viral hepatitis are caused by hepatitis virus A, B and C.

4.1 **Hepatitis A Virus**

Hepatitis A virus (HAV), a picornavirus, causes an acute, self-limiting disease. The viral polyprotein is processed by the 3C proteinase which has been defined as a cysteine proteinase on the basis of mutagenesis and inhibitor studies. The X-ray crystallographic structure of the 3C proteinase, however, reveals it to be structurally related to the chymotrypsin family with a cysteine in place of a serine residue as the active site nucleophile.[21] The substrate specificity of the HAV 3C proteinase has been investigated using natural substrates[22] and potent peptide based aldehyde inhibitors have been generated.[23] The availability of effective prophylactic HAV vaccines may, however, limit the future effort to identify HAV antiviral therapeutics.

4.2 **Hepatitis B Virus**

Hepatitis B virus (HBV) causes polymorphic liver disease and chronic infection is associated with the development of hepatocellular carcinoma and cirrhosis. For several years it was argued that a viral proteinase was probably involved in its replicative cycle. The presence of amino acid sequence motifs in the core protein similar to those found in the active site of retroviral aspartic proteinases[24] and the observation that production of the secretory e antigen by proteolytic processing of the core antigen was inhibited by pepstatin[25] supported this hypothesis. Mutagenesis studies, however, confirmed that the viral sequences with homology to the aspartic proteinases were not involved in this process.[26] No further evidence indicating the role of a viral proteinase in HBV replication has emerged.

4.3 **Hepatitis C Virus**

Hepatitis C virus (HCV) is the major aetiological agent of post-transfusion and sporadic community acquired non-A, non-B hepatitis. Molecular cloning and sequencing studies[27] revealed that the genome of HCV is a positive-strand RNA encoding a single open reading frame capable of expressing a polyprotein of about 3000 amino acids. On the basis of sequence homology with the flaviviruses and pestiviruses, putative functions were assigned to domains of the virus encoded polyprotein. Expression of defined regions of the HCV genome in various heterologous systems, demonstrated that NS3 encodes a proteinase responsible for cleavage of the downstream polyprotein in at least four places.[28-31] Three amino acid residues, predicted to form the catalytic triad of the enzyme's active site, have been identified in the amino terminal portion of NS3. Genetic studies on HCV have been compromised by the difficulties in growing the virus in tissue culture but by analogy with mutagenesis studies undertaken in yellow fever virus,[32] the HCV NS3 proteinase is considered to be essential for the production of infectious virus and represents a target for the development of antiviral drugs.

Analysis of the amino acid residues flanking each NS3 cleavage site and comparison with a number of HCV strains suggests a consensus which may contribute to the substrate specificity of the enzyme.[33] This sequence is Asp/Glu-X-X-X-X-Cys/Thr↓Ser/Ala, where Asp/Glu, Cys/Thr and Ser/Ala represent the P6, P1 and P1′ position relative to the scissile bond. The unusual substrate specificity suggests that it should be possible to generate inhibitors with a high degree of selectivity for the viral enzyme over cellular serine proteinases. A molecular model of the specificity pocket of the NS3 enzyme has been described[34] and further biochemical characterisation and inhibitor studies have lead to the classification of the enzyme as a member of the chymotrypsin-like serine proteinase family.[35]

In vitro NS3 proteinase assays have been described using material generated in *in vitro* transcription/translation reactions[36] and from cells infected with recombinant vaccinia[37] and baculovirus vectors.[38] Good progress in purification of the NS3 proteinase has also been reported[39] and this augers well for crystallographic studies.

The current availability of purified recombinant HCV proteinase, *in vitro* assays and our increasing knowledge about its biochemistry and enzymology will facilitate progress in screening and designing inhibitors of the enzyme's catalytic apparatus. An additional approach to the inhibition of HCV NS3 is also apparent with the observation that NS4A, a 54 amino acid peptide, is an essential co-factor [29,40,41] which activates NS3. Cleavage at the NS3A/4A, NS4A/4B and NS4B/5A junctions requires NS4A and cleavage at NS5A/5B is enhanced in the presence of NS4A. Several studies have shown that NS3 and NS4A can form a stable complex and that the central domain of NS4A is essential for proteinase activation.[42-44] Sequences at the amino terminus of NS3 are involved in the interaction with NS4A.[29,40,45] Engineered NS4A peptides or analogues could, in principle, offer an additional approach to the abrogation of proteinase function.

A second HCV proteinase has also been identified [46, 47] which mediates cleavage at the NS2/NS3 junction and this enzyme has been proposed to be a novel zinc-dependent metalloproteinase. This proteinase is encoded in the C-terminal portion of the NS2 protein and the adjacent NS3 serine proteinase domain. Microsomal membranes are required for this enzyme to achieve maximum efficiency.[48] The molecular mechanism of proteolytic processing mediated by this enzyme is unclear but it does represent an attractive proteinase target in HCV.

5 HERPES VIRUSES

The Herpesviridae are DNA viruses and herpes simplex (HSV) 1 and 2, varicella zoster and cytomegalovirus (CMV) are the main therapeutic targets. The Herpesviruses have a double strand DNA genome of over 100 kb which replicates inside the nucleus of infected cells and is subsequently packaged into an intermediate capsid during virus maturation. A proteinase, called "assemblin" (reviewed by Gisbon *et al* [49]), that functions during capsid assembly and virion maturation is essential for the production of infectious virus[50] and represents a target for therapeutic intervention. The proteinases of herpes simplex (UL26) and cytomegalovirus (UL80) have been the focus of most attention but proteinases from Epstein Barr virus[51] and human herpes virus six[52] have also been cloned and characterised to a limited extent.

Following the initial identification and preliminary characterisation of the HSV and CMV enzymes, more detailed studies ensued. Affinity labeling with diisopropylfluorophosphate confirmed that Ser-129 in HSV-1[53] and Ser-132 in CMV[54] are the catalytic nucleophiles. Site-directed mutagenesis studies also indicate that one of only two histidine residues absolutely conserved in the proteinases of all members of the herpes virus family is a good candidate to be a second component of the active site.[55] The herpes proteinases, however, do not contain the characteristic sequence motifs found near the active site serine residue of the serine proteinase family members and it appears that they represent a new subclass of the serine proteinase superfamily. The prospect of designing highly selective virus specific inhibitors is a realistic prospect. The substrate specificity of the herpes proteinases has been investigated and *in vitro* assays using peptide substrates have been established for the HSV[56] and CMV[57] enzymes. It is apparent that these enzymes cleave peptide substrates slowly and turnover rates are orders of magnitude lower than those observed for other viral proteinases eg HIV-1 and rhinovirus.

Although the herpesvirus proteinases share remarkable similarities there are also intriguing differences. The CMV proteinase, for example, has an "internal" cleavage site near the catalytic centre of the enzyme.[58] Cleavage at this site, however, has no apparent effect on the kinetic properties of the enzyme [59,60] and the precise function of this cleavage event remains to be resolved.

Good progress in purification of the HSV-1[61] and CMV[62] enzymes has been described and efforts to generate and solve their X-ray structure are underway. The availability of *in vitro* assays using quenched fluorogenic peptide substrates[63] will facilitate high throughput screening to identify inhibitors.

6 VIRAL PROTEINASE INHIBITORS - FUTURE PROSPECTS

As we learn more about the various virus proteinases we can ask: what are the prospects of identifying potent inhibitors that will be the antiviral drugs of tomorrow? The availability of recombinant proteinases and appropriate *in vitro* assays facilitates high throughput "random" screening of diverse compound libraries and natural products. Screening strategies will often give rise to "hits" which are worthy of further investigation and evaluation as proteinase inhibitors. The availability of an X-ray crystallographic structure of the target proteinase, preferably with a bound inhibitor, enables true rational design of more potent inhibitors. In the absence of an X-ray structure, however, a "rational" approach can still be adopted.[64] The substrate specificity of the target enzyme is characterised and peptides with similar features as the substrate but with hydrolysable amide bonds replaced by non-reactive "isosteres" can be a starting point. Potency can be enhanced by incorporating an electrophile capable of forming a transition state analogue complex with the active site nucleophile. Optimisation of the peptidic side chains and backbone can lead to further improvements in the properties of the inhibitors. More recently, parallel synthesis and combinatorial chemistry is being utilised to identify moieties that can replace peptidic features. We can expect to see the results of these approaches to inhibitor design as applied to several of the viral proteinases discussed in this paper in the not too distant future.

The excellent progress in identifying and developing the HIV proteinase inhibitor drugs confirms that viral proteinases are valid molecular targets. We can remain optimistic about the prospects for identifying proteinase inhibitor drugs for other human viruses of clinical significance.

References

1. L. Ratner, W. Haseltine, R. Patarca, K. J. Livak, B. Starcich, S. F. Josephs, E. R. Doran, J. A Rafalski, E. A. Whitehorn, K. Baumeister, L. Ivanoff, S. F.. Petteway, Jr., M. L. Pearson, J. A. Lautenberger, J. Ghrayeb, N. T. Chang, R.C. Gallo and F. Wong-Staal, *Nature*, 1985, **313**, 277.
2. N. A. Roberts, J. A. Martin, D. Kinchington, A. V. Broadhurst, J. C. Craig, I. B. Duncan, S. A. Galpin, B. K. Handa, J. Kay, A. Kröhn, R. W. Lambert, J. H. Merrett, J. S. Mills, K. E. B. Parkes, S. Redshaw, A. J. Ritchie, D. L. Taylor, G. J. Thomas and P. J. Machin, *Science*, 1990, **248**, 358.
3. J. Mills, *International Antiviral News*, 1993, **1**, 18.
4. W. G. Dougherty and B. L. Semler, *Microbiol. Revs*, 1993, **57**, 781.
5. D. A. Matthews, W. W. Smith, R. A. Ferre, B. Condon, G. Budahazi, W. Sisson, J. E. Villafranca, C. A. Janson, H. E. McElroy, C. L. Gribskov and S. Worland, *Cell*, 1994, **77**, 761.
6. G. M. Birch, T. Black, S. K. Malcolm, M. T. Lai, R. E. Zimmerman and S.R.Jaskunas, *Protein Expression and Purification,* 1995, **6**, 609.
7. B. D. Korant, '*Biological Functions of Proteases and Inhibitors*', Katunuma et al (Ed), Karger, Basle, 1994 p149
8. S. W. Kaldor, M. Hammond, B. A. Dressman, J. M. Labus, F. W. Chadwell, A. D. Kline and B. A. Heinz, *Bioorganic and Medicinal Chemistry Letters*, 1995, **5**, 2021.

9. J. O McCall, S. Kadam and L. Katz, *Biotechnology*, 1994, **12**, 1012.
10. H. H. Lu, X. Li, A. Cuconati and E. Wimmer, *J. Virol.,* 1995, **69**, 7445.
11. J. Ziebuhr, J. Herold and S. G. Siddel, *J. Virol.*, 1995, **69**, 4331.
12. M. Cotten and J. M. Weber, *Virology*, 1995, **213**, 494.
13. A. Webster, I. R. Leith and R. T. Hay, *J. Virol.*, 1994, **68**, 7292.
14. D. A. Matthews and W. C. Russell, *J. Gen. Virol.*, 1995, **76**, 1959.
15. A. W. Grierson, R. Nicholson, P. Talbot, A. Webster and G. Kemp, *J. Gen. Virol.*, 1994, **75**, 2761.
16. C. Rancourt, K. Tihanyi, M. Bourbonniere and J. M. Weber, *Proc. Natl Acad. Sci. USA*, 1994, **91**, 844.
17. J. A. Cornish, H. Murray, G.D. Kemp and D. Gani, *Bioorganic and Medicinal Chemistry Letters*, 1995, **5**, 25.
18. L. J. Keefe, S. L. Ginell, E. M. Westbrook and C. W. Anderson, *Protein Sci.*, 1995, **4**, 1658.
19. L. A. Schiff, M. L. Nilbert, M. Sung Co, E. G. Brown and B. N. Fields, *Mol. Cell. Biol.*, 1988, **8**, 273.
20. T. Mabrouk and G. Lemay, *Virology*, 1994, **202**, 615
21. M. Allaire, M. M. Chernaia, B. A. Malcolm and M. N. G. James, *Nature*, 1994, **369**, 72.
22. T. Schultheiss, W. Sommergruber, Y. Kusov and V. Gauss-Müller, *J. Virol*, 1995, **69**, 1727.
23. B. A. Malcolm, C. Lowe, S. Shechosky, R.T. McKay, C. C. Yang, V. J. Shah, R. J Simon, J. C. Vederas and D. V. Santi, *Biochemistry*, 1995, **34**, 8172.
24. O. Jean-Jean, M. Levrero, H. Will, M. Perricaudet and J. -M. Rossignol, *Virology*, 1989, **170**, 99.
25. R. H. Miller, *Science*, 1987, **236**, 722.
26. M. Nassal, P. R. Galle and H. Schaller, *J. Virol.*, 1989, **63**, 2598.
27. Q. -L. Choo, G. Kuo, A. J. Weiner, L. R. Overby, D. W. Bradley and M. Houghton, *Science*, 1989, **244**, 359.
28. R. Bartenschlager, L. Ahlborn-Laake, J. Mous and H. Jacobsen, *J. Virol.,* 1993, **67**, 3835.
29. R. Bartenschlager, L. Ahlborn-Laake, J. Mous and H. Jacobsen, *J. Virol.,* 1994, **68**, 5045.
30. A. Grakoui, C. Wychowski, C. Lin, S. M. Feinstone and C. M. Rice, *J. Virol.*, 1993, **67**, 1385.
31. L. Tomei, C. Failla, E. Santolini, R. De Francesco and N. La Monica, *J. Virol.*, 1993, **67**, *J. Virol.*, 4017.
32. T. J. Chambers, R. C. Weir, A. Grakoui, D. W. McCourt, J. F. Bazan, R. J. Fletterick and C. M. Rice, *Proc. Natl. Acad. Sci. USA,* 1990, **87**, 8898.
33. A. Grakoui, D. W. McCourt, C. Wychowski, S. M. Feinstone and C. M. Rice, *J. Virol.*, 1993, **67**, 2832.
34. E. Pizzi, A, Tramontano, L. Tomei, N. La Monica, C. Failla, M. Sardana, T. Wood and R. De Francesco, *Proc. Natl. Acad. Sci. USA*, 1994, **91**, 888.
35. B. Hahm, D. S. Han, S. H. Back, O. -K. Song, M. -J. Cho, C. -J. Kim, K. Shimotohno and S. K. Jang, *J. Virol.*, 1995, **69**, 2534.
36. E. D. A. D'Souza, E. O'Sullivan, E. M. Amphlett, D. J. Rowlands, D. V. Sangar and B. E. Clarke, *J. Gen. Virol.*, 1994, **75**, 3469.

37. P. Bouffard, R. Bartenschlager, L. Ahlborn-Laake, J. Mous, N. Roberts and H. Jacobsen, *Virology*, 1995, **209**, 52.
38. H. Overton, D. McMillan, F. Gillespie and J. Mills, *J. Gen. Virol.*, 1995, **76**, 3009.
39. C. Steinkühler, L. Tomei and R. De Francesco, *J. Biol. Chem.*, 1996, **271**, 6367.
40. C. Failla, L. Tomei and R. DeFrancesco, *J. Virol.*, 1995, **69**, 1769.
41. C. Lin and C. M. Rice, *Proc. Natl. Acad. Sci. USA*, 1995, **92**, 7622.
42. C. Lin, J. A. Thomson and C. M. Rice, *J. Virol.*, 1995, **69**, 4373.
43. Y. Tanji, M. Hijikata, S. Satoh, T. Kaneko and K. Shimotohno, *J. Virol.*, 1995, **69**, 1575.
44. Y. Shimizu, K. Yamaji, Y. Masuho, T. Yokota, H. Inoue, K. Sudo, S. Satoh and K. Shimotohno, *J. Virol.*, 1996, **70**, 127.
45. S. Satoh, Y. Tanji, M. Hijikata, K. Kimura and K. Shimotohno, *J. Virol.*, 1995, **69**, 4255.
46. A. Grakoui, D. W. McCourt, C. Wychowski, S. M. Feinstone and C. M. Rice, *Proc. Natl Acad. Sci. USA.*, 1993, **90**, 10583.
47. M. Hijikata, H. Mizushima, T. Akagi, S. Mori, N. Kakiuchi, N. Kato, T. Tanaka, K. Kimura and K. Shimotohno, *J. Virol.*, 1993, **67**, 4665.
48. E. Santolini, L. Pacini, C. Fipaldini, G. Migliaccio, N. La Monica, *J. Virol.*, 1995, **69**, 7461.
49. W. Gibson, A. R. Welch and M. R. T. Hall, *Perspectives in Drug Discovery and Design*, 1994, **2**, 413.
50. M. Gao, L. Matusick-Kumar, W. Hurlburt, S. F. DiTusa, W. W. Newcomb, J. C. Brown, P. J. McCann III, I. Deckman and R. J. Colonno, *J. Virol.*, 1994, **68**, 3702.
51. G. Donaghy and R. Jupp, *J. Virol.*, 1995, **69**, 1265.
52. N. J. Tigue, P. J. Matharu, N. A. Roberts, J. S. Mills, J. Kay and R. Jupp, *J. Virol.*, 1996, **70**, 4136.
53. C. L. Dilanni, D. A. Drier, I. C. Deckman, P. J. McCann III, F. Liu, B. Roizman, R. J. Colonno and M. G. Cordingley, *J. Biol. Chem.*, 1993, **268**, 2048.
54. J. T. Stevens, C. Mapelli, J. Tsao, M. Hail, D. R. O'Boyle II, S. P. Weinheimer and C. L. Dilanni, *European Journal of Biochemistry,* 1994, **226**, 361.
55. A. R. Welch, L. M. McNally, M. R. T. Hall and W. Gibson, *J. Virol.*, 1993, **67**, 7360.
56. C. L. Dilanni, C. Mapelli, D. A. Drier, J. Tsao, S. Natarajan, D. Riexinger, S. M. Festin, M. Bolgar, G. Yamanaka, S. P. Weinheimer, C. A. Meyers, R. J. Colonno and M. G. Cordingley, *J. Biol. Chem.*, 1993, **268**, 25449.
57. V. V. Sardana, J. A. Wolfgang, C. A. Veloski, W. J. Long, K. LeGrow, B. Wolanski, E. A. Emini and R. L. LaFemina, *J. Biol. Chem.*, 1994, **269**, 14337.
58. E. Z. Baum, G. A. Bebernitz, J. D. Hulmes, V. P. Muzithras, T. R. Jones and Y. Gluzman, *J. Virol.*, 1993, **67**, 497.
59. B. C. Holwerda, A. J. Wittwer, K. L. Duffin, C. Smith, M. V. Toth, L. S. Carr, R. C. Wiegand and M. L. Bryant, *J. Biol.Chem.*, 1994, **269**, 25911.
60. D. R. O'Boyle II, K. Wager-Smith, J. T. Stevens III, and S. P. Weinheimer, *J. Biol. Chem.*, 1995, **270**, 4753.
61. P. L. Darke, E. Chen, D. L. Hall, M. K. Sardana, C. A. Veloski, R. L. LaFemina, J. A. Shafer and L. C. Kuo, *J. Biol. Chem.*, 1994, **269**, 18708.

62. C. Pinko, S. A. Margosiak, D. Vanderpool, J. C. Gutowski, B. Condon and C. -C. Kan, *J. Biol. Chem*, 1995, **270**, 23634.
63. B. K. Handa, E. Keech, E. A. Conway, A. V. Broadhurst and A. J. Ritchie, *Antiviral Chemistry and Chemotherapy*, 1995, **6**, 255.
64. I. D. Kuntz, *Science*, 1992, **257**, 1078.

21
Design, Synthesis and Testing of Inhibitors of Viral Cysteine Proteinases

Sven Frormann, Yanting Huang, Manjinder Lall, Christopher Lowe and John C. Vederas
DEPARTMENT OF CHEMISTRY, UNIVERSITY OF ALBERTA, EDMONTON, ALBERTA, CANADA T6G 2G2

1 INTRODUCTION

1.1 Picornaviruses and Their 3C Proteinases as Potential Therapeutic Targets

Picornaviruses are important human and animal pathogens whose family includes such members as poliovirus, rhinovirus, hepatitis A virus, encephalomyocarditis virus, and foot and mouth disease virus.[1-4] They possess a small, positive strand RNA genome that functions as messenger RNA to ultimately produce a single 250 kilodalton protein. Within this protein sequence are domains corresponding to proteinases that are necessary to cut the parent into the smaller fragments which are required for the virus. In some cases, the protein is initially cleaved by the 2A proteinase to generate two fragments that are then further processed by the 3C proteinase into the mature proteins (Figure 1). In hepatitis A virus (HAV), the 3C proteinase does the first cleavages.[4] In vitro work indicates that the 3C proteinases are active while still part of precursor polyprotein,[5] and the enzyme may cut itself out of the precursor at an early stage of processing. Studies in a number of laboratories have shown that the activity of the 3C proteinase is critical for viral maturation and infectivity.[1-3] Hence it presents an attractive target for development of new antiviral therapeutic agents. This is especially true for picornaviruses such as rhinovirus where the variation in the coat structure leads to a large number of serotypes and makes vaccine development problematic. Work in our laboratories, done in collaboration with the groups of Professors Michael James and Bruce Malcolm (Biochemistry Department, University of Alberta), has focussed on mechanism-based inhibition of hepatitis A virus (HAV) 3C proteinase.

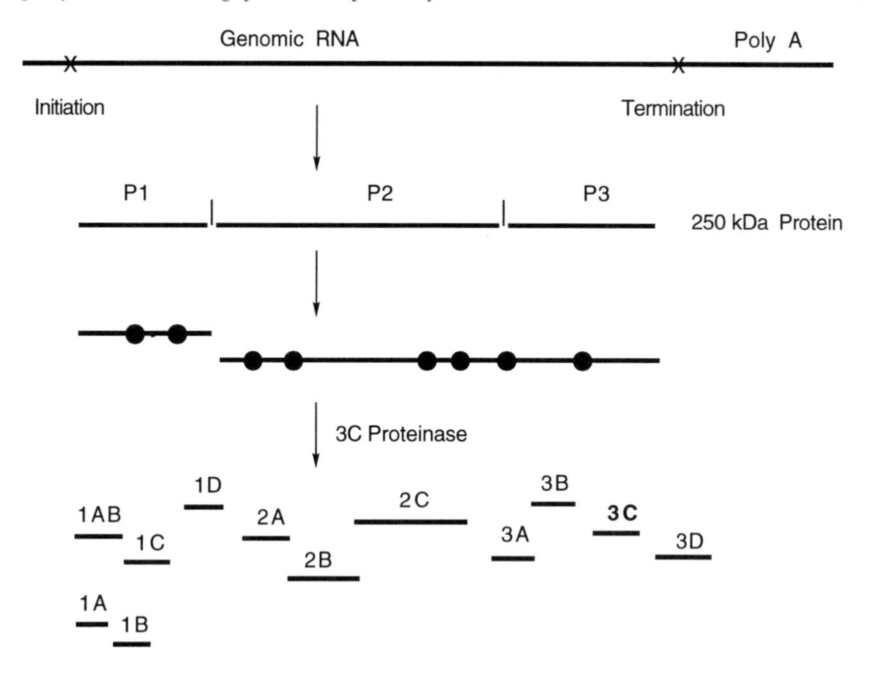

Figure 1 *Generation of Picornaviral Proteins from a 250 KDa Precursor*

1.2 Properties of Hepatitis A (HAV) 3C Proteinase

The HAV 3C enzyme is typical of the picornaviral family of 3C proteinases. Detailed examination of its substrate specificity using synthetic peptides that correspond to the 2B/2C and 2C/3A junctions in the large precursor protein showed that the P_4 to P_2' amino acids are necessary for maximal efficiency of cleavage.[6] In addition, these studies demonstrated that a larger hydrophobic amino acid residue (e.g. leucine) at P_4 is especially favorable, and that there is high specificity for a glutamine residue at the P_1 cleavage site (Figure 2). Although the nitrogens of the glutamine can be alkylated (e.g. N,N-dimethylglutamine is 96% as effective), alterations of the chain length or introduction of charge drastically lowers the recognition by the HAV 3C proteinase.[6]

The wild type HAV 3C enzyme is a cysteine proteinase of 219 amino acids with a molecular weight of 24 kilodaltons and exists as an active monomer. For ideal peptide substrates mimicking the 2B/2C junction, the k_{cat} is typically about 1.8 sec^{-1} with an approximate K_m of 2.1 mM at a pH of 7.5. Attempts to crystallize the wild type HAV 3C for structural studies were unsuccessful due to the presence of two cysteines in the molecule (at position 24 and at the active site position 172), but site-specific mutagenesis generated a C24S-C172A mutant which crystallized and whose structure could thus be

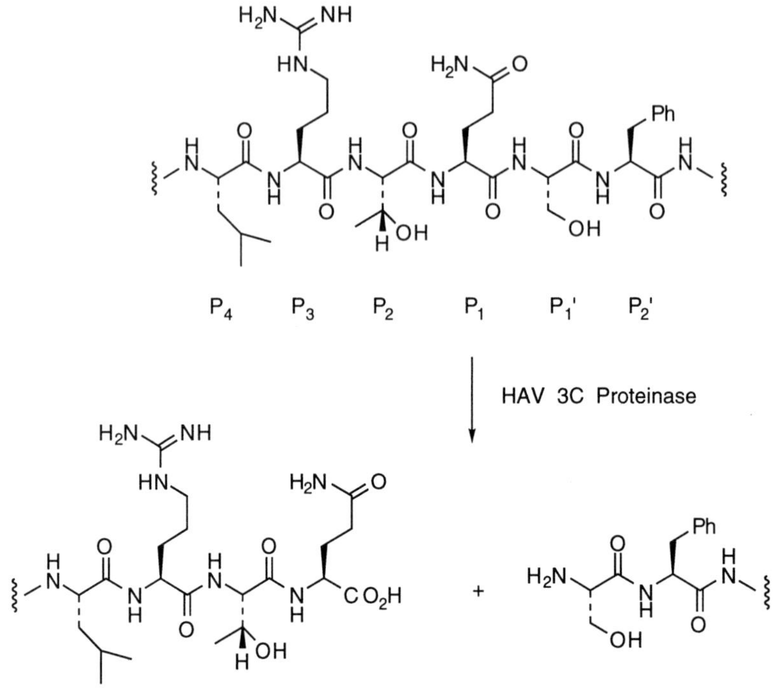

P4 P3 P2 P1 P1' P2'

HAV 3C Proteinase

Figure 2 *A Preferred Cleavage Site of HAV 3C Proteinase*

obtained at 2.3 Å resolution.[7] At about the same time, the group at Agouron Pharmaceuticals reported the crystal structure of the human rhinovirus-14 (HRV) 3C proteinase.[8] More recently, the Alberta group has obtained crystal structures of active HAV 3C mutants wherein only the external cysteine-24 has been replaced (unpublished). These structural studies show that topologies of the HAV and HRV 3C proteinases resemble β-barrel fold serine proteinases such as chymotrypsin and trypsin, respectively, rather than the papain family of cysteine proteinases. The active site cysteine is in close proximity to a histidine residue (His-44 in HAV 3C) which is part of a catalytic diad, or possibly triad.[1] His-191 is believed to be involved in hydrogen bond formation with the distal glutamine carbonyl at P_1 in the substrates and is probably critical for recognition of this amino acid residue.

1.3 Objectives and Strategy for HAV 3C Inhibition

The initial goal of this study was to generate small (≤ four amino acid residues) peptidic inhibitors bearing fairly selective thiol-reactive groups for potent inhibition of HAV 3C proteinase. Peptide-based compounds are likely to show rapid metabolism and poor bioavailability, and will therefore generally not be suitable drug candidates. However, a key objective of our work was to occupy the active site of HAV 3C with a potent, possibly irreversible, inhibitor for crystallographic studies. The resulting information about the inhibitor conformation in the enzyme active site could provide valuable assistance for design of second generation non-peptidic or peptidomimetic inactivators. It would also help guide molecular modeling efforts for drug design and provide a better understanding of enzyme mechanism. In addition, development of new warheads for highly selective reaction with the active site thiol should be useful for production of inhibitors for other types of cysteine proteases.[9,10]

2. RESULTS AND DISCUSSION

2.1 Peptide Aldehyde Inhibitors of HAV 3C

Peptide aldehydes in which the P_1 residue is replaced with an aldehyde are well-established inhibitors of cysteine proteinases as well as serine proteinases.[11-15] Hence, based on the substrate specificity studies,[6] it appeared that a short modified tetrapeptide such as Ac-Leu-Ala-Ala-(glutaminal) having a C-terminal aldehyde could be a potent inhibitor of HAV 3C proteinase. However, it soon became apparent that glutamine aldehydes have a strong tendency to cyclize to internal hemi-acetals or hemi-aminals. This has since been confirmed in independent work on such compounds by researchers at Lilly, who showed that glutamine aldehdyes exist in tautomeric equilibrium with their six-membered hemi-aminals resulting from intramolecular cyclization of the side chain nitrogen onto the aldehyde carbon.[17] To circumvent such cyclizations, we synthesized the corresponding N,N-dimethylglutamine aldehyde (Figure 3), both in unlabelled form and bearing ^{13}C-label at the aldehyde carbon, and this was reported last year.[18]

The tetrapeptide aldehyde, Ac-Leu-Ala-Ala-(N,N-dimethylglutaminal), is a reversible slow-binding inhibitor of HAV 3C proteinase with a K_i^* of 42 nM.[18] The half life of the tight complex is 3 hours. It is about 50 fold less potent as an inhibitor of the highly homologous human rhinovirus (HRV) (strain 14) 3C proteinase whose substrate specificity is somewhat different. Although other aldehydes can interact with the HAV 3C enzyme, the inhibition is highly specific, and the derivative with only the modified parent amino acid, Ac-(N,N-dimethylglutaminal) is a simple competitive inhibitor (not slow-binding) with a K_i of only 2.5 mM (i.e. about 9000 times less effective). The synthetic intermediate ethyl thioester of Ac-Leu-Ala-Ala-(N,N-dimethylglutamine) shows no significant inhibition. Replacement of the leucine residue at P_4 in the key inhibitor with an alanine results in about a five-fold loss of efficiency against HAV 3C.[18] Very recently the Lilly group has demonstrated that a peptide aldehyde inhibits rhinovirus replication.[19]

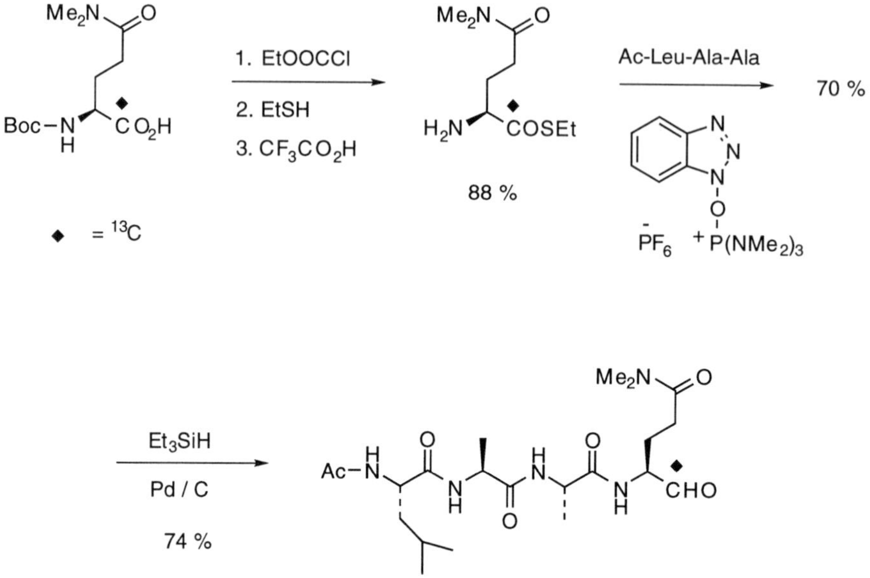

Me2N O
Boc–N(H) CO2H
1. EtOOCCl
2. EtSH
3. CF3CO2H
Me2N O
H2N COSEt
88 %
◆ = 13C
Ac-Leu-Ala-Ala
PF6 + P(NMe2)3
70 %
Et3SiH
Pd / C
74 %
Me2N O
Ac–N(H) ... CHO

Figure 3 *Synthesis of Tetrapeptide Aldehyde for HAV 3C Inhibition*

Electrospray mass spectrometry of HAV 3C inhibited with Ac-Leu-Ala-Ala-(N,N-dimethylglutaminal) clearly shows the appearance of another peak with a mass corresponding to that of the enzyme-inhibitor complex.[18] To confirm that the expected[13] hemithioacetal adduct was indeed being formed, HMQC NMR spectra[20] of the enzyme-inhibitor complex were acquired on a Varian 600 MHz spectrometer. It immediately became apparent that the situation is complicated by the dithiothreitol-containing buffer, which is used to prevent oxidation of the active site sulfhydryl of HAV 3C. In addition, the tetrapeptide aldehyde inhibitor exists primarily as a hydrate (ca. 95 %) in aqueous solution (Figure 4). Dithiothreitol forms a pair of diastereomeric adducts with the labelled inhibitor, and these as well as the aldehyde hydrate are visible in the spectra even after dialysis. However, the stereospecifically-formed enzyme-inhibitor thioacetal can be clearly seen from the correlation signal at approximately 80 ppm on the carbon axis and 5.7 ppm on the proton axis.[18] Attempts to crystallize such enzyme-inhibitor adducts or to soak peptide aldehydes into pre-formed HAV 3C crystals for structural analysis have not been successful. The probable metabolic instability of peptide aldehydes and poor bioavailablity combined with our desire for an enzyme-inhibitor adduct suitable for crystallographic analysis led to investigation of other classes of proteinase inhibitors.

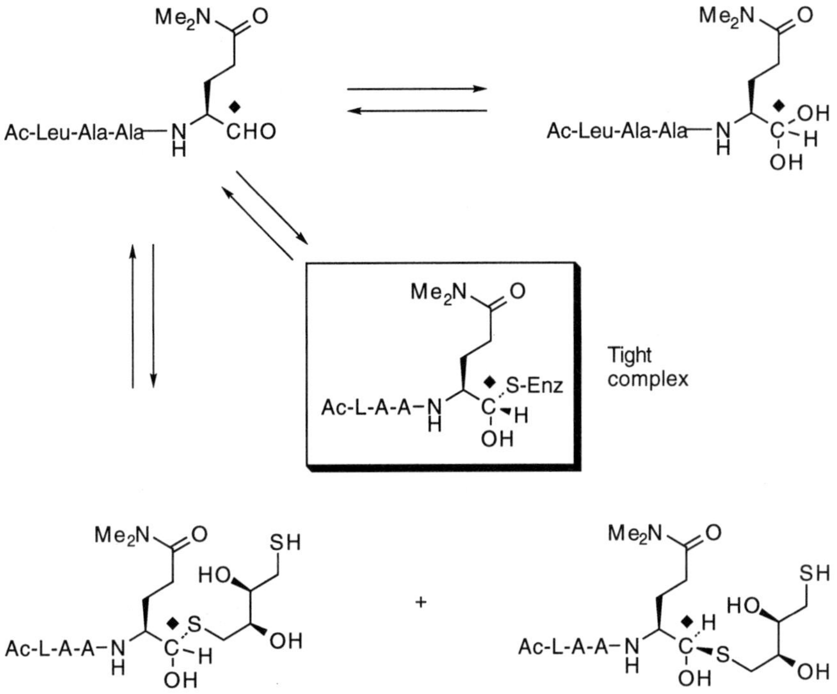

Figure 4 *Labelled Species Observed by NMR (HMQC) in Aqueous Buffer Containing Dithiothreitol and HAV 3C*

2.2 Fluoromethyl Ketone Inhibitors of HAV 3C

Peptidyl fluoromethyl ketones have been shown to be highly potent proteinase inhibitors[21,22] and also work *in vivo* in a several systems.[23-25] Since such compounds are generally more stable than peptide aldehydes, it appeared that attachment of a fluoromethyl ketone functionality to the C-terminus of the Ac-Leu-Ala-Ala-(N,N-dimethylglutamine) tetrapetide could afford a potent inactivator of the HAV 3C proteinase. The synthesis of this target is shown below (Figure 5), and utilizes a versatile strategy in which the terminal amino acid analogue is synthesized via fluoromalonate condensation[26] and coupled in its reduced form (fluoro alcohol) to the Ac-Leu-Ala-Ala tripeptide before reoxidation with Dess-Martin periodinane.[27] Attempts to deprotect the amino acid analogue and directly couple the resulting fluoromethyl ketone of N,N-dimethylglutamine without prior reduction were unsuccessful because of the highly electrophilic nature of the carbonyl group.

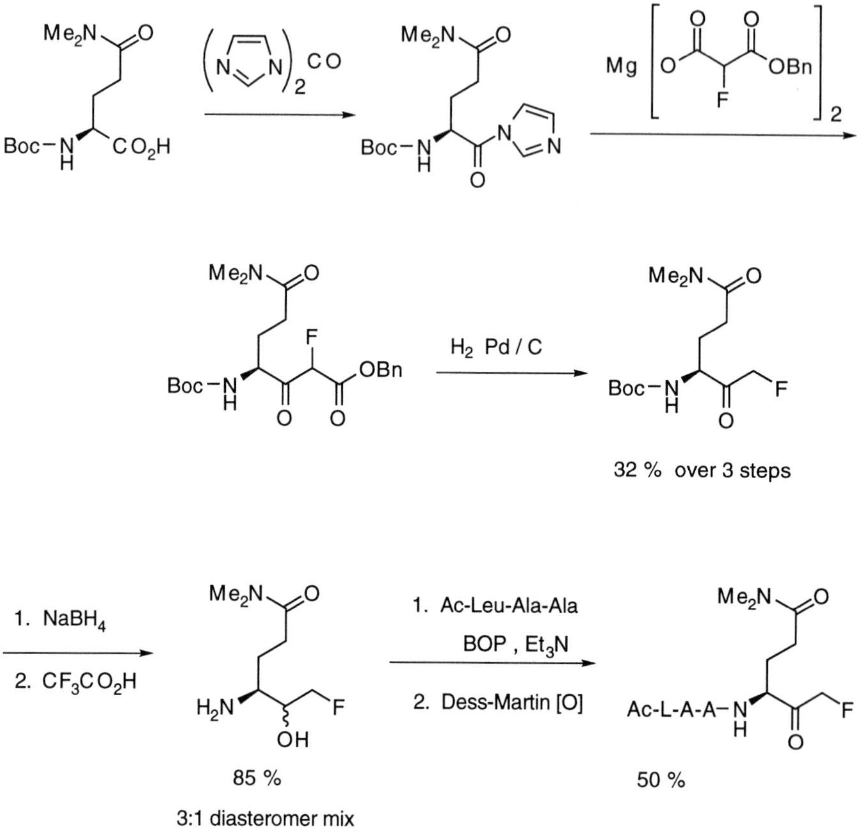

Figure 5 *Synthesis of Tetrapeptide Fluoromethyl Ketone for HAV 3C Inhibition Studies*

The tetrapeptide fluoro*alcohol* precursor displays no significant inhibition of the HAV 3C enzyme over 3 hours at concentrations of 20 μM. However, the corresponding fluoroketone tetrapeptide shows irreversible inactivation of this proteinase with a second order rate constant ($k_{obs}/[I]$) = 3.3x 10^2 M^{-1} s^{-1} ([E]=0.07μM, [I]=1.0μM). The inhibition of HAV 3C (a cysteine protease with a serine proteinase three dimensional structure) is ca 40 fold slower than corresponding inactivation of cathepsin B (a papain-type cysteine proteinase) by a dipeptide fluoromethyl ketone ($k_{obs}/[I]$ = 1.4 x 10^4 M^{-1} s^{-1}).[28] However, attack on HAV 3C is much faster than corresponding inactivation of the structurally-related serine protease, α-chymotrypsin ($k_{obs}/[I]$ = 1.7 M^{-1} s^{-1}).[21] Interestingly, the fluoroketone shows significant antiviral activity in preliminary cell culture tests.[29]

Studies with peptidyl fluoromethyl ketones and their interactions with other cysteine and serine proteinases demonstrate that the final reaction outcomes are different for the two enzyme types. The serine proteinase, α-chymotrypsin, releases fluoride (observed by ^{19}F NMR spectrometry) upon irreversible inactivation by fluoromethyl ketones and is alkylated on an active site histidine residue.[21] The mechanism is probably analogous to that with a α-chloroethyl ketone, namely attack of the serine hydroxyl on the haloketone carbonyl, internal displacement of halide to form an epoxide, followed by oxirane ring opening at the less hindered position via attack of the histidine nitrogen (Figure 6).[30] This results in an N-alkylated acetal as shown below. However, crystallographic studies of the papain-like proteinase cruzain[23] inhibited by a flurometyl ketone and NMR studies of papain inactivated by a chloromethyl ketone[31] clearly demonstrate that these cysteine proteinases have the active site sulfhydryl replacing the halogen to form an α-keto sulfide. This could in principle occur by a mechanism similar to that proposed for chymotrypsin and α-chloroethyl ketone,[30] namely attack of sulfhydryl on the carbonyl and generation of an epoxide, which would then have to be followed by sulfur migration. However, direct halogen displacement by the thiolate, which is much more nucleophilic than the serine hydroxyl, is likely.

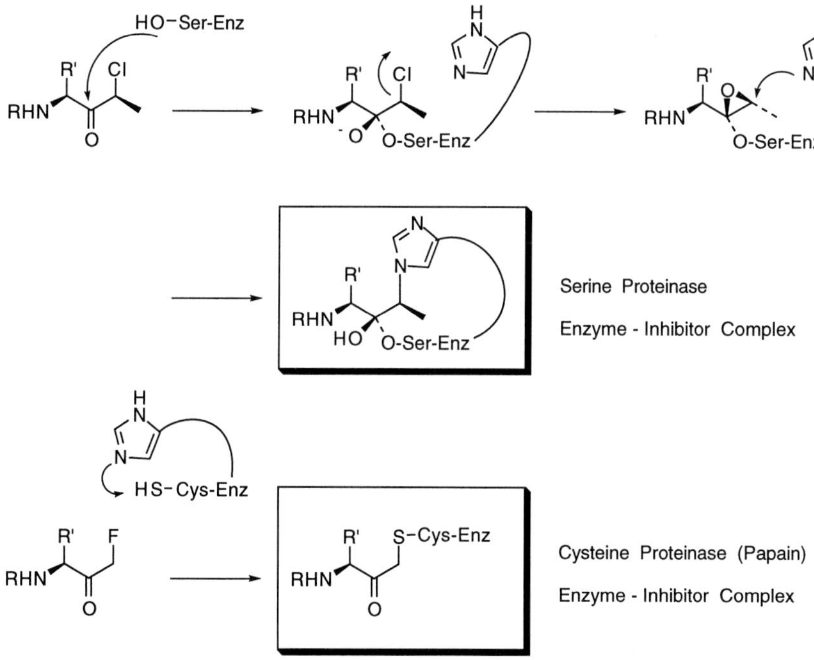

Figure 6 *Modes of Inhibition of Serine and Cysteine Proteinases by Haloalkyl Ketones*

Since HAV 3C possesses a geometry much more similar to serine proteinases such as chymotrypsin than to papain, we examined the inactivation process using the tetrapeptide fluoromethyl ketone labelled with ^{13}C at the methylene carbon. This compound was synthesized using a variation of the halogen exchange method of Kolb (Figure 7).[32] Model compounds having a histidine or a sulfide in place of the fluorine were also generated to assist in ascertaining whether alkylation of HAV 3C occurs with an active site nitrogen (histidine-44) or sulfur (cysteine-172) nucleophile. The chemical shifts of the methylene carbons adjacent to the heteroatoms in the model compounds are 51 ppm for the imidazoloketone and 38 ppm for the (alkylthio)ketone; the latter value is in good agreement with literature values for such sulfides.[31] The methylene hydrogens adjacent to

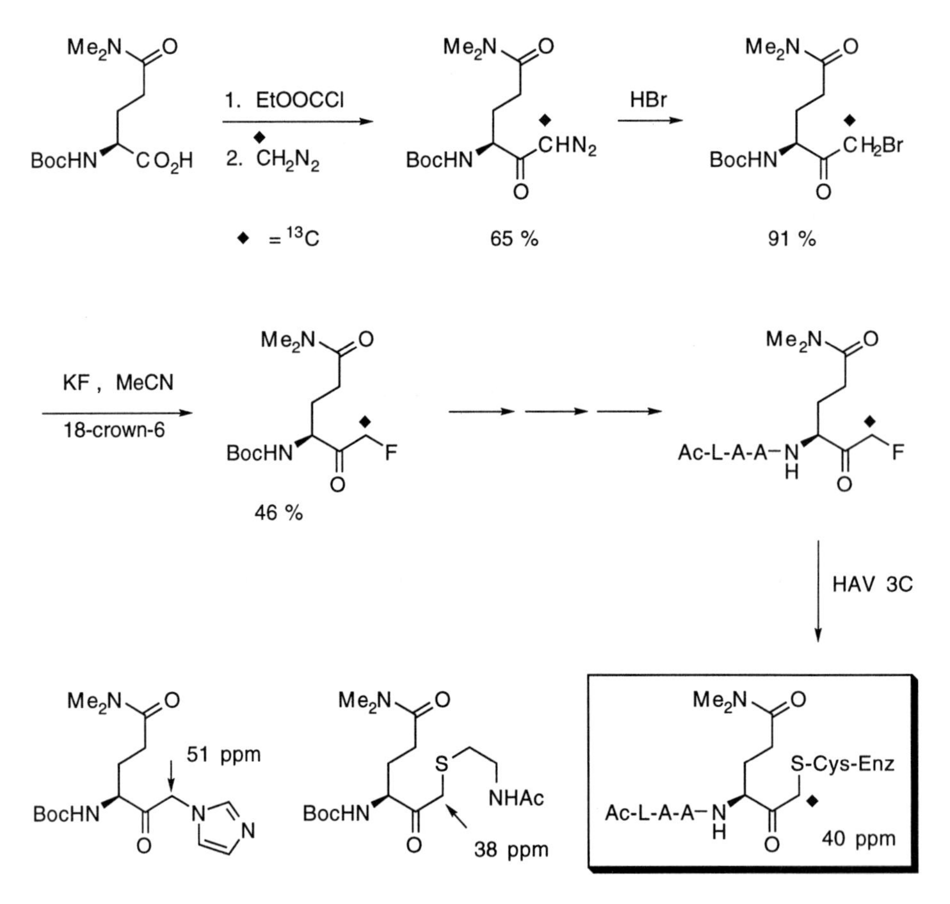

Figure 7 *Synthesis of ^{13}C-Labelled Fluoromethyl Ketone, Its Interaction with HAV 3C, and Chemical Shifts of Model Compounds*

the alkylated heteroatoms in both model compounds exchange rapidly in D_2O and consequently spectra were acquired in water. Reaction of HAV 3C with the ^{13}C-labelled tetrapeptide fluoromethyl ketone rapidly releases fluoride ion (^{19}F NMR chemical shift -120 ppm)[21] and produces an irreversible adduct whose mass spectrum is shifted to higher mass by the expected 471 Da. The ^{13}C-NMR spectrum of the enzyme-inhibitor complex displays a new peak at 40 ppm with a line width at half height of 78 Hz (corrected for line broadening), clearly indicating the formation of a (alkylthio)ketone. Hence, despite an active site geometry which resembles serine proteinases, the active site sulfhydryl of HAV 3C dominates the inactivation chemistry with the fluoromethyl ketone and the process is analogous to that of other cysteine proteinases.

2.3 Azathiopeptide Analogues as Potential HAV 3C Inhibitors

As part of an effort to investigate new types of potential warheads for selective cysteine protease inactivation, especially HAV 3C, three types of modification of the P_1 residue were examined in which the α-carbon is replaced by a nitrogen with an adjacent sulfonamide or sulfenamide moiety (Figure 8). Synthetic investigations on sulfonamide

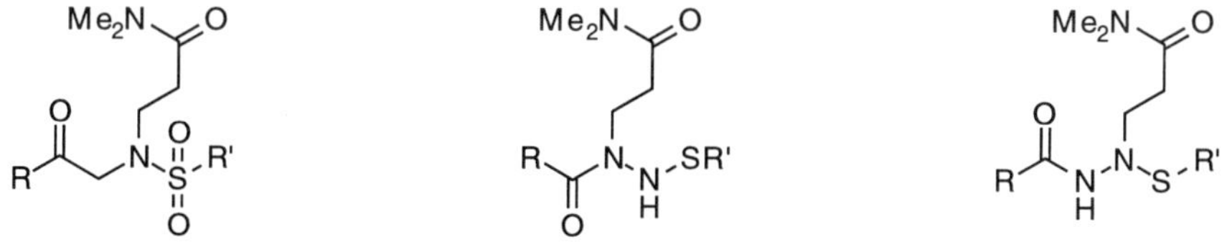

Figure 8 *Azathiopeptide Targets for Potential Cysteine Proteinase Inhibition*

transition state isosteres have previously focussed on their potential inhibition of aspartic proteases such as HIV protease and their use to generate catalytic antibodies.[33,34] To determine whether such compounds could inhibit HAV 3C, we assembled a tetrapeptide sulfonamide analogue as shown below (Figure 9). The key step involves condensation of the anion derived from the N,N-dimethylamide of N'-(methylsulfonyl)-β-alanine with the Cbz-protected bromomethyl ketone derivative of L-alanine (synthesized via the diazoketone analogously to the procedure in Figure 7). Deprotection of the resulting dipeptide analogue and coupling to Ac-Leu-Ala generates the desired tetrapeptide mimic. This compound proved to be a competitive inhibitor of HAV 3C with an IC_{50} of about 75 μM, which is significantly lower than the K_m for an ideal hexapeptide substrate (2.1 mM), but not potent enough to warrant extensive study in view of the aldehyde and fluoromethyl ketone results.

Incorporation of hydrazino functionality into peptide backbones is well-established for generation of proteinase inhibitors[35-39] and has recently seen renewed interest.[40-43] The amide character of sulfenamides coupled with their tendency to be cleaved by thiols under acidic conditions with concomitant formation of disulfides[45,46] suggested that

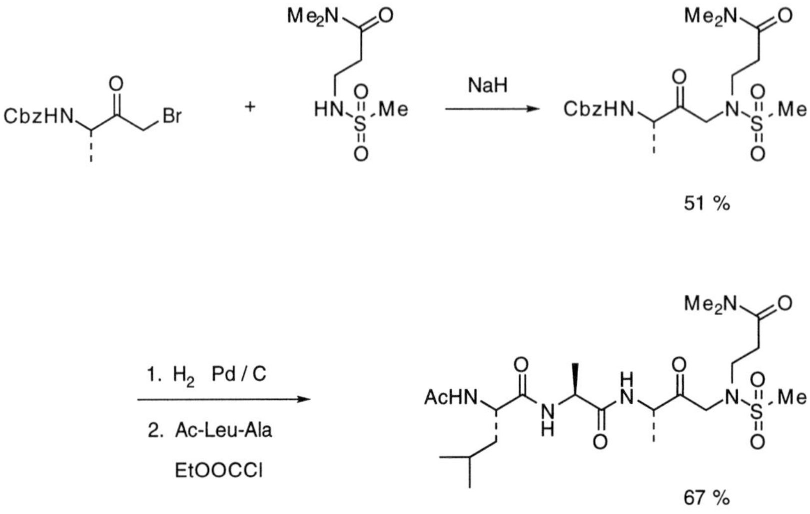

Figure 9 *Synthesis of Sulfonamide Analogue for Inhibition of HAV 3C*

replacement of the P_1 carbonyl with the sulfur of a sulfenamide could generate an irreversible inactivator of a cysteine proteinase. Hence, an aza analogue of N,N-dimethyl-glutamine was synthesized as shown below (Figure 10) through Michael addition of hydrazine to N,N-dimethylacrylamide followed by protection of the more nucleophilic secondary nitrogen with benzyl chloroformate and condensation to the tripeptide, Ac-Leu-Ala-Ala. Hydrogenolytic deprotection frees the central nitrogen which then reacts with o-nitrophenylsulfenyl chloride to produce the target compound **A**. This molecule shows no significant inhibition of HAV 3C proteinase despite the presence of the P_4-P_1 backbone. Apparently the aromatic moiety is too bulky for the enzyme site which binds to the P_1' residue. Attempts to replace the sulfur substituent with other groups, such as methyl or trifluoroethyl gave compounds with very poor stability in water.

Alteration of the synthetic sequence by direct coupling of the tripeptide to the Michael adduct followed with subsequent sulfenamide formation gives a structural isomer **B** wherein the hydrazo nitrogen is placed at the nominal position of the P_1 carbonyl and the sulfur occupies the P_1' nitrogen site. This compound does inhibit HAV 3C with an IC_{50} of ca 100 μM and eventual formation of an enzyme-nitrophenyl disulfide, as demonstrated by electrospray mass spectra. The relatively weak inhibition by **B** may again be due to unfavorable interactions of the aromatic portion at the P_1' binding pocket in the enzyme active site. Although in the present case these sulfenamides are not as potent as the peptidic aldehydes or fluoromethyl ketones, they may be useful structural types for inhibition of other cysteine proteinases.

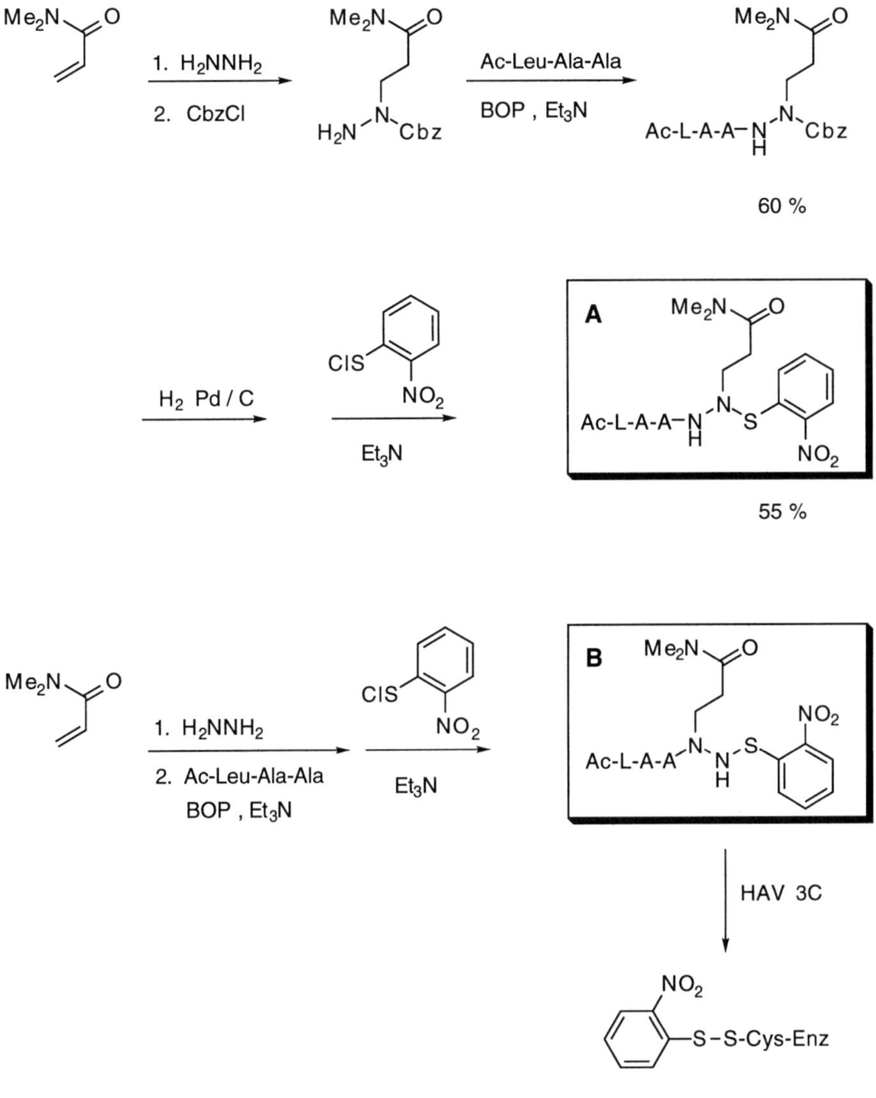

Me$_2$N O

1. H$_2$NNH$_2$

2. CbzCl

Me$_2$N O

H$_2$N N Cbz

Ac-Leu-Ala-Ala

BOP , Et$_3$N

Me$_2$N O

Ac-L-A-A-N N Cbz H

60 %

H$_2$ Pd / C

ClS NO$_2$

Et$_3$N

A Me$_2$N O

Ac-L-A-A-N N S H NO$_2$

55 %

Me$_2$N O

1. H$_2$NNH$_2$

2. Ac-Leu-Ala-Ala BOP , Et$_3$N

ClS NO$_2$

Et$_3$N

B Me$_2$N O

NO$_2$

Ac-L-A-A N N S H

HAV 3C

NO$_2$

S-S-Cys-Enz

Figure 10 *Syntheses of Hydrazino Sulfenamides for HAV 3C Inhibition*

2.4 β-Lactone-Containing Peptides as Potential HAV 3C Inhibitors

β-Lactones occur naturally in a variety of organisms and many possess potent biological activity.[47] Our earlier studies have shown that α-amino-β-lactones are useful intermediates for synthesis of stereochemically pure β-substituted α-amino acids[47-50] and are easily available by cyclization of the appropriately protected β-hydroxy-α-amino acids.[49-51] The ability of thiols to open β-lactones by nucleophilic attack at either the carbonyl or at the β-position[49-50] suggests that cysteine proteinases could be irreversibly inactivated by β-lactones having correct substiutents and stereochemistry. Since α-amino-β-lactones bearing a β-substituent display improved stability in aqueous media, initial investigations focussed on synthesis of tetrapeptide analogues bearing threonine or allo-threonine β-lactone functionality at the C-terminus (Figure 11). Although such compounds are missing the glutamine side chain important for recognition, their easy synthetic accessibility make them ideal as probes for preliminary studies of HAV 3C inhibition.

The L- and D-threonine-β-lactone peptides were synthesized by the general approach that we had developed earlier for the naturally occurring antibiotic obafluorin and its analogues.[51] In contrast to serine derivatives,[49,50] cyclization of β-alkyl substituted β-hydroxy-α-amino acids (e.g. threonine) requires carboxyl group activation to avoid decarboxylative elimination.[52] In addition, a nitrogen protecting group that does not have a carbonyl directly attached is essential to circumvent competing azlactone formation.[52] The *ortho*-nitrophenylsulfenyl group fulfills this function and can be readily exchanged for an N-acyl group (e.g. the P_4-P_2 peptide) after β-lactone formation. Hence, the known[51] L-threonine β-lactone *ortho*-nitrophenylsulfenamide was deprotected and condensed with Ac-Leu-Ala-Ala as shown in Figure 11. The D-threonine derivative can be made similarly. However the analogous L-allo-threonine and D-allo-threonine β-lactone salts could not be N-acylated with the tripeptide in the same fashion, apparently because of the increased sensitivity of the intermediate free amines. Therefore a trans-acylation approach developed by Rao et al.[53] was employed to make these trans-substituted lactone derivatives.

Inhibition studies with the four diastereomeric β-lactones show that none are irreversible inactivators of HAV 3C proteinase under the assay conditions, but the D-threonine analogue displays competitive inhibition with an IC_{50} of ca 200 μM, whereas the corresponding L-threonine derivative has an IC_{50} of 500 μM. These compounds are surprisingly potent given that the glutamine side chain necessary for recognition of the P_1 residue is absent. Analogous studies with related β-lactone peptides and the classical cysteine protease, papain, display similar results. Although the kinetic behaviour of some of these β-lactones with cysteine is complex and is currently under study, it is clear that the desired sulfur alkylation does not occur. To circumvent this difficulty we have recently targeted amino acid-derived β-lactones possessing the correct side chain with the reactive oxetanone functionality replacing the carboxyl group further toward the P' side.

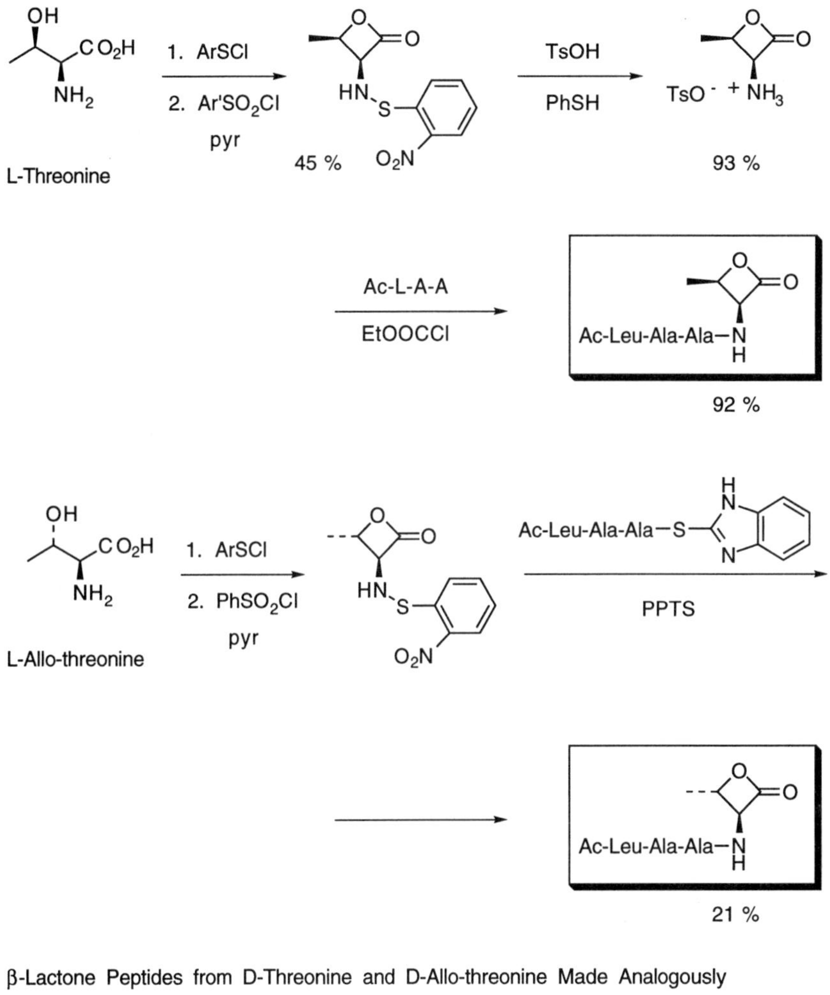

L-Threonine

1. ArSCl
2. Ar'SO$_2$Cl
pyr

45 %

TsOH
PhSH

TsO$^-$ $^+$NH$_3$

93 %

Ac-L-A-A
EtOOCCl

Ac-Leu-Ala-Ala—NH

92 %

L-Allo-threonine

1. ArSCl
2. PhSO$_2$Cl
pyr

Ac-Leu-Ala-Ala—S—(benzimidazol-2-yl)
PPTS

Ac-Leu-Ala-Ala—NH

21 %

β-Lactone Peptides from D-Threonine and D-Allo-threonine Made Analogously

Figure 11 *Syntheses of β-Lactone Peptides*

3. SUMMARY AND CONCLUSIONS

We have synthesized a series of tetrapeptide substrate analogues bearing aldehyde, fluoromethylketone, sulfonamide, hydrazasulfenamide, and β-lactone functionalities for inhibition of the picornaviral 3C cysteine proteinases, in particular, the hepatitis A virus (HAV) 3C enzyme. This enzyme, which prefers glutamine in the P_1 site of its substrate, has an active site thiol nucleophile and proximal histidine as part of a catalytic diad or triad, but the overall geometry resembles that of the serine protease, chymotrypsin. Nevertheless, the inhibition chemistry is dominated by the active site sulfhydryl and appears to be analogous to that of the papain family of cysteine proteinases. The aldehyde, Ac-Leu-Ala-Ala-(N,N-dimethylglutaminal), is a potent (42 nM) specific, slow-binding inhibitor of this enzyme, and is shown by mass spectral and NMR studies (on ^{13}C-labelled inhibitor) to stereospecifically form a hemithioacetal in the active site.[18] The corresponding fluromethyl ketone shows irreversible inactivation of this proteinase, and ^{13}C NMR studies using ^{13}C-labelled inhibitor demonstrate that this occurs by formation of a thiomethyl ketone with the active site cysteine-172. This fluoromethyl ketone peptide is very effective at inhibiting viral replication in cell culture at low concentrations. Although the tetrapeptide sulfonamide analogue is not a good inhibitor of HAV 3C, an aza-peptide sulfenamide derivative irreversibly inactivates the target enzyme by mixed disulfide formation with the active site sulfhydryl group. The work establishes that potent inhibitors of HAV 3C with various warheads can be readily synthesized, and has provided compounds for on-going crystallographic studies to determine preferred conformation of peptide/protein substrates in the enzyme active site. Current studies on modified peptidomimetic inhibitors will be reported later.

Acknowledgement

The authors gratefully acknowledge the on-going contributions of Professors Bruce Malcolm and Michael James and their research groups in the Biochemistry Department of the University of Alberta. These investigations were supported by the Natural Sciences and Engineering Research Council of Canada and by the U.S. National Institutes of Health (Grant AI 38249).

4. REFERENCES

1. B. A. Malcolm, *Protein Sci.*, 1995, **4**, 1439-1445.
2. H. G. Kräusslich and E. Wimmer, *Annu. Rev. Biochem.*, 1988, **57**, 701-754.
3. W. G. Dougherty and B. L. Semler, *Microbiol. Rev.*, 1993, *57*, 781-822.
4. T. Schultheiss, W. Sommergruber, Y. Kusov and V. Gauss-Müller, *J. Virology* 1995, **69**, 1727-1733.
5. X.-Y. Jia, E. Ehrenfeld and D. F. Summers, *J. Virol.*, 1991, **65**, 2595.
6. D. A. Jewell, W. Swietnicki, B. M. Dunn and B. A. Malcolm, *Biochemistry*, 1992, **31**, 7862-7869.

7. M. Allaire, M. Chernaia, B. A. Malcolm and M. N. G. James, *Nature (London)*, 1994, **369**, 72-77
8. D. A. Matthews, W. W. Smith, R. A. Ferre, B. Condon, G. Budahazi, W. Sisson, J. E. Villafranca, C. A. Janson, H. E. McElroy, C. L. Gribskow, and S. Worland, *Cell*, 1994, **77**, 761-771.
9. S. Mehdi, *Bioorg. Chem.*, 1993, **21**, 249-259.
10. E. Shaw, *Adv. Enzymol.*, 1990, 271-346.
11. J. O. Westerik and R. Wolfenden, *J. Biol. Chem.*, 1972, **247**, 8195-8197.
12. R. C. Thompson, *Biochemistry*, 1973, **12**, 47-51.
13. N. E. Mackenzie, S. K. Grant, A. I. Scott and P. G. Malthouse, *Biochemistry*, 1986, **25**, 2293-2298.
14. R. P. Hanzlik, S. P. Jacober and J. Zygmunt, *Biochim. Biophys. Acta*, 1991, **1073**, 33-42.
15. M. D. Mullican, D. J. Laufer, R. J. Gillespie, S. S. Matharau, D. Kay, G. M. Porritt, P. L. Evans, J. M. C. Golec, M. A. Murcko, Y.-P. Luong, S. A. Raybuck and D. J. Livingston, *Bioorg. Med. Chem. Lett.*, 1994, **4**, 2359-2364.
16. E. Dufour, A. C. Storer and R. Ménard, *Biochemistry*, 1995, **34**, 9136-9143.
17. S. W. Kaldor, M. Hammond, B. A. Dressman, J. M. Labus, F. W. Chadwell, A. D. Kline and B. A. Hunt, *Bioorg. Med. Chem. Lett.*, 1995, **5**, 2021-2026.
18. B. A. Malcolm, C. Lowe, S. Shechosky, R. T. McKay, C. C. Yang, V. S. Shah, R. J. Simon, J. C. Vederas and D. V. Santi, *Biochemistry*, 1995, **34**, 8172-8178.
19. B. A. Heinz, J. Tang, J. M. Labus, F. W. Chadwell, S. W. Kaldor and M. Hammond, *Antimicrob. Agents Chemother.*, 1996, **40**, 267-270.
20. M. F. Summers, L. G. Marzilli and A. Bax, *J. Am. Chem. Soc.*, 1986, **108**, 4285-4294.
21. B. Imperiali and R. H. Abeles, *Biochemistry*, 1986, **25**, 3760-3767.
22. B. Imperiali, *Advances in Biotechnological Processes*, 1988, **10**, 97-229.
23. M. E. McGrath, A. E. Eakin, J. C. Engel, J. H. McKerrow, C. S. Craik and R. J. Fletterick, *J. Mol. Biol.*, 1995, **247**, 251-259.
24. R. E. Esser, R. A. Angelo, M. D. Murphey, L. M. Watts, L. P. Thornburg, J. T. Palmer, J. W. Talhouk and R. E. Smith, *Arthritis Rheumatism*, 1994, **37**, 236-247.
25. J. K. Richer, W. G. Hunt, J. A. Sakanari and R. B. Grieve, *Exper. Parasitol.*, 1993, **76**, 221-231.
26. J. T. Palmer, *Eur. Pat. Appl. No. 91301237.3*, 1991, *Reagents and Methods for Stereospecific Fluoromethylation.*
27. R. J. Linderman and D. M. Graves, D. M., *Tetrahedron Lett.*, 1987, **28**, 4259-4262.
28. D. Rasnick, *Anal. Biochem.*, 1985, **149**, 461-465.

29. Unpublished results, Dr. Tina Schultheiss, National Institute of Allergies and Infectious Diseases, National Institutes of Health, Bethesda, MD.
30. K. Kreutter, A. C. U. Steinmetz, T.-C. Liang, M. Prorok, R. H. Abeles and D. Ringe, *Biochemistry* , 1994, **33**, 13792-13800.
31. V. J. Robinson, H. W. Pauls, P. J. Coles, R. A. Smith and A. Krantz, *Bioorg. Chem.*, 1992, **20**, 42-54.
32. M. Kolb, B. Neises and F. Gerhart, *Liebigs Ann. Chem.*, 1990, 1- 6.
33. W. J. Moree, L. C. Van Gent, G. Van der Marel and R. M. J. Liskamp, *Tetrahedron*, 1993, **49**, 1133-1150.
34. W. J. Moree, L. C. Van Gent, G. Van der Marel and R. M. J. Liskamp, *J. Org. Chem.*, 1995, **60**, 5157-5169.
35. A. S. Dutta and J. S. Morley, *J. Chem. Soc., Perkin Trans. 1*, 1975, 1712-1720.
36. C. P. Dorn, M. Zimmerman, S. S. Yang, E. C. Yurewicz, B. M. Ashe, R. Frankshun and H. J. Jones, *J. Med. Chem.*, 1977, **20**, 1464-1468.
37. A. S. Dutta, M. B. Giles and J. C. Williams, *J. Chem. Soc., Perkin Trans. 1*, 1986, 1655-1664.
38. A. S. Dutta, M. B. Giles, J. J. Gormley, J. C. Williams and E. J. Kusner, *J. Chem. Soc., Perkin Trans. 1*, 1987, 111-120.
39. U. Neumann, T. Steinmetzer, A. Barth and H. U. Demuth, *J. Enzyme Inhib.*, 1991, **4**, 213-226.
40. C. Giordano, R. Calabretta, C. Gallina, V. Consalvi and R. Scandurra, *Il Farmaco*, 1991, **46**, 1497-1516.
41. R. H. Abeles and J. Magrath, *J. Med. Chem.*, 1992, **35**, 4279-4283.
42. C. Giordano, R. Calabretta, C. Gallina, V. Consalvi, R. Scandurra, F. C. Noya and C. Franchini, *Eur. J. Med. Chem.*, 1993, **28**, 297-311.
43. E. Peisach, D. Casebier, S. L. Gallion, P. Furth, G. A. Petsko, J. C. Hogan Jr. and D. Ringe, *Science*, 1995, **269**, 66-69.
44. H. L. Sham, W. Rosenbrook, W. Kati, D. A. Betebenner, N. E. Wideburg, A. Saldivar, J. J. Plattner and D. W. Norbeck, *J. Chem. Soc., Perkin Trans. 1*, 1995, 1081-1082.
45. J. L. Kice and A. G. Kutateladze, *J. Org. Chem.*, 1992, **57**, 3298-3303.
46. Y. Pu, C. Lowe, M. Sailer and J. C. Vederas, *J. Org. Chem.*, 1994, **59**, 3642-3655.
47. For a review see: C. Lowe and J. C. Vederas, *Org. Prep. Proc. Int.*, 1995, **27**, 305-346.
48. E. S. Ratemi and J. C. Vederas, *Tetrahedron Lett.*, 1994, **35**, 7605-7608.
49. S. V. Pansare, G. Huyer, L. D. Arnold and J. C. Vederas, *Org. Syn.*, 1991, **70**, 1-9.
50. S. V. Pansare, L. D. Arnold and J. C. Vederas, *Org. Syn.*, 1991, **70**, 10-17.

51. Y. Pu, C. Lowe, M. Sailer, M. and J. C. Vederas, *J. Org. Chem.*, 1994, **59**, 3642-3655.

52. S. V. Pansare and J. C. Vederas, J.C., *J. Org. Chem.*, 1989, **54**, 2311-2316.

53. M. N. Rao, A. G. Holkar and N. R. Ayyangar, *J. Chem Soc., Chem Commun.*, 1991, 1007-1008.

22

Nucleosides with a Six-membered Carbohydrate Moiety: Structural Requirements for Antiviral Activity

P. Herdewijn

LABORATORY OF MEDICINAL CHEMISTRY, REGA INSTITUTE FOR MEDICAL RESEARCH, KATHOLIEKE UNIVERSITEIT LEUVEN, B-3000 LEUVEN, BELGIUM

Modifications of the sugar part of nucleosides has been the most successful strategy to discover antiviral compounds. The research on antiviral nucleosides with a six-membered carbohydrate moiety has been, however, far behind the research on biological active nucleosides with modified five-membered rings. This is due to the fact that six-membered rings are conformationally less flexible than five-membered rings and that conformational flexibility of a nucleoside is believed to be important for the metabolic activation and for the interaction with the target enzyme. With non-flexible molecules, a perfect steric and electric fit is necessary between enzyme and substrate and this is difficult to achieve. Recently, however, we discovered a new series of nucleoside analogues with a six-membered carbohydrate moiety.[1,2] These nucleosides have a 1,5-anhydro-2,3-dideoxy-D-mannitol moiety with the heterocyclic base situated in the 2-position. Here, we describe the further structure-activity relationship of this class of nucleoside analogues.

1 1,5-ANHYDROHEXITOL NUCLEOSIDES AND ANALOGUES

1,5-Anhydrohexitol nucleosides were initially synthesized as ring-expanded analogues of 2'-deoxynucleosides. The insertion of a methylene group between the ring oxygen and the carbon atom linked to the base moiety removes the anomeric center. These compounds could have the advantage of carbocyclic nucleosides while a ring oxygen atom is still present, should this be necessary for recognition by enzymes involved with nucleoside metabolism. The starting material (3-deoxy-D-glucitol) for synthesizing these compounds was first obtained from D-glucose according to following reaction scheme (scheme 1).[1] The base moiety can be introduced either by nucleophilic substitution on the tosylate or triflate (using the lithium or sodium salt of the nucleoside base) or using Mitsunobu conditions. The nucleophilic substitution reaction is preferred, although substantial formation of O^2-alkylated compounds when using pyrimidine bases can not be avoided.[3]

A second scheme to 1,5-anhydro-4,6-*O*-benzylidene-3-deoxy-D-glucitol, more apt to large scale synthesis as it does not make use of chromatographical purification, starts from 3-isopropylidene protected 3-deoxy-D-glucofuranose (scheme 2) yielding the final compound in 30-40% overall yield.[4]

From the many compounds tested, two of them (5-iodouracil and 5-ethyluracil) proved to be highly selective, viral TK-dependent inhibitors of HSV-1 and HSV-2.[1,2]

Scheme 1

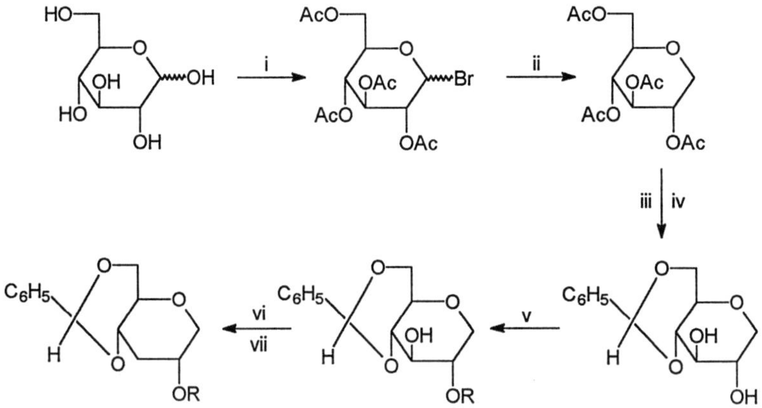

i) Ac_2O, HBr/HOAc; ii) Bu_3SnH, Et_2O; KF/H_2O; iii) NaOMe; iv) C_6H_5CHO, $ZnCl_2$; v) Bu_2SnO, C_6H_6, TosCl or TolCl (R:Tos or Tol); vi) $CSCl_2$, DMAP, 2,4-$Cl_2C_6H_3OH$, CH_2Cl_2; vii) Bu_3SnH, AIBN, toluene.

Scheme 2

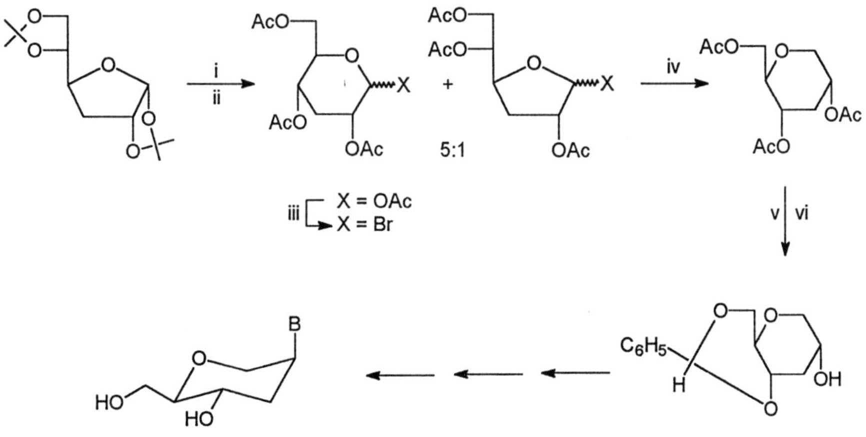

i) IRA-120(H^+) resins, EtOH, H_2O; ii) Ac_2O, C_5H_5N; iii) HBr, HOAc; iv) Bu_3SnH, Et_2O; v) NaOMe, MeOH; vi) $C_6H_5CH(OCH_3)_2$, dioxane, TsOH.

The 5-fluorocytosine and the 2,6-diaminopurine derivatives showed the broadest anti-DNA virus activity spectrum[2] and are also active against TK^- strains of HSV. It may be indicated that the L-series of the anhydrohexitol nucleosides with the four natural bases were also synthesized but they lack antiviral activity.[4] The solution and solid phase conformation of these compound indicate an axial position of the heterocyclic moiety in an anti conformation.[1,2] This conformational preference may be explained by the influence of steric factors, avoiding as many as possible unfavorable 1,3-diaxial interactions. The introduction of a supplementary hydroxyl group in the 3'-position leads to D-altritol and D-mannitol nucleosides. The introduction of this hydroxyl group may have a further stabilization effect on the aforementioned conformation because of the presence of an additional gauche effect in both examples. Therefore it should be expected that these nucleosides adopt the same conformation as D-arabino-hexitol nucleosides and, hence, demonstrate antiviral activity, should this conformational preference be important for biological activity. Synthetic procedures leading to these nucleoside analogues is based on trans-diaxial opening of epoxides and represented in scheme 3[5,6], as exemplified with uracil base.

Scheme 3

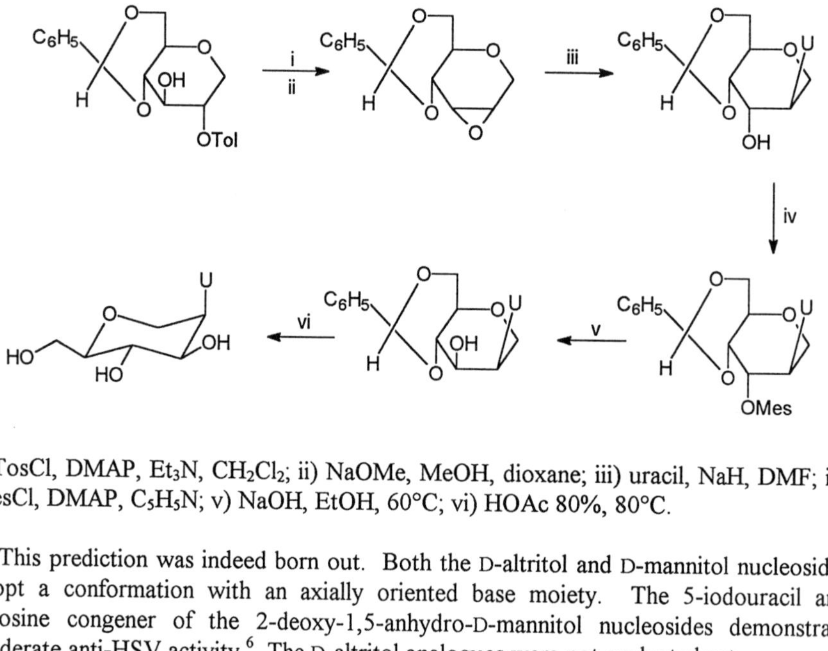

i) TosCl, DMAP, Et_3N, CH_2Cl_2; ii) NaOMe, MeOH, dioxane; iii) uracil, NaH, DMF; iv) MesCl, DMAP, C_5H_5N; v) NaOH, EtOH, 60°C; vi) HOAc 80%, 80°C.

This prediction was indeed born out. Both the D-altritol and D-mannitol nucleosides adopt a conformation with an axially oriented base moiety. The 5-iodouracil and cytosine congener of the 2-deoxy-1,5-anhydro-D-mannitol nucleosides demonstrate moderate anti-HSV activity.[6] The D-altritol analogues were not evaluated yet.

We also synthesized and evaluated 1,5-anhydro-2,3,4-trideoxy-β-D-threo-hex-3-enose compounds together with the 4-azido and the reduced congener.[7,8] These molecules can be considered as mimics of the anti-HIV compounds AZT, D4T and DDC. The synthetic strategy used to synthesize the unsaturated hexitols is based on a Mitsunobu type condensation of nucleoside bases with an allylic positioned alcohol function[8] represented in scheme 4. The starting material 1,5-anhydro-4,6-*O*-benzylidene-2-*O*-p-toluoyl-D-glucitol, is the same as used in scheme 4.

The conformation ($^{0}H_{1}$) was deduced from ^{1}H NMR data and shows a pseudoaxially orientation of the base moiety as well with purine as pyridimine nucleosides. This can be explained mainly by the influence of stereoelectronic effects (π-σ^* interactions, gauche effects, hydrogen bonds) and by the absence of unfavorable steric interactions (between the CH_2OH group and the 1'-hydrogen atoms in the $^1H_0$ conformation). The compounds with an adenine and thymine base moiety demonstrate low anti-HIV activity, again indicating the importance of an axial oriented heterocyclic for biological activity.

Scheme 4

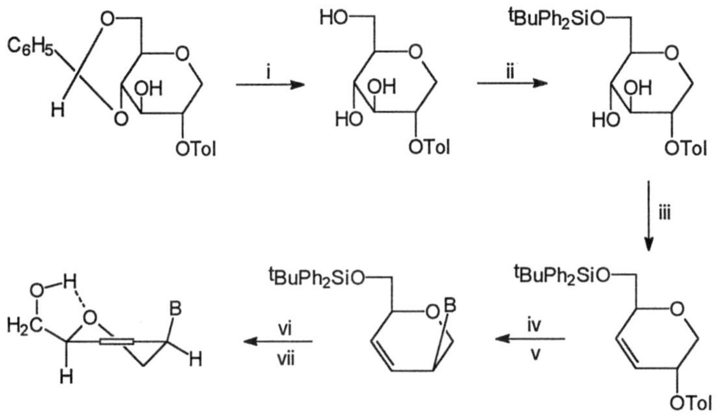

$^0H_1$

i) 80% HOAc; ii) $tBuPh_2SiCl$, imidazole, CH_2Cl_2; iii) Ph_2PCl, I_2, imidazole, CH_3CN, C_7H_8; Zn dust; iv) NaOMe, MeOH; v) nucleoside base, Ph_3P, DEAD, dioxane; vi) base deprotection; vii) Bu_4NF, THF.

2 1,3,5-SUBSTITUTED PYRANOSE NUCLEOSIDES WITH ANOMERIC CENTER

Synthesis of the ring expanded analogues of 2'-deoxynucleosides with a methylene group inserted between C4' and the endocyclic oxygen atom may give us an insight in the conformational preference of pyranose nucleosides with anomeric positioned base moiety and lacking the usual 6'-hydroxymethyl substituent. This hydroxymethyl functionality directs sugar-base condensation reactions giving mainly the anomer with an equatorial oriented CH_2OR group.[9] Four types of compounds were considered : 3'-deoxy-3'-C-hydroxymethyl-α-L-lyxo-pyranosyl nucleosides, 4'-deoxy-4'-C-hydroxy-methyl-α-L-lyxo-pyranosyl nucleosides and their 2'-deoxygenated analogues. Three of the four molecules were yet synthesized. The synthesis of the C-3' branched pyranosyl nucleosides starts from β-D-glycero-pent-2'-enopyranosyl nucleosides and is based on a radical cyclization reaction, a Tamao oxidation and a Mitsunobu type inversion of configuration (scheme 5). While this reaction sequence works well with thymine, uracil and adenine bases[10,11], application with cytosine and guanine bases gives much lower yield.[11] Minimizing steric repulsion could explain the equatorial orientation of the base moiety in the 3'-deoxy-3'-hydroxymethyl-aldopentopyranosyl series ($^4C_1$ conformation). As purine bases give less steric hindrance than pyrimidine bases, the 2,3-deoxy-3-C-hydroxymethyl-β-D-erythro-pentopyranosyl adenine adopts a $^1C_4$ conformation.[11] The analogues with a pyrimidine base has a predominant $^4C_1$ conformation.[10] So, when the adenine base is present, epimerization at the 4'-position induces a conformational change which can be explained by the gauche effect ($OCH_2C(4')O$).[11] This difference in conformational behaviour of purine and pyrimidine nucleosides was also observed with β-D-glycero-pento-2'-enopyranosyl nucleosides[12] where purine bases are oriented pseudoaxially and pyrimidine bases more pseudoequatorially (scheme 6). The driving force for the orientation of the pyrimidine bases is predominant steric, the driving force for the conformation of the purine nucleosides is more stereoelectronic.

Scheme 5

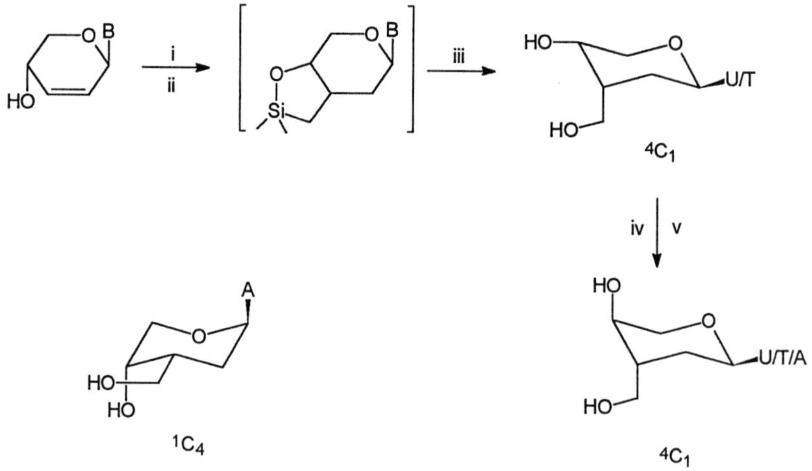

i) $ClSi(CH_3)_2CH_2Br$, imidazole, DMF; ii) Bu_3SnH, AIBN, toluene; iii) KF, $KHCO_3$, H_2O_2, DMF; iv) DEAD, Ph_3P, BzOH, dioxane, xylene; v) NH_3, MeOH.

Scheme 6

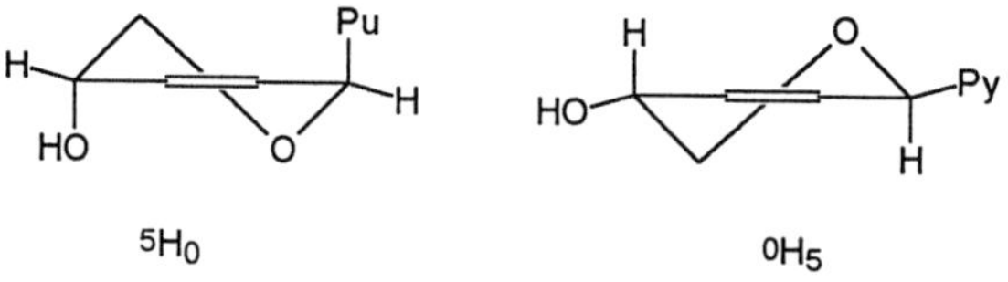

These unsaturated compounds were synthesized as starting material for the preparation of the C-3' branched nucleosides (scheme 5). Normally the acid mediated Ferrier rearrangement of di-*O*-acyl-D-xylal and nucleoside bases affords a mixture of compounds (α/β nucleosides and 3'-substituted 2',3'-unsaturated derivatives). Looking for optimized conditions to use this reaction on preparative scale, we found out that the fusion of 3,4-di-*O*-p-nitrobenzoyl-D-xylal with purine and pyrimidine bases (their sodium salt or trimethylsilylated bases) in boiling DMF in the absence of externally added acid catalyst avoids formation of the 3'-substituted side compounds.[12,13] The exclusive formation of the α/β mixture in preparative quantities may be explained by the lack of anchimeric stabilization of the intermediate cation (scheme 7) and the fact that liberated *p*-nitrobenzoic acid is too weak to force anomerization reaction.[12]

Scheme 7

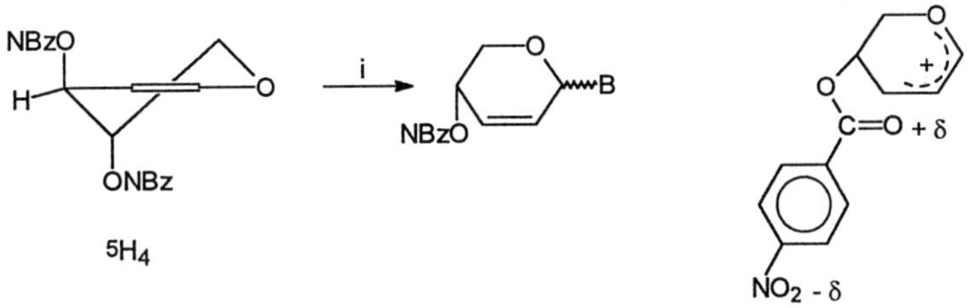

NBzO H ONBz $^5H_4$ i NBzO B O C=O + δ NO_2 - δ

i) B or TMS(B) or B^-Na^+, DMF, Δ

The synthesis of 3'-deoxy-3'-C-hydroxymethyl-α-L-lyxopyranosyl thymine (scheme 8) starts from 1,2-isopropylidene-3-deoxy-3-C-hydroxymethyl-L-lyxofuranose. The approach makes use of a furanose → pyranose conversion and the formation of both furanose and pyranose nucleosides during a Vörbruggen sugar-base condensation reaction starting from tetra-*O*-acetyl-3-deoxy-3-C-hydroxymethyl-L-lyxo-(1,6)-furanose. The compound adopts a $^4C_1$ conformation in solution.

Scheme 8

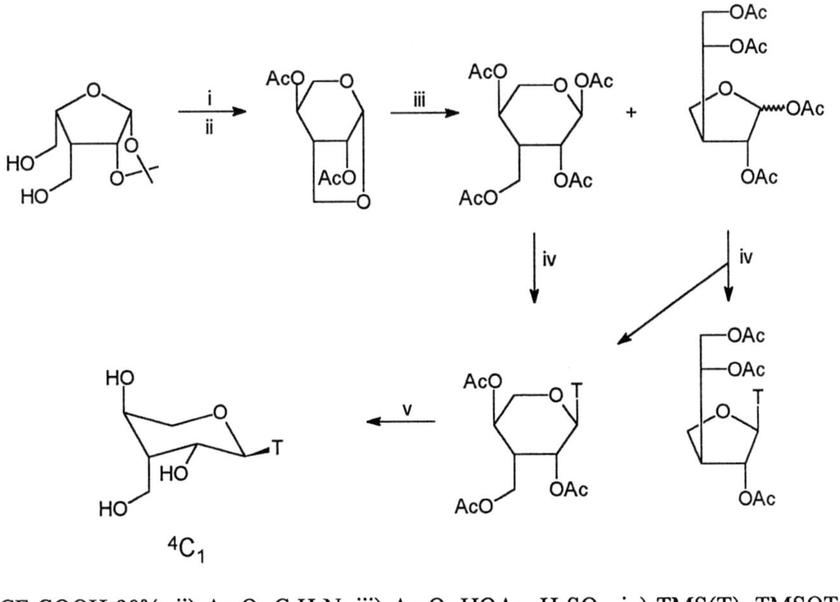

HO HO i ii AcO AcO iii AcO OAc OAc AcO + OAc OAc OAc OAc iv iv OAc OAc T OAc AcO T OAc AcO v HO HO HO T $^4C_1$

i) CF_3COOH 90%; ii) Ac_2O, C_5H_5N; iii) Ac_2O, HOAc, H_2SO_4; iv) TMS(T), TMSOTfl, $C_2H_4Cl_2$; v) NaOMe, MeOH.

The regioisomeric 4-deoxy-4-C-hydroxymethyl-α-L-lyxo-pyranosyl thymine also adopts a $^4C_1$ conformation. Its synthesis (given in scheme 9) makes use of the soft nucleophilic character of malonyl anions and of ozonolytic cleavage of enol ether.

Scheme 9

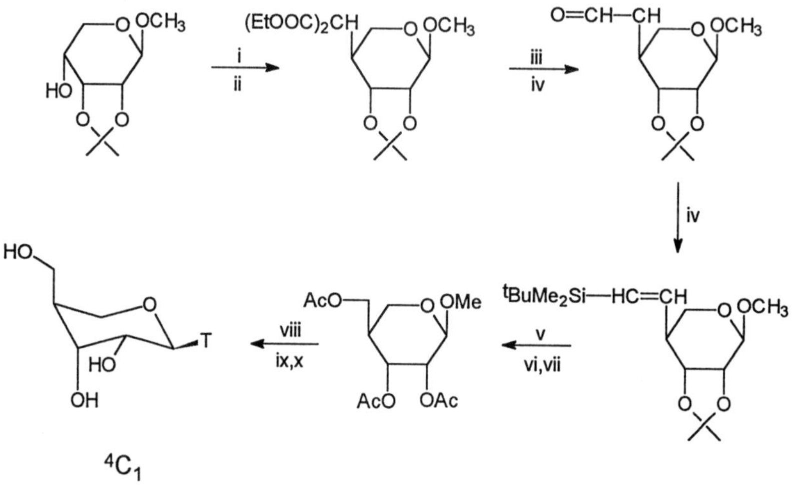

i) Tfl_2O, pyridine; ii) $NaCH(COOEt)_2$, DMF; iii) LiCl, H_2O, DMSO, bp; iv) DIBAL, CH_2Cl_2; iv) $^tBuMe_2Si$-OTfl, CH_2Cl_2, Et_3N; v) O_3, MeOH; $NaBH_4$; vi) CF_3COOH 90%; vii) Ac_2O, C_5H_5N; viii) AcOH, Ac_2O, H_2SO_4; ix) TMS(T), TMSOTfl, CH_2Cl_2; x) NaOMe, MeOH.

All compounds described in this chapter were not active when tested against herpes simplex virus infections, pointing to their configurational and/or conformational incompatibility with recognition by viral enzymes.

3 PHOSPHONATE ANALOGUES OF PENTOPYRANOSYL NUCLEOSIDES

As nucleosides have to be phosphorylated in a cell to exert an antiviral activity and as this reaction can be circumvented by synthesizing phosphonate analogues, we started synthesis of several pentopyranosyl nucleosides with a 1,4-relationship between base moiety and phosphonomethoxy substituent. The compounds are unsaturated (to mimic D4T-monophosphate and D4A-monophosphate), and the oxygen is positioned on the four possible positions. Two different routes were developed leading to the ring expanded nucleoside monophosphonates[16,17] from which only one is presented here (scheme 10). This scheme is based on the aforementioned condensation of protected D-xylal with heterocyclic bases and the same scheme leads to two isomeric nucleosides.

The two 2,5-cis-substituted dihydro-2-H-pyranosyl nucleosides were obtained as represented in scheme 11. The key step here is the introduction of the phosphono-methoxy moiety on pentopyranosyl glycals through an acid catalyzed Ferrier-type rearrangement.[18,19] The attack of the alcohol function occurs preferentially anti to the C-4 substituent.

Scheme 10

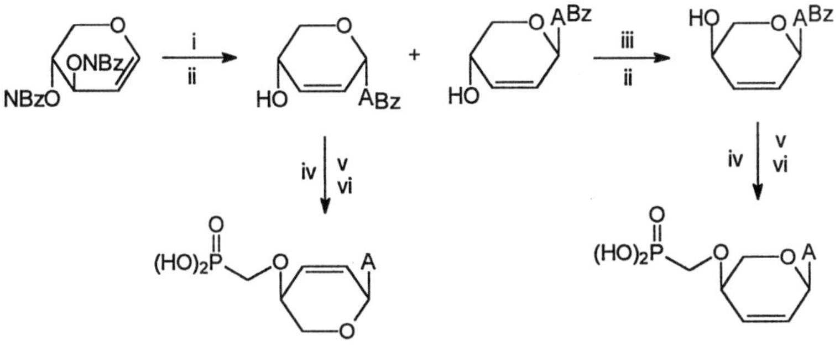

i) N[6]benzoylA, DMF; ii) NaOMe, MeOH; iii) pNBzOH, Ph_3P, DEAD dioxane; iv) NaH, $(iPrO)_2P(O)CH_2OH$, DMF; v) MeOH, NaOH; vi) TMSiBr, DMF, lutidine; NH_3, H_2O.

The diastereoselectivity of this reaction is also determined by attaining an anomeric equilibrium during acid-catalyzed Ferrier reaction of glycals with alcohols leading to the energetically most stable trans isomer. Because a phosphonomethoxy moiety is a poorer leaving group than a heterocyclic base, the compounds presented in scheme 11 are more stable towards acid degradation than the compounds of scheme 10. The compounds were also converted into their saturated congeners.

Scheme 11

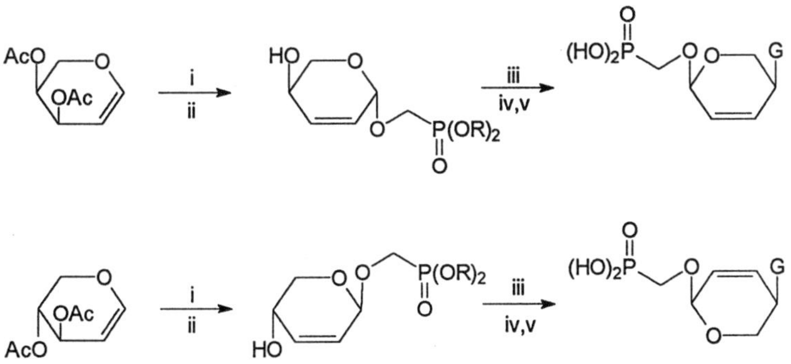

i) $(iPrO)_2P(O)CH_2OH$, TMSOTfl, CH_3CN; ii) NH_3, MeOH; iii) 2-amino-6-chloropurine, Ph_3P, DEAD, dioxane; iv) Me_3N, H_2O, DBU; v) TMSiBr, DMF, lutidine, NH_3, H_2O.

All the phosphonate derivatives were found inactive as antiviral compounds. One of the reasons is that the phosphonate compounds (at least those presented in scheme 11) are not substrates for the GMP kinase and, hence, are most probably badly phosphorylated in cells.[18]

4 ACYCLIC NUCLEOSIDES DERIVED FROM ANHYDROHEXITOLS

Acyclic nucleosides have been a favourite topic of nucleoside chemists during almost two decades following the discovery of acyclovir. The available data on acyclic analogues of anhydrohexitols are rather limited. An upper-part mimic can be easily obtained from diethyleneglycol and the guanine congener ($HO(CH_2)_2O(CH_2)_2G$) shows moderate anti-HSV-1 activity.[20] Recently also acyclic analogues representing all functionalities of the anhydrohexitol nucleosides were synthesized as well in the dideoxy-D-mannityl series (scheme 12) as in the dideoxy-D-glucityl series.[21] The reaction scheme starts from 2-deoxyribose and involves a stereoselective Sharpless dihydroxylation reaction and a Mitsunobu type alkylation of nucleoside bases. It should be mentioned that dihydroxylation of the unsaturated compound using AD-mix-β did not improve the stereoselectivity of the reaction while AD-mix-α does. The same scheme can also be used as an alternative way to synthesize anhydrohexitol nucleosides with a pyrimidine base moiety by ring closure reaction of the 4,6-di-*O*-benzyl-2,3-dideoxy-2-D-mannityl compound. These acyclic nucleosides, however, do not show antiherpes activity.

Scheme 12

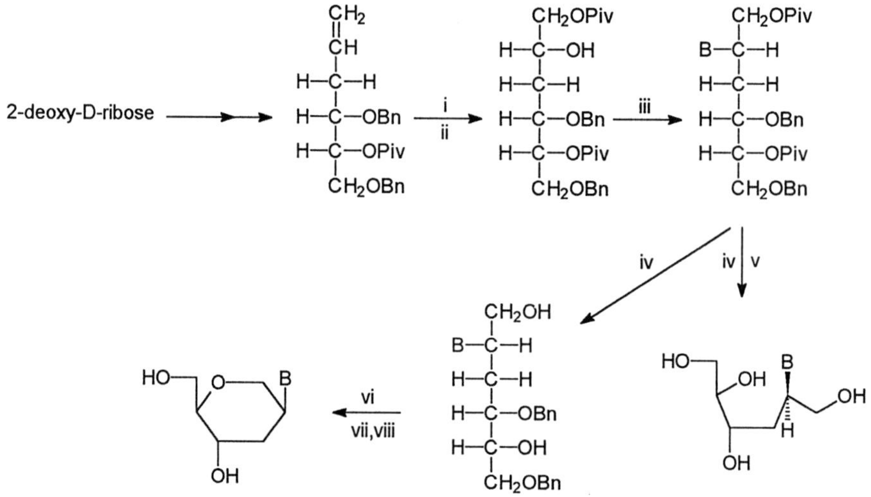

i) AD-mix-α, ^{t}BuOH, H_2O; ii) Piv.Cl, C_5H_5N; iii) protected base, Ph_3P, DEAD, dioxane; iv) NaOH, H_2O, dioxane; v) $Pd(OH_2)$, MeOH, cyclohexene; vi) pTosCl, C_5H_5N; vii) NaH, DMF; viii) $Pd(OH)_2$, MeOH, cyclohexene.

5 CARBOCYCLIC NUCLEOSIDES WITH A SIX-MEMBERED RING

Nucleophilic substitution on a cyclohexane ring is more difficult than on a cyclopentane ring because of steric hindrance and this hampers introduction of nucleoside bases on the six-membered ring structure. This was experienced during the synthesis of 4,4-dihydroxymethyl cyclohexane nucleosides[22] and during introduction of nucleoside bases on saturated six membered carbocycles using Mitsunobu reactions.[23]

Several solutions to this problem are possible. A Pd(O) catalyzed alkylation of heterocyclic bases by allylic epoxide was low yielded.[23] Better results were obtained using a Mitsunobu-type condensation of nucleoside bases with an unsaturated alcohol. The allylic alcohol has higher reactivity in a nucleophilic substitution reaction and the saturated compound could be obtained after catalytic hydrogenation. This method was used for the synthesis of carbocyclic phosphonate nucleosides[23] (scheme 13). The phosphonomethoxy moiety is introduced prior to coupling with the base to avoid protection or undesired alkylation of the bases. These compounds, however, were found to be inactive against HSV and HIV replication.

Scheme 13

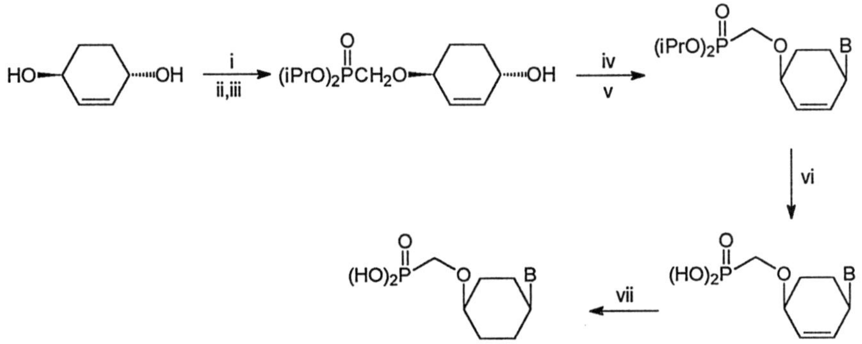

i) TritylCl, DMAP, Et_3N, CH_2Cl_2; ii) $(iPrO)_2P(O)CH_2OH$, NaH, DMF; iii) HOAc 80%; iv) protected nucleoside base, Ph_3P, DEAD, dioxane; v) NH_3, MeOH; vi) TMSiBr, DMF; NH_4OH; vii) H_2, Pd, EtOH.

A third possibility is to introduce the heterocyclic moiety by a Michael-type addition reaction which likewise works efficiently with purine bases.[24] This method was used to synthesize the carbocyclic analogues of the anhydrohexitol nucleosides following scheme 14. Hydroboration gives a separable mixture of compounds. These carbocyclic nucleosides have an equatorial oriented base part.

Scheme 14

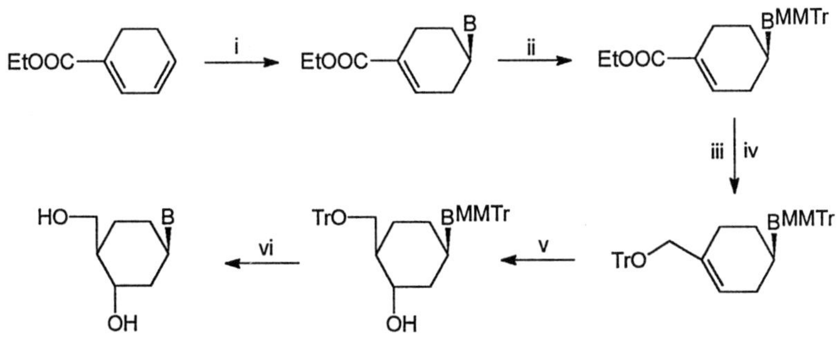

i) nucleoside base, DBU, DMF; ii) MMTrCl, C_5H_5N; iii) DIBAL, CH_2Cl_2; iv) TrCl, C_5H_5N; v) BH_3, THF; MeOH, EtOH, H_2O_2; vi) HOAc 80%.

6 CONCLUSIONS

In the series of pyranosyl-like nucleosides, antiviral activity is observed when a 1,4-substitution pattern between the hydroxymethyl group and the base moiety is present and when the base adopts an axial position. The molecules are forced into this geometry, when more steric hindrance is created (unfavourable 1,3-diaxial interactions) around the hydroxymethylgroup than in the neighbourhood of the base moiety. The heterocyclic substituent should therefore be placed in a 1,3-relationship to the ring oxygen atom. These anhydrohexitol nucleosides geometrically fit to a furanose nucleoside in the 2'-exo, 3'-endo conformation and into an A-type nucleic acid duplex after incorporation into the genome, while natural dsDNA is of the B-type. These data suggest a similar conformational preference of furanose nucleosides during one of the crucial metabolic steps leading to antiviral activity and to a disruption of dsDNA structure in case incorporation of hexitol nucleosides in DNA would occur.

7 REFERENCES

1. I. Verheggen, A. Van Aerschot, S. Toppet, R. Snoeck, G. Janssen, P. Claes, J. Balzarini, E. De Clercq and P. Herdewijn, *J. Med. Chem.*, 1993, **36**, 2033.
2. I. Verheggen, A. Van Aerschot, L. Van Meervelt, J. Rozenski, L. Wiebe, R. Snoeck, G. Andrei, J. Balzarini, E. De Clercq and P. Herdewijn, *J. Med. Chem.*, 1995, **38**, 826.
3. B. De Bouvere, L. Kerremans and P. Herdewijn, in preparation.
4. M. Andersen, S. Daluge, L. Kerremans and P. Herdewijn, *Tetrahedron Lett.* (submitted).
5. I. Verheggen, A. Van Aerschot, N. Pillet, E.M. van der Wenden, A. IJzerman and P. Herdewijn, *Nucleosides and Nucleotides*, 1995, **14**, 321.
6. M.-J. Pérez-Pérez, E. De Clercq and P. Herdewijn, *Bioorg. Med. Chem. Lett.*, 1996, in press.
7. I. Verheggen, A. Van Aerschot, J. Rozenski, G. Janssen, E. De Clercq and P. Herdewijn, 1996, *Nucleosides and Nucleotides*, 1996, **15**, 325.
8. I. Luyten and P. Herdewijn, *Tetrahedron* **1996**, in press.
9. K. Augustyns, J. Rozenski, A. Van Aerschot, P. Claes and P. Herdewijn, *Tetrahedron*, 1994, **50**, 1189.
10. B. Doboszewski, N. Blaton, J. Rozenski, A. De Bruyn and P. Herdewijn, *Tetrahedron*, 1995, **51**, 5381.
11. B. Doboszewski, H. De Winter, A. van Aerschot and P. Herdewijn, *Tetrahedron*, 1995, **51**, 12319.
12. B. Doboszewski, N. Blaton and P. Herdewijn, *J. Org. Chem.*, 1995, **60**, 7909.
13. B. Doboszewski and P. Herdewijn, *Tetrahedron Lett.*, 1995, **36**, 1321.
14. B. Doboszewski and P. Herdewijn, *Tetrahedron*, 1996, **52**, 1651.
15. B. Doboszewski and P. Herdewijn, *Nucleosides and Nucleotides* **1996**, in press.
16. M.-J. Pérez-Pérez, B. Doboszewski, E. De Clercq and P. Herdewijn, *Nucleosides and Nucleotides*, 1995, **14**, 707.
17. M.-J. Pérez-Pérez, J. Rozenski and P. Herdewijn, *Bioorg. Med. Chem. Lett.*, 1994, **4**, 1199.

18. M.-J. Pérez-Pérez, J. Balzarini, J. Rozenski, E. De Clercq and P. Herdewijn, *Bioorg. Med. Chem. Lett.*, 1995, **5**, 1115.
19. M.-J. Pérez-Pérez, B. Doboszewski, J. Rozenski and P. Herdewijn, *Tetrahedron Assymmetry*, 1995, **6**, 973.
20. A. Van Aerschot, N. Zhigang, J. Rozenski, P. Claes, E. De Clercq and P. Herdewijn, *Nucleosides and Nucleotides*, 1994, **13**, 1791.
21. N. Hossain, J. Rozenski, E. De Clercq and P. Herdewijn, *Tetrahedron*, submitted.
22. S.N. Mikhailov, N. Blaton, J. Rozenski, J. Balzarini, E. De Clercq and P. Herdewijn, *Nucleosides and Nucleotides*, 1996, **15**, 867.
23. M.-J. Pérez-Pérez, J. Rozenski, R. Busson and P. Herdewijn, *J. Org. Chem.*, 1995, **60**, 1531.
24. Y. Maurinsh and P. Herdewijn, in preparation.

23

Current Perspectives on Mechanisms of HIV-1 Reverse Transcriptase Inhibition by Nonnucleoside Inhibitors

Stefan G. Sarafianos,[1] Kalyan Das,[1] Jianping Ding,[1] Yu Hsiou,[1] Stephen H. Hughes,[2] and Edward Arnold[1,*]

[1]CENTER FOR ADVANCED BIOTECHNOLOGY AND MEDICINE (CABM) AND DEPARTMENT OF CHEMISTRY, RUTGERS UNIVERSITY, 679 HOES LANE, PISCATAWAY, NJ 08854-5638, USA
[2]ABL-BASIC RESEARCH PROGRAM, NCI-FREDERICK CANCER RESEARCH AND DEVELOPMENT CENTER, PO BOX B, FREDERICK, MD 21701-1201, USA

1 INTRODUCTION

HIV-1 reverse transcriptase (RT) is the target of several clinically important therapeutic agents. Recently, a number of promising therapies, using RT inhibitors in combination with protease inhibitors, have substantially reduced the viral load in HIV-infected patients, in some cases to undetectable levels (data presented at the International AIDS Conference, Vancouver, July 1996). It is important to understand how the various inhibitors block RT function. This knowledge might be useful in the development of drugs (or drug combinations) that are more efficient and less susceptible to the development of resistance.

The vast majority of RT inhibitors now available can be divided into two classes: nucleoside RT inhibitors (NRTIs) and nonnucleoside RT inhibitors (NNRTIs). NRTIs are analogs of the normal nucleotide substrates of RT. NRTIs lack a 3'-OH and act as chain terminators when incorporated into the growing DNA chain. Despite extensive biochemical and structural studies of NNRTIs, it is unclear how they inhibit RT. In this paper, we analyze the currently available biochemical (particularly from pre-steady state kinetic studies) and structural data in light of possible mechanisms of NNRTI action.

2 AVAILABLE EXPERIMENTAL DATA

2.1 Comparison of HIV-1 RT Structures

Crystal structures of RT/NNRTI complexes show that NNRTIs bind in a hydrophobic pocket at the base of the p66 thumb subdomain[1-5]. This nonnucleoside inhibitor binding pocket (NNIBP) is formed primarily by amino acid residues of the β5-β6 loop, β6, the β9-β10 hairpin, the β12-β13 hairpin, and β15 of p66, and the β7-β8 connecting loop of p51[2]. It is proximal to (~10 Å), but distinct from, the polymerase active site which is defined by three aspartates (110, 185, and 186). The β12-β13 hairpin is involved in positioning the primer terminus at the catalytic site, and has been called the "primer grip."[6]

Comparison of the structures of a number of RT/NNRTI complexes with those of unliganded HIV-1 RT or of an RT complex with DNA shows that binding of an NNRTI causes a number of major structural changes in HIV-1 RT: First, the p66 thumb moves from a position where it nearly touches the fingers[7, 8] to an upright configuration[1] which is even more extended[8] than that seen in the RT/DNA structure[6] (Figure 1); second, a hinge-like movement takes place at the base of the p66 thumb and is accompanied by displacement of the p66 connection subdomain and the RNase H domain[8]; and third, NNRTI binding causes the β12-β13-β14 sheet, which contains the primer grip, to undergo a relatively large displacement[2, 7-10] (Figure 2).

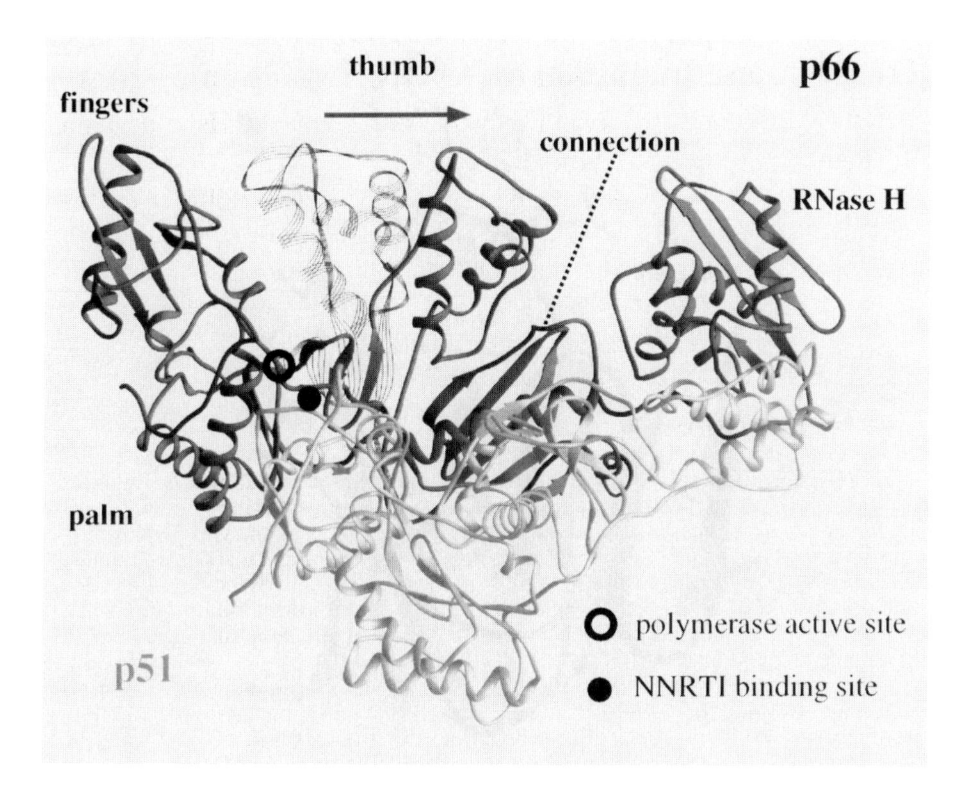

Figure 1 *Ribbon diagram of the HIV-1 RT p66/p51 heterodimer showing the different positions of the p66 thumb subdomain in structures of unliganded HIV-1 RT (light dashed line) and an HIV-1 RT/NNRTI complex (black). The individual subdomains of the p66 subunit are named fingers, palm, thumb, connection, and RNase H. The p51 subunit is shown in gray. The position of the polymerase active site is indicated by a circle and the NNRTI-binding site by a filled sphere.*

In addition to changes that involve repositioning of the primer grip and the p66 thumb, NNRTI binding also causes a number of smaller changes. In the unliganded structures[7, 8] the side chains of Tyr181 and Tyr188 occupy the region corresponding to the NNIBP in the RT/NNRTI structures[1-4]. However, in the RT/NNRTI structures, the side chains of Tyr181 and Tyr188 reorient toward the polymerase active site, creating space to accommodate the incoming NNRTI[2, 7, 9]. A comparison of the structures of the RT/DNA and RT/NNRTI complexes shows significant conformational differences of the YMDD motif in the β6-β10-β9 sheet at the catalytic site (as reflected by displacements of Cα positions of 2 Å or more)[2]. Even though these changes are smaller in magnitude than those of the thumb or primer grip they involve the catalytic site. Esnouf *et al.* (1995) reported a movement of the β6-β10-β9 sheet (up to 2 Å) (which contains the active site aspartates) as a major change caused by NNRTI binding[11]. There are, however, substantial differences between the structure of unliganded RT reported by Esnouf *et al.*[11] and all other unliganded RT structures determined to date (references 7, 8, and E. Arnold *et al.*, unpublished) which can be attributed to the method used to prepare the crystals,

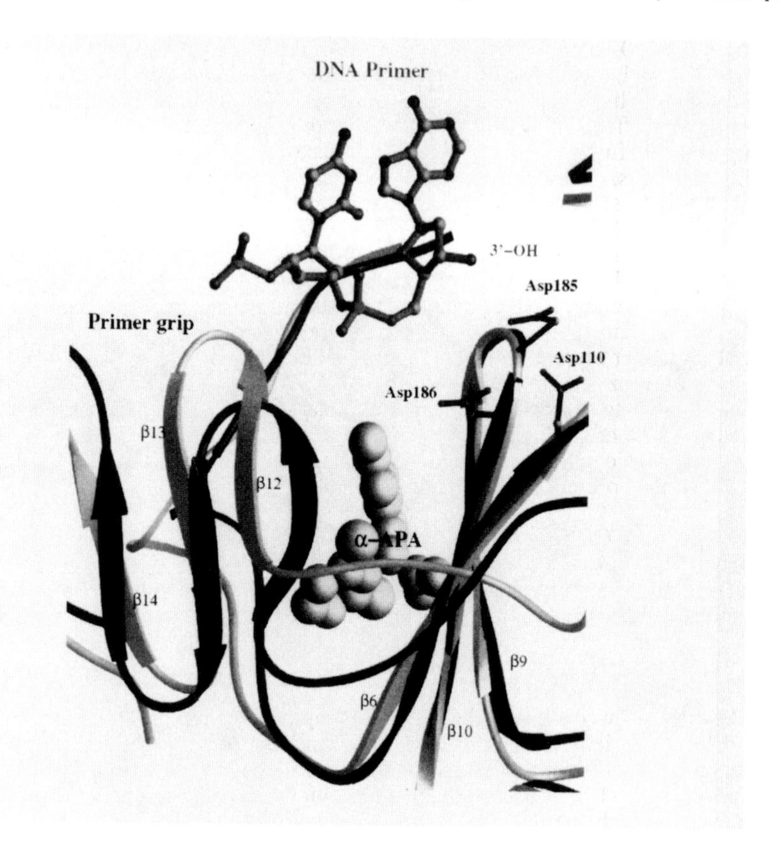

Figure 2 *Superposition of the HIV-1 RT/DNA/Fab complex structure (black) on the HIV-1 RT/α-APA complex structure (gray) in the region near the NNIBP and the polymerase active site. The displacement of the β12-β13 hairpin (primer grip) in the RT/α-APA structure, which makes up part of the NNIBP, is shown relative to the primer strand. For the sake of clarity, the template strand is omitted.*

which suggests that this structure is not representative of "unliganded RT" (see detailed discussion in reference 8). A very serious limitation of all of the currently available NNRTI-bound RT structures is that none contain template-primer or dNTP. The lack of a biologically relevant NNRTI-bound HIV1 RT structure limits our ability to clearly understand the mechanism of NNRTI inhibition.

2.2 Biochemical Evidence

There is a growing body of biochemical data that are relevant to the mechanism of NNRTI action. Several steady-state kinetic studies suggested that inhibition is non-competitive with respect to both template-primer and dNTP substrates[12-18]. However, these studies did not identify the step of the polymerization reaction that is affected by NNRTIs.

Pre-steady state kinetic analysis has been used to investigate which step is blocked[19-21]. In the uninhibited reaction, a two-step dNTP binding mechanism has been proposed, with initial ground state nucleotide binding followed by a rate-limiting conformational change that precedes formation of the phosphodiester bond (chemical step). NNRTIs do not substantially interfere with dNTP binding (increase in binding affinity ranged from negligible to 10-fold[19-21]) or with the conformational change induced by nucleotide binding. Instead, NNRTIs block the chemical step of the reaction.

Although there has been a large number of additional biochemical analyses, measuring the effects of NNRTIs on a number of properties of HIV-1 RT, including (but not limited to) effects on processivity[22, 23], effects on the ability of RT to copy different template-primers, different sequences[13, 15, 17], and elements of secondary structure in the template[23], it is useful to differentiate those assays where results can be simply explained as some sort of secondary effects of interference with the chemical step (like processivity) from those assays that provide additional insights into the effects of NNRTI binding. NNRTI binding has also been reported to moderately affect template-primer binding. Specifically, NNRTI binding alters the fluorescence quenching seen when RT interacts with template-primer[24, 25] (These differences might involve Trp229 of the primer grip among other protein residues); and, NNRTI binding increases (up to 10-fold) the affinity of the enzyme for template-primer[21, 26]. Although this change in DNA binding may not contribute to the inhibition *per se*, it may lead to the formation of a complex that can be more effectively inhibited by NNRTIs. Finally, NNRTI binding has also been reported to induce some changes in the cleavage specificity of the RNase H activity of HIV-1 RT in the presence of nevirapine (an NNRTI)[27].

3 MODELS OF THE MECHANISM OF RT INHIBITION BY NNRTIs

Based on the structural and biochemical results discussed above, three major types of models have been proposed to explain the mechanism of RT inhibition by NNRTIs.

3.1 Molecular Arthritis Model

Steitz and coworkers postulated that nevirapine may act like "sand in the gears of a machine,"[1] potentially inhibiting HIV-1 RT by restricting the mobility of the p66 thumb subdomain. We will refer to this hypothesis as the "molecular arthritis model." The p66 thumb subdomain has extensive interactions with the template-primer[6]. Thus, restriction of the thumb mobility could affect template-primer handling by HIV-1 RT. Conceivably, limiting mobility of the p66 thumb could interfere with processivity of polymerization[22] which has been suggested to involve the motion of this subdomain[8, 28]. However, the molecular arthritis model cannot account for the dramatic decrease in enzyme activity caused by NNRTI binding. Specifically, the pre-steady state experiments clearly show that the chemical step of the reaction is affected and there is no experimental evidence linking restriction of thumb mobility to the chemical step of polymerization. Hence, from the point of view of kinetics of single nucleotide incorporation, the molecular arthritis model appears to be inadequate.

3.2 Primer Grip Model

The crystal structure of RT complexed with a dsDNA template-primer showed the interactions of RT with DNA, and permitted identification of the primer grip (the β12-β13 hairpin) as an element that positions the primer terminus near the polymerase active site[6]. The position of the primer grip in the HIV-1 RT/DNA and unliganded HIV-1 RT structures is similar[7, 8]. However, comparison of the RT/DNA complex and unliganded HIV-1 RT structures with HIV-1 RT/NNRTI complex structures has shown that a significant displacement of the primer grip accompanies NNRTI binding[2, 7-9].

These observations led to the "primer grip model," which proposes that NNRTI binding distorts the primer grip, interfering with the correct placement of the 3'-OH of the primer strand relative to the polymerase active site[2, 7-9, 29].

Hsiou *et al.* (1996) recently showed that displacements of the p66 thumb (including movement that extends its position beyond the position it has in the RT/DNA complex) and the primer grip that accompany NNRTI binding appear to be coupled. The concerted effects of both of these movements could explain experimental observations linking NNRTI binding to changes in template-primer affinity, processivity of polymerization, and changes in RNase H cleavage specificity. More importantly, the primer grip model may be able to explain the data obtained from pre-steady state kinetic experiments that show binding an NNRTI blocks the chemical step of polymerization by HIV-1 RT[19, 21]. According to this model, NNRTI binding displaces the primer grip, which either disturbs the position and/or mobility of the primer terminus at the active site or alters the stereochemical relationship between the primer terminus and a bound dNTP and associated divalent metal cation(s). Thus, an incoming dNTP may still bind with similar affinity with and without bound NNRTIs, but the presence of an NNRTI could alter the relative positions of the incoming substrate (dNTP) and the 3'-OH to which it would be linked. Indirect support for this model is provided by recent studies showing that mutations across the DNA-binding cleft (which may directly interact with the template strand) have the ability to affect NNRTI binding[30] or polymerase impairment caused by mutations in the NNRTI-binding region[31]; in both of these cases, the communication between distant sites is likely to be mediated through the interactions of the primer grip and the template-primer.

3.3. Active Site Distortion Model

The proximity of residues involved in NNRTI binding[32] and in NNRTI resistance[33, 34] to the catalytically essential aspartate residues of HIV-1 RT suggested that NNRTI binding could affect the polymerase active site. Structure determinations of RT/NNRTI complexes confirmed that the bound inhibitors are located reasonably close to the aspartic acid residues[1-4]. The idea that NNRTI binding could result in an alteration of the position of the catalytic aspartates is referred to here as the "active site distortion model."

Some indirect structural evidence that the catalytic site was disturbed in the presence of NNRTI was provided by the comparison of crystal structure of an RT/NNRTI[2] with the structure of the RT/DNA complex[6]. As has already been discussed there is distortion of the primer grip region (Figure 1) and reorientation of the side chains of Tyr181 and Tyr188 toward the polymerase active site. These movements could alter the electrochemical environment and solvent accessibility of the catalytic aspartates, the coordination geometry of the divalent metal(s), and/or coordination of dNTP/metal ion complexes. The smaller, yet significant, distortion of the YMDD motif observed between the RT/DNA and RT/NNRTI complex structures[2] could potentially impair the chemical step of polymerization by NNRTIs. However, the position of the YMDD motif during inhibition of polymerization is not directly known, as no NNRTI-bound HIV-1 RT structure containing DNA has been determined.

The active site distortion model is favored by Stuart and coworkers[4, 11]. As mentioned earlier, they report movements of the β6-β10-β9 sheet (using the conventional nomenclature for HIV-1 RT structural elements) based on the comparison of an unliganded RT structure with structures of RT/NNRTI complexes[4, 11]. However, as has been discussed, this unliganded RT structure differs from other reported unliganded HIV-1 RT structures[7, 8] (detailed discussion in reference 8), suggesting that the reference structure used in their comparisons is not appropriate.

Although the active site distortion model offers an attractive explanation of the dramatic reduction of polymerase activity upon NNRTI binding, the available structural data are not sufficient to resolve the issue.

4 CONCLUSIONS

There is solid biochemical evidence suggesting that NNRTIs inhibit HIV-1 RT by blocking the chemical step of the polymerization reaction. The available structural evidence from crystal structures of HIV-1 RT complexes containing NNRTIs demonstrates that NNRTI binding is accompanied by displacement of the primer grip; this change could interfere with the correct placement of the 3'-OH of the primer strand.

Although the binding of NNRTIs appears to limit movement of the p66 thumb, it seems unlikely that the restriction of mobility can account for NNRTI inhibition. Smaller distortions of key aspartate residues at the polymerase active site have been observed, but their significance is unclear. Clearly, more biochemical and structural data about HIV-1 RT will help to clarify the relative importance of each of these mechanisms, which need not be mutually exclusive. In that respect, the key structures that will directly and unambiguously identify the structural elements involved in the mechanism of inhibition by NNRTI are the structures of HIV-1 RT in complexes with NNRTI and both dNTP and template-primer substrates.

Acknowledgments

We thank the other members of the Arnold and Hughes laboratories as well as our other collaborators for their assistance and helpful discussions. The research in EA's laboratory has been supported by NIH grants (AI27960 and AI36144), Janssen Research Foundation, Hoechst Pharma AG, a grant from NCI-ABL and the Keck Foundation; SGS was supported by an NIH NRSA postdoctoral fellowship (F32 AI09578) and SHH's laboratory is sponsored by the National Cancer Institute, DHHS, under contract with ABL and by NIGMS.

References

1. Kohlstaedt, L. A., Wang, J., Friedman, J. M., Rice, P. A. & Steitz, T. A. (1992) *Science* **256**, 1783-1790.
2. Ding, J., Das, K., Tantillo, C., Zhang, W., Clark, A. D., Jr., Jessen, S., Lu, X., Hsiou, Y., Jacobo-Molina, A., Andries, K., Pauwels, R., Moereels, H., Koymans, L., Janssen, P. A. J., Smith, R., Koepke, M. K., Michejda, C., Hughes, S. H. & Arnold, E. (1995) *Structure* **3**, 365-379.
3. Ding, J., Das, K., Moereels, H., Koymans, L., Andries, K., Janssen, P. A. J., Hughes, S. H. & Arnold, E. (1995) *Nature Struct. Biol.* **2**, 407-415.
4. Ren, J., Esnouf, R., Garman, E., Somers, D., Ross, C., Kirby, I., Keeling, J., Darby, G., Jones, Y., Stuart, D. & Stammers, D. (1995) *Nature Struct. Biol.* **2**, 293-302.
5. Ren, J., Esnouf, R., Hopkins, A., Ross, C., Jones, Y., Stammers, D. & Stuart, D. (1995) *Structure* **3**, 915-926.
6. Jacobo-Molina, A., Ding, J., Nanni, R. G., Clark, A. D., Jr., Lu, X., Tantillo, C., Williams, R. L., Kamer, G., Ferris, A. L., Clark, P., Hizi, A., Hughes, S. H. & Arnold, E. (1993) *Proc. Natl. Acad. Sci. USA* **90**, 6320-6324.
7. Rodgers, D. W., Gamblin, S. J., Harris, B. A., Ray, S., Culp, J. S., Hellmig, B., Woolf, D. J., Debouck, C. & Harrison, S. C. (1995) *Proc. Natl. Acad. Sci. USA* **92**, 1222-1226.
8. Hsiou, Y., Ding, J., Das, K., Clark, A. D., Jr., Hughes, S. H. & Arnold, E. (1996) *Structure* **4**, 853-860.
9. Tantillo, C., Ding, J., Jacobo-Molina, A., Nanni, R. G., Boyer, P. L., Hughes, S. H., Pauwels, R., Andries, K., Janssen, P. A. J. & Arnold, E. (1994) *J. Mol. Biol.* **243**, 369-387.
10. Das, K., Ding, J., Hsiou, Y., Clark, A. D., Jr., Moereels, H., Koymans, L., Andries, K., Pauwels, R., Janssen, P. A. J., Boyer, P. L., Clark, P., Smith, R. H., Jr, Kroeger-Smith, M. B., Michejda, C., Hughes, S. H. & Arnold, E. (1996). Submitted.
11. Esnouf, R., Ren, J., Ross, R., Jones, Y., Stammers, D. & Stuart, D. (1995) *Nature Struct. Biol.* **2**, 303-308.
12. Merluzzi, V. J., Hargrave, K. D., Labadia, M., Grozinger, K., Skoog, M., Wu, J. C., Shih, C.-K., Eckner, K., Hattox, S., Adams, J., Rosenthal, A. S., Faanes, R., Eckner, R. J., Koup, R. A. & Sullivan, J. L. (1990) *Science* **250**, 1411-1413.

13. Carroll, S. S., Olsen, D. B., Bennett, C. D., Gotlib, L., Graham, D. J., Condra, J. H., Stern, A. M., Shafer, J. A. & Kuo, L. C. (1993) *J. Biol. Chem.* **268**, 276-281.
14. Althaus, I. W., Chou, J. J., Gonzalez, A. J., Deibel, M. R., Chou, K.-C., Kezdy, F. J., Romero, D. L., Aristoff, P. A., Tarpley, W. G. & Reusser, F. (1993) *J. Biol. Chem.* **268**, 6119-6124.
15. Debyser, Z., Pauwels, R., Andries, K., Desmyter, J., Kukla, M., Janssen, P. A. J. & De Clercq, E. (1991) *Proc. Natl. Acad. Sci. USA* **88**, 1451-1455.
16. Frank, K. B., Noll, G. J., Connell, E. V. & Sim, I. S. (1991) *J. Biol. Chem.* **266**, 14232-14236.
17. Goldman, M. E., Nunberg, J. H., O'Brien, J. A., Quintero, J. C., Schleif, W. A., Freund, K. F., Gaul, S. L., Saari, W. S., Wai, J. S., Hoffman, J. M., Anderson, P. S., Hupe, D. J., Emini, E. A. & Stern, A. M. (1991) *Proc. Natl. Acad. Sci. USA* **88**, 6863-6867.
18. Taylor, P. B., Culp, J. S., Debouck, C., Johnson, R. K., Patil, A. D., Woolf, D. J., Brooks, I. & Hertzberg, R. P. (1994) *J. Biol. Chem.* **269**, 6325-6331.
19. Spence, R. A., Kati, W. M., Anderson, K. S. & Johnson, K. A. (1995) *Science* **267**, 988-993.
20. Spence, R. A., Anderson, K. S. & Johnson, K. A. (1996) *Biochemistry* **35**, 1054-1063.
21. Rittinger, K., Divita, G. & Goody, R. (1995) *Proc. Natl. Acad. Sci. USA* **92**, 8046-8049.
22. Kopp, E. B., Miglietta, J. J., Shrutkowski, A. G., Shih, C.-K., Grob, P. M. & Skoog, M. T. (1991) *Nucleic Acids Res.* **19**, 3035-3039.
23. Olsen, D. B., Caroll, S. S., Culberson, C. J., Shafer, J. A. a. & Kuo, L., C. (1994) *Nucleic Acids Res.* **22**, 1437-1443.
24. Bacolla, A., Shih, C.-K., Rose, J. M., Piras, G., Warren, T. C., Grygon, C. A., Ingraham, R. H., Cousins, R. C., Greenwood, D. J., Richman, D., Cheng, Y.-C. & Griffin, J. A. (1993) *J. Biol. Chem.* **268**, 16571-16577.
25. Divita, G., Muller, B., Immendorfer, U., Gautel, M., Rittinger, K., Restle, T. & Goody, R. S. (1993) *Biochemistry* **32**, 7966-7971.
26. DeStefano, J. J., Bambara, R. A. & Fay, P. J. (1993) *Biochemistry* **32**, 6908-6915.
27. Palaniappan, C., Fay, P. J. & Bambara, R. A. (1995) *J. Biol. Chem.* **270**, 4861-4869.
28. Patel, P. H., Jacobo-Molina, A., Ding, J., Tantillo, C., Clark, A. D., Jr., Raag, R., Nanni, R. G., Hughes, S. H. & Arnold, E. (1995) *Biochemistry* **34**, 5351-5363.
29. Smith, M. B. K., Rouzer, C. A., Taneyhill, L. A., Smith, N. A., Hughes, S. H., Boyer, P. L., Janssen, P. A. J., Moereels, H., Koymans, L., Arnold, E., Ding, J., Das, K., Zhang, W., Michejda, C. J. & Smith, R. H., Jr. (1995) *Protein Science* **4**, 2203-2222.
30. Kew, Y., Salomon, H., Olsen, L. R., Wainberg, M. A. & Prasad, V. R. (1996) *Antimicrob. Agents and Chemother.* **40**, 1711-1714.
31. Kleim, J.-P., Rosner, M., Winkler, I., Paessens, A., Kirsch, R., Hsiou, Y., Arnold, E. & Riess, G. (1996) *Proc. Natl. Acad. Sci. USA* **93**, 34-38.
32. Wu, J. C., Warren, T. C., Adams, J., Proudfoot, J., Skiles, J., Raghaven, P., Perry, C., Potocki, I., Farina, P. R. & Grob, P. M. (1991) *Biochemistry* **30**, 2022-2026.
33. Nunberg, J. H., Schleif, W. A., Boots, E. J., O'Brien, J. A., Quintero, J. C., Hoffman, J. M., Emini, E. A. & Goldman, M. E. (1991) *J. Virol.* **65**, 4887-4892.
34. Richman, D., Shih, C. K., Lowy, I., Rose, J., Prodanovich, P., Goff, S. & Griffin, J. (1991) *Proc. Natl. Acad. Sci. USA* **88**, 11241-11245.

Subject Index